# ATMOSPHERIC SULFUR AND NITROGEN OXIDES

# ATMOSPHERIC SULFUR AND NITROGEN OXIDES

## Eastern North American Source–Receptor Relationships

**George M. Hidy**
*Environmental Division*
*Electric Power Research Institute*
*Palo Alto, California*

ACADEMIC PRESS
A Division of Harcourt Brace & Company
San Diego New York Boston
London Sydney Tokyo Toronto

This book is printed on acid-free paper. ♾



Academic Press, Inc.
525 B Street, Suite 1900, San Diego, California 92101-4495

*United Kingdom Edition published by*
Academic Press Limited
24–28 Oval Road, London NW1 7DX

Library of Congress Cataloging-in-Publication Data

Hidy, George M.
Atmospheric sulfur and nitrogen oxides : eastern North American source-receptor relationships / George M. Hidy.
p. cm.
Includes index.
ISBN 0-12-347255-5
1. Atmospheric chemistry. 2. Sulphur oxides--Environmental aspects--North America. 3. Nitrogen oxides--Environmental aspects--North America. 4. Acid deposition--Environmental aspects--North America. 5. North America--Climate. I. Title.
QC879.6.H54 1994
551.5'11--dc20 93-25313
CIP

PRINTED IN THE UNITED STATES OF AMERICA
94 95 96 97 98 99 QW 9 8 7 6 5 4 3 2 1

To my good friend, Brynjulf Ottar, of the Norwegian Institute for Air Research. Twenty years ago, he introduced me to the many facets of long-range transport of air pollutants from the Scandinavian perspective, and followed closely with me the parallel, later developments in North American air chemistry.

# CONTENTS

## 2 CLIMATOLOGICAL FACTORS IN EASTERN NORTH AMERICA



## 3 CHARACTERIZATION OF REGIONAL AIR AND DEPOSITION CHEMISTRY

## 4 AIR CHEMISTRY AND ATMOSPHERIC PROCESSES



## 5 INHERENT VARIABILITY IN DEPOSITION CHEMISTRY

## 6 THEORETICAL INFERENCE OF SOURCE–RECEPTOR RELATIONSHIPS



## 7 RETROSPECTIVE

## APPENDIX A CHEMICAL CONVERSION AND DEPOSITION PROCESSES



## APPENDIX B ATMOSPHERIC VISIBILITY

# PREFACE

Although the knowledge of regional-scale atmospheric chemistry has increased considerably in the past 25 years, no consolidation of this information is yet available. *Atmospheric Sulfur and Nitrogen Oxides* is intended to provide such a synthesis. Its specific objectives are to (1) assess the knowledge of sulfur oxides ($SO_x$) and nitrogen oxides ($NO_x$) in ambient air; (2) assess knowledge about sulfate ($SO_4^{2-}$) and nitrate ($NO_3^-$) deposition in relation to environmental effects to limit acidity exposure and visibility impairment; (3) relate present $SO_x$ and $NO_x$ emission distributions to regional pollution indexes; (4) examine knowledge of source–receptor relationships and limitations or uncertainties in estimates of such relationships; and (5) trace the progress made in developing a knowledge base of regional air pollution over eastern North America, noting areas that need continued research to better define the impact on human activity.

This book summarizes climatology and meteorology relevant to regional pollution in eastern North Amerca and discusses phenomenological aspects of chemical measurements for $SO_x$ and $NO_x$ chemistry over time scales from daily to multiyear averages in terms of $SO_2$ and $NO_x$ emission distributions. It provides a summary of current knowledge on spatial and temporal variability of ambient concentrations. Visibility impairment, as well as dry and wet deposition patterns, are also discussed as important "exposure" factors associated with air chemistry.

In addition, this monograph characterizes regional-scale pollution events based on the dynamics of particulate $SO_4^{2-}$ events. This discussion leads to important aspects of synoptic meteorological classification of multiday events and long-term average (wet) deposition behavior. Data analysis and segregation or factor analysis help in understanding the classification of pollution phenomena. The application of such analyses to temporal and spatial variability of $SO_4^{2-}$ and other aerometric constituents reveals important relationships that exist between the constituents' parameters. The relationships are interpreted in terms of the interactions of atmospheric processes anticipated over North America.

This volume also discusses aspects of the theory of atmospheric processes that are synthesized into simulation models to describe individual regional phenomena. Ideally, theoretical models describe chemical distributions while paralleling the view given by the measurements. Initial attempts to simulate regional processes were based on simplistic schemes that emphasized emission distributions and air mass transport. New, more sophisticated approaches detail atmospheric fluid dynamics as well as chemical transformation and deposition processes. The discussion focuses on these latter approaches to illustrate the complexity of phenomena and to provide insight for interpreting the observations. Particular emphasis is placed on establishing a relationship between sources and emissions, and conditions at selected receptor locations.

Finally, this book covers exposure levels and air pollution index documentation, including visibility impairment and atmospheric concentration or deposition data. Such a compendium will be of interest not only to atmospheric scientists but also to those concerned with environmental effects on humans and natural or managed ecosystems.

*Atmospheric Sulfur and Nitrogen Oxides* is written at the graduate level for practicing scientists and engineers, as well as students entering the field of atmospheric chemistry. It is intended for chemists, atmospheric scientists, and engineers who want a summary of a large body of recent literature on regional-scale air chemistry.

George M. Hidy

# ACKNOWLEDGMENTS

This monograph was first completed in 1987 under a contract to the Desert Research Institute from the Electric Power Research Institute (EPRI). However, I have delayed its publication for several years to take account of the rapid expansion of knowledge acquired as part of the National Acid Precipitation Assessment Program (NAPAP), and the Electric Power Research Institute's cooperative program with NAPAP. These were completed in 1990. This book represents a compilation and interpretation of knowledge reported through 1992, recognizing that active research is continuing on this subject for the forseeable future.

I am indebted to the Electric Power Research Institute for sponsorship of this writing, both in its initial phases at the Desert Research Institute, and later at EPRI itself. I give my sincere thanks to Glenn Hilst, Peter Mueller, Alan Hansen, A. Venkatram, and Charles Hakkarinen for sharing their knowledge with me; and my appreciation to many other colleagues, too numerous to list here, for their energetic pursuit of knowledge that has led to a far deeper understanding of $SO_2$ and $NO_x$ behavior in the atmosphere than was available just 20 years ago. Further, I am grateful to Ms. Juana Blackburn (DRI), Ms. Kitty Vroom, and Dianne Zarcone for their effort to prepare the manuscript for publication. I also acknowledge Janet Miller's assistance in supervising the graphics preparation.

Finally, I recognize C. E. Junge, Walter Orr Roberts, and James P. Lodge, who led me into this world of atmospheric phenomena three decades ago. Without their wisdom and sense of excitement about atmospheric chemistry, I would not have become so taken with this important subject.

# ABBREVIATIONS

| | |
|---|---|
| **AES** | (Canadian) Atmospheric Environment Service |
| **AH** | Absolute humidity |
| **ANATEX** | Across North America Tracer Experiment |
| **APN** | (Canadian) Air Pollution Network |
| **APOS** | Air Pollution in Ontario Study |
| **CAAA** | Clean Air Act Amendments (of 1990) |
| **CANSAP** | Canadian Network for Sampling Precipitation |
| **CAPMON** | Canadian Air and Precipitation Monitoring Network |
| **CAPTEX** | Cross Appalachian Pollutant Tracer Experiment |
| **CMB** | Chemical Mass Balance |
| **DOE** | U.S. Department of Energy |
| **EMEFS** | Eulerian Model Evaluation and Field Study |
| **ENA** | Eastern North Ameirca |
| **EOF** | Empirical orthogonal function |
| **EOFA** | Empirical orthogonal function analysis |
| **EPA** | U.S. Environmental Protection Agency |
| **EPRI** | Electric Power Research Institute |
| **ERAQS** | Eastern Regional Air Quality Study |
| **EUS** | Eastern United States |
| **IPM** | Inhalable particulate matter ($<15$ $\mu$m diameter) |
| **MAP3S** | Multistate Atmospheric Pollution and Power Production Study |
| **MATEX** | Massive Aerometric Tracer Experiment |

| | |
|---|---|
| **MCCP** | Mountain and Cloud Chemistry Project |
| **MOE** | Ontario Ministry of Environment |
| **NADP** | National Acid Deposition Program |
| **NAMS** | National Air Monitoring System |
| **NAPAP** | National Acid Precipitation Assessment Program |
| **NASN** | National Air Sampling Network |
| **NOAA** | National Oceanic and Atmospheric Administration |
| **NWS** | U.S. National Weather Service (NOAA) |
| **OECD** | Organization for Economic and Cooperative Development |
| **ORBES** | Ohio River Basin Environmental Study |
| **OSCAR** | Oxidizing and Scavenging Characteristics of April Rains |
| **PCA** | Principal component analysis |
| **PEPE** | Prolonged Episodic Pollution Event Study |
| **RH** | Relative humidity |
| **RSP** | Fine (or regional) particle concentration ($>\mu$m diameter) |
| **SCF** | Spatial correlation function |
| **SURE** | Sulfate Regional Experiment |
| **TONC** | Total organic noncarbonate carbon |
| **TSP** | Total suspended particle concentration ($\mu g/m^3$) |
| **UAPSP** | Utility Acid Precipitation Study Program |
| **USGS** | U.S. Geological Survey |
| **WMO** | World Meteorological Organization |

# 1 INTRODUCTION

## OVERVIEW

Scientific knowledge of the physical processes within the atmosphere that produce weather has accumulated over the past century. In contrast, understanding of chemical constituents in the atmosphere has emerged only within the past three decades. Acquisition of this knowledge has accelerated in part because of public concern regarding air pollution and also in part because of a greatly expanded interest in chemical interactions with the biosphere. Today, a number of species are of concern, including carbon monoxide (CO); sulfur dioxide ($SO_2$); nitric oxide (NO) and nitrogen dioxide ($NO_2$)[1]; ozone ($O_3$) gases; and suspended sulfate, lead, and carbonaceous particulate matter.

Sulfur oxides ($SO_x$) and nitrogen oxides ($NO_y$) have received particular attention since the 1970s because of concern about not only human health effects and atmospheric visibility impairment but also a number of other effects, including material corrosion and ecosystem interference. Human health effects and material damage from corrosive acidic species like $SO_x$ and $NO_y$ are confined to local regions around pollution sources. Ecological effects, including acidification of lakes and streams with effects on fish and interference with forest productivity, are believed to be geographi-

[1] $NO_x$ refers to NO and $NO_2$; $NO_y$ is adopted for all N species, NO, $NO_x$, $N_2O_3$, $N_2O_4$, $NO_3$, $N_2O_5$, HONO, $HONO_2$, $NO_3^-$, and others.

cally widespread. The vulnerability of aquatic and terrestrial ecosystems to deterioration resulting from exposure to acidic pollution depends on a complicated set of circumstances. These involve the natural waters, soil, and bedrock chemistry of acid neutralization, as well as the presence of leachable, potentially toxic chemical species. Although little direct evidence of damage has been found for North America, areas in Europe or Scandinavia have reported widespread, apparent degradation of freshwater aquatic systems and forests attributed to acid exposure. These important issues of environmental stress are beyond the scope of this book; for details, the reader is referred to the extensive literature reviewed in the reports of OTA (1984), EPA (1984), and Irving (1991). Cowling (1980) has summarized early fears about environmental effects in North America, and Irving (1991) provides a valuable perspective on the most recent consensus about the nature and effects of exposure to and surface deposition of atmospheric $SO_x$ and $NO_y$.

EPRI's (1991) environmental impact assessment of the regional effects of acidic air pollutants complements other evaluations and adds perspective on emission control strategies that would optimize reduction in pollutant exposure at minimum cost to industrial sources. This work further notes that the aesthetic aspects of regional visibility impairment in the eastern United States appeared to be economically the most motivating effect of regional $SO_x$ pollution. Visibility reduction in the East is believed to be strongly influenced by airborne sulfate particles. The valuation of visibility impairment as a public issue is discussed by EPA (1979) and in EPRI's (1991) evaluation. Only the physical and chemical aspects of regional visibility change are covered here.

Given the motivation for environmental protection, this book concentrates on the atmospheric chemistry of $SO_x$ and $NO_y$ species that have the potential to produce contamination of large geographic regions.

The ability of atmospheric processes to disperse contaminants over great distances has been appreciated for a long time. Philosophers of science in the 18th and 19th centuries speculated about many phenomena, for example, the impact of dust fallout associated with volcanic eruptions thousands of kilometers away. The analysis of radioactive fallout from atomic weapons testing after World War II added to evidence for the long-distance transport of contaminants. Only in the past few decades, however, has the public been sufficiently conscious of public health issues to give priority to large industrial sources of pollutants and urban air conditions. In the 1960s, Junge (1963) reported speculations that abnormally high concentrations of sulfate ($SO_4^{2-}$) in southern Scandinavian rainwater resulted from scavenging of sulfur oxides ($SO_x$) emitted from distant industrialized parts of Europe (Sweden's Case Study for the U.N., 1972).

Since the 1960s, scientists have identified the symptoms of large-scale (regional) dispersion of pollutants in the form of haze and widespread elevated concentrations of aerosol particles and pollutant gases. Carbon monoxide, for example, a by-product of combustion, is found in high concentrations in the Northern Hemisphere compared with the Southern Hemisphere. Other species that appear to be globally dispersed are $SO_x$, including airborne $SO_4^{2-}$. In the early 1970s, experts suspected geographically widespread pollution over eastern North America (ENA) based on regionally elevated pollutant concentrations, particularly of $O_3$ and airborne particles containing $SO_4^{2-}$.

Research on regionwide dispersion of $SO_x$ accelerated in the 1960s in Europe, where a consensus arose about the ecological consequences of the fallout of acidifying substances, such as $SO_4^{2-}$, in precipitation (Oden, 1968). The Europeans launched a major international program to study the long-range transport of air pollution under the sponsorship of the Organization for Economic and Cooperative Development (OECD). Results of this research characterized regional-scale $SO_x$ pollution in Europe in relation to acid precipitation (OECD, 1977). As part of this project, researchers systematically measured $SO_4^{2-}$ in the air and in rainwater. Others developed a European emissions inventory (Semb, 1978) and a source–receptor simulation capability. Although the early simulation models were relatively simple in concept, they enhanced appreciation of the nearby versus distant source contributions to deposition in remote areas averaged over periods of a year or more (e.g., Eliassen and Saltbones, 1975).

Shortly thereafter, historical data analyses showed the potential for acid deposition in ENA (Cowling, 1980). North American scientists became more interested in regional $SO_x$ pollution because of allegations of health effects associated with airborne $SO_4^{2-}$ particles (U.S. Environmental Protection Agency, 1975) and the existence and importance of long-range transport of air pollutants (LRTAP) in Europe inferred from measurements and computer model simulations (e.g., OECD, 1977). As a result of investigations such as the Sulfate Regional Experiment (SURE) and later field studies, such as the Eastern Regional Air Quality Study (ERAQS), the Cumberland Experiment, the Prolonged Episodic Pollution Event Study (PEPE) (Gillani and Pooler, 1980), and satellite photograph analysis (e.g., Lyons *et al.*, 1978), scientists had defined the role of LRTAP in ENA by the early 1980s. These studies indicated that in the presence of volatile organic compounds (VOC or HC), $SO_x$, $NO_x$, and their atmospheric end products (e.g., $SO_2$, nitrate [$NO_3^-$], peroxyacetyl nitrate [PAN], and $O_3$) are implicated in regional pollution.

Concern about cumulative deposition effects on lake water quality and forest ecosystems caused European scientists to work on long-term average calculations rather than on individual pollution events. U.S. research in the 1970s, however, focused on exposure to, or dosage from, ambient air so that researchers could describe pollution events and their frequency of occurrence. More recently, North American scientists shifted their concern to long-term deposition effects on surface waters and terrestrial ecosystems, following the lead of Europeans (e.g., U.S. Environmental Protection Agency, 1984).

Experts speculated about the regional extent of air pollution in North America based on their analysis of many years of annual average data on airport visibility. Visibility is much lower in ENA than in the West, and lowest visibility occurs in the heavily industrialized region of the Midwest (Frank and Posseill, 1976; Trijonis, 1982). Meteorological factors, such as humidity, reduce regional visibility. However, the prevailing levels of particulate $SO_4^{2-}$ in 1970s air-monitoring data suggested a circumstantial relationship between that species and visibility over a large portion of ENA (Trijonis, 1982). The extent of regional exposure to airborne $SO_4^{2-}$ particles was uncertain. Furthermore, whether this phenomenon was related to the historic geographic distribution of large $SO_2$ sources in the East or to the influence of large-scale meteorological processes that permit pollution accumulation was unclear.

The occurrence of regional pollution "events" lasting one to several days as identified by elevated $SO_4^{2-}$ concentrations measured simultaneously over distances of hundreds of kilometers was documented in early studies of rural air quality by Hidy *et al.* (1976). Other researchers studied the manifestations of regionally elevated $SO_4^{2-}$ concentrations, including reduced visibility over large areas covered with haze (e.g., Lyons *et al.* 1978) and high $SO_4^{2-}$ levels in rainwater in areas of elevated ambient concentrations (e.g., Lodge, 1968). Examples of the qualitative association between visibility degradation and elevated $SO_4^{2-}$ concentrations in the northeastern United States were reported by Hidy *et al.* (1978). Similar case studies have been reported by Husar and Patterson (1980). These results implicated airborne $SO_4^{2-}$ and other finely divided particulates, which are known to be efficient scatterers of light.

The results also helped establish a spatial and temporal scale for regional events. Haze accumulation typically extends into the planetary boundary layer to heights of 1.5 km or more, covers a geographic area of more than $10^6$ km$^2$, and has a linear scale of 10 to 1000 km or more. Regional phenomena tend to evolve from several hours to several days (Hidy *et al.*, 1976, 1978). From these scales, synoptic meteorological processes appear to be those relevant to regional pollution behavior. The behavior of $SO_4^{2-}$, as seen from data collected in the mid-1970s (Hidy *et al.*, 1976), gives an estimate of scale that is consistent with expected regional phenomena in the eastern United States. The data were obtained at eight nonurban sites ranging from the lower Ohio River Valley to New

York State. Their orientation roughly followed the prevailing eastward movement of air masses. The observations showed both conditions of localized, mutually independent $SO_4^{2-}$ concentration variation and conditions in which concentrations measured at several stations were highly correlated.

The fate of airborne material depends on atmospheric processes and surface factors that influence the material's deposition. If gases or airborne particles are soluble or reactive in aqueous media, cloud processes followed by precipitation provide a major vehicle for deposition. The regional extent of $SO_4^{2-}$ and $NO_3^-$ in the atmosphere should then be reflected in precipitation data. Indeed, researchers found this to be the case for both species. Although regional-scale phenomena were not defined in the 1950s, spatial characterization from surveys by Junge and his associates (Junge, 1963) showed elevated $SO_4^{2-}$ in precipitation over a broad area of ENA. This region was later confirmed by observations in the 1960s reported by Lodge and his associates (1968).

Measurements in the 1980s have better characterized the regional extent of $SO_4^{2-}$ and $NO_3^-$ in wet deposition. Data from the National Acid Deposition Program (NADP) and the Canadian networks (CANSAP) indicate that $SO_4^{2-}$ and $NO_3^-$ concentrations in precipitation are elevated over a large part of ENA. The distributions are qualitatively similar to the visual range distribution (Hidy *et al.*, 1976) and the airborne $SO_4^{2-}$ distribution reported by Frank and Posseill (1976). The geographic dispersal and regional extent of elevated $SO_4^{2-}$ concentration in precipitation revealed a broad area of weak concentration gradients associated with the regional distribution of large $SO_2$ sources. The area of maximum concentration is elongated along the direction taken by the prevailing air mass movement. This area extends horizontally over 1000 km from the upper Ohio River Valley in the West to Quebec and Nova Scotia in the East. From south to north, the region extends hundreds of kilometers from Tennessee and the Carolinas to Quebec. Thus, the geographic scales for both pollution events and long-term average changes in ENA parallel one another.

Although regional-scale pollution phenomena have been reported in various parts of the world, the area by far the most thoroughly studied is eastern North America, particularly the United States. In this region, atmospheric chemistry research has a richness of extensive measurements, theoretical modeling, and interpretation results not reported elsewhere. Regional meteorology varies from region to region, but the linkages between atmospheric physical processes and $SO_x$ or $NO_y$ chemistry should be well exemplified by the experience in the eastern United States. Thus, I have elected to concentrate on the eastern United States to illustrate regional atmospheric chemistry of $SO_x$ and some components of

$NO_y$. The examples should serve as a general basis for interpretation of these chemical species elsewhere.

## DEFINITIONS

### Source–Receptor Relationships

A source–receptor relationship (SRR) is the contribution of emissions from a given source to conditions at a given receptor either by exposure to ambient air or by deposition. A wide range of sources may inject contaminants into the atmosphere. Once injected, these materials may be influenced by a variety of atmospheric processes before their return to the earth. Important processes include transport away from the source by airflow, as well as dilution. If the contaminants are reactive, atmospheric chemical reactions can convert them to related products in the ambient air or in aqueous media (e.g., cloud droplets, ice crystals, snowflakes, graupel, or hailstones). For example, $SO_2$ may oxidize to airborne $SO_4^{2-}$ or NO may convert to $NO_2$, which, in turn, may convert to nitric acid ($HNO_3$). Material is removed from the air either by absorption at the ground or on vegetation, as in the case of gases, or by collection on obstacles, as in the case of particles. Removal also occurs by collection (scavenging) during precipitation. Airborne material is transported, diluted, and chemically modified as it travels away from a source until it arrives at a receptor, where it can remain airborne or be deposited. The area surrounding a source in which ambient concentrations or deposition conditions are measurably or "significantly" affected by the source is often called the *zone of influence* of that source. On the average, atmospheric processes tend to decrease exponentially the ground level concentration with distance from a source. Therefore, an effective zone can be defined heuristically through measurement or measurement analysis.

### Chemical Constituents

The atmospheric constituents of concern in this monograph are $SO_x$ and $NO_x$, oxidants (including $O_3$ and hydrogen peroxide [$H_2O_2$]), carbon-containing compounds, and certain cationic species, such as ammonia or ammonium (Table 1.1). These species dominate acid deposition and visibility impairment in ENA. Table 1.1 does not list reactive volatile organic compounds (VOCs), which are key to oxidant formation. For reliable predictions of source–receptor relationships, VOCs have to be accounted for, at least implicitly, in dealing with S and N chemistry; additional details of the chemical linkages are given in Appendix A.

**TABLE 1.1 Constituents Considered for Characterization of Regional-Scale Pollution Phenomena in Eastern North America**

| Constituent | | Remarks |
|---|---|---|
| Sulfur | | |
| Sulfur dioxide | $SO_2$ | Principal emitted precursor gas for $SO_4^{2-}$ |
| Sulfate | $SO_4^{2-}$ | Airborne particle component formed by oxidation of $SO_2$, found in precipitation as an acid-forming anion |
| Hydrogen sulfide, dimethyl sulfide, and carbonyl sulfide | $H_2S$, $(CH_3)_2S$, COS | Predominantly natural sulfur constituents that oxidize to $SO_4^{2-}$ |
| Nitrogen | | |
| Nitric oxide | NO | Principal emitted precursor gas for $NO_x$ cycle |
| Nitrogen dioxide | $NO_2$ | Minor emitted precursor gas for $NO_x$ cycle |
| Nitric acid | $HNO_3$ | Acid vapor produced by oxidation of $NO_x$ (NO + $NO_2$) |
| Peroxyacetyl nitrate (PAN) | $CH_3CH_2ONO_3$ | An organic nitrate that is a major product of $NO_3^-$ chemistry |
| Nitrate | $NO_3^-$ | A minor airborne particle constituent, a major acid former in precipitation |
| Nitrogen pentoxide, nitrogen trioxide | $N_2O_5$, $NO_3$, HONO | Short-lived products of $NO_3^-$ chemistry |
| Oxidants | | |
| Ozone | $O_3$ | Major gas-phase product of $NO_x$-VOC photochemical system |
| Hydrogen peroxide | $H_2O_2$ | Gas-phase product of $NO_x$-HC system, important oxidizer in aqueous media |
| Hydroxyl radical | OH | Major free radical oxidizer in $NO_x$-VOC system |
| Cationic species | | |
| Hydronium ion | $H_3O^+(H^+)$ | Ubiquitous acid cation |
| Ammonia, ammonium | $NH_3(g)$, $NH_4^+$ | Major cationic species in deposition, biologically active in soil as an $NH_4^+$ producer |
| Calcium, potassium, magnesium, sodium | $Ca^{2+}$, $K^+$, $Mg^{2+}$, $Na^+$ | Suite of soil- or sea salt–derived cationic species |
| Carbon | | |
| Elemental carbon | C | Principal light-absorbing species in particles found in urban air |
| Reactive hydrocarbons | HC | A large variety of nonmethane hydrocarbon species including paraffins, olefins, and aromatics found in trace quantities in the atmosphere |
| Volatile organic compounds | VOC | A large number of organic vapors including HC, oxygenated species, and organic nitrogen species found in the atmosphere |
| Condensed organics | Paraffins, acids, esters | Major components of finely divided particles in air over ENA |

## Deposition Processes

Wet or dry deposition results from physical and chemical processes that occur at or near the ground or from precipitation in clouds. *Dry deposition* of gases and suspended particles depends on the rate of vertical transport in air toward a surface, which, in turn, depends on the surface roughness and the air velocity near the surface. Gases are absorbed at the surface with different rates, depending on the nature and wetness of the surface. Particle collection results from gravitational settling, interception (the collision of neighboring bodies of finite size), impaction (the collection of particles as they depart from gas flow paths around bodies), and diffusion by Brownian motion or small-scale (microscale) air turbulence. Brownian motion is the thermal agitation of particles caused by bombardment of gas molecules.

In *wet deposition*, gases and particles are transported to hydrometeors, where they collide and/or are absorbed and sometimes undergo aqueous-phase chemical reactions. They are scavenged in clouds (*rainout*) and below clouds (*washout*) and deposited through precipitation. Interception of cloud or fog particles by vegetation surfaces or other objects or dew formation can also result in wet deposition. Although collection occurs in the presence of condensed moisture, fog droplet collection and dew formation are sometimes called *occult deposition*.

The vernacular term *acid rain* refers to the presence of acid-forming species in precipitation, especially $SO_4^{2-}$, $NO_3^-$, carbonate, and organic acid anions. In recent literature this term is synonymous with all forms of deposition of acid-forming material.

## Visibility

Visibility is a measure of the ability of radiant energy to evoke visual sensation in an observer. Atmospheric constituents affect visible light transmission through the atmosphere by absorbing or scattering wavelengths. Visibility impairment by suspended material depends on absorption and scattering of light along a nearly horizontal path. Optical effects depend on the burden of suspended material or the product of ambient concentration and the path length of a light beam (Appendix B). In ENA visibility generally ranges from $<10$ to 40 km, in contrast to a "natural" (Rayleigh) limit from scattering by gas molecules of about 300 km. Thus, the "receptor" for source influence represents a volume of air extending horizontally to 40 km or more from an observer.

## Scale of Phenomena

The spatial and temporal behavior of materials in the atmosphere can be classified according to scales as summarized in Table 1.2. The *local*

TABLE 1.2 **Spatial and Temporal Scales for Contaminant Episode Behavior in the Atmosphere**

| Pollution class | Meteorological class | Horizontal spatial range (km) | Temporal range (days) | Comments |
|---|---|---|---|---|
| Local | Microscale | 0–10 | 1 | Recognized as classic pollution impact, basis for noncreative pollutant reactions |
| Urban | Mesoscale | 10–100 | 1–2 | Identified in the 1950s largely with reactive pollutant products, such as photochemical oxidants |
| Regional | Mesoscale synoptic scale | 100–1000 | 1–5 | Recognized by problems related to haze and visibility degradation, acid rain, $SO_4^{2-}$ |
| Global | Planetary scale | 1000 | 5–10 | Surveyed extensively in relation to dispersal of radionuclides from bomb tests |

Source: Hidy *et al.*, 1978.

*scale* is characterized by highly variable exposures to ambient concentrations with hourly or 24-hour extrema much different from annual means, which are at least a factor of 2 lower than shorter-term averages. Nearby sources dominate ambient concentrations. Chemical processes that produce products of reactive species, such as $SO_4^{2-}$, and $NO_3^-$, yield ambient concentrations that are generally lower and more widely distributed from sources than their precursors, $SO_2$ and NO (or $NO_2$).

The *urban scale* covers a regime of multiple sources exemplified by a moderate to large metropolitan center. Phenomena occurring on this scale involve complex mixtures of pollutants, including partially reacted emissions, and their atmospheric end products, such as particulate $SO_4^{2-}$, $O_3$, and $HNO_3$.

The *regional scale* encompasses much larger geographic areas and time periods. Meteorological phenomena distribute material over 24 hours or more; dispersion is controlled by regional (synoptic) wind patterns and turbulent mixing throughout the atmospheric layers near the earth's surface, called the *planetary boundary layer* (height to about 4 km). Concentrations are low, and their gradients are relatively weak. The chemical conversion approaches steady state or equilibrium. Both dry deposition and wet deposition become important for reducing airborne pollutant levels.

On the *global scale*, atmospheric processes reduce concentrations of contaminants and their reaction products to very low levels, particularly the more reactive contaminants, because of the distance between sources.

The spatial and temporal scales of atmospheric gaseous and particulate species differ. Residence time reflects chemical reactivity, loss by deposition, and scavenging potential in clouds or precipitation (e.g., water solubility). The shorter the residence time, the more reactive the species. Acid deposition can occur in very small areas—individual lakes or ponds, buildings, or small stands of crops or trees. In keeping with concepts of human population exposure, however, receptors for deposition are treated on a coarse areal extent. These are normally associated with susceptible lake populations, such as those in the Adirondack Mountains, or sensitive forest areas of the same size. As a practical definition, a source area is also considered to be the same areal resolution for regional studies.

This monograph focuses on the regional scale in which $SO_x$ and $NO_y$ chemistry, $O_3$ and other oxidants, and suspended particles less than 20 mm in diameter play an important role. For regional-scale phenomena, the residence time regime is $10^3$–$10^7$ seconds (s) and the transport distance regimes are >10 to $10^4$ km. SRRs extending over a regional scale have been difficult to estimate reliably because the atmospheric processes involved on such scales have been poorly characterized. There are few measurements through which sources can be traced for distances beyond a few tens of kilometers. The difficulties of establishing regional SRRs have been reviewed by Hidy (1984a).

## DESCRIPTIVE FACTORS

Certain key factors have to be accounted for in characterizing regional-scale processes. Distribution and areal density are important as basic factors for the $SO_x$ and $NO_y$ behavior in ENA. At the same time, meteorological processes are important in reinforcing pollutant accumulation.

### Emission Distributions

The East is characterized by a geographically broad distribution of $SO_2$ and $NO_x$ sources located in heavily industrialized areas of the Midwest and in different large metropolitan areas, including the New York–Philadelphia area, Chicago, Detroit, Atlanta, and Cincinnati. In many areas, emission sources are closely packed for several tens of kilometers. The distributions of $SO_x$ and $NO_x$ sources are similar in extent throughout the East and are aligned roughly along a direction parallel to the prevailing winds. There is a large potential for accumulation of pollution with air movement across ENA from the midcontinent to the Atlantic Coast.

### Meteorological Factors

Earlier studies describing conditions favoring air pollution suggest that three dynamic elements should be important in characterizing regional-scale performance.

- The strength and persistence of airflow leading to the potential for *long-range transport* of polluted air masses over distances of several hundred kilometers or more
- The intensity and depth of *vertical mixing* by air eddying or turbulence, modulated by diurnal changes over periods of 24 hours or more
- The *chemical reactivity* of species to form transformed products and the coupling with *dry and wet removal processes* to deplete atmospheric concentrations of emitted contaminants

One of the frequently discussed elements of regional pollution is the potential for long-range transport of air masses. Conditions for weak or intensified airflow (advection) play as important a role in regional pollution behavior as they do in events occurring in cities or near large sources. King and Vukovich (1982) have described pollution buildup in poorly ventilated conditions near large air masses that circulate slowly (anticyclonic–clockwise flow disturbances) around areas of high barometric pressure. They hypothesized that concentrations of pollutants should build up in anticyclonic disturbances where air motion is weak and residence time of material is longest. The potential for these conditions appears to be strongest in the southwest sector of an anticyclonic circulation. Persistence of wind direction or speed over several hundred kilometers for more than a day is needed for direct observational evidence of long-range transport. This phenomenon was largely unexplored in ENA until the mid-1970s, when Hidy *et al.* (1978) and Lioy and Samson (1979) described the evolution of particulate $SO_4^{2-}$ concentration distributions shaped by persistent winds along summer frontal systems moving northeastward across the midwestern and eastern United States. They speculated that such conditions could provide direct evidence for long-range pollutant transport. This hypothesis has subsequently been tested and is shown to be characteristic of regional pollution conditions over ENA.

The conditions for intensity of advection (horizontal transport and mixing) are coupled with vertical mixing processes in the boundary layer; there is a question about the height of representative transport winds for regional-scale phenomena. Surface winds do not appear to be a justifiable choice for transport winds; winds nominally in the middle of the planetary boundary at a height of about 2 km are generally believed (but have not been proved) to be most relevant. Diurnal changes in the boundary layer can strongly influence advection, particularly in stable conditions. The diurnal sequence is accompanied by the appearance at night of a shallow layer of high-speed winds (nocturnal jet) at a height of 1 to 2 km. The jet may affect advective flow in different air layers, particularly at night.

Turbulent mixing of air plays a critical role in the dilution and dispersal of atmospheric contaminants. Pollution levels generally increase when mixing is restricted in buoyantly stable conditions. Regional-scale accumulation of pollution depends on constrained mixing over a large area persisting for more than a day. Such conditions are known to occur in stagnating air surrounding anticyclones associated with zones of high barometric pressure (e.g., Korshover, 1976). When pollution accumulates for more than a day, the area covered may expand over that locally influenced by sources. In addition, the diurnal cycle of mixing in the boundary layer becomes important. During the day, solar heating at the ground and in clouds stimulates vertical mixing (convection). At night, air near the surface cools as the surface cools, and vertical mixing is suppressed by stratification in stable air. This sequence is a factor that results in vertical distribution of pollutants to several kilometers in height, as well as at the ground. Furthermore, the potential for residual layers of polluted air to fumigate or sink to the ground overnight is a unique feature of the regional scale. The possibility for sinking is enhanced by the fact that large sources in parts of ENA emit pollutants at heights in excess of 300 m with buoyant plumes from tall stacks. The gases from these stacks are hot and exit with high velocities. Once aloft, polluted air may travel considerable distances with the winds before cooling and sinking to the ground.

A third element influencing regional behavior of pollutant species is how long they reside in the atmosphere. *Residence time* is determined by the combined influence of chemical reactions and deposition processes. Strictly speaking, the notion of residence time is defined in terms of an airborne constituent that decays or decreases according to a simple first-order rate. Then the residence time is defined as the ratio of the concentration in air to the (volumetric) removal rate. The residence time is the time required for a material to be reduced to $1/e$ of its original concentration. This implies that a residual concentration of any material may exist for very long times in the atmosphere. Thus, a small but significant fraction of contaminant emissions may be present for many months (or longer) despite the fact that its original concentration has decayed rapidly to a "background" level in a few days.

The residence time of water-soluble, reactive species such as $SO_x$, $NO_x$, or $NO_y$ is relatively short in the atmosphere. For regional-scale phenomena, the residence time is generally limited to a few days, compared with trace materials that are globally important. Examples of the latter include carbon dioxide, methane, nitrous oxide, and the chlorofluorocarbons (e.g., the Freons), all of which are quite nonreactive in the lower atmosphere. The residence time in the lower atmosphere for $SO_2$ is 1–3 days and for sulfate particles is 2–4 days. Residence times for $NO_x$ and $NO_y$ species are somewhat more complex than for $SO_x$ because of their atmospheric chemistry. For $NO_x$ the residence time is believed to range from less than a day to 2–3 days, and the residence time for nitrate is also said to range from 1–3 days.

The atmospheric chemical reactions of $SO_x$ are often considered to be quasi-first order in character, whereas the $NO_x$ and $NO_y$ reactions are more nonlinear in character. $SO_x$ and $NO_x$ readily oxidize in the air either by homogeneous gas-phase reactions, by reactions in aqueous droplets, or in the presence of liquid or solid aerosol particles. The transformation products are found either in the condensed phase, as in the case of $SO_4^{2-}$, or in the gas phase, as in the case of $HNO_3$ vapor. Dry and wet deposition processes are also important for regional-scale phenomena. Dry processes vary in rate, depending on the gas-absorbing capability of the underlying surface and the presence of surface moisture. Where vegetation is the accumulating medium, biological assimilation is also important in gas removal. Different reactive or water-soluble gases are thought to have similar removal rates at the ground, which are also thought to be substantially larger than the rates for finely divided particles. Deposition rates for particles vary widely, depending on particle size, collecting surface roughness, and surface wind and mixing conditions. Wet removal is generally thought to be most influenced by in-cloud scavenging rather than by below-cloud activity. Cloud processes are intermittent in nature but, when present, are efficient scavengers of $SO_2$ and particulate $SO_4^{2-}$. Scavenging $SO_2$ is enhanced by the rapid oxidation of dissolved $SO_2$ in the presence of $H_2O_2$. $SO_4^{2-}$ particles are effective nuclei for condensation and serve as centers for new droplet formation in a cloud containing supersaturated water vapor. Neither NO nor $NO_2$ appears to be efficiently scavenged in clouds. However, their oxidation products, nitrous acid (HONO) and nitric acid ($HONO_2$ or $HNO_3$), are very soluble and are removed effectively by cloud droplet scavenging.

The combination of dry and wet deposition provides an efficient means of cleansing the atmosphere. Dry processes take place continuously. They depend on material concentration near the ground, so they achieve high rates at high ambient concentration or near sources. Wet deposition depends on precipitation but is efficient when ice is not a major factor. Evidence suggests that ice in clouds and snowfall may inhibit removal because of poor scavenging rates and low chemical transformation rates in the ice phase. The removal processes, in effect, tend to reduce the characteristic distance of dispersal of reactive pollutants to subcontinental proportions.

The interactions between meteorological processes that contribute to regional-scale chemical phenomena have been studied only recently. Since the mid-1970s a number of studies have produced considerable information. Nevertheless, important questions about atmospheric chemical processes and regional air dynamics remain. This monograph provides a summary and analysis of the knowledge in this field. The broad features of atmospheric $SO_x$, $NO_x$, and oxidant behavior in ENA are now considered reasonably well established.

## ROLE OF MAJOR EXPERIMENTS

This monograph relies heavily on the analysis of a large collection of data taken in the SURE and the conceptual development of the Massive Aerometric Tracer Experiment (MATEX), as well as a large body of wet deposition measurements taken since 1979 under the sponsorship of the Electric Power Research Institute (EPRI) in the U.S. and Canadian governments. The SURE collection includes a wide rate of aerometric observations ranging from basic meteorological parameters to pollutant concentration distributions based on nonurban conditions (e.g., Mueller and Hidy, 1983). The data included measurements taken at the ground at selected locations, representing regional conditions, and taken by aircraft to heights of approximately 3 km over certain sites. The data are particularly useful in that they provide a daily picture of $SO_x$ behavior for meteorological conditions extending over the greater northeastern United States in 1977 and 1978. At a few locations the observations were continuous in certain months; this sampling indicates the short duration in $SO_x$ variability as well as longer-term variability at the same sites. The acquisition of a regionally segregated $SO_x$ emissions inventory for both anthropogenic sources (Klemm and Brennan, 1981) and estimated biogenic sources (Adams *et al.*, 1980) enhanced the field observations in the SURE. The inventory permitted a far more in-depth study of the relation between source emissions and ambient conditions than would otherwise have been possible.

The SURE proved to be a valuable pioneering study; it was however, limited in certain ways. First, it did not compare information about the fine structure of pollutant distributions with information about meteorological processes on a temporal scale of hours and over distances less than 100 km. Second, the auxiliary observations of $NO_x$, VOCs, and $O_3$ were limited in spatial and temporal coverage. Third, instrumentation available in the late 1970s did not permit routine, direct measurements of dry deposition of either $SO_2$, $NO_y$, or particulate $SO_4^{2-}$. Fourth, accelerated interest in precipitation chemistry observations lagged by approximately a year so that wet deposition was not well documented during the 1977–1978 period. Despite its limitations, the SURE data set was the only comprehensive one available for many years to describe the range of atmospheric processes taking place over ENA. Since the SURE, long-term field acquisition has been carried out at existing state and local air-monitoring stations and wet deposition sites. And a major new field program to generate a data set to evaluate large models was initiated in 1987 (Hansen, 1989).

The MATEX was to be a very large field experiment in which researchers would release multiple aerometric tracers and use deposition observations to infer SRRs empirically. Although the MATEX has not been implemented, scientific investigations conducted in 1983–1984 to support its design provided important insights into regional-scale phenomena (e.g., Hidy, *et al.*, 1985). These studies integrated the knowledge of ambient $SO_x$ and $NO_x$ concentration patterns with observations of wet deposition. Furthermore, they explored the inherent variability in emissions data as well as interpretation of that data.

The work of these major projects is supplemented and enriched by a large body of literature, developed from the mid-1970s until 1986, that describes aspects of regional-scale pollution in ENA. Important field studies to be noted for their contributions include the Ohio River Basin Environmental Study (ORBES) (e.g., Nieman, 1980), the PEPE Study (e.g., Gillani, 1985), and the Oxidizing and Scavenging Characteristics of April Rains (OSCAR) (e.g., Easter *et al.*, 1984). Various short-term specialized studies also describe the regional aspects of $SO_x$ and $NO_y$ chemistry in ENA.

The regional-scale field experiments of the 1970s and 1980s culminated with a large and elaborate experiment aimed at advancing the SURE and collecting a very large data base to evaluate predictive air quality models. This project is sometimes called the Eulerian Model Evaluation and Field Study (EMEFS). It was a cooperative venture between the U.S. and Canadian governments, the Electric Power Research Institute and the Florida Electric Power Coordinating Group (Hansen, 1989). The project involved a large aerometric and (wet) deposition monitoring network of more than 90 stations operating over more than two years, 1987–1989, supplemented with aircraft surveys and special experiments. The project employed improved observational instrumentation and addressed the weaknesses of the SURE by collecting a full suite of data for $SO_x$, $NO_x$-HC, and oxidant chemistry, as well as wet deposition. This project was superimposed on the routine monitoring operations for wet and dry deposition (e.g., Interagency Task Force on Acid Precipitation, 1983).

The results of the EMEFS project are in the initial stages of reporting; analysis and interpretation of results will follow for many years to come. Although the early analysis has focused on model evaluation (e.g., Dennis *et al.*, 1990), the data will ultimately produce a broadened understanding of regional phenomena in years to come. So far the results of EMEFS confirm the results described in this monograph.

# 2 CLIMATOLOGICAL FACTORS IN EASTERN NORTH AMERICA

## ROLE OF CLIMATE FACTORS

### Factors Influencing Exposure

Biota are exposed to a variety of materials deposited from the atmosphere. The acid or acid forming compounds include carbonic acid, a naturally derived acid; $SO_x$; $NO_y$; sulfuric acid ($H_2SO_4$); $HNO_3$; and organic acids. Potentially harmful metals, such as chromium (Cr), vanadium (V), nickel (Ni), and lead (Pb), may also be deposited. Ammonia ($NH_3$) or its cation, ammonium ($NH_4^+$), is of interest because although it is a base in the atmosphere, it can react in soils after deposition to produce hydrogen ions.

Chemical reactions complicate the characterization of emitted contaminants. $SO_2$ follows a complicated path from emission to deposition. The gas and its chemical products are transported and diluted by the winds and turbulent eddies. The gas-phase reactions are photochemically stimulated and believed to be dominated by hydroxyl radical (OH) interaction as an intermediate product of $NO_x$-HC or $NO_x$-VOC reactions (Appendix A). A secondary reaction of $SO_2$ and an intermediate in the ozone reaction sequence may also be a factor. The aqueous-phase chemistry is believed to be dominated by an oxidation reaction involving dissolved $H_2O_2$. Other possible reactions may be dissolved oxygen interactions, catalyzed by such trace metals as manganous ion ($Mn^{2+}$), ferrous ion ($Fe^{2+}$) and ferric ion ($Fe^{3+}$), and by dissolved $O_3$.

The photochemical processes of oxidation are complex interactions between $NO_x$ and HCs emitted from both human activities and natural sources. The HC emissions in the East are widely distributed in a geographic pattern similar to those of $SO_x$ and $NO_x$. Major metropolitan areas are considered principal locations of photochemical activity.

The importance of the state of condensed water as a factor in atmospheric oxidation reactions is not known. That $SO_2$ absorption and oxidation occur in supercooled water, but not in ice crystals or other frozen hydrometeors, has been hypothesized.[1]

Dry deposition of $SO_2$ and $SO_4^{2-}$ occurs by gas absorption on surfaces or by particle scavenging on those surfaces. The rate of deposition differs between gases and particles and depends on the nature of the surface. Vegetation is a particularly variable scavenger. Wet deposition depends on scavenging of $SO_2$ and particulate $SO_4^{2-}$ within and below clouds. Scavenging takes place at varying rates, depending on absorption, nucleation, and collision processes (Appendix A). These processes, in turn, depend on the size of the cloud and rain drops, the presence of falling frozen water as ice crystals or graupel, and the updraft intensity (or vertical velocity) in the clouds.

$NO_y$ pathways to deposition are more complicated. The applicable physical processes of dry and wet deposition are the same for $NO_y$ and $SO_x$; however, processes at the surface, including biological assimilation, are probably different. $NO_x$ oxidation reactions result in several products, including $HNO_2$ and $HNO_3$ vapor as well as other $NO_y$ species (e.g., nitrogen pentoxide [$N_2O_5$] and $NO_3$). $NO_3^-$ is believed to be low in concentration in aerosol particles compared with $SO_4^{2-}$ because of the acid volatility and the volatility of salts, such as ammonium nitrate ($NH_4NO_3$).

A complex blend of source and atmospheric factors influences deposition patterns expected in the East.

- Distribution and relative geographic proximity of $SO_2$ and $NO_x$ sources
- Wind patterns for dry or storm conditions in the eastern terrain, including the surface-roughened mountainous areas
- Mixing conditions with summer and winter climatological regimes, including vertical structuring by convection
- Prevailing conditions that influence dry or occult deposition, including vegetation, soils, and other ground or surface cover (snow)
- Nature of atmospheric chemical activity with variation in altitude—for example, sunny conditions contrasted with cloudy conditions

[1] Although this hypothesis has not been disproved, it has been questioned with the results of laboratory studies showing that chemical reactions can occur in microscopic, supercooled liquid layers on ice surfaces (e.g., Lee and Kinsley, 1986).

- The ratio of ice and water in clouds
- Precipitation, especially in receptor locations with susceptible visual or biological material

## Factors Influencing Visibility

Visibility is a useful index of climatologically influenced moisture conditions and suspended particulate matter. Change in visibility is a direct consequence of change in the extinction coefficient for visible light. The extinction coefficient in pristine air approaches the so-called Rayleigh molecular scattering limit, which is the light scattering associated with the principal gas molecules in air. Light scattering and absorption from trace gases and suspended particles in the air are superimposed on this baseline. The Rayleigh scattering limit is rare in tropospheric conditions, but it is believed to occur occasionally in the arid southwestern United States. In most circumstances light extinction is far greater than the Rayleigh theory would predict. In ENA it is dominated by suspended-particle light scattering and absorption. For practical purposes, visibility depends mainly on the concentration, size–volume distribution, and chemical composition of suspended particles. The principal climatological factor affecting visibility is relative humidity. Because particles are often hygroscopic, their effective mass (and volume) is much larger in moist air than in dry conditions. Figure 2.1 shows that hygroscopic particles, typical of

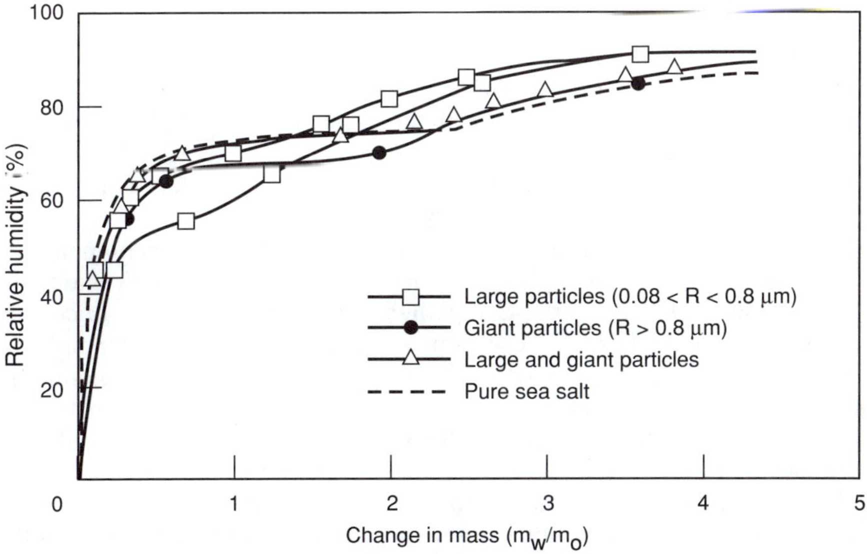


**FIGURE 2.1** Change in mass of suspended particles with relative humidity expressed as the ratio of wet mass ($m_w$) to dry mass ($m_o$). Curves correspond to collections of atmospheric particles in different radius ($R$) ranges. (From Winkler, 1967.)

atmospheric aerosol, absorb large quantities of water with increased humidity. Thus, visibility over ENA will be strongly linked to the moisture content of a large air mass. Workers often attribute reduced visibility mainly to airborne $SO_4^{2-}$ particles. However, humidity (or condensed water and other particulate matter) is also potentially important.

## CLIMATE OF EASTERN NORTH AMERICA

Climatic conditions in North America are characteristic of midlatitude continental conditions mediated by airflow from three marine regions. These conditions create a relatively dry climate in the middle of the continent, with a moist, variable climate in ENA.

### Airflow

The mean airflow across North America is from west to east. This flow may be distorted substantially at any given time by the north–south flow associated with large-scale wavelike disturbances traveling around the globe. Within the crests or troughs of these waves lie circulatory airflows that tend to rotate either counterclockwise (cyclones) or clockwise (anticyclones). The former are associated with areas of low barometric pressure and stormy weather, the latter with high barometric pressure and fair weather.

Three distinct major airstreams, and a fourth which is a modification of the first three, dominate the climate of ENA (Figure 2.2). Cold, dry arctic air comes from the north; warm, moist tropical air from the south; and mild, dry Pacific air intrudes from the west between the first two. A seasonal, semipermanent circulation, which mixes Pacific or arctic air, occurs over the southeastern United States in fall and winter. This mixing has been designated as a southern cyclonic regime.

The interaction of airstreams produces such weather phenomena as blizzards, thunderstorms, and very slow-moving (stagnating) regions of high pressure. When averaged together, these phenomena produce the various climates of ENA. These climates are strongly latitude dependent, which is modified somewhat by the Appalachians and other mountain ranges and by the Atlantic Ocean. The cold, dry air from the north has more effect at northern latitudes; the warm moist air influences southern latitudes. The relatively flat terrain at the Atlantic Coast has a minimal effect on the flow of air, but the inland mountain ranges are a factor in airflow. Airstream interaction occurs not only at the surface but also aloft. Warm air from tropical regions often overrides cold arctic air from the polar region to bring moisture and precipitation to northern latitudes.

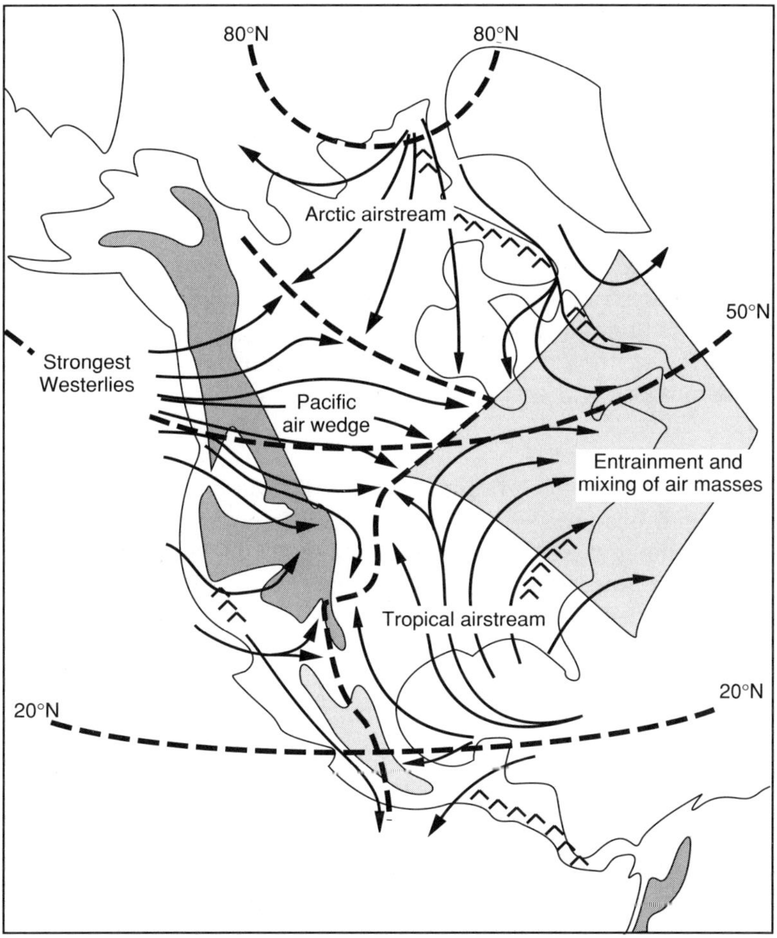


FIGURE 2.2 Schematic representation of the surface flow across North America, illustrating the wedge of Pacific air east of the Cordillera and the downstream entrainment and mixing of airstreams. Based on July resultant surface winds. (From Bryson and Hare, 1974; reproduced with permission of Elsevier Science Publishing Co.)

Similarly, Pacific air can override arctic and tropical air inland over the continent. Boundaries between warm and cold air, termed *fronts*, are often relatively sharp. These boundaries are unstable and are usually regions where stormy weather is prevalent.

Climate conditions are seasonal, controlled by the difference between incoming solar radiation and the earth's outgoing long-wave radiation. The difference in radiation varies with latitude as the solar declination changes. In winter the low angle of the sun results in very cold air at higher latitudes because of strong net outward radiation. A temperature gradient forms between the cold air to the north and the warm air to the south, and a band of westerly winds develops, generally centered around 45°N latitude at the surface and 30°N of the elevation of the jet stream. The high mountain ranges in western North America modify the Pacific air, but the westerly flow remains distinct. Large-scale instabilities develop that set up wave patterns the width of North America, in which very cold arctic air can penetrate to the Gulf of Mexico and moist tropical air can reach southern Canada. Cyclonic storms form along the thermal discontinuities or fronts between the air masses, bringing heavy snowfalls followed by cold polar airflow.

The north–south temperature gradient gradually decreases, and its maximum moves northward in spring and summer. The westerlies diminish in strength, and air from the Pacific Ocean seldom penetrates to the east coast. Convergence of cold, dry arctic air and moist, warm tropical air results in large areas of thunderstorms. The convergence zone moves northward and becomes weaker as summer progresses.

An anticyclonic circulation around a zone of high barometric pressure generally dominates the East during early autumn. The westerlies remain weak and at a more northern latitude. Tropical storms and hurricanes may be steered into the southeastern United States and more rarely into the Northeast. Such storms result in large amounts of precipitation. As insolation decreases at higher latitudes, the westerlies intensify and the winter pattern begins again.

## Mixing

A second factor closely associated with storminess is the mean pattern of mixing height and intensity across ENA. The mixing intensity of the atmosphere is linked with wind speed, the roughness of the earth's surface, and the heating and cooling of air near the ground. Qualitatively, the strength of mixing is characterized by the intensity of eddying (turbulence) and the height of penetration of eddying in the air. Figure 2.3 shows the average morning and midday mixing heights estimated from vertical temperature gradients in the air. Mixing heights are characteristically low at night and in the morning, when air is cooled near the ground relative to warmer air above. The ground is warmed by sunlight during the day. As air is heated by thermal transfer with the ground, mixing heights tend to increase substantially. In general, the summer heights are greater than winter ones, mainly because of the deeper, more intense eddying (convection) that occurs in summer with stronger surface area heating than in winter.

Mixing heights over ENA vary, with greater variation in the Midwest

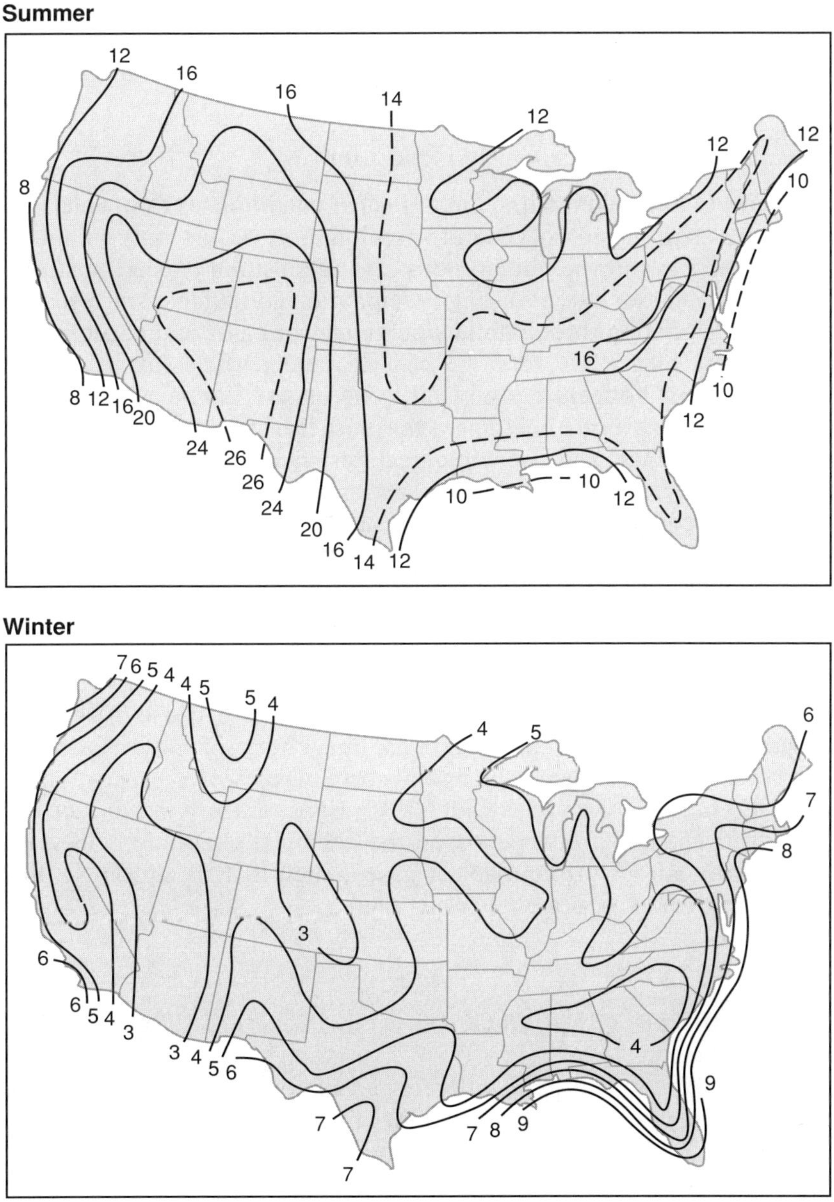


FIGURE 2.3 Mean daytime mixing heights in meters ($\times 10^{-2}$). (From Holzworth, 1972).

and Southeast than elsewhere. Convection spawned by large-scale storm systems is a factor in this distribution, as is the moderating influence of ocean conditions along coastal areas.

### Prevailing Conditions for Dry Deposition

A third factor linked with climatological conditions is that which influences dry deposition. In general, precipitation occurs only about one-third of the days in the midlatitudes. Dry deposition occurs most of the time and to a lesser extent during precipitation conditions, since scavenging for precipitation throughfall is inefficient for gases and particles (e.g., Pruppacher, 1986). Dry removal of gases and particles at ground level depends on the concentration of airborne material in air adjacent to the surface as well as a rate coefficient for mass transfer (deposition velocity) across the surface and the biological resistance to exchange (Appendix A).

Ambient concentrations of contaminants vary widely. However, dry deposition is usually most intense near sources where ambient concentrations are high. The deposition velocity depends on wind speed and roughness of the surface. Estimates of surface roughness are based on terrain, vegetation, and density and height of buildings in large metropolitan areas. Values of roughness based on aerodynamic criteria vary widely from 0 to 800 cm. Surface roughness changes seasonally because of vegetation growth or leaf loss, as well as the blanketing effects of snow cover. Snow cover and moisture also influence absorption of soluble gases, such as $SO_2$. Vegetation cover varies widely over ENA. Underlying surface resistance to exchange also depends on the chemical interactions in vegetation. Plants can absorb and reject gases, such as $SO_2$ and $NO_x$, during their biological cycles (e.g., Hicks, 1986).

## MOISTURE CONTENT, CLOUDS, AND PRECIPITATION

Water vapor in the atmosphere is an important ingredient of atmospheric chemistry. It acts as a reactant in photochemical processes and is a factor in creating wet particles as hosts for reactions well below the saturation point for droplet formation (e.g., Figure 2.1). Condensed water in various forms represents a productive aqueous regime for both chemical reactions and physical scavenging processes (Appendix A).

Relative humidity is an important measure of water vapor availability and the potential to form clouds. Over ENA the annual mean relative humidities are generally 70% or higher. Annual levels of 80% are found in parts of the Northeast and coastal Southeast. These high humidities at the surface are found at higher altitudes, too, so that the potential for fog and cloud formation is generally high.

In many locations fog or dew may enhance deposition. The interception of fog droplets and the condensation of moisture on surfaces provide a means of concentrating contaminants. Fog and cloud droplets are said to have higher concentrations of scavenged pollutant than precipitation. Although few measurements of dew chemistry exist, they show high concentrations of contaminants (e.g., Pierson *et al.*, 1987). Fog can occur through radiative cooling of air near the ground or by mixing of cool air with warm, moist air. Both can result in supersaturation of water vapor, leading to droplet formation. Fog is likely to occur in mountain valleys. A cold valley floor is conducive to radiation fogs, whereas fog in the form of low clouds may envelop the surrounding mountains. Fog can as much as double the amount of precipitation on mountaintops. In extreme conditions, for example, weather records at Mount Washington, NH, indicate 6–30 times as much fog at this location than at nearby lower elevations.

Figure 2.4 shows the distribution of fog in the United States. The mean annual occurrence of fog in ENA is most frequent in the coastal areas of New England, extending into the mountains of New Hampshire. Other areas of frequent fog are in the central Appalachians, extending from

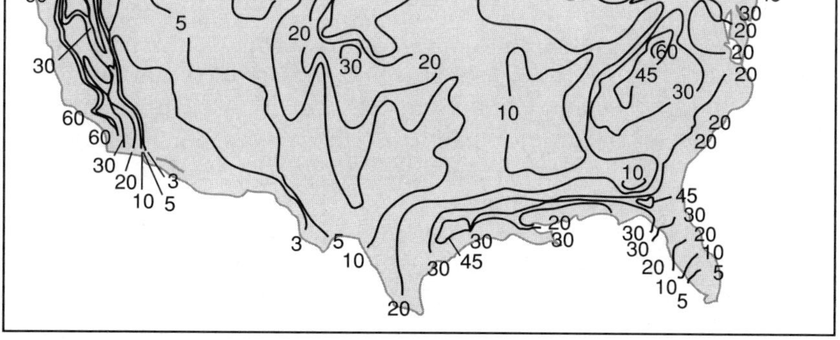


**FIGURE 2.4** Mean annual number of days with fog (visibility less than 300 m) from records of 251 stations through 1960. (From Court and Gerston, 1966, by permission of the American Geographical Society.)

West Virginia and Virginia into the Smoky Mountains of Tennessee, and the area along the Florida coast of the Gulf of Mexico. If concentrations of sulfate or nitrate are similar in fogs, one would expect enhanced deposition from fog interception to be high in the areas where frequency of occurrence of fog is high.

## Cloud Cover and Microphysics

The presence of clouds has a complex influence on chemical processes occurring in the air. These processes have been identified and characterized qualitatively. However, their significance in shaping the distribution of chemical reactants and products is not characterized regionally. Clouds influence homogeneous gas–phase processes and permit in-cloud processes. Layers of clouds substantially modify the amount of sunlight as a function of height, inhibiting photochemical activity. Cloud absorption of $SO_2$ and oxidants, such as $H_2O_2$, accelerate $SO_4^{2-}$ production and remove $HNO_3$ efficiently from the air. Clouds also provide hydrometeors for aerosol particle scavenging. They also act as distributors of suspended material after scavenging is followed by evaporation. If the clouds are formed with ice crystals or if the clouds are mixed with supercooled water droplets and ice, chemical reactions and in-cloud absorption rates are expected to be modified or possibly suppressed. Aerosol particle scavenging is affected by collection on crystal surfaces that can be better aerodynamic collectors than spherical particles.

A rough indicator for ''indexing'' cloud processes is average cloud cover. Figure 2.5 shows cloud cover over ENA in summer and winter. There are some differences, with winter showing added cloudiness over most of ENA. Cloud cover, of course, is linked with zones of high humidity. In ENA clouds are present on the average more than 30% of the time. Cloudiness tends to increase in the coastal areas and is also influenced by large inland water bodies, such as the Great Lakes. Differences are also relative to height. At mountaintop level in New Hampshire, for example, NOAA climate records indicate that it is cloudy most of the time. However, at both Concord and Burlington, NH, it is clear 1.5 to 2 times more frequently than over the mountains. At these sites precipitation often takes the form of snow and sleet, which reflects the possible significance of ice in clouds at these locations.

Because pollutants are concentrated near the ground, the effectiveness of the cloud environment on scavenging and chemical conversion will also depend, in part, on cloud height. Figure 2.6 shows an example of the portion of daily cloud cover with height. Cloud cover clearly varies diurnally and with height. Only about 20% of the cover is associated with clouds below 2 km in this winter example. If this is characteristic of most

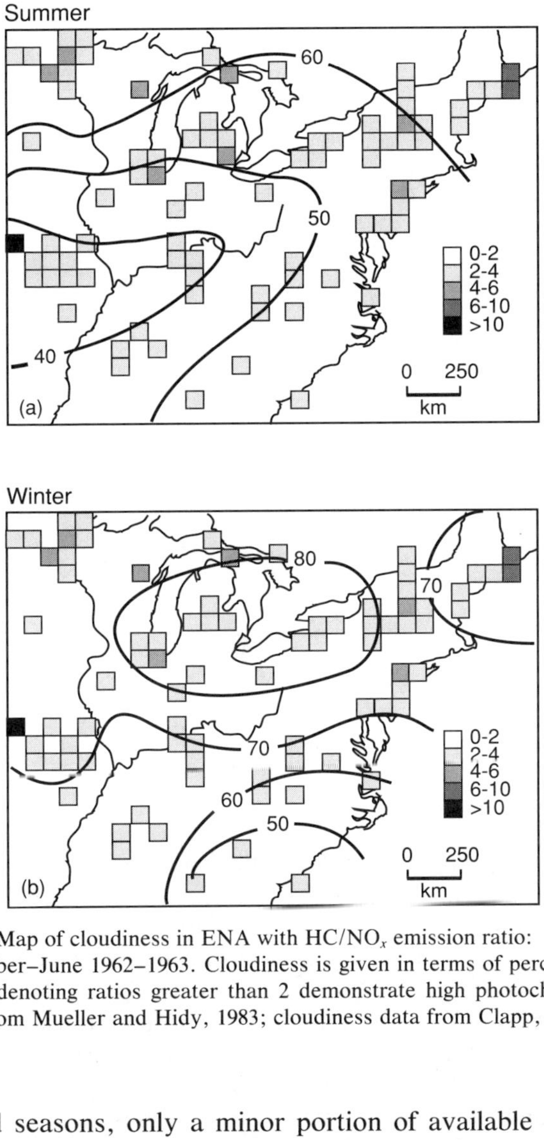


FIGURE 2.5 Map of cloudiness in ENA with $HC/NO_x$ emission ratio: (a) June–August 1962; (b) December–June 1962–1963. Cloudiness is given in terms of percentage covered. Shaded regions denoting ratios greater than 2 demonstrate high photochemical activity. ($HC/NO_x$ data from Mueller and Hidy, 1983; cloudiness data from Clapp, 1964.)

locations and seasons, only a minor portion of available clouds would come into contact with polluted air near the ground.

A simple index of air reactivity over ENA for oxidation rates of $SO_x$ and $NO_x$ and for production of oxidants (e.g., $O_3$, $H_2O_2$, and $HNO_3$) can be obtained by comparing cloudiness with the distribution of HC to $NO_x$ emissions. Cloudiness influences the actinometric conditions by selective absorption or scattering of solar radiation. The $HC/NO_x$ ratio is often used as an index for photochemical oxidant production (e.g., Hidy, 1984b).

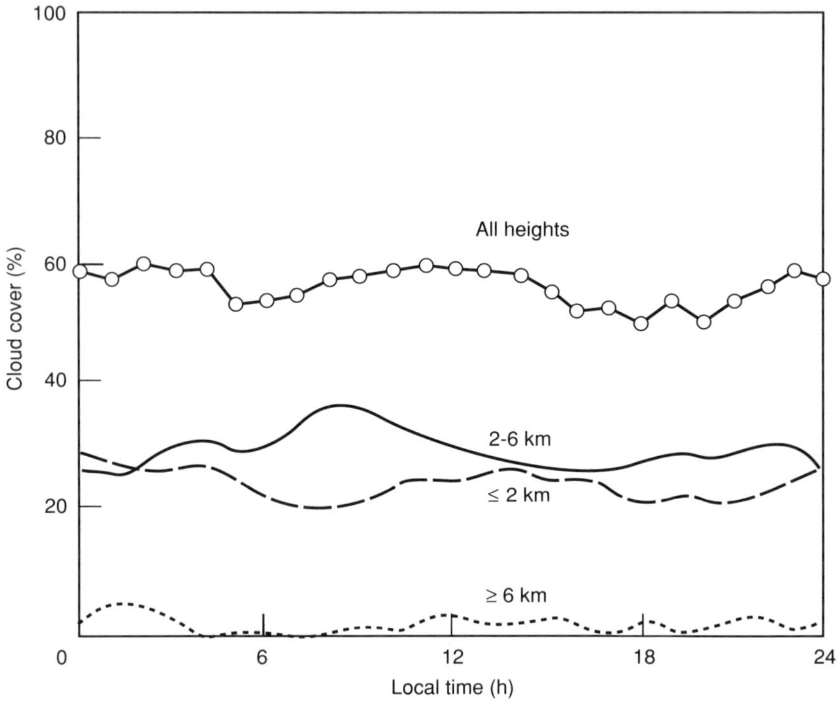


FIGURE 2.6 Example of total, low, middle, and high cloud cover, derived from geosynchronous orbiting earth satellite (GOES) data, for November 1978 at 39.4°N, 88.5°W (ca., Mattoon, IL). (From Minnis and Harrison, 1984.)

Ambient ratios of approximately 3 reflect possible active photochemical activity (Figure 2.5). High photochemical activity would be located around certain metropolitan areas, including New York, Philadelphia, Detroit, St. Louis, Atlanta, and cities in the Ohio River Valley. With high photochemical activity one observes relatively high oxidant production in these areas, reflected in elevated $O_3$ concentrations. The photochemical activity may be suppressed in cloudy areas around the Appalachian Mountains and the Great Lakes, while aqueous-phase reactions may dominate production of sulfate and nitrate in these areas.

## Precipitation

Because it is relatively easy to measure, the removal of airborne chemicals by precipitation is widely reported. Precipitation concentrations vary spatially less than precipitation quantity within storms, and consequently chemical deposition varies widely over ENA. Average precipitation patterns provide a climatology of areas where net deposition may be important to total chemical deposition. Climatologists sometimes divide Canada and the United States into several regions by considering the annual cycle of rainfall at climatological stations. Table 2.1 describes the regions of ENA. The precipitation patterns reflect the areas where boundaries between air mass types tend to occur.

Seasonal variation of rainfall at a particular location can be attributed to large-scale changes in the air mass types that influence the site throughout the year. Seasonal changes in airstream influence account for large-scale climatic regions. Each region also has local variations, which depend, for the most part, on local terrain character, including elevation, prevailing winds and insolation, and local sources of water vapor, such as large lakes. Mountainous terrain tends to promote strong nocturnal inver-

TABLE 2.1 **Climatic Regimes of Eastern North America Based on Annual Cycle of Precipitation Amounts**

| **Location** | **Description** | **Approximate annual precipitation amount (cm × $10^{-1}$)** |
|---|---|---|
| Ohio River Valley | Warm season maximum, cold season minimum | 10–12 |
| Tennessee | Maximum in winter/early spring, secondary maximum in late summer | 13–15 |
| Arkansas | Maximum in winter/early spring, weak secondary maximum in later summer | 12 |
| Oklahoma, Texas, Arkansas | Maximum in later spring, secondary maximum in early fall | 11 |
| Southeast Texas | Transition between Arkansas, Texas, Oklahoma, and the northern Gulf Coast | 11–13 |
| Northern Gulf Coast | Maximum in summer, secondary maximum in March | 16 |
| Florida | Maximum in summer, slight secondary maximum in March | 13 |
| Virginia, North Carolina, South Carolina | Maximum in summer, secondary maximum in March | 12 |
| Middle Atlantic | Slight summer maximum, partly uniform rest of year | 11 |
| New England | Fairly uniform | 10–11 |
| Maritime Canada | Winter exceeds summer | 10 |
| Newfoundland, Labrador | Fairly uniform | 10 |
| Quebec, Ontario, northern New York | Maximum in warm season | 9 |

sions in the valleys with decoupling of local airflow from the large-scale motion aloft. However, daytime heating breaks up inversion conditions, and air from the valleys mixes with the air aloft. Precipitation generally increases with elevation, and snow is more likely at higher elevations. This increase is often the result of orographic (terrain-forced) lifting of air to the water saturation level, where clouds can form and allow precipitation to develop.

Figure 2.7 shows annual mean levels of precipitation over ENA. The

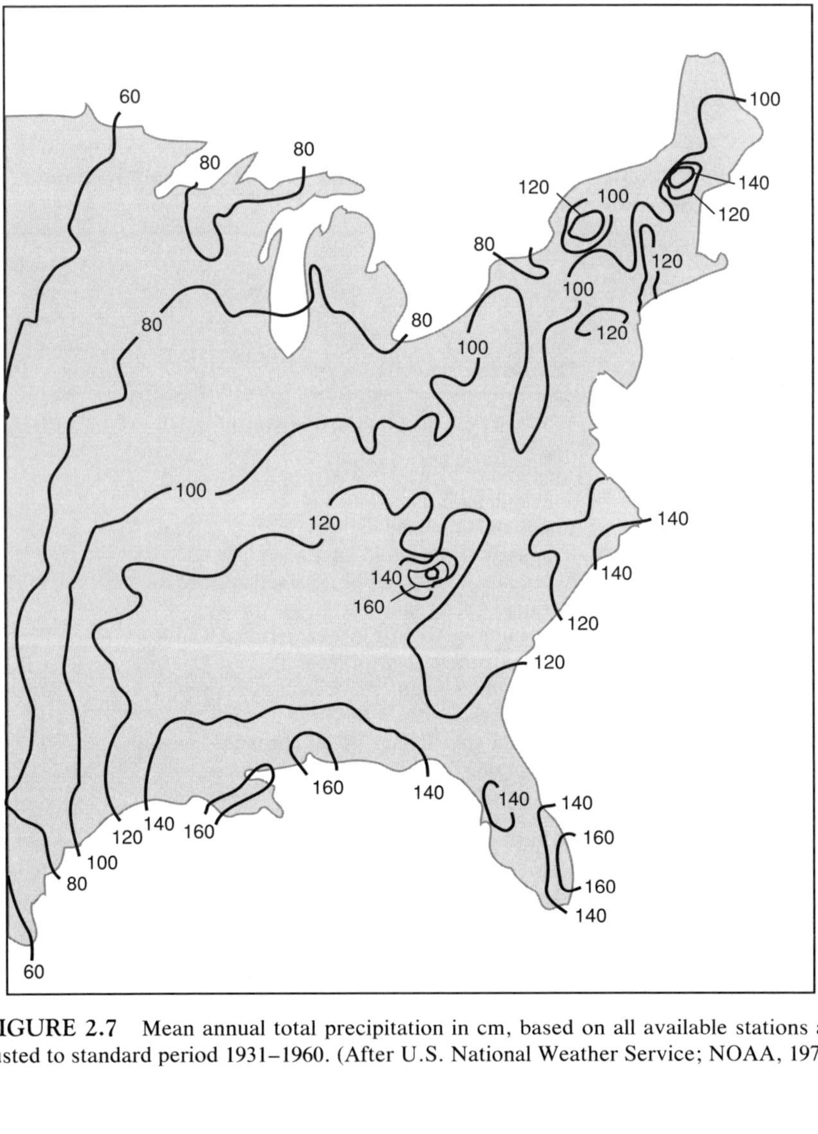


FIGURE 2.7 Mean annual total precipitation in cm, based on all available stations adjusted to standard period 1931–1960. (After U.S. National Weather Service; NOAA, 1974.)

strongest superficial gradients in precipitation are located in mountainous regions. The figure shows the year-to-year variation in precipitation quantity compared with the storm-by-storm change; the year-to-year variability in precipitation quantity over ENA is small. Areas of high precipitation also reflect the areas where persistent encounters between different classes of air mass tend to occur. There is a tendency for high levels of precipitation where warm moist air mixes with cool air. In the easternmost region of North America, numerous storms evolve from disturbances along a nominal boundary between air from the arctic and moist air from the Gulf of Mexico or the Pacific.

Climatological maps of precipitation patterns in North America tend to mask the influence of terrain. For the most part, measurements are made in valleys, and terrain features are only weakly evident. One exception is the Mount Washington Observatory in New Hampshire, which is on an exposed mountaintop more than 2000 m above sea level. Mount Washing-

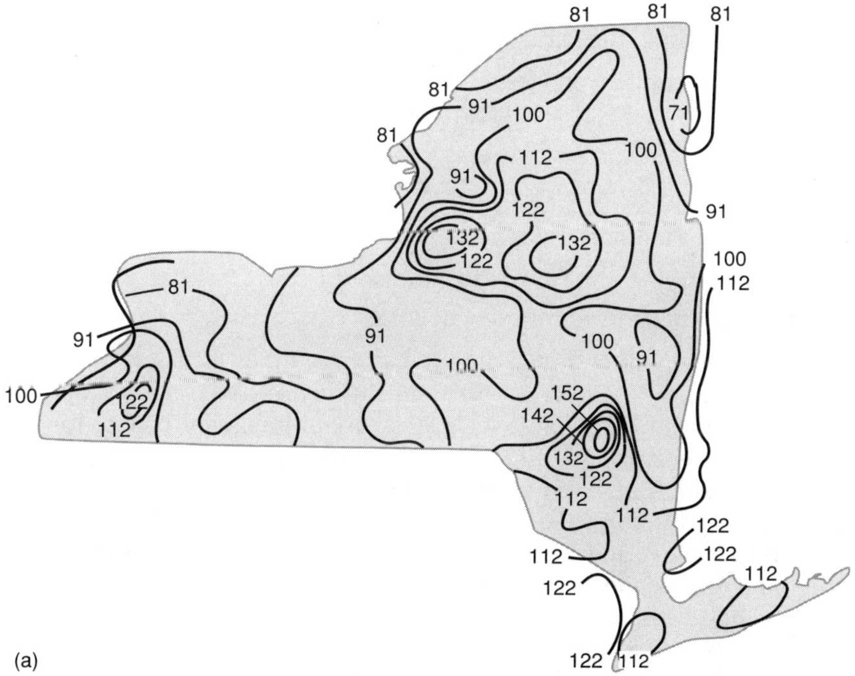


(a)

**FIGURE 2.8** (a) Mean annual precipitation in cm for New York. Data taken from *Climate of the States* (U.S. National Weather Service). Isolines are drawn through points of approximately equal value. (b) Mean annual precipitation in cm for New England. Data taken from *Climate of the States* (U.S. National Weather Service). Isolines are drawn through points of approximately equal value.

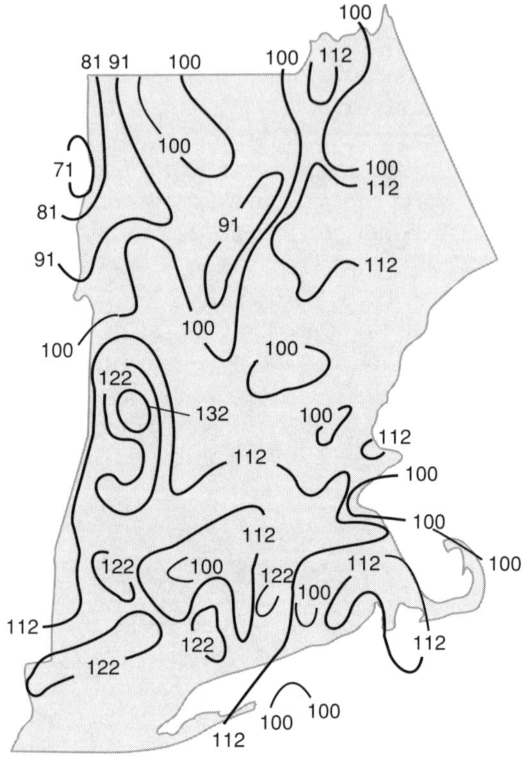


(b)

FIGURE 2.8 *(continued)*

ton has twice as much precipitation and three times as much snowfall as nearby locations at Concord, NH and Burlington, VT.

Figure 2.8 shows the mesoscale structure of precipitation rate for New York and New England, which have a generally mountainous character. The orographic influence on precipitation in the Adirondacks, the lake effect around Buffalo, the marine and orographic effect of the mid-Hudson River Valley, and the orographic influence around the Green Mountains and the White Mountains are apparent. The marine influence is perhaps best seen north of Long Island, NY along the southern coast of Connecticut.

Figure 2.9 shows the effect of terrain on precipitation in the Great Smoky Mountains of North Carolina, where a finer-scale network of stations was set up by the Tennessee Valley Authority (TVA). The contours of equal precipitation levels show complicated patterns of precipitation, which depend on the mountainous terrain.

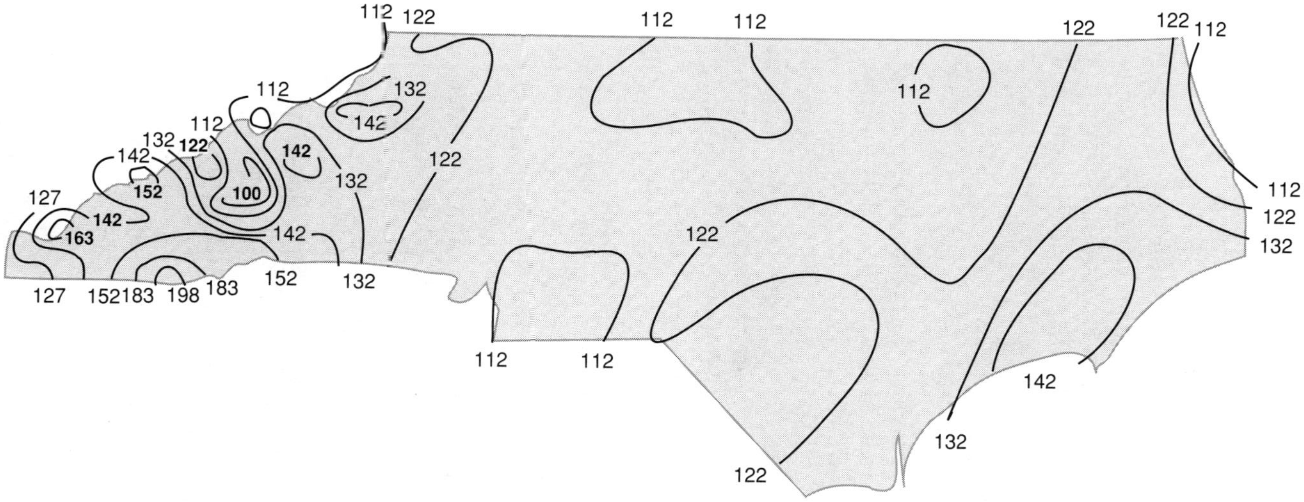


**FIGURE 2.9** Mean annual precipitation in cm for North Carolina. Isolines are drawn through points of approximately equal value. Caution should be used in interpolating on these maps, particularly in mountainous areas. Data taken from Climate of the States.

## VARIABILITY IN AEROMETRIC FACTORS

An important aspect of climatological phenomena is their variability. Variability can be measured spatially or in terms of time. Temporal observations can be arranged to show seasonal oscillations or cover several years. Relatively little is known about variability in atmospheric chemical processes, because observations have been acquired in a highly irregular manner. However, long-term meteorological records are helpful in delineating the variability associated with physical processes.

### Airflow and Wind Trajectories

Mean flow patterns provide information about atmospheric motion, but they lose the daily variability that typifies midlatitude weather. There are two main types of mean airflow disturbance, depending on their mode and location of formation. Cyclones and anticyclones form as perturbations of the westerly flow at middle to upper latitudes. Cyclonic disturbances described curvilinear airflow around zones of low or high (anticyclones) barometric pressure. These disturbances are associated with large-scale wavelike wind patterns that propagate around the globe. At low latitudes, tropical storms and hurricanes form as disturbances of the easterly trade winds that involve strong curvilinear flow, often described as a "vortex."

Cyclones and anticyclones have a strong effect on the precipitation of ENA. Because of their association with the fronts of the westerly flow, they vary seasonally in location and behavior. The latitudes of greatest occurrence were 45°N in winter, 43°N in spring, 54°N in summer, and 48°N in autumn. More disturbances occur over a wider area in winter and spring than in summer and autumn. Cyclones bring precipitation in the form of rain in summer and snow in winter. Thunderstorms and other violent weather are also associated with cyclones. Anticyclones generally bring fair weather but can be associated in very cold or highly stagnating conditions with little air movement. The latter is conducive to pollution accumulation over a large region.

### Air Mass Trajectories

The flow of air from one location to another is often described by the concept of *air mass trajectory*. Trajectories are calculated from the wind field by following (coordinates on) a parcel of air traveling across the ground over time, a *Lagrangian frame of reference*. In reality, air masses do not travel for great distances intact but mix with surrounding air. Thus the picture of air parcels is an ideal conceptual construct. Nevertheless, calculated trajectories of air parcels on a large scale are useful indicators of pathways between air over a source and air reaching a particular receptor location, because horizontal mixing is relatively weak in many circumstances. Air mass trajectories, which can be calculated in different ways (Chapter 6), are rarely straight paths between locations. Beginning and end positions often needed to establish source–receptor relationships are only coarse indicators of actual patterns.

Repetitive calculations can be made for dry or wet conditions within a year or from year to year using meteorological data (e.g., Wilson, 1984). Figure 2.10 shows the seasonal variability for dry conditions in such calculations and the relationship between sources and receptors (SRRs) for a receptor area in the Adirondack Mountains. Calculations were made of 24-hour trajectories backward (upwind) from this location by Heffter's (1981) method. In the example in Figure 2.10a and b, winter "exposure" is influenced more heavily by airflow from the north and west. In summer, airflow to the Adirondacks comes more uniformly from all directions.

The calculations also indicate the variability in pathways of air motion representing dry and wet conditions (Figure 2.10c and d). The trajectory origins to the south are biased for wet conditions compared with dry. Comparisons between the years 1978 and 1982 suggest a range of difference in transport origins that would be significant in tracing sources responsible for exposure at a given location. The factors influencing SRRs are variable. Clearly, the wind trajectory pathways are a primary factor, and perhaps the single most critical factor, influencing SRRs. Trajectory pathways indicate whether airborne material can reach a receptor. Their variability is detected by the cycle of large-scale storms and fair weather. In any time period, winds can shift, creating potential exposure from a wide range of distances and directions. The level of exposure depends on the dilution and mixing of material from different sources upwind, in addition to loss in transit by deposition.

## Mixing

The mixing process varies with air density stratification and wind speed. Its variability is reflected in the range of scales of turbulence in the atmosphere. A *turbulent energy spectrum* (wind variability as a function of inverse time scale) provides a measure of the intensity of mixing. The higher the energy, the more intense the mixing rate. Typically a diurnal ($\frac{1}{12}$ h$^{-1}$) peak occurs in the frequency spectrum. At higher frequencies, rapid wind variations exist that show intensification of mixing with heating and wind increase during the day, followed by calmer conditions at night. These variations generally are the ones identified with the long-term mixing. Evidence suggests a multiday cycle at lower frequencies, consistent with the cyclone development and passage in the wind field.

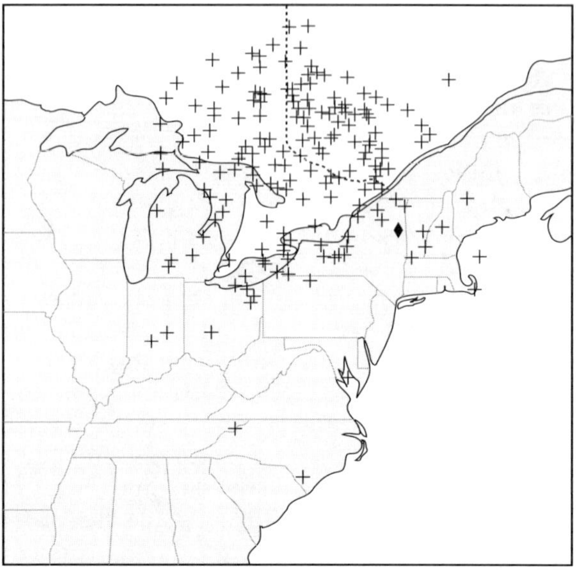

(a) January-March 1978; Nonprecipitation

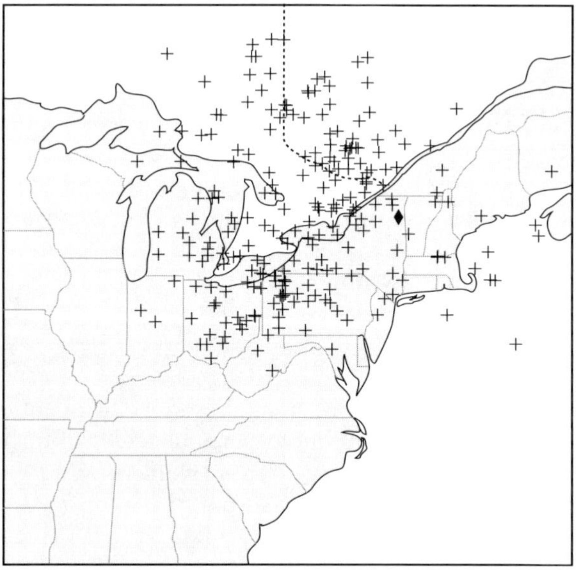

(b) August-September 1977, July 1978; Nonprecipitation

FIGURE 2.10 Calculation of 24-hour upwind air mass trajectory end points from a receptor in the Adirondacks for summer and winter conditions in 1978. (a and b) Nonprecipitation; (c and d) precipitation. Calculations based on the Heffter (1981) model, assuming winds representation of the planetary boundary layer. (Courtesy of J. Wilson; graphs from Hidy *et al.*, 1985; reproduced with permission, Electric Power Research Institute.)

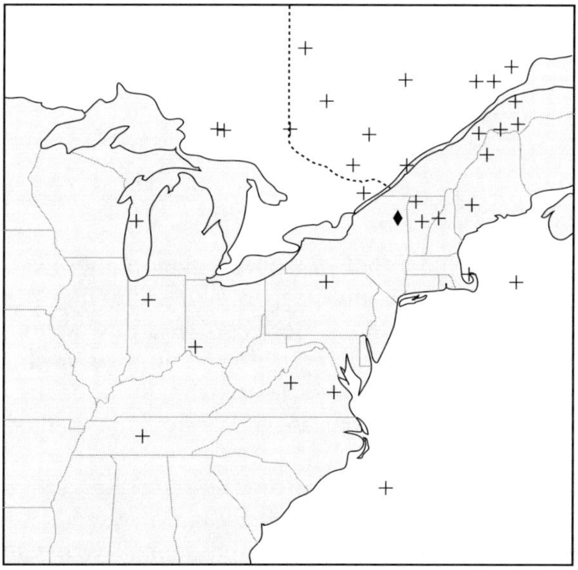

(c) January-March 1978; Precipitation

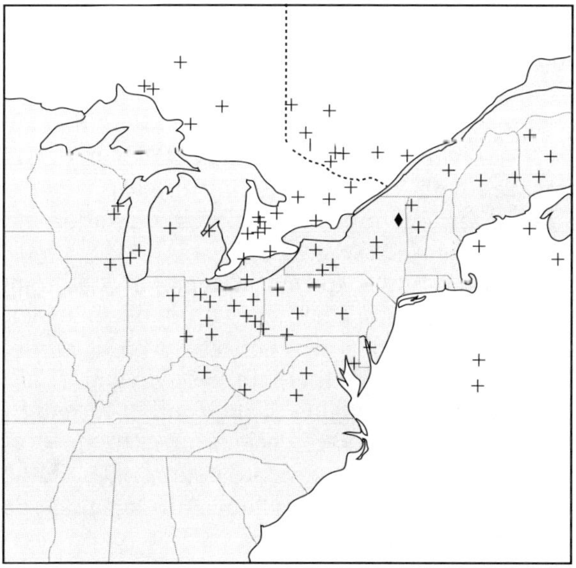

(d) August-September 1977, July 1978; Precipitation

FIGURE 2.10 *(continued)*

Observations have documented diurnal changes in mixing height with formation of temperature inversions at night, which break up in daylight. Although the seasonal changes in mixing height have been observed and recorded, little information is available about weekly variability with storm passage.

### Cloudiness and Precipitation

Climatological data show that sunshine and cloudiness vary geographically. One can infer seasonal and year-to-year variability from these records; however, such variability is not reported. Few summaries of seasonal and year-to-year variability in clouds, fog, or mist exist. In the mountains, the occurrence of fog varies seasonally; it occurs less in summer. Data summaries of this type are available from analysis of the surface observations.

Microphysical conditions vary seasonally and from year to year. Much greater ice content is expected in winter than in warmer seasons. Summaries of ice and liquid water content and year-to-year variability from satellite remote sensing are obtainable, but no reports of observational analysis are available.

Other than wind fields, perhaps the best documented climatological variability is precipitation. Precipitation changes in character seasonally from rain to snow in the midlatitudes. The year-to-year variability is reflected on a large scale by the coefficient of variation in annual precipitation (Figure 2.11). Over much of ENA the coefficient of variation is less than 20%, compared with larger values in the West.

Data from a long-term national record of precipitation have been analyzed for dry and wet periods (e.g., Diaz and Quayle, 1980; Diaz, 1983). Long dry and wet periods give extremes in regional conditions for wet and dry deposition. As discussed in Chapter 3, droughts may be identifiable in acidity chemistry. In particular, dry periods have signs of elevated calcium and other soil dust material that imply scavenging of high suspended dust concentrations. The scavenging of such cation-forming species tends to reduce acidity in precipitation. Periods of extreme drought vary in regional areas over a long period of record (Diaz, 1983). Although drought varies by subregion, in some circumstances it can be widespread over portions of the continent and persist for several years.

Atmospheric meteorological and climatological processes vary over a wide range of space and time scales. The time scales relevant to chemical exposures can range from hours to decades and beyond. The measures of atmospheric chemical parameters are generally too brief in record to separate meteorological variability from other factors, such as emissions variability. An indication of the variability in chemical factors can be

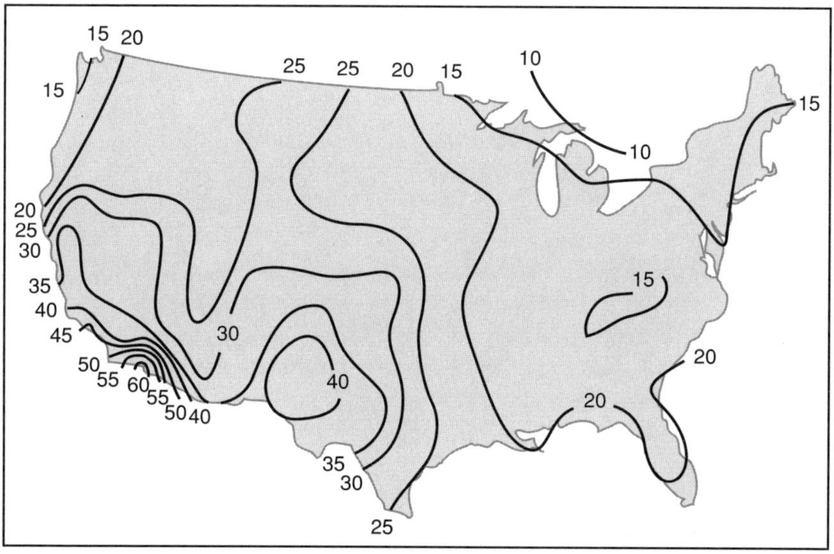


FIGURE 2.11 Coefficient of variation (%) of annual precipitation, 1931–1960, based on 220 stations in the United States (from Hershfield, 1962; courtesy of the American Meteorological Society.)

extracted from precipitation data in a few locations, as well as from regionally representative ambient air quality data. Comparison of variations in $SO_x$ concentrations with variations of estimated emissions for a period of a year or more suggests that meteorological (or aerometric) factors dominate the observed variability in both ambient concentrations and precipitation (see Mueller and Hidy, 1983; Hidy *et al.*, 1985). Thus, variability is important for interpreting data collected for limited periods. The analysis of variability provides critical information about both meteorological and chemical factors that are involved in regional-scale processes.

# 3 CHARACTERIZATION OF REGIONAL AIR AND DEPOSITION CHEMISTRY

## OVERVIEW

$SO_x$ chemistry concerns mainly two constituents: $SO_2$ and its oxidation product, $SO_4^{2-}$. Anthropogenic $SO_2$ is mainly produced as a by-product of fossil fuel combustion; some $SO_3$ or $H_2SO_4$ mist is also produced from combustion processes. $SO_3$ is generally only 1–2 percent of total $SO_x$ emissions. Sulfate produced in the atmosphere or from direct emissions takes the form of suspended particles or is contained in cloud and precipitation water. Other S constituents, including reduced S gases and sulfite, $SO_3^{2-}$, are of interest, are unstable to oxidation, and are considerably lower in atmospheric concentration.

$NO_x$ chemistry and $NO_y$ chemistry are more complicated. Two ($NO_x$) species, NO and $NO_2$, are produced by combustion and are emitted. These oxidize and react to form $NO_y$ products that include ($N_2O_5$), $HNO_3$, and organic nitrates and peroxynitrates, such as PAN. In the presence of aerosol particles, nitrate salts, including $NH_4NO_3$, may also form.

The geographic distributions of $SO_x$ emissions in North America are shown in Figure 3.1. These patterns indicate that the main concentrations of emissions are in the eastern United States and the eastern U.S.–Canadian border region of Ontario and Quebec. In the West, sources are generally absent or widely spaced (isolated), whereas the area east of the Mississippi River has more closely packed sources, associated with major industrialized areas and large population centers. The geographic distri-

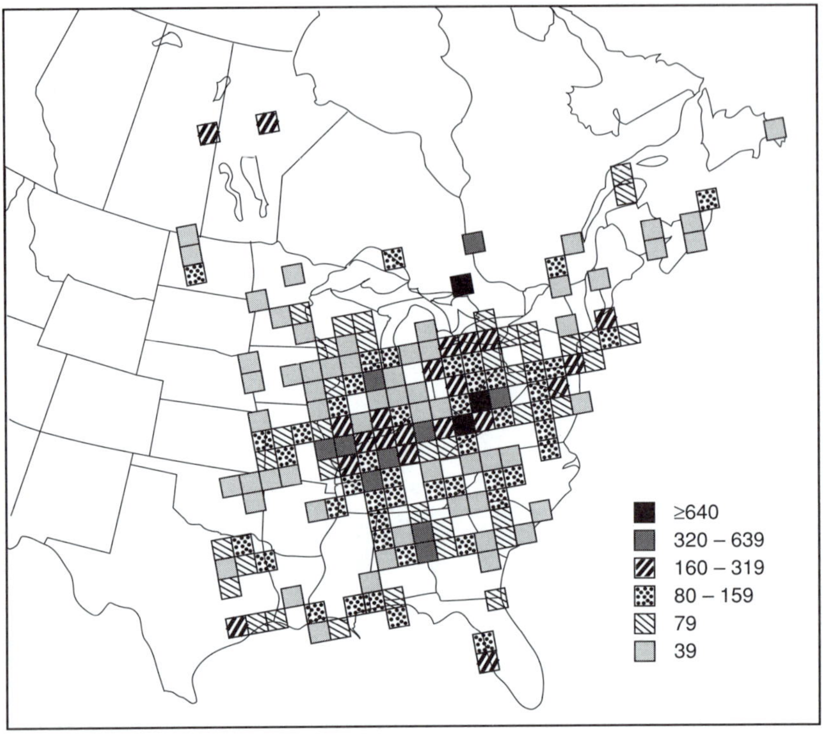


**FIGURE 3.1** 1985 $SO_2$ emissions distribution in eastern North America (kilotonne per 127 km × 127 km) grid cell. (From Canadian Federal/Provincial Research and Monitoring Coordinating Committee, 1990.)

butions of $SO_x$ and $NO_x$ are similar in ENA. In the United States, about 90% of the 1985 $SO_x$ emissions were concentrated east of the Mississippi. Similarly, the majority of the $NO_x$ emissions are in the eastern half of the continent.[1] The $SO_x$ and $NO_x$ emissions in ENA are also similar in density and size to those found in western Europe.

There are also emissions of sulfur components and nitrogen oxides, as well as reactive hydrocarbons, from natural sources throughout ENA. The sulfur compounds are biogenic in origin or sulfate associated with sea spray. The former include compounds such as hydrogen sulfide, dimethyl sulfide, or carbonyl sulfide. These are readily oxidized in the atmosphere, ultimately to form sulfate. The natural sulfur sources are believed to derive mainly from biogenic processes in standing water, wetlands, and

[1] Details of anthropogenic $SO_x$, $NO_x$, and HC emissions inventories from the mid-1970s can be found in Mueller and Hidy (1983), Heisler *et al.* (1988), and NAPAP (1990a).

intertidal zones. Nitrogen oxides derive from soil biogenic processes, forming nitrous oxide ($N_2O$) and NO, as well as from lightning discharges, producing NO and $NO_2$. $N_2O$ is considered reactive in the lower atmosphere. Natural production of $SO_x$ and $NO_x$ species is estimated generally to be small relative to anthropogenic sources in ENA. They are larger in summer than in winter. NAPAP (1990a) reviews natural sources over ENA.

Natural HCs or VOCs are emitted from both terrestrial and marine sources. Most reactive are terpene compounds from trees and brush. Biogenic methane ($CH_4$) is by far the most prevalent natural hydrocarbon species in the air, but is far less reactive than other components. There is a continuing debate about natural HC emissions and production of regional $O_3$ in ENA. However, most estimates suggest that anthropogenic HC levels are generally higher than natural HC concentrations in cities, probably dominating oxidant production in urban areas. But estimated regional natural VOC emissions are comparable to or larger than anthropogenic rates, particularly in summer. Thus, natural VOCs may influence the regional oxidant production in summer, especially in rural areas.

$SO_4^{2-}$ and $NO_3^-$ chemistries are closely linked with oxidation processes. The latter depends on the photochemistry of atmospheric $NO_x$ and HC mixtures (Appendix A). Measures of oxidants are important for the characterization of regional S and N chemistry. Aside from oxygen ($O_2$) itself, the oxidants observed are $O_3$ and $H_2O_2$, as well as $HNO_3$ and other $NO_y$ species. These are produced with other oxidants in HC-$NO_x$ mixtures found in the atmosphere. HC mixtures in the atmosphere deriving from anthropogenic sources are numerous, including most commonly $C_2$–$C_{12}$ paraffins and olefins, as well as simple aromatic compounds such as benzene and toluene. Oxygenated organics include aldehydes and acids. The olefins and oxygenates are generally the more reactive species.

Any analysis of the regional air chemistry of ENA must also consider a global continental background. Atmospheric conditions in remote areas are variable but are considered representative of a background. The residual material found in remote areas comes from both natural and anthropogenic sources. The residue represents a result of aging, mixing, and loss processes balanced against sources over the earth. Knowledge of background conditions in the Northern Hemisphere and on the North American continent is limited but sufficient to provide a basis for discussion. Table 3.1 gives typical values for ambient concentrations of $SO_2$, $NO_2$, $HNO_3$, and $O_3$ at remote sites. Table 3.2 lists values of particle sulfate and nitrate concentrations. The S and N gases of concern are in the parts per billion by volume (ppbv) range and below. These levels are consistent with measurements of S and N gases at midlatitude sites over the western Atlantic Ocean (e.g., Dredgen *et al.*, 1986; Bottenheim and Gallant, 1986).

TABLE 3.1 **Typical Concentrations of Atmospheric Gas Constituents Representative of North American Continental Baseline Conditions**

| **Location** | **Elevation (m)** | **Measurement period** | **Mean concentrations (ppb)** $SO_2$ | $NO_x$ | $HNO_3$ | $O_3$ | $NH_3$ | **Reference** |
|---|---|---|---|---|---|---|---|---|
| Boundary layer over western United States | 3000 | 1977–1978 | — | — | 0.06–0.08 | — | — | Huebert and Lazrus (1980) |
| Niwot Ridge, Colorado (80 km NW of Denver) | 3050 | 2/78–8/79 | — | 0.2–0.25 | <0.03–0.1 | — | — | Kelly *et al.* (1980) |
| Thurber Ranch, Arizona (55 km SW of Tucson) | 1600 | Summer 1980 | 0.3 | — | 0.3 | — | 0.8 | Dawson and Farmer (1984) |
| | | Summer 1981 | 2.1 | — | 0.6 | — | 1.1 | |
| Sells, Arizona (100 km SW of Tucson) | | Winter 1980 | 0.8 | — | 0.4 | — | 0.9 | Dawson and Farmer (1984) |
| Mt. Lemmon, Arizona (32 km NMW of Tucson) | 2800 | Summer 1980 | 0.2 | — | 0.2 | — | 1.0 | Dawson and Farmer (1984) |
| Saguaro National Monument, Arizona (29 km W of Tucson) | | Summer 1980 | 0.2 | — | 0.2 | — | 1.0 | Dawson and Farmer (1984) |
| | | Winter 1980 | 0.4 | — | 0.6 | — | 1.9 | |
| El Nido, California (200 km SE of San Francisco) | | Spring 1981 | 2.0 | — | 1.2 | — | 10.0 | Dawson and Farmer (1984) |
| Niwot Ridge, Colorado | 3050 | January 1981 | — | — | — | 40 | — | Fehsenfeld *et al.* (1983) |
| | | July 1981 | — | — | — | 48 | — | Fehsenfeld *et al.* (1983) |
| | | 1981–1982 | — | 0.9 | — | — | — | Fehsenfeld (1985), unpublished |
| | | 6–10/81 | — | 0.5 | — | 43 | — | Roberts *et al.* (1984) |

| | | | | | | | | |
|---|---|---|---|---|---|---|---|---|
| Clear Creek, Colorado | 2600 | 11/80–12/81 | ≤1.7 | 1.3 | — | 58 | — | Young (1982) |
| Gasquet, California (near Crescent City) | 100 | 1989 | 0.02 | 1.7 | 0.01 | — | 1.2 | California Air Resources Board (1991) |
| Rock Creek, California (34 km SE of Sacramento) | 3020 | 7–12/84 | 3.0 | 5.0 | — | 42 | — | Science Applications International (1985a,b) |
| Chico, California (150 km N of Sacramento) | 50 | 1983 | — | 11.0 | — | 20 | — | California Air Resources Board (1983) |
| Sequoia NP, California | 2000 | 1983 | — | — | — | 51 | — | Calfiornia Air Resources Board (1983, 1991) |
| | | 1988–1989 | 0.3 | 0.005 | 0.23 | — | 2.3 | |
| Gothic, Colorado (200 km WSW of Denver) | 2900 | 8–9/84 | — | — | — | 22 | — | Bruck *et al.* (1985) |
| Scobey, Montana (northwestern Montana) | | 1983 | 1.0 | — | — | — | — | Sternberg (1985) |
| Devils Tower, Wyoming (east central Wyoming) | | 1983 | — | 2 | — | — | — | WDEQ (1984) |
| Cree Lake, NWT Canada | | 1983–1987 | 0.4 | — | 0.04 | — | — | Sirois and Fricke (1992) |
| Experimental Lakes Area Manitoba, Canada | | 1983–1987 | 0.3 | — | 0.1 | — | — | Sirois and Fricke (1992) |
| Yampa, Colorado | 2000 | 1987–1988 | 0.36 | 1.0 | 0.22 | 52.0 | 0.23 | EMEFS (1991)[a] |

[a] EMEFS data summaries, summer 1991, courtesy of M. A. Allan.

TABLE 3.2 **Typical Concentrations of Atmospheric Particle Constituents Representative of North American Continental Baseline Conditions**

| Location | Measurement period | Concentrations in fine particle ($\mu g/m^3$) $SO_4^{2-}$ | $NO_3^-$ | Reference |
|---|---|---|---|---|
| Clear Creek, CO | 11/80–12/81 | 1.1 | 0.08 | Young (1982)[a] |
| Yampa, CO | 1988–89 | 1.0 | 0.2 | EMEFS (1991) |
| Rock Creek, CA | 7–12/84 | 0.9 | 0.2 | SAI (1985a,b)[a] |
| Chico, CA | 1983 | 2.1 | 3.1 | CARB (1983)[a] |
| Sequoia NP, CA | 1983 | 4.7 | 3.1 | CARB (1983)[a] |
| Gasquet, CA | 1989 | 0.84 | 0.32 | CARB (1991)[b] |
| Scobey, MT (northwestern MT) | 1983 | 3.4 | 0.8 | Sternberg (1985)[a] |
| Polebridge, MT (northwestern MT) | 1983 | 2.6 | 0.8 | Sternberg (1985)[a] |
| Ranan, MT | 1983 | 1.9 | 0.9 | Sternberg (1985)[a] |
| Experimental Lakes Area, CA | 1989 | 1.46 | 0.3 | Sirois and Fiske (1991) |
| Cree Lake, CA | 1989 | 0.93 | 0.06 | |
| WRAQS[c] Data | 9/81–8/82 | | | Tombach *et al.* (1985) |
| Harlton, Montana | | 0.83 | 0.06 | |
| Little Butte, Idaho | | 1.06 | 0.21 | |
| Fish Pathfinder, Wyoming | | 0.97 | 0.05 | |
| Creek Ranch, Nevada | | 0.68 | 0.06 | |
| Delta, Utah | | 1.16 | 0.13 | |
| Ouray, Colorado | | 0.81 | 0.13 | |
| Walsenburg, Colorado | | 0.96 | 0.05 | |
| Blythe, California | | 1.59 | 0.19 | |
| Sunshine, Arizona | | 1.50 | 0.08 | |
| Encino, New Mexico | | 1.41 | 0.05 | |
| Sierra Vista, Arizona | | 2.68 | 0.07 | |

[a] Total suspended particulates ($NO_3^-$ probably overestimated).
[b] Daytime data particles less than 2.5 μm in diameter.
[c] Western Regional Air Quality Study.

$O_3$ is in the range of 10 to 50 ppbv. Concentrations of particles smaller than 10 $\mu$m in diameter are 15–20 $\mu g/m^3$ or lower, with $SO_4^{2-}$ concentrations less than 5 $\mu g/m^3$ and $NO_3^-$ less than 1 $\mu g/m^3$.

Table 3.3 lists major ion concentrations for five remote sites over the world. The locations outside the United States were chosen to be as free from pollution as possible. The two California sites included a remote coastal/redwood site (Gasquet) and a rural northern Sierra Nevada community (Quincy). Volume-weighted pH levels at the world sites were approximately 5 ($H^+$ concentrations of 11–16.6 $\mu$mol/L). Sulfate concentrations resulting from sea salt injections into the air range from 5.4 to 36.6

TABLE 3.3 **Annual Volume-Weighted Precipitation Chemistry Data from Remote Sites around the World**

| Component | Amsterdam Is., Indian Ocean[a,c] pH: 4.92 | Poker Flat, Alaska[a,d] 4.96 | Katherine, Australia[a,d] 4.78 | San Carols, Venezuela[a,d] 4.81 | St. George's, Bermuda[a,c] 4.79 | Gasquet, California[c,e] 5.34 | Quincy, California[d,e] 5.27 |
|---|---|---|---|---|---|---|---|
| | | | | Concentration (μmol/L) | | | |
| $H^+$ | 12.0 | 11.0 | 16.6 | 15.5 | 16.2 | — | — |
| $SO_4^{2-}$ | 30.6 | 14.4 | 12.6 | 5.8 | 72.6 | 14.4 | 7.8 |
| $SO_4^{2-}$ (ex)[b] | 17.6 | 14.2 | 11.0 | 5.4 | 36.6 | 5.8 | 6.6 |
| $NO_3^-$ | 1.7 | 1.9 | 4.3 | 2.6 | 5.5 | 1.9 | 4.7 |
| $NH_4^+$ | 2.1 | 1.1 | 2.0 | 2.3 | 3.8 | 2.4 | 4.2 |
| $Ca^{2+}$ | 14.8 | 0.2 | 5.0 | 0.6 | 19.4 | 4.6 | 2.7 |
| $Mg^{2+}$ | 77.4 | 0.4 | 4.0 | 1.0 | 69.0 | 8.3 | 1.3 |
| $Na^+$ | 177.0 | 1.0 | 7.0 | 1.8 | 147.0 | 35.4 | 4.9 |
| $K^+$ | 3.7 | 0.6 | 0.9 | 0.8 | 4.3 | 1.2 | 1.0 |
| $Cl^-$ | 208.0 | 2.5 | 11.8 | 2.5 | 175.0 | 41.8 | 5.8 |
| $SO_4^{2-}/NO_3^-$ | 10.3 | 7.5 | 2.6 | 2.1 | 6.7 | 3.1 | 1.4 |

[a] Reprinted from Galloway *et al.* (1982). Copyright by the American Geophysical Union.

[b] Estimated $SO_4^{2-}$ in excess of seawater calculated assuming $Na^+$ or $Mg^{2+}$ originates from ocean spray.

[c] Marine sites.

[d] Continental sites.

[e] Data from California Air Resources Board, 1991.

μmol/L. The coastal California site compares well with the remote sites elsewhere. The rural mountain location in the Sierra Nevada is also similar. For comparison, annual precipitation chemistry data for rural sites in Colorado are given in Table 3.4. Inspection of the concentration data suggests similar $H^+$ concentrations between continental locations, as in the case of $Na^+$, $Mg^{2+}$, and $Cl^-$. However, concentrations of other ions such as $Ca^{2+}$ and $NH_4^+$ are higher in Colorado than the remote sites. Both $SO_4^{2-}$ and $NO_3^-$ are higher in Colorado than at remote sites. Although a continental baseline is expected to be variable, the Quincy, CA site is probably more representative of continental baseline conditions than the Colorado sites. The latter would be considered clean conditions, however, for the midcontinental United States.

Wet deposition levels for Colorado are also listed in Table 3.4. These are considered representative of deposition expected in the midcontinental United States. For comparison, calculated values of dry deposition are listed as typical upper limit values for the West in Table 3.5. Also included

TABLE 3.4 **Annual Precipitation Chemistry Data for Colorado; Considered Typical of Continental Baseline Conditions**[a]

| | **Colorado average**[b] | | | |
|---|---|---|---|---|
| | **1981** | **1982** | **1983** | **1984** |
| | **pH: 5.04** | **4.99** | **5.09** | **5.15** |
| | **Concentration (μmol/L)** | | | |
| $H^+$ | 9.1 | 10.2 | 8.1 | 7.8 |
| $SO_4^{2-}$ | 35.7 | 19.4 | 20.3 | 26.0 |
| $NO_3^-$ | 22.6 | 15.1 | 16.3 | 19.9 |
| $NH_4^+$ | 20.1 | 11.7 | 13.9 | 16.6 |
| $Ca^{2+}$ | 20.9 | 11.1 | 12.9 | 17.0 |
| $Mg^{2+}$ | 6.7 | 3.1 | 3.5 | 4.7 |
| $Na^+$ | 5.1 | 2.8 | 4.0 | 5.9 |
| $K^+$ | 1.2 | 1.2 | 1.3 | 1.1 |
| $Cl^-$ | 4.1 | 2.7 | 3.1 | 3.7 |
| $SO_4^{2-}/NO_3^-$ | 1.62 | 1.30 | 1.25 | 1.31 |
| Deposition (kg/ha) | | | | |
| $H^+$ | 0.033 | 0.035 | 0.033 | 0.034 |
| $SO_4^{2-}$ | 6.4 | 3.8 | 4.3 | 5.3 |
| $NO_3^-$ | 5.2 | 3.8 | 4.4 | 5.3 |
| $NH_4^+$ | 1.3 | 0.9 | 1.1 | 1.5 |
| $Ca^{2+}$ | 1.6 | 0.9 | 1.1 | 1.5 |
| Precipitation amount (cm) | 37.5 | 40.3 | 43.5 | 43.1 |

[a] From Young *et al.*, 1988.

[b] Average of volume-weighted mean for three sites with data for 1981–1984 (Sand Spring, Pawnee, and Manitou).

TABLE 3.5 **Annual Dry Deposition for Rural/Background Areas in the West**

| Species | Typical concentration | Deposition velocity (cm/s) | | Calculated dry deposition rate (kg/ha-yr)[a,b] | Inferred dry deposit ion rate (kg/ha-yr)[c] |
|---|---|---|---|---|---|
| $SO_2$ | 0.5 ppb | 0.5 | | 2 | 0.9 |
| $SO_4^{2-}$ | 1 $\mu g/m^3$ | 0.2 | | 0.6 | 0.9 |
| | | | Subtotal | 2.6 | 1.8 |
| $NO_2$[a] | 1 ppb | 0.1 | | 0.5 | |
| $HNO_3$ | 0.2 ppb | 1.0 | | 2.0 | 2.2 |
| $NO_3^-$ | 0.5 $\mu g/m^3$ | 0.2 | | 0.3 | 0.04 |
| | | | Subtotal | 2.8 | 2.2 |

[a] Estimated from the product of deposition velocity and typical concentration.
[b] $SO_4^{2-}$ and $NO_3^-$.
[c] Based on measurements at Pawnee, Grassland, CO; reported by Meyers *et al.*, 1991. Reprinted with permission from Pergamon Press Ltd., Oxford, England.

are limited observations of (inferred) dry deposition in eastern Colorado. The measurement-determined values are of the same order as the calculated values but are the best estimates for a dry deposition baseline. Comparisons of wet and dry deposition at Pawnee suggest that about one-third of sulfate and nitrate is dry deposited under conditions in Colorado, where precipitation is lower than in the East. Note from Table 3.4 that wet deposition levels appear to be higher than dry deposition in Colorado for both $SO_4^{2-}$ and $NO_3^-$. Comparison of the $SO_4^{2-}/NO_3^-$ ratios for wet and dry deposition suggests that the ratio is lower for dry deposition than for wet deposition.

The levels of the urbanized and industrialized ENA are superimposed on the western midcontinental levels exemplified by Colorado conditions and the areas farther west.

## MEASUREMENT METHODS

Methods for sampling atmospheric S and N species and oxidants have improved considerably since the 1960s. Before the mid-1970s, measurements were very limited and largely confined to urban or suburban areas, where higher ambient concentrations exist. Increased interest in regional-scale processes and conditions resulted in observational programs. These programs have pressed for improvements in both sampling and analytical methods to provide data near the detection limits of instrumentation. As a result, the reader should be aware of the limitations of data reported in the literature. In the case of ambient concentrations, Mueller and Hidy (1983) and Mueller and Watson (1983) report limitations in measurement technology through 1980. Limitations in measurement of dry deposition are discussed by Hicks *et al.* (1980). Topol (1984) describes the limitations in wet deposition monitoring practice, and Hansen and Hidy (1982) discuss historical measurements of precipitation. Tanner (1984) summarizes available sampling and observational methods accessible for field programs in the 1980s. A compendium of methods, instrumentation, intercomparisons, and data treatment for the 1980s has been released (Sisterson *et al.*, 1989). Table 3.6 summarizes techniques adapted for most North American observational programs from which most of the knowledge of regional air chemistry derives.

The observations discussed rely heavily on the data from SURE, ORBES, and PEPE (Nieman, 1980; Gillani, 1985). Account is also taken of measurements reported from the EMEFS project, a major regional observational network supporting model evaluation. These data are becoming available but have not been examined extensively for additional insight on atmospheric processes (Hansen, 1989). The design specifications for the EMEFS listed in Table 3.7 provide an illustration of the achievable standards of measurement that can be expected from 1980s air-monitoring instrumentation. The design criteria for precision, accuracy, and lower quantifiable limit of measurements are specified for regional-scale field experiments.

Other data are available from the National Aerometric Sampling Network (NASN). Precipitation chemistry measurements from a number of networks operating since 1978 are available: event sampling from the MAP3S/RAINE project (MAP3S/RAINE, 1982); the EPRI Utility Acid Precipitation Studies Program (UAPSP) (Mueller *et al.*, 1984) and its predecessor derived from the SURE; the Canadian networks, CANSAP and Air Precipitation Network (APN) (Whelpdale and Barrie, 1982); the weekly sampling networks; NADP and the National Trends Network (NTN) (ITFAP, 1983). Additional experimental measurements are integrated with long-term monitoring for perspective on other airborne chemical species, such as $HNO_3$, $H_2O_2$, $NH_3$, and C. Notable for a continuing air chemistry record are stations at Whiteface Mountain, NY (Kadlecek *et al.*, 1983) and for wet deposition are Hubbard Brook, NH (Likens *et al.*, 1980) and the USGS New York State network (Peters and Bonelli, 1982). The first extensive cloud chemistry observations are reported (Mountain Cloud Chemistry Project, MCCP; Mohnen and Kadlecek, 1989), and historical records at Mount Washington, NH are reported.

Complementing the ground network are data taken from aircraft sampling, including surveys made in the SURE and cloud sampling conducted in the EPA Acid Precipitation Experiment (APEX) (Lazrus *et al.*, 1981) and EPRI's programs (Blumenthal *et al.*, 1981, 1984; Daum *et al.*, 1984).

TABLE 3.6 **Sampling and Measurement Techniques for Key Chemical Parameters**[a]

| Constituent | Method |
|---|---|
| $SO_2$ | Continuous, flame photometric, or pulse fluorescence; time averaged, filter pack (particle filter followed by KOH absorber-filter) |
| NO, $NO_2$ | Continuous modified chemiluminescent analysis |
| $HNO_3$ | Continuous, infrared spectroscopy; semicontinuous, two-channel ozone chemiluminescent with nylon filter in one inlet, denuder followed by thermal release to an $O_3$ chemiluminescent device |
| Alkyl nitrates, peroxyacetyl nitrate (PAN) | Semicontinuous, chromatographic separation with electron capture detector |
| $O_3$ | Ultraviolet spectrometer |
| $H_2O_2$ | Time-averaged, absorber with enzyme-catalyzed fluorimetry |
| Particle mass | (Size selective) filter sample, gravimetric |
| Particle chemistry | |
| Sulfate | Ion chromatography |
| Nitrate | |
| Chloride | |
| Ammonium | |
| Metals | Neutron activation, plasma, emission spectroscopy, X-ray fluorescence |
| Organic carbon and elemental carbon | Thermal oxidation followed by ionization detector |
| Precipitation and cloud water chemistry | Precipitation sample, polyethylene bucket or snow scoop. Cloud-hydrometeor, inertial impactors |
| Sulfate, nitrate, ammonium, chloride | Ion chromatography |
| Heavy metals | Atomic absorption, plasma emission spectroscopy, or ion chromatography |
| Sulfite, phosphate | Fixation with $Hg(Cl)_4^{2-}$, colorimetry, automated calorimetry |
| Hydrogen peroxide | Enzyme-catalyzed fluorimetry |
| Organic acids | Fixation with biocide, liquid chromatography |
| Aldehydes | Derivatization, liquid chromatography |

[a] Based on Tanner, 1984.

## AMBIENT CONCENTRATIONS

A discussion of the basic character of regional $SO_x$ behavior in air over the ENA should examine different features of the spatial and temporal distributions and ambient concentrations compared with emissions distributions. This information is available from the SURE and the EMEFS.

TABLE 3.7 **Achievable Data Quality for Precision, Accuracy, Lower Quantifiable Limit Based on EMEFS**[a,b]

| Air quality observable (interval) | Methods (units) | Precision (the larger of) | Accuracy | Lower quantifiable limits | |
|---|---|---|---|---|---|
| | | | | 3 × sd | 10 × sd |
| Particulate mass (24 h) | FP/G ($\mu g/m^3$) | ±15% $\mu g/m^3$ | ±10% | 2 ($\mu g/m^3$) | 6 ($\mu g/m^3$) |
| | FPC/G ($\mu g/m^3$) | ±15% $\mu g/m^3$ | ±10% | 2 ($\mu g/m^3$) | 6 ($\mu g/m^3$) |
| Particulate sulfate (24 h) | FP/AC ($\mu g/m^3$) FP/IC ($\mu g/m^3$) | ±15% 0.4 $\mu g/m^3$ | ±10% | 0.3 ($\mu g/m^3$) | 0.8 ($\mu g/m^3$) |
| Particulate nitrate (24 h) | FP/AC ($\mu g/m^3$) FP/IC ($\mu g/m^3$) | ±15% 0.3 $\mu g/m^3$ | ±10% | 0.2 ($\mu g/m^3$) | 0.5 ($\mu g/m^3$) |
| Particulate ammonium (24 h) | FP/AC ($\mu g/m^3$) | ±15% 0.03 $\mu g/m^3$ | ±10% | 0.02 ($\mu g/m^3$) | 0.05 ($\mu g/m^3$) |
| Sulfur dioxide (24 h) | FP/AC ($\mu g/m^3$) FP/IC ($\mu g/m^3$) | ±15% 0.4 $\mu g/m^3$ | ±10% | 0.3 ($\mu g/m^3$) | 0.8 ($\mu g/m^3$) |
| Nitric acid (24 h) | TFR/IC ($\mu g/m^3$) TFR/AC ($\mu g/m^3$) FP/AC ($\mu g/m^3$) FP/IC ($\mu g/m^3$) | ±15% 0.4 $\mu g/m^3$ | ±10% | 0.3 ($\mu g/m^3$) | 0.8 ($\mu g/m^3$) |

| | | | | | |
|---|---|---|---|---|---|
| Ammonia (24 h) | TFR/AC ($\mu g/m^3$) FP/AC ($\mu g/m^3$) | ±15% 0.1 $\mu g/m^3$ | ±10% | 0.07 ($\mu g/m^3$) | 0.2 ($\mu g/m^3$) |
| Ozone (1 h) | Photometry ($\mu g/m^3$) | ±10% or 10 $\mu g/m^3$ | ±10% | 8 ($\mu g/m^3$) | 25 ($\mu g/m^3$) |
| Nitrogen dioxide (1 h) | Luminol CL ($\mu g/m^3$) | ±10% 0.2 $\mu g/m^3$ | ±10% | 0.17 ($\mu g/m^3$) | 0.5 ($\mu g/m^3$) |
| (24 h) | TEA FF/IC ($\mu g/m^3$) | ±15% 0.4 $\mu g/m^3$ | ±10% | 0.3 ($\mu g/m^3$) | 0.8 ($\mu g/m^3$) |
| PAN (24 h) | FS/IC ($\mu g/m^3$) | ±15% 4 $\mu g/m^3$ | ±10% | 4 ($\mu g/m^3$) | 14 ($\mu g/m^3$) |
| Hydrogen peroxide (1 h) | E/F (ppb) | | | | |
| Hydrocarbons | GC/FID (ppbc) | (Not Specified) | | | |
| Aldehydes | Der/HPLC (ppb) | | | | |

[a] From Hansen, 1989.

[b] Abbreviations: E/F, enzyme-catalyzed fluorimetry; FP/AC, filter pack, automated colorimetric analysis; FP/IC, filter pack, ion chromatographic analysis; Luminol CL, luminol chemiluminescence; TFR/IC, transition flow reactor, ion chromatographic analysis; TFR/AC, transition flow reactor, automated colorimetric analysis; FS/IC, filter sampler, ion chromatographic analysis; FP/G, filter pack, gravimetry; FPC/G, fine particle collector, gravimetry; PAN, peroxyacetyl nitrate; GC/FID, gas chromatography analysis with flame ionization detection; Der/HPLC, derivatization with high-performance liquid chromatographic analysis; sd, standard deviation; NS, not specified.

The former is described by Mueller and Hidy (1983), the latter by Hansen (1989).

The geographic region covered in the SURE extended from west of the Mississippi eastward to the Atlantic Ocean; it encompassed parts of states just north of the Gulf of Mexico northward to the Great Lakes and southeastern Canada. This area was chosen because it included essentially all of the industrialized ENA as of the mid-1970s, bordered by areas with low $SO_x$ and $NO_x$ emissions, so that influx from the boundaries of the region would be minimized.

The region is also believed to be minimally influenced by natural emissions. Natural emissions of gaseous S compounds such as hydrogen sulfide ($H_2S$) and dimethyl sulfide (DMS) were estimated in the SURE region based on measurements. Adams *et al.* (1980) calculated that, in aggregate, the natural S component is less than 2% of the total S emissions in the region. Only in a few localized areas—coastal Virginia and the Carolinas, where marshy or wetland areas exist—could the biogenic S emissions flux approach that of anthropogenic emissions. Both $H_2S$ and DMS oxidize in the atmosphere, ultimately forming $SO_4^{2-}$. Natural $SO_4^{2-}$ is also derived from injections of sea spray. Although this source may be large in coastal areas, it does not appear to be significant inland in the Northeast.

The regional behavior of $SO_x$ was first characterized mainly by the SURE results for the late 1970s (1977–1978); later, they were obtained for the late 1980s in EMEFS (1987–1989). Spatially representative measurements of $NO_x$ have been reported from the EMEFS; some systematic $NO_x$ data were acquired earlier at a few locations in the SURE, as well as the ORBES. $O_3$ concentrations have been measured for a number of years at several locations through the National Air Monitoring System (NAMS) (Vukovich and Fishman, 1986). Limited temporal average observations are available from the SURE and EMEFS. These data provide a useful geographic and temporal record, more complete than that for S and N.

The SURE network of 54 ground stations was established and operated between August 1977 and October 1978. The network included nine stations (Class I, or CI) distributed throughout the region, which were operated continually during this period. Their operation was extended in ERAQS through June 1979. The network also had 45 additional (Class II, CII) stations distributed over the region. The nine CI stations were operated continuously, measuring $O_3$, $SO_2$, NO, $NO_2$, temperature, and dew point. In addition, 24-hour samples of airborne particles were collected for chemical analysis using a standard high-volume filter unit in which inlet size cutoff is nominally 30 $\mu$m (aerodynamic) mass median diameter (AMMD).[2] The CII stations were run continuously in six seasonal

[2] AMMD is defined in terms of the settling or deposition rate of nonspherical and irregularly shaped particles.

months, August 1977, October 1977, 10 January to 9 February 1978, April 1978, July 1978, and October 1978. These stations were equipped with instruments to measure $SO_2$ and to collect 24-hour particulate samples for analysis. During periods when all stations were operating, high-volume filter samplers were run at the nine CI stations to collect 3-hour averaged particulate samples. These samples were obtained for the size ranges $\lesssim 11$ $\mu$m AMMD (TSP) and $\lesssim 2.5$ $\mu$m AMMD (RSP). The sites were carefully chosen for their rural nature; they were at least 16 km from identifiable $SO_2$ and particulate sources and were representative of regional conditions. The details of the network, including instrumentation, operations, and siting procedures, are given by Mueller and Hidy (1983).

Supplementing the SURE ground network was an aerial sampling program that used two aircraft (Blumenthal *et al.*, 1981, 1984). The aircraft were equipped to measure continuously $SO_2$, $O_3$, NO, and $NO_2$ concentrations, temperature, humidity, pressure, turbulence intensity, condensation nuclei (total aerosol particle) number concentration, and light scattering coefficient. Samples of aerosol particles less than 3 mm AMMD were also collected for chemical analysis using high-volume filters and a tape sampler with a cellulose filter tape substrate. The main purpose of the aircraft program was to obtain routine, repetitive measurements of the vertical structure of aerometric parameters over ground stations, so that pollution behavior in the atmospheric boundary layer could be assessed for comparison with the ground observations. The aerometric data set was completed by obtaining surface and upper level meteorological data from the National Weather Service (NWS) and other sources. These data included wind, temperature, humidity, visual range, cloud cover, and precipitation.

The EMEFS involved coordination of 10 different measurement network operations, including

(a) 10 stations from APIOS (Ontario Ministry of Environment)
(b) 10 stations of CAPMON (Atmospheric Environment Service)
(c) The 25-station Operational Evaluation Network, OEN (EPRI)
(d) 39 stations of ME-35 (EPA)
(e) 7-station EPA-TVA supplementary sites
(f) 15-station Gradient Resolution Network, GRAD (EPA)
(g) 45-station subgrid Variability Network, VAR (EPA)
(h) 45-station Florida Acid Deposition Monitoring Program, FADMP (FCG).

The sites were located in a region larger than covered by the SURE, extending from southern Alberta and Quebec in the north to Florida in the south and from South Dakota and Texas in the west to the Atlantic Coast and Nova Scotia in the east. In general, the spacing between EMEFS stations was designed to characterize regional models with resolution of about 100 km by 100 km; however, the GRAD network in Pennsylvania concentrated stations in a more closely spaced network to allow examination of conditions within the designated regional grid of 100 km by 100 km.

The EMEFS network was embedded in NAMS and NADP monitoring programs and was supplemented by U.S. and Canadian aircraft observations during a key intensive period in summer 1988. Not all sites were operated for the entire planned operation period. However, the full coverage involved selected stations representative of the region.

The archiving and validation of the extremely large EMEFS data set have been completed. The data and results of their interpretation generally will not be available for the entire sampling period for some time. However, preliminary results reported for the operations indicate spatial patterns generally similar to those found in the earlier SURE work.

The SURE results indicated the relative contribution of $SO_x$, $NO_x$, and $O_3$ to the atmospheric burden of constituents over ENA. Table 3.8 indicates the aggregated concentrations of $SO_2$ and particulate $SO_4^{2-}$ relative to other constituents in nonurban air. Of the constituents measured in the SURE, the observations indicate that $O_3$ is present in by far the largest concentration at both the median (50th percentile of occurrence frequency) and extreme concentration (90th percentile) level. At the median level $SO_2$ and $SO_4^{2-}$ (as ammonium bisulfate, $NH_4HSO_4$) are at approxi-

TABLE 3.8 **Average Concentrations of Air Pollutants Observed at SURE CI Stations from August 1977 through October 1978**[a,b]

| | At 50th percentile | | At 90th percentile | |
|---|---|---|---|---|
| **Pollutant** | **Concentration ($\mu g/m^3$) at STP** | **% of Total** | **Concentration ($\mu g/m^3$) at STP** | **% of Total** |
| $SO_2$ | 9.6 | 7.8 | 47.0 | 15.2 |
| $NH_4HSO_4$ | 7.8 | 6.4 | 20.0 | 6.5 |
| $NH_4NO_3$ | 0.3 | 0.2 | 1.2 | 0.4 |
| $NO_x$ (1NO + 9$NO_2$) | 13.0 | 10.6 | 50.0 | 16.1 |
| $HNO_3$[c] | 2.0 | 1.6 | 6.0 | 1.9 |
| Residual particulate mass[d] | 26.0 | 21.2 | 51.0 | 16.4 |
| $O_3$ | 64.0 | 52.2 | 135.0 | 43.5 |
| Total | 122.7 | 100.0 | 310.2 | 100.0 |

[a] From Mueller and Hidy, 1983; reproduced with permission of the Electric Power Research Institute.

[b] Estimated idealized composition based on average stoichiometry of field observations.

[c] Estimated from indirect measurements.

[d] Determined by difference between total mass concentration and the sum of ammonium bisulfate and ammonium nitrate.

mately the same concentration. Furthermore, the total $SO_x$ constituents at the median are roughly the same in concentration as the $NO_x$ and $NO_3^-$ compounds. However, most of the $NO_x$ and $NO_3^-$ are found in the gas phase, at the 90th percentile. At this extreme, particle concentrations excluding $NH_4HSO_4$ and $NH_4NO_3$ are smaller in relative concentration than at the median level.

Inspection of data reported from EMEFS in 1989 and 1990 shows a similar distribution of pollutants as illustrated in Table 3.8. However, ozone concentrations over the EMEFS region are a larger fraction (~70%) and $NO_x$ and $HNO_3$ are substantially smaller portions of the total (combined ~2.4%).

The general pattern shown in Table 3.8 gives the distribution of the major constituents in the S and N chemistry. The ambient concentration levels reported in the SURE are similar to those of the EMEFS. However, in the latter case the median levels (90th percentile) of $SO_2$, $SO_4^2$, $NO_x$, and $HNO_3$ are generally lower but $O_3$ levels are reported to be higher for this network. Additional direct observations of $HNO_3$ and $H_2O_2$ indicate concentrations in the range of 0.1–1.0 ppbv (Mohnen, 1986; for $HNO_3$, M. A. Allan, personal communication) and 0.5–4 ppbv, respectively (Heikes *et al.*, 1987). If these are representative of ENA, $HNO_3$ vapor concentrations are lower than $SO_4^{2-}$, and $H_2O_2$ concentrations in air are much lower than $O_3$. Although $SO_4^{2-}$ in the SURE was identified mainly as $NH_4HSO_4$, $SO_4^{2-}$ is also found as $H_2SO_4$ and other salts.

The acidity of ambient aerosol has been studied at Allegheny Mountain, Pennsylvania. Pierson *et al.* (1980) found that the aerosol was quite acidic during a period in August 1978, equivalent to a mix of dilute $H_2SO_4$ and ammonium sulfate [$(NH_4)_2SO_4$] salts. Later measurements also indicate acidity that is in excess of ammonium bisulfate; there is evidence that much of the $SO_4^{2-}$ in particles is $H_2SO_4$ during some sampling periods.

Table 3.9 lists concentrations of components in aerosol particles measured at three eastern rural locations. Data from two of the sites are segregated into coarse and fine particle fractions. There are similarities at all three locations, namely in $SO_4^{2-}$, $NH_4^+$, acidity, and calcium level. Differences appear in carbon content, silicon, and certain other metals.

Figure 3.2 illustrates the identified contributions by mass of the finely divided particle fraction during January and July 1978, based on aggregated data obtained from selected SURE sites. Although $SO_4^{2-}$ represents a large fraction of the fine particles, total organic noncarbonate carbon (TONC) and other unidentified compounds also represent large mass fractions of sampled particles. The unidentified constituents are believed to be largely soil dust, metal salts, or metal oxide debris and condensed or adsorbed water. The fraction apportioned to soil dust and to heavy metals is identified in work of Ferman *et al.* (1981) and Stevens *et al.* (1984). The

TABLE 3.9 **Comparison of Average Aerosol Concentrations from Three Studies**[a,b,c]

| | **Smoky Mountains, Tennessee 20–26 September 1978** | | **Shenandoah Valley, Virginia 23 July–5 August 1980** | | **Allegheny Mountains, Pennsylvania 24 July–11 August 1977** |
|---|---|---|---|---|---|
| | **Fine** | **Coarse** | **Fine** | **Coarse** | **Total (<30 $\mu$m)** |
| Mass | 24000 ± 3000 | 5600 ± 3000 | 27000 ± 5000 | 7400 ± 5080 | 38000 |
| $SO_4^{2-}$ | 12000 ± 1300 | NA[d] | 13600 ± 1300 | 780 ± 200 | 14060 |
| $NO_3^-$ particle | 300 ± 300 | NA | NA | NA | — |
| $NO_3^-$ total | 900 ± 300 | NA | 1271 ± 127 | NA | 530[e] |
| $NH_4^+$ | 2280 ± 390 | NA | 3356 ± 390 | NA | 2310 |
| $H^+$ | 114 | NA | 96 | NA | 157 |
| C organic | 2220 ± 400 | 1200 ± 440 | 440 ± 420 | NA | — |
| C elemental | 1100 ± 800 | <100 | 1120 ± 780 | NA | — |
| C total | 3300 ± 600 | 1300 ± 600 | 1560 ± 660 | 1680 | +780[b] |
| Al | 20 ± 18 | 195 ± 101 | 44 ± 35 | 311 ± 104 | +700[b] |
| Si | 38 ± 10 | 580 ± 262 | 166 ± 23 | 813 ± 288 | +1000[b] |
| S | 3744 ± 218 | 204 ± 187 | 4539 ± 273 | 260 ± 237 | +1300[b] |
| Cl | <10 | 7 ± 4 | 10 ± 5 | 179 ± 63 | — |
| K | 40 ± 3 | 108 ± 30 | 61 ± 5 | 129 ± 30 | 70 |
| Ca | 16 ± 1 | 332 ± 73 | 35 ± 3 | 304 ± 54 | 330 |

| | | | | | |
|---|---|---|---|---|---|
| Ti | <6 | 18 ± 5 | <10 | 17 ± 10 | +40[f] |
| V | <4 | <5 | <10 | <10 | 3 |
| Mn | NA | NA | 3 ± 2 | 3 ± 2 | 9[c] |
| Fe | 28 ± 2 | 118 ± 9 | 54 ± 5 | 158 ± 12 | 320 |
| Ni | 1 ± 0.5 | 1 ± 0.5 | 1 ± 1 | 1 ± 1 | — |
| Cu | 3 ± 0.7 | <5 | 5 ± 2 | 6 ± 2 | — |
| Zn | 9 ± 1 | <4 | 11 ± 2 | 6 ± 2 | 20 |
| As | 2.2 ± 1 | <1 | 1 ± 1 | <2 | — |
| Se | 1.4 ± 0.3 | 0.2 ± 0.2 | 1.2 ± 0.5 | <1 | 3 |
| Br | 18 ± 1 | 5 ± 0.4 | 8 ± 1 | 3 ± 1 | 20 |
| Pb | 97 ± 5 | 14 ± 1 | 52 ± 4 | 9 ± 2 | 90 |
| Na | — | — | — | — | 200 |
| Mg | — | — | — | — | 70 |
| Cd | — | — | — | — | 3 |
| Ba | — | — | — | — | 4 |

[a] From Stevens *et al.*, 1984, Pierson *et al.*, 1980; data reprinted with permission of the New York Academy of Science and Pergamon Press.

[b] Results are expressed in units of ng/m$^3$; overall measurement errors are shown after the ± symbols and are based on estimates of uncertainties in flow rate determination, instrument calibration, and X-ray attenuation corrections.

[c] The fine and coarse fractions pertain to particles having aerodynamic diameters in the ranges 0–2.5 and 2.5–15 $\mu$m, respectively.

[d] NA, not analyzed.

[e] On fluorocarbon filter (data questionable).

[f] The uncertainties for Mn are ±10% of the concentration reported. The uncertainties in the other species are the same as those reported for the Shenandoah Valley.

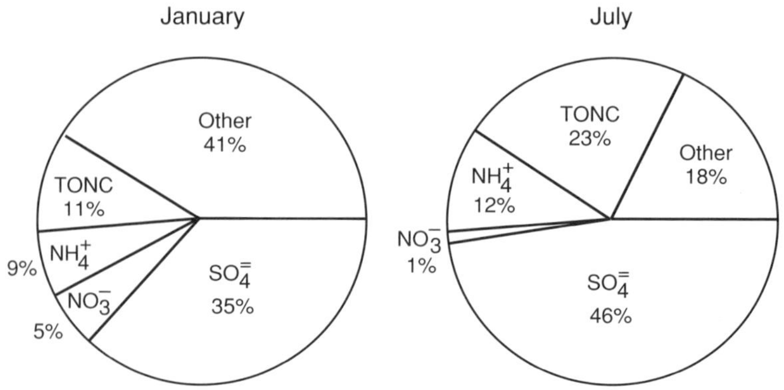


FIGURE 3.2 Relative composition of small particles (≤2.5 μm AMMD) based on special sampling days at selected SURE sites in January and July 1978. (From Mueller and Hidy, 1983; reproduced with permission of the Electric Power Research Institute.)

unidentified fraction was found to be particularly large in winter. The fractional makeup of the fine particles varied somewhat from site to site and by season but should be of a generally similar character by site. The fine particles make up between 40 and 54% of the total suspended particle concentration obtained from the high-volume filter samples.

The fine-particle composition is important for interpretation of visibility conditions in the ENA. Visibility change depends on the total extinction coefficient for visible light (see also Appendix A). The coefficient is usually dominated by fine-particle concentrations (Covert *et al.*, 1980). The particle extinction coefficient depends on both a scattering component and an absorption coefficient. Essentially all the fine particles in the atmosphere scatter light, but few particulate materials absorb light because of their low absorptivity (expressed in terms of the imaginary part of the complex index of refraction). The principal absorber is believed to be elemental (black) carbon. The fraction of total organic noncarbonate carbon in nonurban air is not known but is believed to be small compared with the light-scattering particulate fraction. Thus, the association between particulate $SO_4^{2-}$ changes and regional visibility changes is circumstantial and depends on $SO_4^{2-}$ being a surrogate for the total mass concentration of fine particles.

The SURE data provide a basic picture of the average spatial distributions of $SO_4^{2-}$ and total $NO_3^-$ collected on a Teflon-coated filter, minimizing ambiguities of $SO_2$ and $NO_y$ absorption. The data also give important insight into the mean seasonal and diurnal time sequences in rural conditions. In general, rural $SO_4^{2-}$ tends to be elevated in summer over winter and is elevated in midday relative to the rest of the day. The spatial distributions of concentration show the apparent extent of the geographic area persistently influenced by similar or coherent atmospheric processes in relation to the geographically fixed emission distribution.

## Spatial Distributions

A composite distribution from five representative seasonal months between August 1977 and July 1978 provides an approximation for an annual mean distribution of $SO_2$ and $SO_4^{2-}$ concentrations (Figure 3.3). This seasonal composite shows that $SO_4^{2-}$ concentrations are generally high in the regime of high $SO_2$ concentrations. However, they extend northeastward from the lower Ohio River Valley across Pennsylvania to the Atlantic Coast. The zone of high $SO_4^{2-}$ concentration also covers an area farther south of the high-$SO_2$ regime. The zones of high $SO_2$ and $SO_4^{2-}$ roughly correspond to the subregions where high $SO_2$ emissions are found. Figure 3.4 shows annual median $SO_2$ and $SO_4^{2-}$ distributions a decade later than the SURE composite. Both annual median $SO_2$ and $SO_4^{2-}$ levels over the eastern United States are lower in the late 1980s than in the earlier sample, based on averages of seasonal 1-month composites. The $SO_2$ spatial distributions are similar in both periods. The distribution of elevated $SO_4^{2-}$ appears to have shifted southward by 1989 compared with the earlier period.

The distribution of particulate $SO_4^{2-}$ and total suspended particles (TSP), measured by air filtration, was found to be distinctly different in 1978. This difference is attributed to source differences, as well as the size distribution of material. TSP is dominated by large particles, whereas $SO_4^{2-}$ is found primarily in the diameter range less than 1 $\mu$m (Hidy, 1984a). $SO_4^{2-}$ accounts for about 20% of TSP. Total $NO_3^-$ recovered on the SURE filters also shows a somewhat different spatial distribution than $SO_4^{2-}$. $NO_3^-$ is more confined to sources and is much lower in concentration than $SO_4^{2-}$. The localized character of $NO_3^-$ was believed to be related to the filter absorption of $NO_y$ but may be a characteristic of local atmospheric reactions to produce $NO_3^-$ relative to $SO_4^{2-}$. The results for $NO_3^-$ and $HNO_3$ from the EMEFS indicate that these species are present in similar concentrations in particles and in the vapor phase. The EMEFS data suggest that particulate $NO_3^-$ has a local character, as seen in SURE, but $NO_2$ and $HNO_3$ have a broader regional distribution of concentration.

The distributions of $SO_2$ averaged by month from the SURE and later the EMEFS data generally show a localized concentration around major sources. A notable exception in the SURE was January 1978. In this period elevated concentrations of $SO_2$ were geographically more widespread than in other months. Particulate $SO_4^{2-}$ averaged by month was regional in summer, spring, and winter but more local in fall. The maxi-

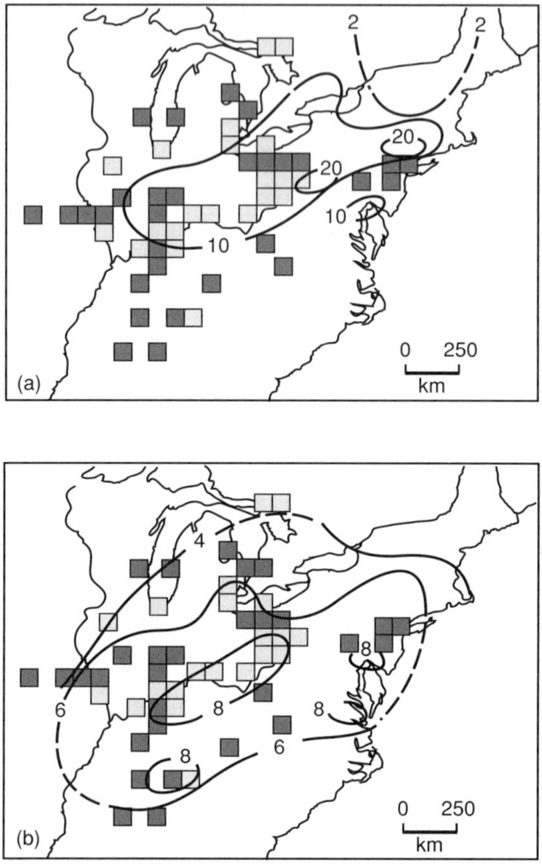


FIGURE 3.3 (a) Composite spatial distribution of 1-hour average concentrations (ppb) of $SO_2$ from one representative month in each season between 1977 and 1978. This average is approximately equivalent to an annual average. Dashed isopleth is based on limited quantities of SURE data from 1977–1978 and on data taken at Whiteface Mountain, New York, after 1978. (b) Composite spatial distribution of 24-hour concentrations $SO_4^{2-}$ ($\mu g/m^3$) from one representative month in each season between 1977 and 1978. This average is approximately equivalent to an annual average. Light 80 km × 80 km squares are $SO_2$ emissions in mT/d areas >1000; darker areas are 500–1000 mT/d emissions. (From Hidy, 1984a; courtesy of the Air and Waste Management Association.)

mum in $SO_4^{2-}$ concentration shifted southward in seasons other than summer.

The ratio of $SO_4^{2-}$ S to total airborne S is sometimes a useful indicator of the age of air masses, the efficiency of chemical conversion to $SO_4^{2-}$, or deposition of $SO_2$ relative to $SO_4^{2-}$. The geographic distribution of the

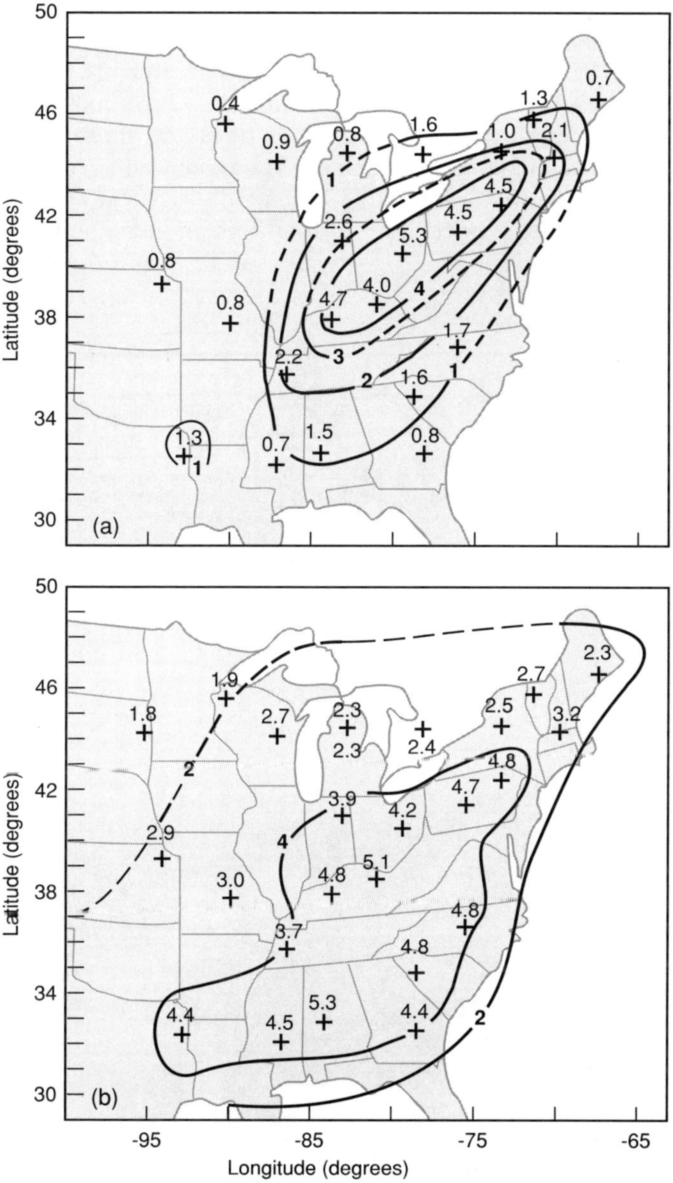


**FIGURE 3.4** (a) Annual median $SO_2$ (ppb) concentration from the EMEFS network, based on 1988–1989 data. (b) Annual median $SO_4^{2-}$ particle concentration ($\mu g/m^3$) from the EMEFS network, based on 1988–1989 data. (Data from the Electric Power Research Institute.)

ratio of $SO_4^{2-}$ S to total S is shown in Figure 3.5 with 1975 emissions. The ratio is generally highest in the summer months and lowest in winter. Dry deposition of $SO_2$ and $SO_4^{2-}$ is believed to be greater in summer than in winter because of increased gas absorption on vegetation and increased particle capture on grasses and deciduous trees and shrubs. Wet deposition also increases in summer, because $SO_2$ oxidation is enhanced, $SO_4^{2-}$ is scavenged efficiently, and rainfall is generally greatest in the East. The increased particulate S in relation to total S is an indication of enhanced

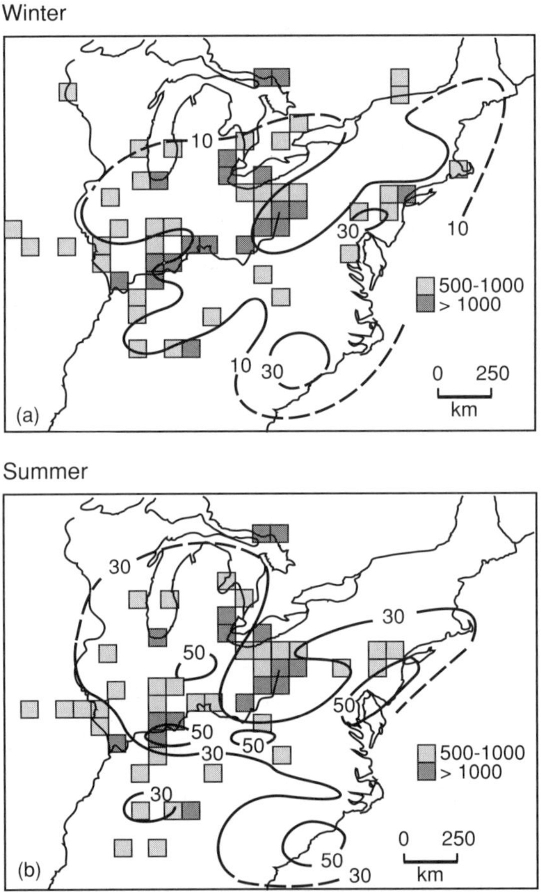


FIGURE 3.5 Geographic distribution of the ratio of $SO_4^{2-}$ sulfur to total airborne sulfur for different seasonal periods (%). (a) Winter (January–February 1978); (b) summer (July 1978). The squares represent areas of 1977 high $SO_2$ emission rate, given in metric tons/day (mT/d). (From Mueller and Hidy, 1983; reproduced with permission of the Electric Power Research Institute.)

oxidation chemistry in summer as well as less intense air mass transport with light winds (see Appendix A).

Higher ratios of particulate S to total S occur near large complexes of $SO_2$ emissions, except for the coastal areas of the Carolinas and Virginia. The reason for a persistently high ratio in this area for all months sampled is not known. However, it may be associated with onshore airflow associated with circulation around a climatologically prevalent region of high barometric pressure offshore, the Bermuda High. Wolff *et al.* (1977) speculated that such an onshore flow would carry with it aged, polluted air from an offshore flow from northern industrialized areas.

The distribution of annual median $NO_2$ and $HNO_3$ concentrations is shown in Figure 3.6. The concentrations are similar in range to $SO_2$, and the distribution of elevated $NO_2$ generally corresponds to a zone of high emission density, similar to $SO_2$. The distribution of median $HNO_3$ concentration from the EMEFS data shows a maximum of about 2 $\mu g/m^3$

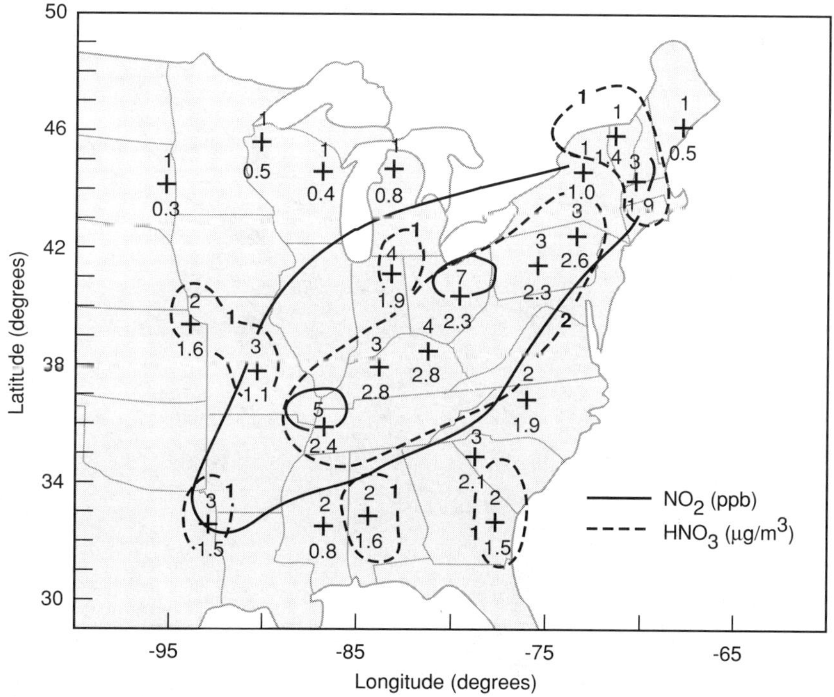


**FIGURE 3.6** Annual median $NO_2$ concentrations in ppb and $HNO_3$ concentration in $\mu g/m^3$ based on 1988–1989 EMEFS data. The top number at each site is $NO_2$ and the bottom number is $HNO_3$. The solid contours are $NO_2$ and the dotted contours are $HNO_3$. (Data from the Electric Power Research Institute.)

present throughout the Ohio River region extending into Pennsylvania and southern New York. Elevated levels of $HNO_3$ of about 1 $\mu g/m^3$ are suggested in several areas around the Ohio River Valley. EMEFS median particulate nitrate concentrations were maximum around urban and industrial areas, especially in the lower and upper Ohio River Valley, west of Chicago, Milwaukee, Detroit, and Kansas City.

The area of maximum $O_3$ concentration is similar to that of particulate $SO_4^{2-}$ (Vuckovich and Fishman, 1986). This maximum is generally located in a region of relatively large HC/$NO_x$ emission ratio. This is illustrated by comparing the HC/$NO_x$ ratios in Figure 2.5 with the summer ozone patterns in Figure 3.7. The elevated ozone concentrations are observed broadly from the southwest to the northwest, with highest concentrations in the region of northern Ohio eastward across New York into New Hampshire. The far eastern area in New England reflects long-range transport of ozone into areas relatively remote from pollutant sources.

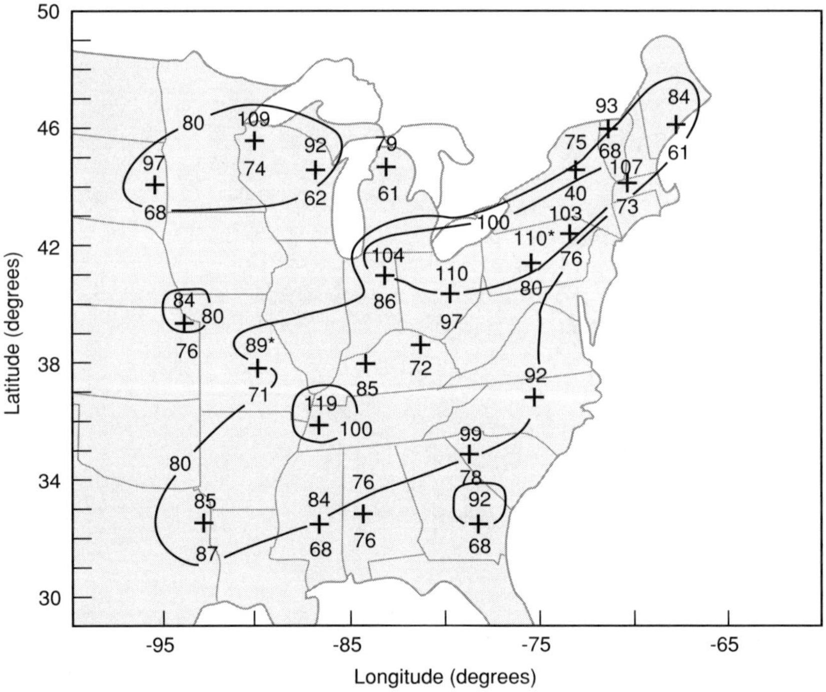


**FIGURE 3.7** Seasonal $O_3$ concentrations (ppb) corresponding to the 90th percentile values for summer 1988 (top numbers); bottom numbers are for summer 1989. (Data from the Electric Power Research Institute.)

The area of elevated $O_3$ reflects an area of high chemical reactivity (see Appendix A). Vukovich and Fishman (1986) show that the monthly patterns depend on trajectories of large-scale anticyclonic systems and differ from year to year compared with $SO_2$ patterns. Another important oxidant, hydrogen peroxide, $H_2O_2$, has been measured in the East (Mohnen, 1986).

The compounds $SO_2$, $NO_2$, $O_3$, and particulate $SO_4^{2-}$ are species governed explicitly or implicitly by the 1968 U.S. Clean Air Act (CAA). Limits on their atmospheric concentrations are set according to National Ambient Air Quality Standards (NAAQS). The NAAQS are reviewed periodically and are established primarily by levels that are considered safe according to known health effects. Secondarily, they are set in terms of certain public welfare considerations, including materials corrosion and visibility impairment. Sulfate, in particular, is currently regulated as a constituent of airborne particulate matter less than or equal to 10 $\mu$m in diameter ($PM_{10}$).[3]

Comparison of annual average NAAQS and regional $SO_2$, $NO_2$, and $O_3$ concentrations indicates that $SO_2$ and $NO_2$ are well below the NAAQS levels of 30 ppb (80 $\mu g/m^3$) and 50 ppb (100 $\mu g/m^3$), respectively. In the case of $O_3$ concentrations, regional concentrations at the 90th percentile and above in Figure 3.7 suggest that regionwide $O_3$ levels may be a major contributor to the NAAQS (120 ppb for 1 h). Assuming that sulfate is 20–40% of $PM_{10}$, the reported regional particulate levels represent a substantial fraction of the NAAQS, 50 $\mu g/m^3$.

The results for regional $O_3$ and $SO_4^{2-}$ concentrations indicate that exclusively local strategies to reduce $O_3$ and $PM_{10}$ may be completely inadequate without parallel management of regional emissions. The regionwide elevated levels seen over the eastern United States will probably require interstate strategies as well as intrastate planning for continued progress in $PM_{10}$ and $O_3$ pollution reduction. Interstate studies have been considered for regional authorities. However, specific guidance for such strategies remains only in the conceptual planning stage.

Observations of $H_2O_2$ concentrations in summer were obtained from aircraft measurements on a north–south pathway and indicate that southern areas have higher concentrations than the Northeast. $H_2O_2$ concentrations were observed at 2 ppbv over Mississippi and decreased to less than 0.5 ppbv over New York.

[3] $PM_{10}$ is distinguished from TSP by the air sampler. The former is well defined by a filter inlet carefully designed to remove particles $<10$ $\mu$m. The old TSP measurement was taken with a high-volume air filter sampler with an ill-defined inlet, permitting particles $>10$ $\mu$m to be collected.

## Temporal Changes

The temporal changes of ground-level gas and particle concentrations were found to range widely, depending on the season and daily conditions at a given station. Figure 3.8 shows changes observed in the monthly averages for three of the SURE sites representing the rural Northeast, a highly industrialized region of the upper Ohio River Valley, and a rural area surrounded by isolated sources of pollution in central Tennessee.

At all three sites, $O_3$ concentrations have a strong seasonal average variation. A minimum was found in early winter, followed by a strong increase beginning in January to a maximum in late spring to early summer. This pattern suggests the importance of photochemical processes that produce oxidant throughout much of the year except early winter. Downward transport of $O_3$ from the upper atmosphere may also be a factor in $O_3$ behavior, particularly in early to middle spring. Logan (1985) summarized monthly average $O_3$ concentrations from the SURE and other sources. Her composites yield essentially the same patterns as observed in the SURE and are consistent with the recent data of EMEFS.

Monthly average $SO_2$ concentrations reach a maximum in midwinter at all three sites shown in Figure 3.8. In contrast, mean concentrations show not only a maximum in winter for particulate $SO_4^{2-}$ but also a second maximum in summer with simultaneous minimum $SO_2$ concentrations. This finding differs somewhat from earlier observations in 1974, which showed a stronger summer maximum in $SO_4^{2-}$ concentrations than in 1977–1978 (Hidy *et al.*, 1976). In the late 1980s, the EMEFS data indicate the summer $SO_4^{2-}$ maximum and $SO_2$ minimum and vice versa in winter.

The monthly mean concentrations of $NO_x$ and NO suggested that most of the nonurban $NO_x$ is $NO_2$ (by difference). The two northern stations showed a maximum in $NO_x$ levels in winter, paralleling $SO_2$ concentration. At Giles, Tennessee, however, $NO_x$ concentrations were higher and more constant between seasons than at the other sites. This is believed to be related to urban and industrial sources of $NO_x$, for example, within 100–200 km of Giles, TN. These sources appear to be distant from the $SO_2$ sources in the area. Mixing is known to be suppressed in winter with the formation of cold air inversions at the ground. The EMEFS data in 1989 and 1990 show the summer minimum in $NO_2$ with higher levels in the other seasons, particularly in winter. Nitric acid tends to be present with a weak summer maximum; particulate nitrate shows a summer minimum and a winter maximum. The elevated $SO_2$ and $NO_x$ concentrations in winter may be a reflection of this effect. The high $SO_4^{2-}$ in summer with low $SO_2$ and high $O_3$ concentrations may reflect the more active $SO_2$ oxidation rates in summer.

The monthly average concentration of suspended particles from SURE

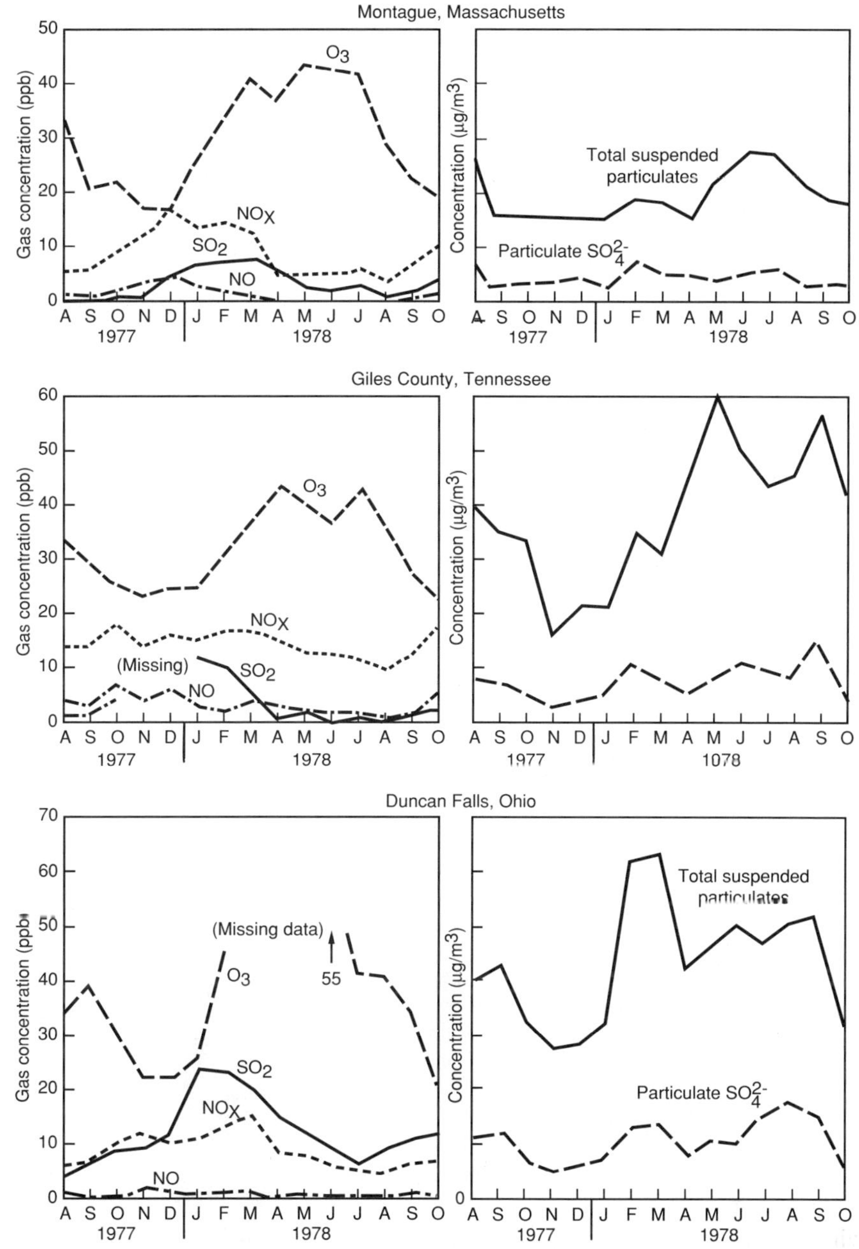


FIGURE 3.8 Monthly average concentrations of species recorded at Montague, Massachusetts, Duncan Falls, Ohio, and Giles County, Tennessee, 1977–1978. Gases are in ppb; particles are in $\mu g/m^3$. (From Mueller and Hidy, 1983; reproduced with permission of the Electric Power Research Institute.)

was minimum in early winter, rising to a maximum in spring to summer. This seasonal pattern was attributed not only to increased $SO_4^{2-}$ concentration but also to soil mobilization with drying and loss of snow cover.

The patterns of seasonal change at the six other SURE sites are similar. However, some differences were found that apparently reflect a subregional source, proximity to the ocean, and climatological factors. The daily variations in gas and particle concentrations were large, dominated by meteorological factors, such as ventilation, the product of mixing height, and wind speed. The fraction of SURE stations in the network simultaneously reporting certain concentrations of $SO_4^{2-}$ shows the degree of daily variability over the SURE region. When all stations were operating, a geographically representative picture of the extent of uniform, regionwide rise and fall of $SO_4^{2-}$ concentration was obtained (Figure 3.9).

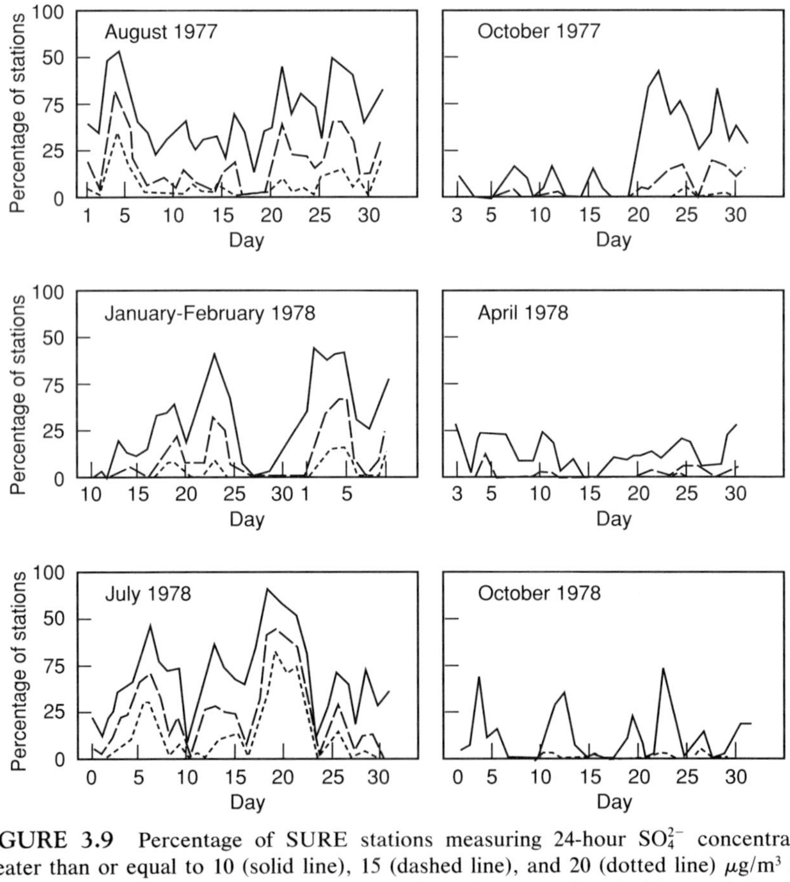


**FIGURE 3.9** Percentage of SURE stations measuring 24-hour $SO_4^{2-}$ concentrations greater than or equal to 10 (solid line), 15 (dashed line), and 20 (dotted line) $\mu g/m^3$ $SO_4^{2-}$ during intensive months. (From Mueller and Hidy, 1983; reproduced with permission of the Electric Power Research Institute.)

The daily sequences show wide variation in $SO_4^{2-}$ concentrations found uniformly at all reporting stations. The episodic nature of $SO_4^{2-}$ accumulation and distribution over a broad area can be readily seen in the results.

The diurnal changes in the results showed highly variable patterns from day to day and site to site. Aggregation of the average values reveals broad diurnal changes that are seasonally distinct (Figure 3.10). The composite diurnal change is stronger in summer than in winter. $O_3$ shows a midday maximum in both cases, but $SO_4^{2-}$ has a midday maximum with $O_3$

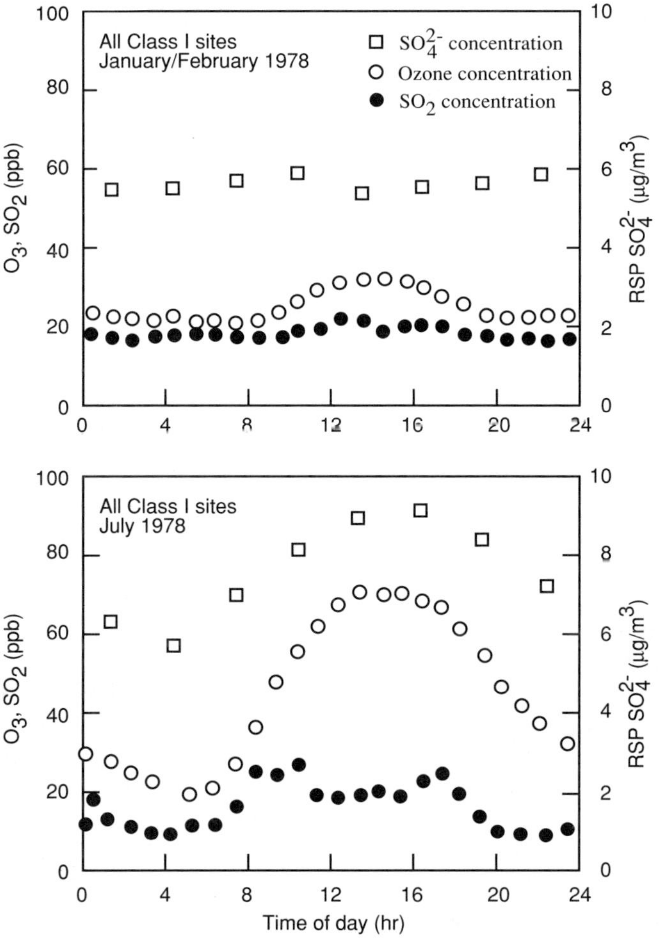


FIGURE 3.10 Diurnal distribution of air quality parameters observed at nine SURE stations in the eastern United States. (□) $SO_4^{2-}$ concentration; (○) ozone concentration; (●) $SO_2$ concentration. (From Mueller and Hidy, 1983; reproduced with permission of the Electric Power Research Institute.)

only in summer. In winter the mean $SO_4^{2-}$ pattern is approximately constant through the day, following the $SO_2$ concentration. The summer $SO_2$ change is distinct from that of $SO_4^{2-}$. The summer $SO_2$ change appears to reflect increased daytime industrial activity or deepened vertical mixing, which brings $SO_2$-rich layers of air to the ground. In contrast, the change in average $SO_4^{2-}$ concentration in summer follows $O_3$ production. $O_3$ levels at a remote, high-altitude site (Whiteface Mountain, NY) differ from those at the SURE stations (Mohnen, 1986). At Whiteface, $O_3$ shows less variation diurnally in summer than at the SURE sites. The concentrations are higher in summer, however, while $O_3$ maxima occur over several hours in late afternoon to evening. The former change evidently reflects $O_3$ transport of polluted air to the site as contrasted with local production, which peaks at midday. The latter is expected with summer seasonal increase in production because of intensified chemical reaction in the air.

## Vertical Structure

The interpretation of $SO_2$ behavior on a regional scale, characterizing conditions throughout the planetary boundary layer, assumes that the ground measurements represent a volume of air 100–300 m thick. Aircraft sampling, which was used to investigate this representativeness, at least in a limited way, was an important part of the SURE. The aircraft observations were made at night, early morning, and midday in the intensive sampling months to explore the extent and uniformity of vertical structure. Figure 3.11 shows results of the observations taken as averages over all samples at a fixed height. Because these samples represent collections integrated over heights nominally from 0.1 to 1.5 km and 1.5 to 3.0 km, they are shown as stepped concentrations.

The observations show vertical $SO_2$ concentrations to be uniform on the average for the lower one-eighth percentile of data. A maximum in average $SO_2$ concentration occurs at about 300 m. The upper one-eighth percentile of $SO_2$ concentrations indicates less vertical uniformity with considerable layering. Midday warming causes vertical mixing and a rise in elevated $SO_2$ concentration to 1 km and above.

Although aircraft observations of SURE (Figure 3.12) and other programs (Daum *et al.*, 1984) have indicated aerial $O_3$ concentrations similar to those at ground level, the $O_3$ concentration is often greater aloft than at the ground (Blumethal *et al.*, 1981), which is related to deposition destruction at the ground or depletion by reactions of $O_3$ with NO.

Gas–phase $H_2O_2$ reactions have been reported for summer conditions (Heikes *et al.*, 1987). Concentrations of $H_2O_2$ at the surface were found to be 0.5–1.5 ppbv; above 3 km in height they were 2–4 ppbv. The data were collected by air sampling in a path from the Gulf Coast at Gulfport,

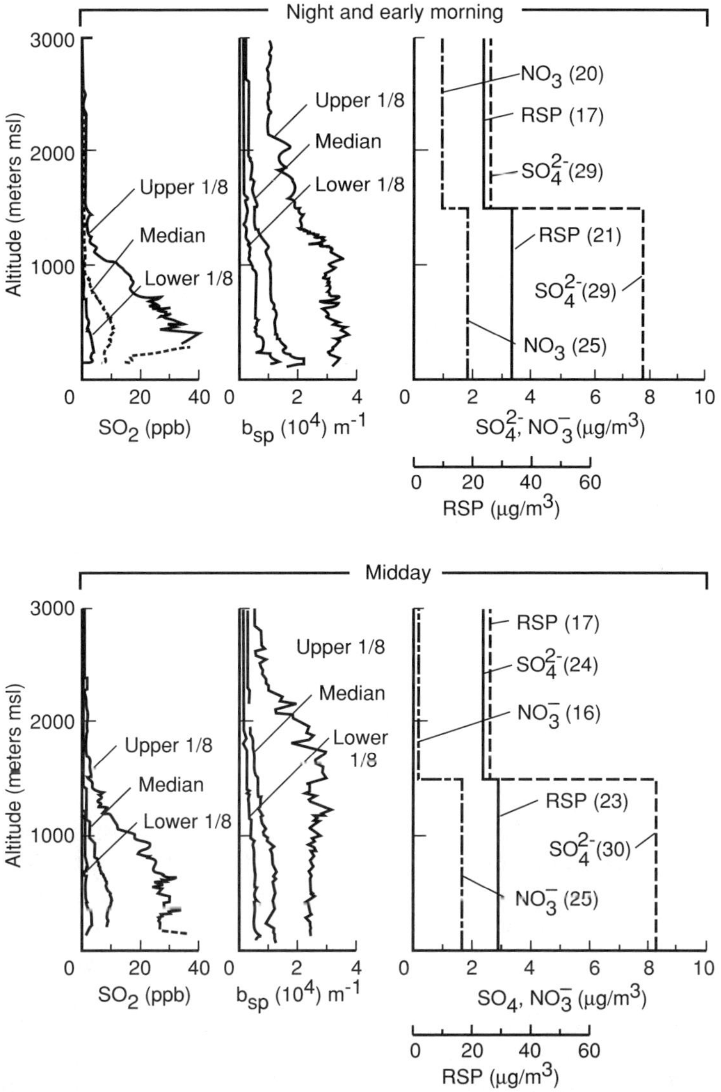


**FIGURE 3.11** Night and day vertical profiles constructed from median $SO_2$ and $b_{sp}$ values for upper and lower one-eighth of data points. Data also shown for composite RSP, $SO_4^{2-}$, and $NO_3^-$ at heights $>$1500 m and $>$1500 m taken at the same time. Observations taken between August 1977 and October 1978 near Duncan Falls, Ohio (number of points in average in parentheses). (Data from Blumenthal *et al.*, 1981, and from SURE data archives.)

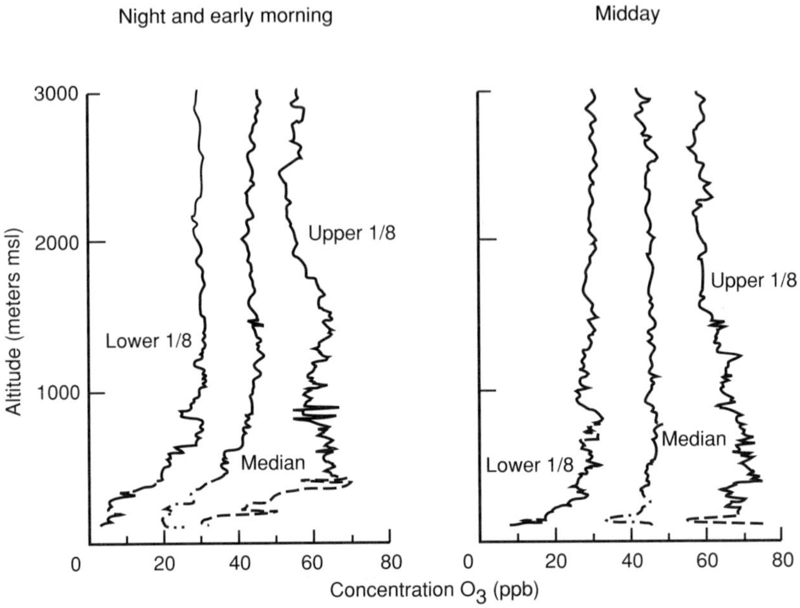


FIGURE 3.12 Night and day vertical profiles constructed from median $O_3$ values in each altitude bin (each 15 m for altitude <1500 m msl, 30 m for altitude >1500 m msl) for MRI data base (minus episode flights). Dividing lines for upper and lower one-eighth of data points also shown. (From Blumenthal *et al.*, 1981; reproduced with permission of the Electric Power Research Institute.)

Mississippi, north to Syracuse, New York, at intermediate locations within the surface boundary layer (<1.5 km) above cloud tops and in the free troposphere (above 3 km in height). Data show that concentrations of $H_2O_2$ are higher at high altitude than near the ground. The observed depletion with travel northeastward is ascribed to chemical reactions with $NO_x$ and HC that deplete $H_2O_2$ production. Depletion may also be associated with increased $SO_2$ with altitude, which is available for oxidation and consequent destruction of $H_2O_2$.

The profile of light-scattering coefficient ($b_{sp}$) in Figure 3.11 indicates a vertical distribution of particles in the light-scattering size range (0.1–1.0 $\mu$m in diameter). The concentrations of these particles (roughly proportional to $b_{sp}$) are uniform with height except in the morning in layers nearest the ground. The upper one-eighth percentile of $b_{sp}$ data shows layering, but the average values of $b_{sp}$ are roughly uniform. The zone of elevated $b_{sp}$ tends to extend uniformly to higher elevations than $SO_2$ concentrations. The difference between night and midday is particularly strong in the upper one-eighth percentile profile. Thus, a profile of a single pollutant does not necessarily indicate the top of the "well-mixed" air layers.

The vertical structure in aerosol particle components is also shown in Figure 3.11. However, the data suggest that the fine particle concentration (RSP) is more uniformly distributed with height than $SO_4^{2-}$ or $NO_3^-$. The uniformity in RSP concentration increase by midday is unexpected because $SO_4^{2-}$ is a large fraction of RSP and is sometimes used as a surrogate for total mass concentration. Both $SO_4^{2-}$ and $NO_3^-$ are much higher in concentration in the lower layer than the upper layer. On the average, $SO_4^{2-}$ tends to increase by midday. However, $NO_3^-$ concentrations either remain constant or decrease aloft by midday. The $NO_3^-$ results are uncertain because a few samples are averaged together, and those available are marginal in analytical detectability.

The aircraft data collected in the lower air layer were compared with ground-level samples for RSP mass and $SO_4^{2-}$ concentration over approximately the same period. Correlation coefficients for RSP mass concentration were calculated for pairs of aircraft samples and the ground station. None of the correlations were significant at the 95% level. The correlations at three sites were significant at 90%. Other sites were correlated at the 50% level of confidence or below.

The average RSP $SO_4^{2-}$ concentrations between 0.3 and 1.5 km were compared as a function of ground-level RSP $SO_4^{2-}$. The correlation coefficient for the 100 pairs of data is 0.682, which is significant at the 99.9% level. Analysis of these comparisons indicated that a one-to-one relationship did not exist between observations at ground level and aloft. Concentrations of $SO_4^{2-}$ tended to be higher aloft. These results suggest that the ground-level observations are useful semiquantitative measures of air chemistry in layers at 300 m and below. However, inference of lower tropospheric behavior from the ground-level data is not necessarily highly reliable.

Correlation coefficients were calculated for pairs of aircraft RSP $SO_4^{2-}$ and ground data, CI sites, and an intensive sampling period. The significance levels range from 1% to greater than 50%. Because the correlation coefficient for all points taken together was significant at the 0.1% level, segregating the data in this manner did not improve the relationship. Thus, differences in the vertical structure from one day to another at a single location were as large as differences from one site to another and from one season to another. The $SO_4^{2-}$ concentrations aloft exceeded the ground-level concentration in 73% (48 of 66) of the cases. Concentrations are mutually predictable within definable but wide limits of precision.

Data for particle concentrations were separated into sets of daytime and nighttime observations to determine whether the overall differences in vertical distribution were influenced by the observed diurnal variation in vertical structure. Tests of significant differences among the three vertical levels by day and night were conducted. Only the result for RSP $SO_4^{2-}$ was substantially different when comparing the 0.3- to 1.5-km level to ground level. In this case the median daytime 0.3- to 1.5-km observation for RSP $SO_4^{2-}$ is 7.7 $\mu g/m^3$, the median nighttime value is 4.3 $\mu g/m^3$, and the observed ground-level median value for either day or night is 6.6 $\mu g/m^3$. A test performed on the 0.3- to 1.5-km RSP $SO_4^{2-}$ identified a significant difference (at 5% and also at 1% between the daytime and nighttime observations). This finding suggests that $SO_4^{2-}$ in fine particles is produced aloft during the day by photochemically accelerated reactions.

## VISIBILITY

Perception of complex visual images through the atmosphere involves an ill-defined combination of physical optics, psychophysical processes in the human eye, and psychological impressions of features. Appendix B summarizes optical and perception principles that affect visibility. The propagation of light through a gaseous medium is well understood. In the absence of specific knowledge linking all three components of perception, studies of visibility have focused first on relating optical properties of air to pollution and second on a somewhat subjective human visual observation, the *visual range*. This term is defined as the greatest distance at which an observer can distinguish a contrast between an object and its background. *Contrast* refers to a color or brightness of an object relative to its background. An ancillary concept, *prevailing visibility*, is important because it indicates the extent of geographic area in which the visual range is restricted. It is the visual range attained or surpassed around at least half of the horizon but not necessarily in continuous sectors. The concept is a useful one because the distances can be determined from visual observations of surrounding landmarks as viewed from a receptor location, such as an airport control tower or a mountain fire-watch tower. As a result of its importance to aviation, a considerable collection of data on prevailing visibility exists for locations in North America.

Figure 3.13 shows data on the frequency of occurrence of prevailing visibility from two rural stations in ENA. These data indicate that the eastern Ohio site has poorer visibility than the site in eastern Pennsylvania; the median values are 17 km and 32 km, respectively.

The visual range (VR) is linked empirically with the extinction coefficient $b_{ext}$ of visible light through a seeing path by the Koschmieder relationship:

$$VR = -\text{constant}/b_{ext} \quad (3.1)$$

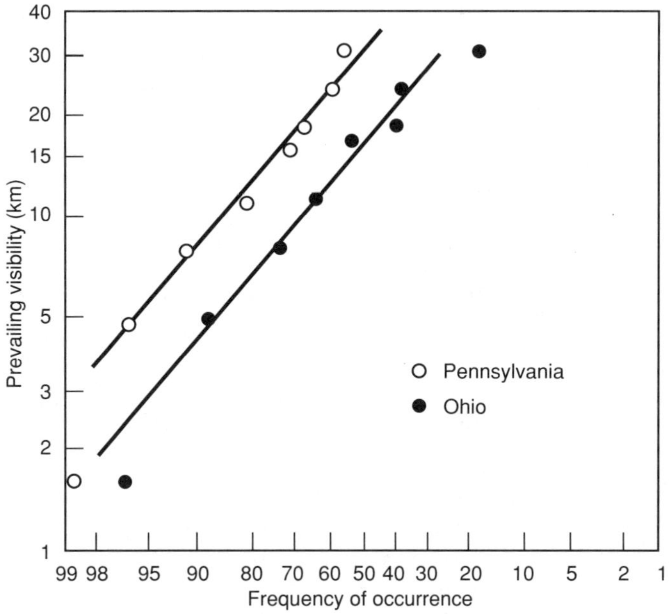


FIGURE 3.13 Cumulative distribution of prevailing visibility ≤ indicated values; total program, June 1979–February 1980. Data from Mehoopany, Pennsylvania, and Blue Rock, Ohio. (From Tombach and Allard, 1983; reproduced with permission of the Electric Power Research Institute.)

where the constant is the logarithm of a threshold contrast for a black object on a white background. The constant is ln(0.02) or −3.912, based on many measurements. The extinction coefficient contains the light scattering and absorption associated with suspended material in the observer's line of sight.

Thus,

$$b_{\text{ext}} = b_{\text{sg}} + b_{\text{ag}} + b_{\text{sp}} + b_{\text{ap}} \tag{3.2}$$

where subscript s refers to scattering, a to absorption, g to gas, and p to particles. The particle component is usually much larger than the gas component. Light scattering from gas molecules is predicted from Rayleigh's theory, with strong dependence on light wavelength. The Mie scattering from particles is much less wavelength dependent but is size dependent. The weak wavelength dependence on the Mie range is said to lead to a blue-white to gray appearance of haze.

Measurements at a rural site near Shenandoah, Virginia, show an empirical relationship between VR and $b_{\text{sp}}$, as determined with a nephelometer (Ferman *et al.*, 1981). The "constant" in Eq. (3.1) is somewhat less than the Koschmieder value for the Shenandoah data.

Because VR and $b_{sp}$ depend on relative humidity, a consistent relationship between $b_{sp}$ and fine-particle concentration is found for eastern samples at a fixed, low relative humidity (~40%). The data also establish the relation between the light-scattering coefficient in dry conditions, $b_{sdb}$, and dry filter collection particle samples. Extending the relationship to chemical constituents in the fine particles, Ferman *et al.* (1981) found that

$$b_{sdb} = 7.8(\pm 0.6)(NH_4HSO_4) + 6.0(\pm 3.0)\,[\text{fine carbon}] + 2.3(\pm 1.0)[R] + 4.1(\pm 1.9) \quad (3.3)$$

where R is the remainder of material in the fine-particle concentration. This indicates that $SO_4^{2-}$ (as ammonium bisulfate) is an efficient light scatterer and is comparable with the carbon fraction. Organic carbon concentration in these samples from rural Virginia was about 4.5 times the elemental (black) carbon concentration, so that the particle absorption coefficient was small relative to the scattering effect.

Water vapor has a dramatic influence on light scattering (Figure 3.14). The scattering in the range above 70% relative humidity (RH) can be reduced by normalization of $b_{sp}$ (as $b_{sw}$) to $SO_4^{2-}$ concentration, which suggests the importance of the hygroscopicity of $SO_4^{2-}$ species in the humidity effect.

The empirical relationships for optical effects illustrated in the Shenandoah example are thought to be representative of regional haze in ENA based on similar results from other studies. Other studies have shown variation of $b_{sdb}$ fine-particle mass ratios from site to site. Wolff *et al.* (1982) hypothesized that this ratio is a function of $SO_4^{2-}$, which is believed to be a more efficient light scatterer than other fine-particle constituents.

Trends have been determined in airport prevailing visibility data collected since 1948. A thorough study of selected data has been supported by Sloane (1984). She attempted to adjust the visual observations for representative local meteorological conditions using relative humidity, dew point, and temperature for indexes of air mass characteristics. Reported relative humidities were in the range of >51–55%, with a standard deviation (sd) of ±13% RH. Temperature was restricted to 25–28°C with an sd of ±4°C. Wind speed and pressure were screened to eliminate extremes, with restriction that they be within 2 sd of their quarterly mean at each site. The 60th percentile values were used as a representative value for visibility. Sloane's results showed that visibility differed substantially over the years from site to site in the East. In some areas visibility decreased; in others it improved. Her results showed a less pronounced period of excellent visibility in the 1950s than other analyses (Trijonis and Yuan, 1978).

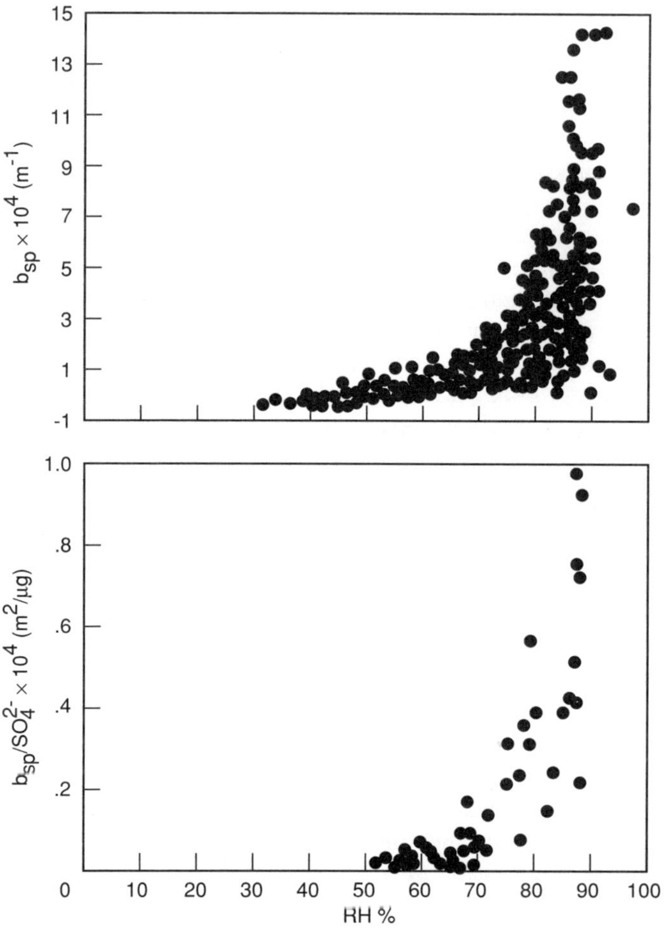


**FIGURE 3.14** Effect of relative humidity on light-scattering coefficient and on $b_{sp}/SO_4^{2-}$ ratio. (From Ferman *et al.*, 1981; courtesy of the Air Pollution Control Association.)

Sloane's results showed a moderate (10–30%) decline in visibility for two Virginia sites but a steady (no trend) condition for Williamsport, PA. The trends at urban or suburban sites in the region were also mixed for summertime conditions, but they generally displayed a moderate decline between 1948 and 1980. Meteorological adjustment of the data showed much less deterioration of visibility than does trend analysis of raw data. The adjusted picture resulted from exclusion of locally atypical conditions. Summertime metropolitan area declines are small during the period, but the trend to declining visibility is apparent in rural and medium-sized urban areas. This decline was attributed to the increases in post-1960, summertime $SO_2$ emissions and their atmospheric conversion to particle $SO_4^{2-}$. The emission increases are identified with increased demand for electric power, partly for air conditioning. Improvement in winter (first and fourth quarter) visibilities, consistent with shifts in fuel use in the 1970s and restrictions in fuel S content, were noted.

In the eastern United States, historical trends in light extinction, estimated from visual range, and $SO_2$ emissions trends apparently correspond well with each other both regionally and seasonally (NAPAP, 1990b). From the late 1940s to the early 1980s, the Northeast has undergone a moderate decline in haziness and emissions during the winter and a moderate increase in both haziness and emissions during the summer. Over the same period, the Southeast has experienced a moderate increase in emissions and haziness during the winter and a strong increase in both during the summer. The increasing trends for both emissions and calculated light extinction in the summer in the Northeast and in the winter and summer in the Southeast were greatest during the 1950s and 1960s, with a leveling off or decrease after the 1970s. The NAPAP authors hypothesize that haziness over the eastern United States has been dominated by sulfur emission sources since the late 1940s.

***Turbidity and Optical Depth*** Although most people rarely see the earth from space or near-space, the principles of visibility apply equally well when viewing the earth's surface or looking straight up at the sky. Viewing in this way involves looking through the entire atmosphere as a shell. This measure of atmospheric murkiness or the observation of surface or sky features is stated in terms of opacity or *turbidity*.

A convenient measure of turbidity or the total particle burden in a vertical column of air is the *optical depth* ($\delta$),

$$\delta = \int_0^\infty b_{sp}\, dz$$

Aside from its importance to astronomical observations, atmospheric turbidity is important in accounting for the radiation fluxes in the atmosphere. The balance of incoming solar radiation to the earth and outgoing radiation from the earth is maintained through the medium of the atmosphere such that the earth's surface temperature is about 20°C.

Concern has developed that the atmospheric accumulation of certain radiation-absorbing gases emitted from human activities may modify the radiation balance to add heat to the surface air. This effect, sometimes called the *greenhouse effect*, involves the gases carbon dioxide ($CO_2$), methane ($CH_4$), nitrous oxide ($N_2O$), and chlorofluorohydrocarbons (CFCs). These gases absorb (outgoing) radiation in infrared wavelengths while being transparent to incoming solar radiation, primarily in shorter wavelengths. The "trapping" of outgoing radiation and reemission at warm temperatures near the earth's surface effectively provide for "stabilizing" the average surface temperature. If the concentration of the greenhouse gases increases, the mean surface air temperature will tend to increase in response.

If clouds or aerosol particles are present, the radiation balance can be modified. The modification can force cooling or warming depending on the optical properties of the cloud of particles and their effective height. The widespread transient effect of aerosols has been observed in connection with widespread injections of volcanic dust into the stratosphere (see also Hidy, 1984b).

Generally, the effect on global temperature of atmospheric aerosol particles was believed to be small relative to the absorbing gas effect. However, calculations by Charlson *et al.* (1991) indicate that the present total sulfate burden in the Northern Hemisphere can change the reflected solar energy flux up to about $-1$ W/m$^2$. The present-day radiative forcing from anthropogenic emissions is $+1.5$ W/m$^2$; the two are comparable but of opposite sign. The spatial variability in sulfate, for example, between western and eastern North America creates large-scale radiation or heating gradients whose effects on weather or climate change have not been studied.

If verified, the effect of anthropogenic aerosols could have a significant influence on the apparent greenhouse effect associated with increases in radiatively absorbing gases in the atmosphere.

## PRECIPITATION CHEMISTRY

Scavenging during cloud formation followed by precipitation represents a major mechanism for loss of material from the atmosphere. The efficiency of scavenging differs with the solubility and reactivity of species (Appendix A). The chemistry of precipitation water gives information on deposited material, as well as clues about the scavenging process for different airborne species.

Both $SO_4^{2-}$ and $NO_3^-$ are found in significant quantities in precipitation. Their presence may be the result of the scavenging of $SO_2$ and $NO_x$ or their reaction products from the air. S and N chemistry in precipitation is often identified with acidity. Acidity is difficult to measure in precipitation water. Therefore, to account for acidity, the major anion and cation species must be specified to ensure that an ion balance is achieved in the chemistry. Historically, this is often overlooked in discussions of precipitation chemistry. Nevertheless, an ion balance is crucial to characterizing precipitation properties with confidence.

Table 3.10 lists typical concentrations of major ions in precipitation. The reported anions include $SO_4^{2-}$ and $NO_3^-$, with minor contributions from chloride, sulfite, and phosphate. The major cations are $H^+$, $NH_4^+$, $Ca^{2+}$, $Na^+$, $Mg^{2+}$, and $K^+$. Total acidity depends on the ionization equilibrium of different acids. At pH levels encountered in precipitation, strong inorganic acids ($H_2SO_4$ and $HNO_3$) are substantial contributors, but the contributions from organic acids are also factors (Likens, 1976). In some locations, hydrochloric and phosphoric acids may also be significant.

Data collected in the early 1980s established the spatial distributions of $SO_4^{2-}$ and $NO_3^-$ concentrations in precipitation (e.g., Endlich *et al.*, 1986) (Figure 3.15). The $SO_4^{2-}$ distribution is elliptical in shape with a maximum in the Ohio River Valley, Indiana, and western Pennsylvania, in the zone of high emission density. This pattern is similar to that of airborne $SO_4^{2-}$

TABLE 3.10 **Precipitation Measurements in the EPRI-UAPSP Network**[a,b]

| Observable | Network medians (μmol/L) 1979, 1980 | 1982 | Lower quantifiable limit (μmol/L)[d] | Precision for duplicate measurements (AD/median)[c] 1979[e] | 1982[f] | Precision for collocated[d] measurements (AD/median)[c] 1979[e] | 1982[g] |
|---|---|---|---|---|---|---|---|
| Precipitation amount: | 0.58 cm | | 0.05 cm | | | 5 | 3 |
| $H^+$ | 60.3 | 43.7 | 25 | 6 | 5 | 10 | 13 |
| $NH_4^+$ | 15.9 | 14.1 | 1.39 | 4 | 2 | 8 | 6 |
| $Na^+$ | 6.63 | 2.49 | 1.09 | 6 | 7 | 25 | 11 |
| $K^+$ | 1.21 | 0.57 | 0.64 | 19 | 18 | 42 | 22 |
| $Mg^{2+}$ | 1.16 | 0.85 | 0.21 | 4 | 4 | 11 | 9 |
| $Ca^{2+}$ | 3.48 | 2.51 | 0.40 | 5 | 5 | 16 | 12 |
| $Cl^-$ | 6.0 | 4.26 | 0.70 | 2 | 1 | 12 | 9 |
| $NO_3^-$ | 25.7 | 21.4 | 0.40 | 2 | <1 | 5 | 3 |
| $SO_4^{2-}$ | 26.5 | 19.7 | 0.26 | 1 | <1 | 5 | 4 |
| $H_2PO_4^-$ | — | 0.11 | 0.13 | — | 28 | — | 22 |

[a] From Mueller *et al.*, 1984; reproduced with permission of the Electric Power Research Institute.

[b] 20 stations operated, 16 of which were east of the Mississippi.

[c] Percent of mean of absolute differences (AD) with respect to median concentration observed in the precipitation.

[d] 10% precision limit based on replicate laboratory runs, except for precipitation amount, which is based on rain gauge sensitivity.

[e] Data from 9 stations (about 1000 sample pairs).

[f] Data from 19 stations.

[g] Data from 2 stations (about 150 sample pairs).

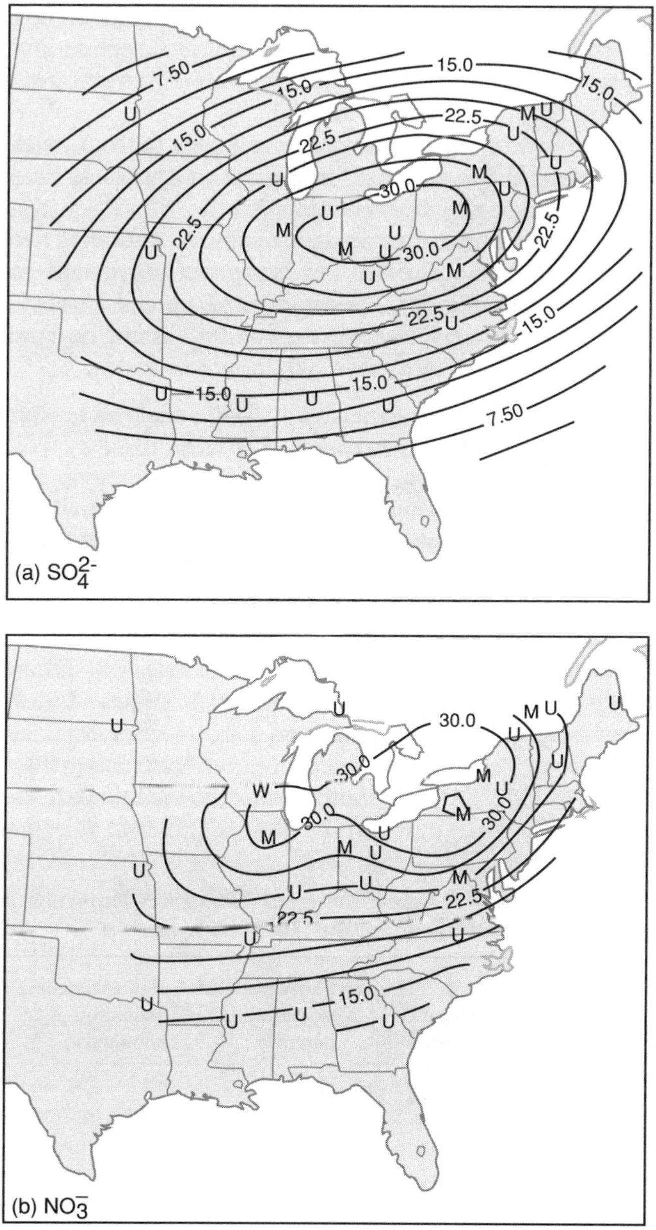


**FIGURE 3.15** (a) Map of median $SO_4^{2-}$ concentration ($\mu$mol/L) for 1982 based on UAPSP stations (U) and MAP3S stations (M). (b) Map of median aq $NO_3^-$ concentration ($\mu$mol/L) for 1982 based on UAPSP stations (U) and MAP3S stations (M). (From Endlich *et al.*, 1986; reproduced with permission of the Electric Power Research Institute.)

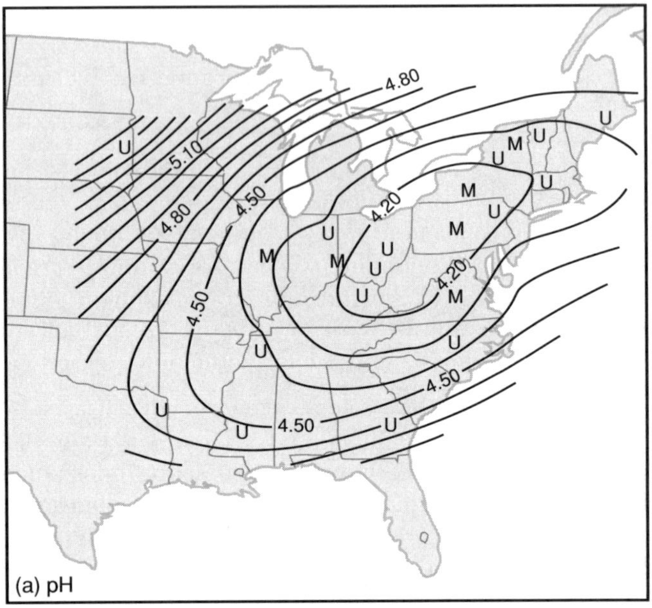


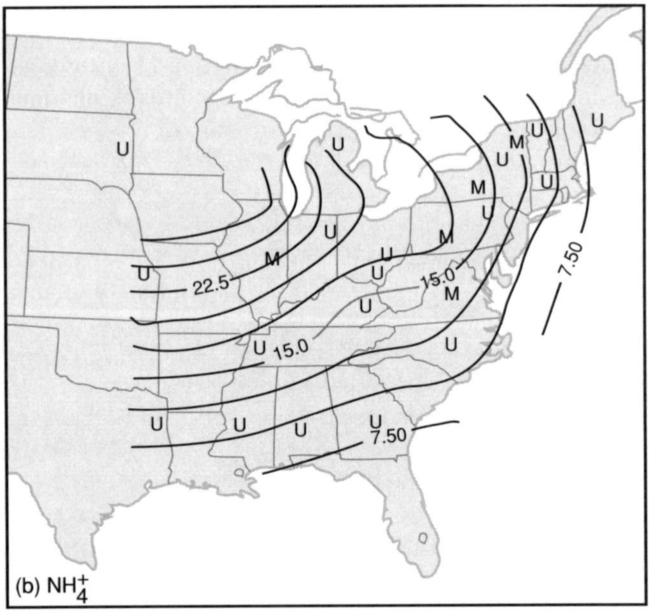


FIGURE 3.16 (a) Map of median laboratory pH values for 1982 and maps of median $NH_4^+$ (b), $Ca^{2+}$ (c), and $Na^+$ (d) concentrations ($\mu$m/L) for 1982. Data based on UAPSP stations (U) and MAP3S stations (M) for (a) and (b), and UAPSP stations for (c) and (d). (From Endlich *et al.*, 1986, and Topol, 1989; reproduced with permission of the Electric Power Research Institute.)

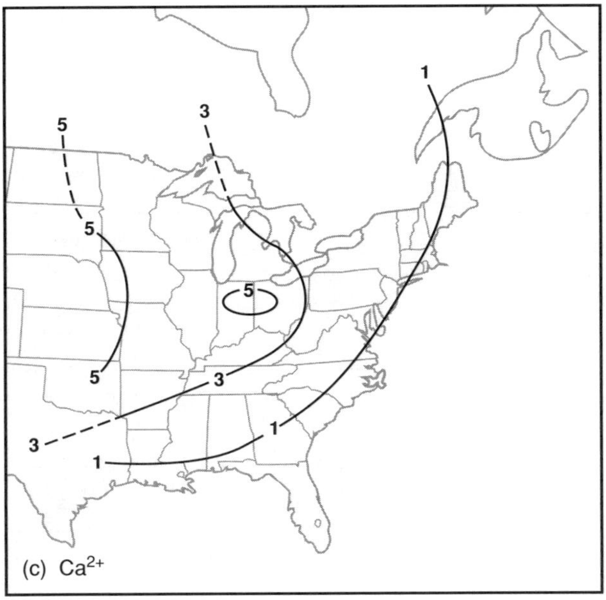


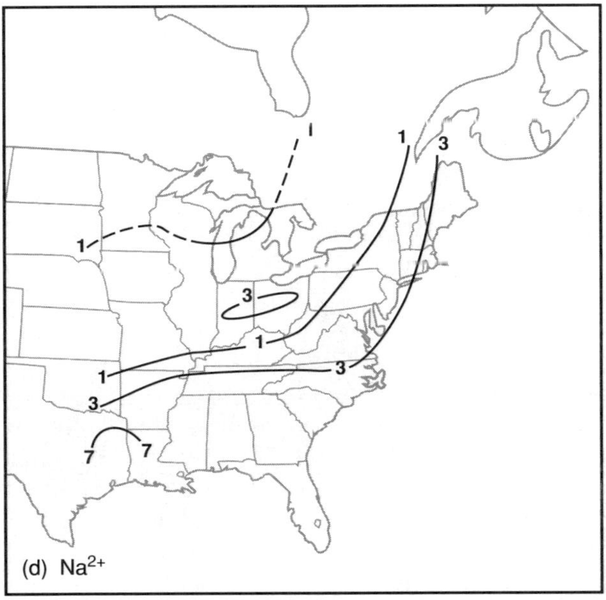


**FIGURE 3.16** (*continued*)

except that the maximum is farther east. The precipitation $NO_3^-$ distribution shows a more asymmetric pattern, with the maximum region farther east than that of $SO_4^{2-}$. The area of maximum concentration encompasses Lake Erie and Ontario.

Precipitation pH values corresponding to the $SO_4^{2-}$ and $NO_3^-$ distributions are shown in Figure 3.16 (Endlich *et al.*, 1986). Generally, the lower pH levels (higher acidity) correspond to the region of high $SO_4^{2-}$ and $NO_3^-$. However, the area of minimum pH occurs over Pennsylvania and New York State. This difference is ascribed to eastward decreases in cation concentrations, particularly $Ca^{2+}$ and $NH_4^+$, relative to $SO_4^{2-}$ and $NO_3^-$. As an example, median $NH_4^+$ and $Ca^{2+}$ concentrations are shown in Figure 3.16b and c. These cation concentrations decrease significantly toward the coastline. In contrast, $Na^+$ (Fig. 3.16d) is high along the coastline and tends to decrease in concentration inland. Note both $Ca^{2+}$ and $Na^+$ have a localized maximum around the Ohio River Valley. Both $H^+$ and $SO_4^{2-}$ generally have a mean seasonal change with a maximum in summer. The exception is a western station where a maximum in $SO_4^{2-}$ occurs in spring but a maximum in $H^+$ occurs in winter. The seasonal pattern of $NO_3^-$ is similar, but three of the six stations have a summer $NO_3^-$ maximum.

Data for changes in precipitation $SO_4^{2-}$, $NO_3^-$, and $H^+$ concentrations from year to year were collected from bulk deposition observations between 1965 and 1983 at Hubbard Brook, NH (NAS, 1983). Year-to-year variations can be substantial in station data, in part because of emission changes in $SO_4^{2-}$ and $NO_3^-$ during the period. The changes in $H^+$ concentration do not correspond to those expected from summing the anions, $SO_4^{2-}$ and $NO_3^-$. The reason for this may be related to sample contamination. In any case, this is a good example of the need to analyze such data for consistency in terms of an ion balance.

## CLOUD CHEMISTRY

In the past decade, data that characterize cloud conditions and link their chemistry with precipitation have increased significantly. Surface observations at Whiteface Mountain, NY, and at Allegheny Mountain, PA, give perspective on in-cloud or fog conditions. The Mountain Cloud Chemistry Project (MCCP) has provided cloud observations at five high-altitude sites in ENA (Mohnen, 1988). In addition to Whiteface Mountain, measurements were made during 1986–1988 at Shenandoah, VA, Whitetop, VA; Mitchell, NC; and Moosilauke, NH. The MCCP measurements add insight into the spatial distribution of chemicals in cloud water. Mean ion concentrations for the five sites are listed in Table 3.11. The highest concentration of the ions listed was found at Mt. Mitchell, NC, and lowest values were observed at Shenandoah and Whiteface. $SO_4^{2-}$ and $NO_3^-$ ion

TABLE 3.11 **Unweighted-Mean Ion Concentrations for Both Precipitating and Nonprecipitating Clouds of 1986–1988 at Six High-Elevation Locations Monitored by MCCP**[a]

| | Concentration (μeq/L) | | | | | |
|---|---|---|---|---|---|---|
| **Site** | $H^+$ | $SO_4^{2-}$ | $NO_3^-$ | $Cl^-$ | $NH_4^+$ | **Elevation (m)** |
| Mitchell, NC | 398 | 489 | 174 | 31 | 184 | 2020 |
| Moosilauke, NH[b] | 263 | 257 | 132 | 16 | 107 | 1000 |
| Shenandoah, VA | 171 | 176 | 94 | 13 | 93 | 1014 |
| Whiteface-1, NY | 171 | 205 | 73 | 5 | 97 | 1483 |
| Whiteface-2, NY[b] | 255 | 352 | 92 | 7 | 157 | 1250 |
| Whitetop, VA | 274 | 321 | 144 | 16 | 152 | 1089 |

[a] From Vong, 1989.
[b] No cloud water collected during 1986.

concentrations are generally substantially higher than in precipitation collected at lower altitudes. The data do not show a clear spatial variation from these sites, except for Mt. Mitchell. Comparisons between results at two heights suggest higher concentrations of ions in the lower levels of clouds. The high elevations of mountains in the Appalachians are frequently in clouds. In the southern mountains 30–40% immersion has been observed, while in the northern Appalachians in-cloud conditions are 40–50%. Cloud water content was generally measured at 0.2 g/L on the average.

Seasonal differences in cloud chemistry appear to be subtle except possibly in the case of sulfate and nitrate. An example of median and maximum levels at Whiteface Mountain for some ions is listed in Table 3.12. Comparison of the winter and summer month concentrations for these years of sampling suggests that hydrogen ion and nitrate ion concentrations are similar in winter and summer. However, sulfate and ammonium are higher in summer than in winter. This appears to reflect the summer role of $H_2O_2$ in oxidizing $SO_2$ and the added biological activity in summer, producing higher atmospheric levels of $NH_4^+$.

Measurements of fog water on Allegheny Mountain (Table 3.13), suggest that the concentrations of acidity, $SO_4^{2-}$, and $NO_3^-$ are larger than accompanying precipitation but that $H^+$ is similar in both precipitation and fog (Pierson *et al.*, 1987).

A few observations of the chemical composition of cloud water samples taken on Mt. Washington in New Hampshire have been reported. A record at this site dates back to 1937 and was renewed in the early 1980s (Kimball *et al.*, 1986). Table 3.14 lists the concentrations of several elements, acid content, and $SO_4^{2-}$ for the 1980s data. Also tabulated are species equivalent ratios to $SO_4^{2-}$ in the samples. Comparison of these

**TABLE 3.12 Comparison of Winter and Summer Cloud Ion Concentrations (μeq/L) at Whiteface Mountain, NY[a]**

| Year | $H^+$ Median | $H^+$ Max | $SO_4^{2-}$ Median | $SO_4^{2-}$ Max | $NH_4^+$ Median | $NH_4^+$ Max | $NO_3^-$ Median | $NO_3^-$ Max | Number of samples | Days with data |
|---|---|---|---|---|---|---|---|---|---|---|
| | | | | | *January–February* | | | | | |
| 1983 | 129 | 589 | 100 | 554 | 54 | 205 | 71 | 258 | 107 | 8 |
| 1984 | 204 | 1148 | 160 | 1632 | 95 | 680 | 140 | 838 | 156 | 11 |
| 1986 | 191 | 1585 | 168 | 972 | 73 | 522 | 86 | 1131 | 247 | 22 |
| | | | | | *June–August* | | | | | |
| 1983 | 166 | 3162 | 224 | 3050 | 130 | 855 | 69 | 1338 | 304 | 29 |
| 1984 | 251 | 977 | 316 | 1202 | 147 | 612 | 83 | 285 | 124 | 21 |
| 1986 | 145 | 1413 | 142 | 1318 | 64 | 770 | 46 | 752 | 162 | 23 |

[a] From Mohnen and Kadlecek, 1989.

TABLE 3.13 **Comparison of Rain Properties at Allegheny and Laurel, August 1983, with Allegheny Dew and Settled Fog Water (Volume-Weighted Averages)**[a]

| | **Rain, Allegheny ($n$ = 12)** | **Rain, Laurel ($n$ = 5)** | **Dew ($n$ = 15)** | **Fog ($n$ = 1)**[b] |
|---|---|---|---|---|
| pH | 3.5 (3.1–3.75) | 3.5 (3.3–3.6) | 4.0 (3.5–5.3) | 3.47 |
| Titratable acid ($\mu$eq/L) | 324 (227–763) | 349 (317–615) | 108 (16–382) | 350 |
| $H^+$ ($\mu$eq/L) | 290 (178–741) | 311 (257–550) | 91 (5.5–347) | 340 |
| $NH_4^+$ ($\mu$eq/L) | 53 (15–116) | 70 (61–119) | 8 (0–55) | 144 |
| $Na^+$, $K^+$, $Mg^{2+}$, $Ca^{2+}$ ($\mu$eq/L) | 39 (8–99) | 70 (48–159) | 41 (14–99) | 155 |
| $NO_2^-$ ($\mu$eq/L) | 0.6–1.0 | 0.6–1.3 | 0.7 (0.1–2.0) | — |
| $NO_3^-$ ($\mu$eq/L) | 68 (41–252) | 86 (63–202) | 32 (3–138) | 230 |
| $SO_4^{2-}$ ($\mu$eq/L) | | | | |
| Prompt[c] | 256 (125–700) | 294 (249–575) | 73 (10–254) | 387 |
| Delayed | 262 (138–700) | 299 (251–575) | 81 (14–251) | 373 |
| $Cl^-$ ($\mu$eq/L) | 8 (3–25) | 8.5 (6–1) | 5 (0.3–16) | 26 |
| Ion balance $\Sigma+/\Sigma-$ | 1.15 | 1.16 | 1.27 | 0.99 |
| Å ($\mu$mho/cm) | 115 (67–271) | 122 (109–220) | 42 (10–137) | 159 |
| Å accounted for | 106% | 111% | 94% | 107% |

[a] Reprinted from Pierson *et al.*, 1987. Copyright (1987) American Chemical Society.

[b] $n$ is the number of samples; does not include all fog events.

[c] Distinction is made in terms of sample storage time.

TABLE 3.14 **Analysis of Cloud Water from Summit of Mount Washington**[a]

| Species | Arithmetic mean (ng/ml) | Standard deviation[b] (%) | Equivalent ratio to $SO_4^{2-}$ |
|---|---|---|---|
| Na | 630 ± 70 | (11) | 0.33 |
| Mg | 74 ± 17 | (23) | 0.07 |
| Al | 14 ± 9 | (65) | 0.02 |
| Cl | 860 ± 9 | (11) | 0.29 |
| Ca | 290 ± 30 | (10) | 0.17 |
| Se | 0.0044 ± 0.0025 | (57) | — |
| V | 0.15 ± 0.06 | (38) | — |
| Mn | 1.36 ± 0.09 | (7) | — |
| Fe | 20 ± 6 | (30) | — |
| Co | 0.016 ± 0.005 | (35) | — |
| Zn | 9.9 ± 1.3 | (13) | — |
| As | 0.48 ± 0.04 | (9) | — |
| Se | 0.095 ± 0.010 | (11) | — |
| Br | 4.5 ± 0.5 | (12) | — |
| Sb | 0.050 ± 0.012 | (23) | — |
| $SO_4^{2-}$ | 4200[c] | — | 1.00 |
| $H^+$ | 100[c] | — | 1.2 |

[a] Unpublished data reported by K. Rahn.
[b] Based on five samples.
[c] From median values of summer month data between 1983 and 1985 (Kimball *et al.*, 1986).

data with cloud and precipitation water samples taken elsewhere in ENA indicates similar concentration levels. The equivalent ratio suggests that sodium is approximately balanced by chloride, indicating a possible marine air influence at Mt. Washington. Furthermore, the $H^+$ ion appears to be largely accounted for in terms of $SO_4^{2-}$ anion. This is considered typical of conditions over the upper Ohio River Valley area.

Kimball *et al.* (1986) have compared summer $SO_4^{2-}$ and pH observations on Mt. Washington between 1939–1940 and 1983–1986. They found that the ranges of pH values observed were similar, yet the frequency of high acidity occurrence in clouds was greater in the 1980s. The $SO_4^{2-}$ concentration patterns were not significantly different.

At Whiteface Mountain $SO_4^{2-}$ and $NO_3^-$ concentrations exhibit seasonal differences. $SO_4^{2-}$ is maximum in summer in precipitation, but the $NO_3^-$ concentration becomes larger in winter. $H^+$ and $NH_4^+$ appear to have an inconsistent seasonal variation.

The absorbed $SO_2$ oxidation process varies with $H_2O_2$ concentration. Aqueous-phase $H_2O_2$ concentration averages 30 $\mu$mol/L in summer but

0.3 $\mu$mol/L in winter (Kadlecek *et al.*, 1986). These seasonal differences are likely to be associated with reduced photochemical activity in winter. Ambient $SO_2$ concentration also shows the difference in $SO_2$ removal by cloud oxidation in winter. $SO_2$ levels are reportedly higher in winter than in summer. Observations of aqueous $H_2O_2$ concentrations for four stations in the MCCP are listed in Table 3.15 for the period between 1986 and 1988. There is considerable variation in cloud water $H_2O_2$, but the maximum mean levels were found at Mitchell, as well as the recorded maximum. The moist atmosphere at this site appears to be most oxidative and highest in acidifiers of the ENA high-elevation sites examined to date.

Aircraft measurements have been reported by Lazrus *et al.* (1981) and Daum *et al.* (1984). Their data suggest that $H^+$, $SO_4^{2-}$, and $NO_3^-$ concentrations are higher in cloud water than in precipitation. Comparison of concentrations in particle, cloud, and precipitation water ratios to $SO_4^{2-}$ suggest enrichment of $NO_3^-$ in cloud water, presumably through scavenging of $HNO_3$ vapor. $Ca^{2+}$, $Na^+$, and $Mg^{2+}$ are depleted in relation to $SO_4^{2-}$, which suggests either a difference in scavenging efficiency in hydrometeors or differences in concentration with height.

Comparison of the $SO_4^{2-}$ ion ratios on Mt. Washington with observations reported by Lazrus *et al.* (1981) suggests that clouds at this site are rich in alkaline metals. The concentrations of species in cloud water at Mt. Washington fall within the range suggested by Pennsylvania data. In this case, there was no evidence of dramatic differences between cloud water and precipitation concentrations.

More extensive measurements have been reported by Daum *et al.* (1984); their aircraft flights targeted warm (>0°C) nonprecipitating stratiform clouds from February 1982 to June 1983. The measurements ranged from South Carolina to Long Island, NY. $H^+$, $NH_4^+$, $NO_3^-$, and $SO_4^{2-}$ dominated the inorganic composition of cloud water. Concentration of interstitial aerosol species and gaseous $HNO_3$ were found to be low compared with concentrations of ionic species in cloud water. In contrast, $NO_x$ and organic $NO_3^-$ were in excess of cloud water $NO_3^-$. In some cases,

TABLE 3.15 **Cloud Water $H_2O_2$ Measurements ($\mu$mol/L) at Four MCCP Summit Sites for 1986–1988**

| Site | Minimum | Maximum | Mean | Number of samples |
|---|---|---|---|---|
| Whiteface, NY | 1.6 | 136.4 | 37.7 | 54 |
| Shenandoah, VA | 3.0 | 27.0 | 13.7 | 10 |
| Whitetop, VA | 0.8 | 246.6 | 44.1 | 141 |
| Mitchell, NC | 1.0 | 219.0 | 64.9 | 94 |

interstitial $SO_2$ concentrations were low (<0.2 ppb), with concentrations of cloud water $SO_4^{2-}$ above 1 ppb equivalent gas-phase $SO_2$. The data suggest substantial conversion to $SO_4^{2-}$ in such clouds. In other cases, plumes of $SO_2$ were found where gas-phase $SO_2$ concentrations exceeded equivalent cloud water $SO_4^{2-}$ levels. Aqueous-phase $H_2O_2$ concentrations were found to be inversely correlated with interstitial $SO_2$. The data suggest that $NO_x$ uptake as cloud water $NO_3^-$ was less complete than $SO_2$ uptake as $SO_4^{2-}$. Evidence for $NO_x$ conversion to $NO_3^-$ in cloud droplets was not found.

## DEPOSITION RATES

The characterization of deposition rates is an important extension of the description of air chemistry because deposition is an important measure of chemical exposure for ecosystems. Because of difficulties in measurement methods, the principal information on deposition focuses on the wet component.

### Wet Deposition

The patterns of wet deposition of $H^+$, $SO_4^{2-}$, and $NO_3^-$ (Figure 3.17) are similar to the precipitation concentration patterns. The deposition levels for $H^+$ range from 19 to 97 mM/m$^2$-yr (0.19–0.97 kg/ha-yr); those of $SO_4^{2-}$ range from 15 to 50 mM/m$^2$-yr (15–50 kg/ha-yr) and those of $NO_3^-$ from 14–40 mM/m$^2$-yr (9–25 kg/ha-yr). The $SO_4^{2-}$ and $NO_3^-$ deposition distributions are linked with the broad distributions of $SO_2$ and $NO_x$ emission density. The (wet) $SO_4^{2-}$ deposition levels are an order of magnitude larger in the East than in the West. The aq $NO_3^-$ deposition levels are as much as a factor of 4 or more larger in the East than in the West.

### Dry Deposition

There is relatively little direct information on dry deposition of S and N compounds. Measurements are based on micrometeorological techniques for the flux of gases or particles to a surface. The direct methods used adapt an eddy correlation scheme in which the covariance between concentration and fluctuating air velocity is observed and averaged over a period of time near the ground. The dry deposition rate involves three components (Appendix A). The first two are turbulent-aerodynamic and viscous-sublayer components, which depend on the air speed near the surface, the aerodynamic roughness of the surface, and the molecular diffusivity of material. The third component involves the surface resistance. For soluble gases, the surface wetness is a major factor for absorp-

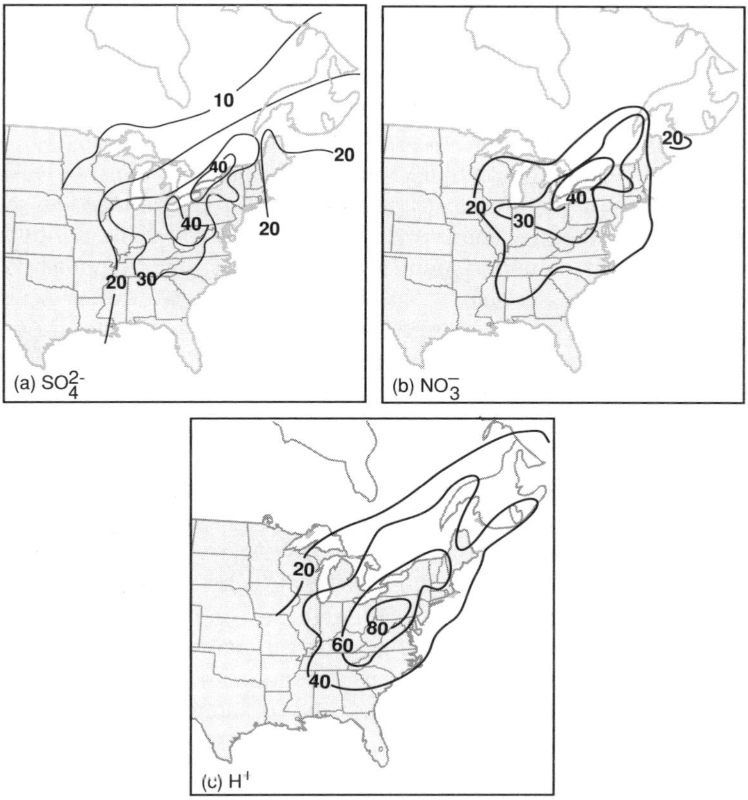


FIGURE 3.17 (a) Spatial distribution of annual wet deposition of $SO_4^{2-}$ ions (mM/m$^2$) in North America in 1980. (b) Spatial distribution of annual wet deposition of $NO_3^-$ ions (mM/m$^2$) in North America in 1980. (c) Spatial distribution of annual wet deposition of $H^+$ ions (mM/m$^2$) in North America in 1980. Data from four networks are plotted: Canada, CANSAP and APN; United States, NADP and MAP3S. (From Barrie and Hales, 1984.)

tion of gases. In the case of $SO_2$, the surface resistance depends on vegetation condition; that is, there is a stomatal resistance. When the stomata are closed, for example, the gas transfer is suppressed. In particle deposition, the rate depends on particle size. Depending on the size range, the rate can differ theoretically by more than an order of magnitude.

Typical measurements of dry deposition of $SO_2$ and submicrometer particles are given in terms of a deposition velocity or air–surface mass transfer coefficient (Hicks *et al.*, 1985). The deposition rate is defined as the product

$$\text{Deposition rate} = (\text{deposition velocity}) \times (\text{ambient concentration}) \tag{3.4}$$

In effect, reporting deposition as a velocity normalizes the rate for ambient concentration differences. Observations of the $SO_2$ deposition velocity show substantial diurnal variation as expected from the interaction with leaf surfaces, where wetness and stomatal conditions change. By convention, a positive velocity is directed downward. A small diurnal variation occurs in the particle deposition velocity for small particles. However, a larger apparent variation occurs for large particles. A negative (upward) velocity suggests reentrainment as well as deposition of material. Other data suggest that $HNO_3$ deposition is rapid and governed by aerodynamic resistance. Particle deposition depends on surface roughness, but there is some evidence that particle removal from surfaces can occur, creating a flux of particles away from the surface at times.

Indirect means of estimating dry deposition also are available. These use an indirect measure of eddy flux relying on the analogy between moisture and heat transfer at the surface (Hicks and McMillen, 1986). Various means for determining deposition velocity have been used. The results depend on the definition of ambient concentration at the ground, which is generally regarded as 10 m in height or less. Typically, daytime summer values for $SO_2$ are cited as 0.5–1.0 cm/s; $NO_x$, 0.2 cm/s; and $HNO_3$, 1.0 cm/s. Deposition of $SO_2$ decreases with snow cover and may be reversible. However, $HNO_3$ deposition apparently does not decrease because of its affinity for ice. Particle deposition velocities are nominally 0.1–0.3 cm/s, but deposition can vary widely, depending on particle size and surface characteristics.

The uncertainty of indirect methods is large, but comparisons with the direct method suggest that agreement can be achieved for micrometeorological tests (Hicks and Matt, 1986). A dilemma then surfaces about extrapolating such results to the much larger spatial scales that correspond to regional conditions. The extrapolations are linked to surface conditions, including land-use practice and urban conditions. Calculations of daily distributions of deposition velocity are incorporated into models (Mueller and Hidy, 1983; Walcek *et al.*, 1986; Dennis *et al.*, 1990). Seasonal averages also have been estimated (Voldner and Sirois, 1986).

Values of dry deposition rate for $SO_x$ and $NO_y$ have been reported, as inferred from concentrations and meteorological flux measurements of the combined routine measurement experiment (CORE) network. This network was formed to test different methods of dry deposition measurement from a point location surrounded by satellite stations. Aerodynamic conditions, including surface roughness, at the sites were carefully selected to establish their representativeness. Eight sites were operated in the eastern United States between late 1984 and 1987.

Table 3.16 lists estimates of annual and seasonal rates along with an average fraction of total deposition at the sites for the sampling period.

TABLE 3.16 **Average Annual and Seasonal Dry Deposition Rates (kg/ha) of Sulfur as $SO_2$ and $SO_4^{2-}$ (in parentheses) and Nitrogen as $HNO_3$ and $NO_3^-$ (in parentheses) at the CORE Sites**[a,b]

| Site | Annual fraction of total | Annual | Winter | Spring | Summer | Fall |
|---|---|---|---|---|---|---|
| A. $SO_2$ and $SO_4^{2-}$ | | | | | | |
| Oak Ridge, TN | 0.50 | 8.7 (1.5) | 1.96 (0.16) | 2.29 (0.46) | 2.38 (0.02) | 2.11 (0.27) |
| Argonne, IL | 0.45 | 12.6 (1.4) | 2.22 (0.21) | 3.33 (0.42) | 4.08 (0.49) | 2.87 (0.26) |
| Whiteface, NY | 0.20 | 1.4 (1.1) | 0.37 (0.13) | 0.28 (0.32) | 0.34 (0.37) | 0.35 (0.25) |
| State College, PA | 0.30 | 3.7 (1.2) | 0.80 (0.13) | 1.35 (0.37) | 1.42 (0.71) | 1.00 (0.26) |
| Bondville, IL | 0.40 | 5.0 (1.2) | 0.70 (0.10) | 0.98 (0.33) | 1.92 (0.48) | 1.10 (0.20) |
| Panola, GA | 0.40 | 3.9 (2.0) | 0.92 (0.24) | 0.83 (0.60) | 1.46 (0.84) | 0.71 (0.56) |
| Shenandoah NP, VA | — | 5.6 (1.7) | 0.45 (0.15) | 0.98 (0.25) | 1.92 (0.75) | — |
| B. $HNO_3$ and $NO_3^-$ | | | | | | |
| Oak Ridge, TN | 0.45 | 1.9 (0.06) | 0.44 (0.02) | 0.70 (0.02) | 0.38 (0.01) | 0.34 (0.01) |
| Argonne, IL | 0.40 | 4.6 (0.08) | 0.95 (0.03) | 1.07 (0.03) | 0.64 (0.01) | 0.75 (0.01) |
| Whiteface Mt., NY | 0.30 | 1.5 (0.04) | 0.27 (0.01) | 0.46 (0.02) | 0.45 (0.01) | 0.36 (0.01) |
| State College, PA | 0.35 | 2.7 (0.04) | 0.56 (0.01) | 0.91 (0.01) | 0.71 (0.01) | 0.72 (0.01) |
| Bondville, IL | 0.50 | 3.0 (0.04) | 0.76 (0.01) | 0.86 (0.01) | 0.71 (0.01) | 0.63 (0.01) |
| Panola, GA | 0.45 | 2.1 (0.04) | 0.50 (0.01) | 0.67 (0.01) | 0.58 (0.01) | 0.33 (0.01) |
| Shenandoah, VA | — | 2.0 (0.04) | 0.57 (0.01) | 0.66 (0.01) | 0.74 (0.01) | — |

[a] From Sisterson *et al.*, 1990.

[b] Also listed for S and N are the average percent of total deposition at the sites. Averages are determined over intermittent periods from Fall 1984 to Summer 1987.

Sampling was done between 1984 and 1987 with observations reported from 1 to 4 years, with the more consistent, year-to-year data on sulfur. The data indicate an annual range of dry sulfur deposition from 1.4 to 12.6 kg/ha, with an apparent seasonal variation yielding a summer maximum, except for Whiteface Mountain. This remote site receives the least dry deposition, which seems to vary little with season. Except for Whiteface, sulfate makes a small contribution to total dry sulfur deposition. Sulfate deposition appears to be surprisingly consistent from site to site both annually and seasonally.

The dry deposition data for $HNO_3$ in Table 3.16 show an annual rate between 1.5 and 4.6 kg/ha; aq $NO_3^-$ at all stations is a negligible fraction of dry $NO_3$ deposition. The $HNO_3$ deposition rates by mass are somewhat lower than those of $SO_x$, except for Whiteface Mountain. A seasonal pattern for $HNO_3$ is not apparent from the results.

## Cloud and Fog Interception

Little is known about the rates of cloud droplet deposition. The estimates of droplet interception representative of high elevation in ENA were summarized by Vong (1989). His summary is shown in Table 3.17. The results reported derive from cloud droplet concentration measurements and water deposition for the sites and different models for cloud deposition. Various investigators give similar results for the same site. Four of the sites reported, Moosilauke, Whiteface, Whitetop, and Mitchell, tend to have higher deposition rates that are roughly independent of cloud water deposition rates. The results are reported in kg/ha-mo because observations were taken primarily during periods when the mountaintops were accessible; the periods represent conditions expected during the growing season for vegetation.

## Relative Contributions

The relative contributions to total deposition can be estimated from recent measurements. This information is very important for characterizing the nature of exposure to terrestrial and aquatic biota. Calculations based on regional mass balances suggest that dry and wet deposition rates are similar over much of ENA, but wet rates are greater than dry in remote areas, such as parts of eastern Canada. Of course, near sources or in arid areas, dry deposition is expected to predominate.

Estimates of deposition rate components have been reported at two sites in ENA. Lindberg *et al.* (1986) have discussed deposition rates in a hardwood forest near Oak Ridge, Tennessee. Their results indicate that dry deposition of particles and vapors accounts for more than half the

TABLE 3.17 **Estimated Chemical and Cloud Water Deposition by Droplet Interception for Impaction for Various Locations; Data Taken Between 1980 and 1988**[a]

| Site | Elevation (km) | Investigator | Droplet concentration (kg/ha/mo) | | | | $H_2O$ (cm/yr) |
|---|---|---|---|---|---|---|---|
| | | | $H^+$ | $NH_4^+$ | $NO_3^-$ | $SO_4^{2-}$ | |
| Mossilauke, NH | 1.22 | Lovett | 0.20 | 1.4 | 8.5 | 11.5 | 68 |
| Whiteface, NY | 1.0–1.48 | Mohnen | 0.04– | 0.24– | 0.8– | 7.0– | 18–127 |
| | | | 0.18 | 2.3 | 4.8 | 11.2 | |
| Whitetop, VA | 1.69 | Mohnen | 0.20– | 2.6– | 7.7– | 13.1– | 25–26 |
| | | | 0.22 | 11.8 | 10.4 | 13.8 | |
| Mt. Mitchell, NC | 2.02 | Saxena | 0.17 | 1.2 | 3.2 | 9.8 | 35–77 |
| Mt. Mitchell, NC | 2.02 | Lin | 0.09– | 0.7– | 1.9– | 5.2– | 32–57 |
| | | | 0.18 | 1.3 | 4.0 | 8.0 | |
| Mt. Mitchell, NC. | 2.02 | Mohnen | 0.09– | 0.1– | 2.2– | 5.5– | 27 |
| | | | 0.13 | 1.1 | 2.6 | 8.9 | |
| Shenandoah, VA | 1.01 | Krovetz | 0.01 | 0.13 | 0.5 | 0.6 | 9 |
| | | Sigmon | 0.02 | 0.21 | 0.8 | 1.2 | 5 |
| | | Mohnen | 0.04 | 0.26 | 1.3 | 1.3 | — |
| Smoky Mountains | 1.74 | Lindberg | 0.05 | 0.5 | 1.1 | 3.3– | 9–80 |
| | | | | | 1.1 | 4.0 | |
| Roundtop, Que. | 0.97 | Schemenauer | | | | 3.3 | 15–80 |

[a] From Vong, 1989.

total deposition of $SO_4^{2-}$ and $NO_3^-$ at that site (56% for $SO_4^{2-}$ and 63% for $NO_3^-$); $H^+$ ion is deposited about 57% by dry processes. Pierson *et al.* (1987) have attempted to estimate deposition partitioning at Allegheny Mountain, Pennsylvania. The estimates suggest that rainfall dominated both $SO_4^{2-}$ and $NO_3^-$ deposition at this location. Fog and dew depositions contributed a small amount to the total deposition.

Additional information on the relationship between dry deposition and total S and N deposition was reported for several locations. The fraction of total S deposition represented by the dry component at the stations in Table 3.16 ranges from 0.2 to 0.5. Whiteface, NY and State College, PA are most strongly dominated by wet S deposition. The fraction of total nitrate depositions that is dry ranges between 0.3 and 0.5, with the lowest fractions again at Whiteface and State College. These results are generally consistent with expectations from earlier work.

The relative contributions to deposition of cloud and precipitation water are also of concern at high altitude. These contributions are summarized from the MCCP sites in Table 3.18. Included for comparison are the frequency of occurrence of in-cloud conditions during the growing season (April through October). The results in the table suggest substantial differences between sites that are not necessarily consistent with the frequency of occurrence of cloud. At Whiteface and Whitetop, cloud droplet deposition tends to dominate all listed ions, whereas Shenandoah and Moosilauke are roughly equally affected by cloud and precipitation deposition. Mitchell is somewhat higher in cloud deposition than precipitation. The

TABLE 3.18 **Ratios of Cloud Droplet Deposition to Wet Deposition via Precipitation for the Growing Season (Usually April–October) of 1987 at Five Mountain Sites in the Eastern United States**[a,b]

| | **MCCP site** | | | | |
|---|---|---|---|---|---|
| | **Whiteface** | **Whitetop** | **Shenandoah** | **Mitchell** | **Moosilauke** |
| Frequency (%) | 37 | 30 | 11 | 29 | 19 |
| Elevation (m) | 1483 | 1689 | 1014 | 2020 | 1000 |
| $SO_4^{2-}$ | 5.0 | 4.0 | 0.5 | 1.4 | 0.6 |
| $NO_3$ | 5.0 | 6.0 | 1.2 | 1.5 | 0.6 |
| $NH_4^+$ | 7.0 | 8.0 | 0.8 | 3.0 | 1.0 |
| $H^+$ | 4.0 | 4.0 | 1.0 | 3.0 | 1.0 |

[a] From Mohnen, 1988.
[b] Frequency (%) is the Fraction of Time the Sampler was in Clouds.

predominance of cloud deposition seems to be related to both frequency of occurrence and liquid water content.

Deposition reaching the ground in a forest is further complicated by chemical interactions with the forest canopy. Lindberg *et al.* (1986) have discussed this effect for different ions for throughfall in a deciduous forest for the dormant and growing season. $SO_4^{2-}$ is largely unaffected by canopy effects, but $NO_3^-$ is absorbed, as are $H^+$, $NH_4^+$, and $NO_3^-$. Interestingly, cations $Ca^{2+}$ and $K^+$ are actually leached from the trees during the process. The effects appear to be more dramatic during the growing season.

Additional studies of throughfall have been reported from the MCCP activities. Throughfall in the presence of cloud and precipitation deposition gives a complex picture of chemical conditions. Comparisons have been made at two sites on Whiteface Mountain. The lower site is influenced by rain only, while the upper site receives additional cloud deposition. Both sites are a mix of conifers with some deciduous trees present at the lower site. Throughfall from both sites included whatever dry deposition had occurred since the last wetting, residue after evaporation, and releases from the tree itself and biota on the tree surfaces. Table 3.19 summarizes enrichment ratios in species deposition, averaged over the summer season. The water received at tree level was similar. However, there are differences in throughfall enrichment for $NH_4^+$, $NO_3^-$, $K^+$, and $Cl^-$. Thus throughfall change in wet deposition cannot be generalized from the available studies.

TABLE 3.19 **Deposition Enrichment Ratios (Throughfall/Bulk Precipitation) Water Deposition (cm) and Chemical Deposition (Dimensionless Ratio) at Whiteface Mountain, NY**

| | **Under canopy (upper site)** | **Open rain (lower site)** |
|---|---|---|
| Water (cm) | 8.51 ± 0.48 | 8.31 ± 0.08 |
| $H^+$ | 0.98 ± 20% | 0.71 ± 13% |
| $NH_4^+$ | <0.35 | 1.11 ± 22% |
| $NO_3^-$ | 0.86 ± 23% | 1.65 ± 19% |
| $SO_4^{2-}$ | 1.45 ± 19% | 1.16 ± 14% |
| $K^+$ | 9.4 ± 44% | >13.0 |
| $Cl^-$ | 2.51 ± 27% | 2.0 ± 25% |

## SOURCE–RECEPTOR RELATIONSHIPS

An extensive body of observations has developed that characterize elements of regional-scale atmospheric chemistry. These observations provide basic data through which a number of hypotheses about the role of atmospheric processes in distributing material over ENA can be tested. The data also reflect information about exposures to ambient concentrations of pollutants, as well as atmospheric chemistry and deposition patterns. In themselves, the observations generally do not offer definitive opportunities to estimate source–receptor relationships (SRRs). However, some useful information can be obtained using the measurements in three ways: (1) comparison of spatial distributions and distribution of source emissions, (2) comparison of temporal changes with estimated emission changes segregated by wind direction, and (3) identification and observation of unique tracer materials (tracers of opportunity) emitted from groups of sources to infer source contributions at receptors.

### Comparison of Spatial Distributions

Comparison of emission distributions and spatial distributions of $SO_x$ and $NO_x$, either as ambient concentrations or wet deposition concentrations, has been attempted. By inspection, a linkage is established between $SO_2$ and $NO_x$ emission patterns and the ambient concentration or deposition data. Average ambient $SO_x$ concentration distributions, for example, show high levels and a gradient structure that roughly corresponds to that of high emission density (Figure 3.4). Such qualitative relations are insufficient to establish SRRs, however. Hilst and Chapman (1990) extended the qualitative comparison of $SO_2$ emissions and national precipitation $SO_4^{2-}$ observations taken between 1982 and 1985 to search for an empirical relationship between $SO_x$ and $NO_x$ emissions, and sulfate and nitrate distributions in precipitation. They explored relationships that would fit a general linear relationship,

$$\text{Deposition} = a + b\ (\text{emission density})$$

and a nonlinear form,

$$\ln(\text{deposition}) = \text{const.} + n \ln(\text{emission density})$$

Calculations were made for the annual average wet deposition rate aggregate from 110 stations reporting in the United States. The estimated relationships were determined after subdividing the stations into low- and high-emission subregions corresponding to the ratio of molar sulfate concentration and molar nitrate concentration, divided by the ratio of molar $SO_2$ emissions and molar $NO_x$ emissions. Use of this parameter allowed geographic segregation into four subregions in the United States. These four roughly separate into regions of low emission density and high emission density. Furthermore, subdivision depended on the $SO_x/NO_x$ emission ratio and climatological conditions.

The investigations found that in areas of relatively low emissions of $SO_2$ and $NO_x$, the associations between wet $SO_4^{2-}$ concentrations and anthropogenic $SO_2$ emissions and between wet $NO_3^-$ concentrations and anthropogenic $NO_x$ emissions within 560 km of each precipitation chemistry station were weak or nonexistent ($r^2 < 0.42$). These land areas included the Gulf states. The remaining land areas in the northeastern United States and southeastern Canada showed moderate to strong associations between $SO_2$ and $SO_4^{2-}$ and $NO_x$ and $NO_3^-$ during the summer, but weak to nonexistent associations during the winter. During the summer, the maximum correlations between emissions and concentrations were approached when the pseudocircular area of emissions averaging centered on each station has a radius of 400 to 600 km.

At the level of aggregation used in this study, the associations between anthropogenic emissions and concentrations, e.g., $SO_2$ and $SO_4^{2-}$, were equally well represented by either a linear or a power law function. However, in all cases examined, the least-squares best-fit linear function produced a nonzero intercept value of concentration at zero emissions. Because of the nonzero intercept, the data do not substantiate a linear-proportional relationship between local concentrations and subregional anthropogenic emissions within areas less than $\sim 4 \times 10^6$ km$^2$ in the Eastern United States (EUS).

Hilst (1992) further noted that the $SO_2$ emissions to wet $SO_4^{2-}$ concentration proportionately over large areas of the EUS apparently depends on a combination of imported $SO_2$ from major source areas and an oxidant limited conversion of $SO_2$ to $SO_4^{2-}$ within major source areas. Thus, a control strategy of rollback on $SO_2$ emissions for ENA would result in a proportional decrease in wet $SO_4^{2-}$ concentrations in minor $SO_2$ source areas, but no appreciable reduction in wet $SO_4^{2-}$ in major source areas until the oxidant limitation is overcome. This conjecture, of course, remains untested.

The local and subregional anthropogenic emissions of $SO_2$ and $NO_x$ are highly correlated, as are the emissions of HC and $NO_x$. Substantial unilateral changes of one of these emissions would produce pollutant mixes that have not previously existed in this geographic area, with unknown consequences for $SO_4^{2-}$ and $NO_3^-$ formation. Thus we cannot use the experience gained from this analysis *a priori* to predict future conditions, changing only $SO_2$ and $NO_x$, unless a "self-similar" system were maintained.

Hilst and Chapman (1990) found that the combined geographic variabilities of precursor emissions and meteorological factors were well reflected in regional and seasonal variations of wet $SO_4^{2-}$ and $NO_3^-$ concentrations observed in their descriptions of the central North American acid rain system. This result reinforces the perception that such a relationship exists from inspection of spatial distributions of precipitation chemistry overlaid on emissions distributions.

## Comparison of Temporal Changes

Temporal changes in deposition chemistry at receptors provide additional evidence about SRRs. Analysis of temporal change has been attempted in two basic ways. The first involves inspection of data segregated by wind direction or by air mass trajectory analysis, linking source location to expected receptor exposure. The second method compares the long-term record of deposition at receptors and the changes in emissions over the same period.

*Wind Vector Segregation* If there is a persistent exposure pattern between a source and receptor, segregation of exposure by wind direction or by air mass trajectories calculated backward in time offers an opportunity to identify SRRs. Several attempts to adopt this approach have been reported. They have yielded ambiguous and inconsistent relationships even with careful studies of isolated source impact. The reasons for the ambiguity are complex, but they are related to atmospheric processes that influence emitted material as it travels downwind.

Wind sector analyses of precipitation chemistry at Whiteface Mountain and Ithaca, NY, have revealed the importance of both wind and precipitation rate (Wilson *et al.*, 1982; Henderson and Weingartner, 1982). In both of these studies, the wind sector analysis was done in terms of calculated air mass trajectories reaching the receptor with precipitation from up to 48 hours upwind. Interestingly, the highest S concentrations were found in airflow from the south to southwest trajectory sectors. However, the heaviest precipitation was also found in these sectors. Thus, the results indicate that wet $SO_4^{2-}$ deposition essentially comes with heaviest precipitation, which coincidentally comes from the south to southwest. These results also indicate that precipitation $SO_4^{2-}$ is essentially derived from contributions from every trajectory direction, depending on the season. However, most of the deposited $SO_4^{2-}$ was from the 160–350° azimuthal directions.

Henderson and Weingartner (1982) have examined MAP3S data from Ithaca by air mass trajectory analysis between October 1977 and October 1979. They found a clear seasonal dependence on precipitation acidity associated with air mass movement. During the summer, when biological activity is high, most of the $NH_4^+$ and $SO_4^{2-}$ deposited by precipitation are carried by the same air masses, which reduces acidity in the precipitation event. Most of the acidity in summer events was associated with trajectories traveling over large $NO_x$ sources. When biological activity was low, most of the acidity per event was associated with air masses passing over large $SO_2$ sources to the southwest of Ithaca.

Superposition of calculated air mass trajectories to sites in Quebec indicates that sources from virtually any azimuthal sector could affect this subregion, as in the case of the Whiteface Mountain area. Desautels (1981), however, noted that there are important seasonal differences in trajectories reaching Quebec, as well as differences between airflows with and without precipitation. Summer experiences stronger flow from the west and southwest, while dry air has a more northerly component than air with precipitation. This result appears to be consistent with findings at the Whiteface Mountain and Hubbard Brook areas.

There are two potentially significant tests of emission change and change in precipitation chemistry that unfortunately have not been attempted (or at least reported). The two cases of interest occurred with the general economic recession in 1982. It is believed (but not verified) that reductions in electricity production resulted in decreases in coal consumption in parts of the Midwest of as much as 25–30% over 1980 and 1981. At the same time, a 1982 labor strike closed smelters in Ontario. These events would have substantially reduced $SO_2$ emissions from Ontario. Reports of studies for these year-to-year changes in the precipitation chemistry have not yet emerged in the literature. But initial analyses of the Canadian Atmospheric Environment Service and Ontario Ministry of Environment suggested difficulty in tracing the smelter effect on either ambient S concentrations or precipitation $SO_4^{2-}$ data between 1981 and 1982. The difficulty arises, in part, from the highly variable nature of precipitation $SO_4^{2-}$ with southerly and northerly airflow over much of southern Ontario. (See also Hidy *et al.*, 1985.)

***Comparisons with Emissions History*** There has been considerable debate about the link between emissions and precipitation chemistry. As we have seen, this link appears in spatial distributions of $SO_4^{2-}$ or $NO_3^-$ when superimposed on distributions of emissions. Much more difficult is the association between emission change and temporal change in deposition rates. To place this association in context, a discussion of historical changes in data is included here, with particular reference to $SO_2$ emissions. The discussion follows from the National Academy of Sciences (1986) assessment, with additional comments from NAPAP (1990b). In much of this discussion, acidity is taken as a surrogate for pollution-derived anions, especially $SO_4^{2-}$. In general, $SO_2$ emissions have increased since 1900 because of increased fuel consumption (especially coal). As a result of increased pollution controls, $SO_2$ emissions tended to decline after 1965–1970, except for the Southeast, where emissions from growing industrialization created conditions for increasing emissions.

$NO_x$ emissions have increased with fuel consumption, particularly in the widespread increased use of the internal combustion engine. Only since the late 1970s has a leveling of $NO_x$ emissions been estimated—as a result of improved combustion systems and automobile emission controls.

Data from three relatively brief periods have been used to investigate changes in precipitation composition, primarily pH, in the period since the mid-1950s: Junge and Werby (1958); Lodge (1968) (also referred to as the Public Health Service–National Center for Atmospheric Research [PHS/NCAR] data) for the early 1960s; a variety of sources for the mid-1980s (see Likens and Butler, 1981); and the NADP, UAPSP, and CANSAP networks in recent years. Analyses of these data sets and the conclusions drawn from them have given rise to the debate in the published scientific literature over the past few years.

On the basis of their analyses, Cogbill and Likens (1974), Likens and Butler (1981) and Cogbill *et al.* (1984) concluded that a large spread and probable intensification of acid precipitation (pH <5.6) in EUS have occurred during the past 25 years and that acid precipitation may not have been widespread in the EUS before 1930. More specifically, they concluded that precipitation has become more acidic, especially in the Midwest and Southeast, between the 1950s and the present, and that many areas of the East show an increase in acidity of greater than 22 $\mu$eq/L (equal to 11 $\mu$eq/L $SO_4^{2-}$) during the past three decades.

The major points of concern about using these data for trend analysis involve (1) the amount of uncertainty and bias introduced by differences in sampling, analytical, and calculational methods and (2) the influence of differing climatological conditions, particularly drought during the periods of data collection.

Hansen and Hidy (1982) reviewed the available post-1955 data and assessed the biases and uncertainties associated with (1) use of different sampling, analytical, and calculational methods to deduce precipitation acidity; (2) use of data from different times and conditions; (3) failure to collect brief or light rainfall and initial precipitation in an event; (4) incomplete chemical analyses; and (5) selection of data for analysis. The outcome of this assessment was to determine a series of uncertainty values in pH by decade and geographic region. Hansen and Hidy (1982) concluded that the historical data are of insufficient quality and quantity to support any reliable long-term trends in precipitation acidity ($SO_4^{2-}$) change in the EUS. Cogbill *et al.* (1984) disputed several of these estimates and conclusions. Hansen and Hidy (1982) noted that acidity of precipitation, on the average, seemed lower in the mid-1950s than in the mid-1960s and thereafter in parts of the EUS; they also concluded that despite limited observations in the Southeast, a region of elevated acidity (pH <5.0) in rainfall from Alabama through Florida appears to exist that was not evident in the mid-1950s.

Stensland and Semonin (1982) evaluated the potential role of elevated levels of calcium and magnesium from crustal sources (soil dust) in the composition of precipitation in the mid-1950s. Drought conditions in the mid-1950s are likely to have increased the concentrations of wind-raised soil components in precipitation samples, possibly resulting in calculated acidity values lower than would have been typical of the period. These researchers also point out that data for the 1960s (PHS/NCAR) and for the 1970s from the World Meteorological Organization/National Oceanic and Atmospheric Administration/Environmental Protection Agency (WMO/NOAA/EPA) network show generally higher ion concentrations than do current NADP data. Their analysis suggests that the earlier measurements—1950s, 1960s, 1970s—were perhaps affected by evaporation and dry deposition (dust leakage) and may have been more like bulk than wet-only samples in many cases. On the basis of a simple pH analysis, Stensland and Semonin (1982) recalculated the 1955–1956 pH distribution and found a smaller pH change between the mid-1950s and the present than in the calculations of Likens and Butler (1981). Stensland and Semonin note the crucial importance of (1) measuring for all major ions in samples to be used for trend analysis and (2) analyzing concurrently meteorological/climatological factors when investigating large-scale temporal and spatial changes.

Investigators have not reached a consensus on the details of precipitation acidity changes since the 1950s; the quality and the quantity of the earlier data make a quantitative evaluation of the change difficult. Nevertheless, workers agree that precipitation acidity has increased since the 1950s in part of the ENA despite disagreement on the amount of the change and the mechanism. More specifically, there is agreement that the acidity of precipitation has increased in the Southeast since the 1950s (e.g., Barrie and Hales, 1984). However, the reason for the change has not been thoroughly explored. A comparison with the Junge data shows that the concentrations of $SO_4^{2-}$ in the southeastern states have generally increased since the 1950s. A similar trend is indicated for $NO_3^-$ and $H^+$ (Barrie and Hales, 1984). However, concentrations of calcium and (calculated) magnesium have decreased significantly. Researchers do not know the relative importance of strong acids that contribute $H^+$ versus reduced soil components that neutralize.

For the northeastern states, data indicate that $SO_4^{2-}$ concentrations have risen in some areas (MA, NY, OH, IN), although generally not to the same degree as the Southeast. The other northeastern states either exhibit no change (PA, MI) or show decreases (ME, MA, VT, and MD).

*Hubbard Brook Data* One of the best long-term records of change in deposition chemistry has been collected at Hubbard Brook, NH. The data consist of analyses of the more than 500 bulk samples[4] collected between 1964 and 1979. The trend analysis was done for the following ions: $H^+$, $SO_4^{2-}$, $NO_3^-$, $Cl^-$, $NH_4^+$, $Ca^{2+}$, $Mg^{2+}$, $K^+$, and $Na^+$; $Cl^-$ data were not available until June 1967 (Oehlert, 1984). Monthly volume-weighted average concentrations and monthly depositions were calculated from the raw data. The data were analyzed for $Cl^-$, $Ca^{2+}$, $Mg^{2+}$, $K^+$, and $Na^+$ on the logarithmic scale. Seasonal influences were eliminated from the variables by subtracting from each monthly value the mean of the time series for that month. The results show that some strong trends exist in the Hubbard Brook data (Oehlert, 1984). In particular, concentrations of all ions but $H^+$ and $NO_3^-$ are decreasing; pH has remained essentially constant. $NO_3^-$ has increased significantly over the period 1964 to 1979, although the result depends somewhat on statistical technique and the period of analysis. Inclusion of data more recent than 1979 tends to indicate no change in $NO_3^-$ concentrations over the entire time span.

One feature of the $SO_4^{2-}$ trend is that it varies seasonally. The trends for all seasons are downward. But spring and winter concentrations have evidently been falling more rapidly than the summer and fall concentrations. There are also declines in $Ca^{2+}$, $Mg^{2+}$, $K^+$, $Na^+$, and $Cl^-$ concentrations. The trends are not related to bias introduced by changes in sampling procedures, as these methods have apparently not changed over the period of record. That the ions exhibit this change suggests changes in local source strengths, because these substances are usually found in particulate matter that is not transported large distances from sources. In one possible scenario, the decrease in $SO_4^{2-}$ (about 20 $\mu$eq/L for the 18-year period) and the decrease in $Ca^{2+}$ and $Mg^{2+}$ (about 18 $\mu$eq/L) could result, for example, from a decrease in the input of calcium sulfate ($CaSO_4$), either as dry deposition ($SO_2$ plus Ca on the funnel surface) or as wet deposition ($CaSO_4$ in precipitation). The portion of the $SO_4^{2-}$ transported from distant sources would then be much less.

Several authors using different statistical methods have examined the Hubbard Brook precipitation chemistry data for trends. The National Academy of Sciences (1986) summarizes the methods and results from

[4] Bulk sampling collects both wet and some dry fallout as a rough estimate of total deposition.

recent publications on trend analysis of the Hubbard Brook data. The following is a summary of the trends since 1964.

1. $H^+$: The record is variable with no significant trend; slight tendency to lower concentrations for parts of the period.
2. $SO_4^{2-}$: Significant decrease of approximately 2%/yr superimposed on periods of drift; decrease in winter and spring is greater than for summer and fall.
3. $NO_3^-$: The trend over the entire record is variously described as "not significant" (1963–1982) and "increasing" (1964–1979); an apparent change in slope in about 1971 may be from chance.
4. $NH_4^+$: Three authors describe the trend as "no change," and two others say "decreasing" (1.9 and 6.4%/yr).
5. $Na^+$, $Cl^-$, $Ca^{2+}$, $Mg^{2+}$, $K^+$, $Al^{3+}$: Agreement among three authors of a strong decrease in all these constituents.
6. Ratios: $SO_4^{2-}/NO_3^-$, $NH_4^+)/NO_3^-$, and $SO_4^{2-}/H^+$ show a consistent, significant decrease; $NO_3^-/H^+$ shows a significant increase.

Certain features of the NAS review stand out. Results from the various authors are generally consistent for $H^+$ and $SO_4^{2-}$ as well as the group including $Na^+$, $Cl^-$, $Ca^{2+}$, $Mg^{2+}$, $K^+$, $Al^{3+}$, and their ratios. For $NO_3^-$ and $Al^{3+}$ there is qualitative agreement on the behavior of the time-series segments but some disagreement on the trend in the overall record.

The different statistical techniques may introduce some differences in the results, but the length of record used in different studies is more important. The end point of the time series varies among studies between 1977 and 1982. In addition, several authors have noted the differences in trends that may occur with an analysis of selected portions of the time series. The differences in estimated rates of change obtained in the various analyses point to the need for caution if quantitative comparisons are to be made between trends in precipitation composition and emissions.

*U.S. Geological Survey Data* A USGS network was established in 1965 to provide data and continuous records on atmospheric contributions for use in hydrologic and geochemical studies (Peters *et al.*, 1982). The network comprises nine stations, eight in New York and one in northern Pennsylvania; six sites are inland rural, two are urban, and two are coastal (one urban-coastal). Samples are monthly bulk samples; they have been collected and analyzed through the early 1980s.

Because this data set provides one of the longest continuous records of precipitation composition and deposition in North America, it has been examined by several authors. Virtually all of them have commented on problems associated with the data (NAS, 1986). Potential problems include changes in sites, collectors, and analytical methods. Authors differ in their assessments of the importance and implications of these factors for trend analysis; however, the various analyses have been instructive and have produced some valuable trend information.

A time-trend regression analysis by Miller and Everett (1981) for pH data showed no significant trend for any of the nine sites but a slight downward trend for the pooled data. Analysis of the $NO_3^-$ data showed slight increases for individual stations and a greater increase for pooled data.

Barnes et al. (1982) analyzed the USGS data for time trends at all nine sites with monthly bulk data since 1965. The site at Athens, PA, was characterized by high levels of $NO_3^-$ and $NH_4^+$, which were perhaps due to a nearby dairy farm. The time-trend values were calculated with three different methods. The time trend by simple linear regression is commonly used, whereas the seasonal (Kendall) test is described as the most robust of the three. The data show that the magnitude of a time trend varies by as much as threefold, depending on which of the three methods is used. The time trend can change sign as the method is changed. Also, the magnitude of the time trend can change by more than a factor of 2 among the five sites for the same method. The Barnes analysis suggests that quantitative time-trend results from the USGS bulk data are imprecise. Qualitatively, the most consistent result is that $SO_4^{2-}$ has decreased by about 1 to 3%/yr at the five nonurban, noncoastal sites. $NO_3^-$ increases by 4 to 14%/yr for significant values ($P < .10$) and decreases by up to 4%/yr for some nonsignificant values. The significant values of $NH_4^+$ trends vary from −4 to +8%. Similarly, values of cation sum ($Ca^{2+}$ + $Mg^{2+}$ + $Na^+$ + $K^+$) trends vary from −5 to +7%, and the significant $H^+$ trends vary from −9 to +10%. The large variability in trends observed at the five USGS bulk sampling sites may be caused by local effects masking the regional patterns.

Bilonik and Nichols (1983) applied the Box–Jenkins time-series analysis method to "cleaned," pooled, monthly deposition data for 1965–1979. They found that the long-term mean levels of $H^+$, $SO_4^{2-}$, $NO_3^-$, and $Ca^{2+}$ ions in bulk samples were essentially constant.

Hidy *et al.* (1984) attempted to establish the reliability of USGS data by comparing time series from the Hinckley and Mays Point stations with the event data from the MAP3S stations at Whiteface Mountain and Ithaca, respectively. They concluded that although the USGS data displayed a systematic positive bias in $SO_4^{2-}$ concentrations on a monthly basis, these data composed a self-consistent, reliable long-term record of $SO_4^{2-}$ deposition that was adequate to reflect areal trends. However, the long-term patterns of $H^+$, $NO_3^-$, and $NH_4^+$ in the USGS data were less reliable and difficult to interpret for long-term trends.

***Consistent Patterns*** Hidy *et al.* (1984) also examined the spatial representativeness of a number of stations in the Northeast. From an examination of intersite correlations for $SO_4^{2-}$ among the Hinckley, Mays Point, Canton, Salamanca, and Athens USGS stations, they found that only the first three tended to have similar and simultaneous month-to-month variations in $SO_4^{2-}$. This finding points out the difficulty of expecting that a single station or a grouping of a few stations will necessarily provide data that will represent regional or subregional patterns without having established the fact from supplementary information beforehand.

Median $SO_4^{2-}$ concentrations showed an overall downward trend, while median $NO_3^-$ concentrations increased in the 1960s and fluctuated thereafter. Median $NH_4^+$ concentrations appear to have increased after 1975, and median pH showed a weak downward trend. Barnes *et al.* (1982) performed analyses of data from the five USGS stations and Hubbard Brook. Two central New York Sites, Hinckley and Canton, showed a downward trend in $SO_4^{2-}$ concentration, which is comparable with that at Hubbard Brook. Nearby but farther west, Mays Point, Salamanca, and Athens did not. Two of the central New York sites, Hinckley and Mays Point, along with Athens, PA, showed a positive $NO_3^-$ trend, and the site at Hubbard Brook showed a slightly positive trend; Canton and Salamanca showed no trend. $NH_4^+$ appeared to have increased steadily at all USGS sites except Salamanca, and the site at Hubbard Brook showed a decreasing trend. The pH trends, although less reliable, are either neutral or downward over the period of record. Hidy *et al.* (1984) also applied an autoregressive, integrated moving-average model to the $SO_4^{2-}$ data from Hinckley and Mays Point, the two sites where the data record was of sufficient length. Results showed a statistically significant decrease in $SO_4^{2-}$ concentration of about 2%/yr between 1965 and 1980 at both sites.

Despite the uncertainties inherent in the USGS data and the different statistical analysis techniques used, some general statements can be made about trends in the data. Although slight downward trends in pH are evident at some stations for some portions of the record, pH generally shows no trend. The $SO_4^{2-}$ records at Hinckley and Mays Point show a downward trend, consistent with the record at Hubbard Brook. Because of problems with the $NO_3^-$ data, the various analyses are equivocal; some show increases, some decreases, some no changes. $NH_4^+$ generally appears to be increasing.

As important as the trend estimates themselves are, several points stand out from the published analyses. Large variability is evident between stations and parameters; even methods of analysis and data preprocessing affect the results. Stations within a limited geographic area may be affected by different processes or climates and therefore may not be assumed to be representative of a region. The trends in $SO_4^{2-}$ at the two stations examined in detail by Hidy *et al.* (1984) appear to be most certain because of their self-consistent, reliable, long-term record. The fact that the trends are similar to that at Hubbard Brook gives some confidence to inferring a regional trend in precipitation $SO_4^{2-}$ concentration.

This discussion documents certain trends in historical data between 1950 and 1980. These trends in precipitation acidity are not consistent with $SO_2$ and $NO_x$ emission trends; $SO_2$ emissions have declined after 1970, and $NO_x$ increases have not been sufficient to offset acidity change expected with $SO_2$ decline. Furthermore, the correlation among $H^+$, $SO_4^{2-}$, and $NO_3^-$ is poor in the early data, based on bulk deposition. The association between historical changes in $SO_x$ and $NO_x$ and precipitation $SO_4^{2-}$ and $NO_3^-$ is stronger. Estimates of annual emissions have been made that date back to the 1900s. With the industrialization and urbanization of the eastern United States, fossil fuel (particularly coal) combustion increased dramatically after 1900. Emissions of $SO_2$ increased at the same time. Later, with improved combustion technology, estimated $NO_x$ emissions increased, especially after World War II. After the mid-1960s, widespread implementation of regulation under the U.S. Clean Air Act caused a general decrease in $SO_2$ emissions, but $NO_x$ emissions continued to increase through the early 1970s. They have since tended to level off.

Perhaps better evidence for establishing a directly proportional relationship between emission change is comparison of long-term trends with emission changes over the same time period. Such a direct relation is apparent in $SO_4^{2-}$ (and possibly $NO_3^-$) data in bulk deposition collected at Hubbard Brook. An average decrease in $SO_4^{2-}$ of about 2%/yr between 1970 and 1982 is essentially the same as that estimated from decrease in regionwide $SO_2$ emissions (Hidy *et al.*, 1984; NAS, 1986).

From the discussion of spatial correlation, it is logical to expect that any relation between emissions and regionally representative deposition measurements should be consistent for several sites as demonstrated with the aggregated 1980s data (Hilst and Chapman, 1990). To test this, the Hubbard Brook results were compared with limited data from the USGS bulk deposition network, primarily located in New York State. The stations selected are rural and are within 550 km of one another. Qualitative comparison of trend indicators in the data are summarized by Hidy *et al.* (1984). These indicators are based on estimated significant change by statistical testing. The summary indicates that the trends are not consistent for $SO_4^{2-}$ or for other constituents.

Spatial and temporal weighting also have been used on the USGS data to obtain yearly averaged $SO_4^{2-}$ deposition for New York State. Bilonik's (1985) results indicate a maximum in $SO_4^{2-}$ deposition in 1971–1973 compared with an opposite direction in the ENA $SO_2$ emissions maximum between 1970 and 1975. Because deposition rate is a dimensionally consistent parameter with emission rate, these results qualitatively tend to support an $SO_2$ emission–deposition relationship for ENA.

The lack of spatial consistency in relating $SO_4^{2-}$ concentration to $SO_2$ emissions may be the result of differences in the influence of sources near a given site, or of an inadequate quantitative data base, or of meteorological differences in exposure. A test of the first factor was attempted, calculating the effect of regionally different emission changes on different receptors, accounting for reduction in source effect with distance from the University of Michigan Acid Deposition (UMACID) model (Hidy *et al.*, 1984). The calculations show clear differences in expected changes with $SO_2$ emissions after 1965, and the $SO_4^{2-}$ deposition change should not necessarily be linear because the apparent emissions were not linear over the period. Nevertheless, a downward trend in $SO_4^{2-}$ should have been observed consistently if meteorological variability was similar from site to site. The results are ambiguous. The ambiguity derives, in part, from subregional differences in emission change over the EUS. The Hubbard Brook data are considered of higher quality than the USGS set. Thus, the Hubbard Brook data are relied on for supporting a linear SRR.

***Evaluation of Post-1980 Data*** The NAPAP (1990b) assessment has reported an evaluation of wet deposition trends after 1980 using the national sampling networks. The data are likely to be of well-documented quality and are homogeneous in methodology in space and time for the period between 1979 and 1987 (8 years). Trends in the wet deposition data by subregion have been compared with emission trends in those subregions. The results showed essentially no change in nitrate, corresponding to little or no change in $NO_x$ emissions throughout the EUS. The results by subregion for $SO_4^{2-}$ and $SO_2$ emissions are summarized in Figure 3.18. There is a trend to a linear proportionality within the estimated uncertainty of the data for the Northeast, the Atlantic coast and the upper Midwest. A nonproportional relationship (i.e., *less* $SO_4^{2-}$ deposition reduction than emissions reduction) exists in the Ohio River Valley region, the deep South, and the Southeast. The latter are more source "rich" than the former.

NAPAP (1990b) also reported annual (linearized) trends for concentrations of several major ions in rainwater. These are shown in summary in Figure 3.19. These results in aggregate over the ENA indicate that hydrogen ion concentrations showed essentially no change between 1979 and 1987, as shown also by nitrate. In contrast, sulfate concentrations declined significantly, with smaller decreases in the cations except for $NH_4^+$. The decline in sulfate, combined with $NH_4^+$ increase and smaller declines in other cations, partially accounts for the lack of change in $H^+$ concentration.

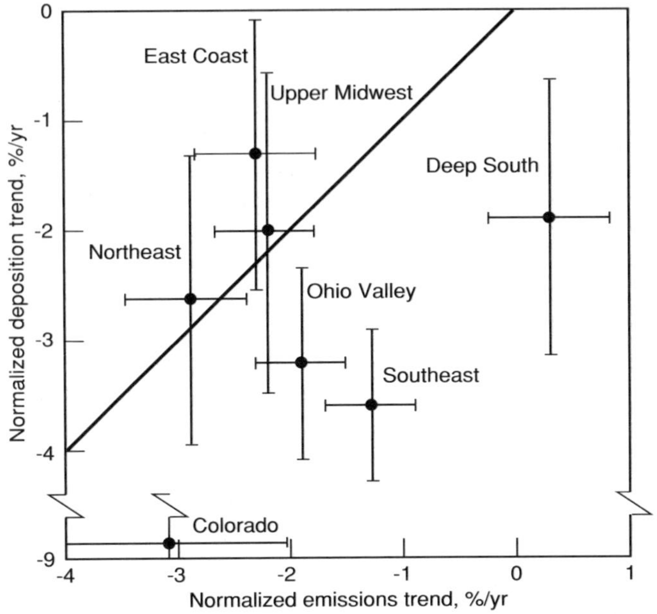


**FIGURE 3.18** Regional 1979–1987 trends in normalized sulfate wet deposition and normalized $SO_2$ emissions. The horizontal and vertical bars show the estimated uncertainty of the trend. The diagonal is the line of equal deposition and emission trends. (From NAPAP, 1990b.)

Analysis of a 13-station group from the UAPSP network data covering 1978 to 1986 showed trends generally similar to those reported by NAPAP. However, the UAPSP observations indicated that changes in major anion concentrations were generally not statistically significant. UAPSP trends suggested decreasing $H^+$ deposition at the midwestern sites (OH, IN, KY, AL), with increased sulfate and nitrate deposition at the (Tunkhannock) Pennsylvania and Big Moose, NY sites. Most other sites indicated a decrease in anion deposition. At the former locations $SO_4^{2-}$ and $NO_3^-$ deposition appeared to have increased with rainfall change because significant changes in concentration were not found.

Qualitatively, the declines in $SO_4^{2-}$ and $NO_3^-$ would be expected with emission reductions. It is not known why the cations would change across ENA. The recent data again show clearly that $H^+$ behavior in wet deposition depends on the variations in both anions and cations. Thus the "management" of acidity in deposition is not a simple function of a single acid-producing air contaminant.

The results reported by NAPAP are somewhat consistent with an analysis of data in the same period by Hilst (1992). Hilst's calculations suggest

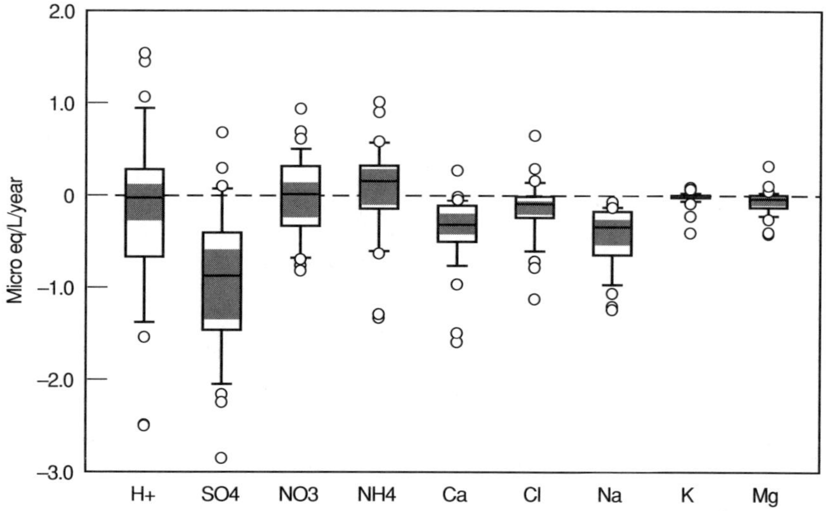


**FIGURE 3.19** Distribution of average annual trends in major ion concentration in precipitation for the period 1979–1987. Plots show the 10th, 25th, 50th, 75th, and 90th percentiles of sites in the United States and Canada that qualified for trends analysis. (From NAPAP, 1990b.)

that level $SO_4^{2-}$ deposition generally was not proportional to local $SO_2$ emissions. For the eastern half of central North America (an area of $4 \times 10^6$ km$^2$), Hilst found a "global" proportionality between areawide average wet $SO_4^{2-}$ deposition and $SO_2$ emissions. For subregions of ENA, the internal structure does not necessarily follow the regionwide association. In subregions, importation of pollutants from major source areas and oxidant limitations within major source areas appear to dictate the local response of $SO_4^{2-}$ wet deposition to emission reductions.

Part of the difficulty in interpreting a regional history is related to the potential for local source impact. An example has been cited in relation to the Hubbard Brook record. Coffey (1984) speculated about the possible influence of a local paper mill on the Hubbard Brook data for $SO_4^{2-}$, although data on actual emissions from the plant were lacking. Pinkerton (see Coffey, 1984) estimated $SO_2$ emissions from the paper mill from 1963 to 1980 based on available information on strikes and closings at the plant, pulp and paper production rates, energy consumption, and knowledge of the boilers and processes. The $SO_2$ emissions from the plant were roughly constant at about 4200 tons/yr from 1963 until the plant shut down in August 1970. After that, the plant operated sporadically with emissions during operational periods at about one-third the pre-1970 rate. The longest period of total shutdown was from July 1976 to November 1978, during which there were no $SO_2$ emissions from the mill.

No step changes are evident in the monthly $SO_4^{2-}$ concentration in precipitation data at Hubbard Brook. Qualitatively, the Hubbard Brook $SO_4^{2-}$ data show short-term increases and decreases superimposed on a linear decrease from 1964 through 1980. The large step decrease in emissions in 1970 occurs at a time when the Hubbard Brook $SO_4^{2-}$ record is at the end of a short-term (4-year) decrease and the beginning of a short-term (3-year) increase. Analysis of the record in detail, however, indicates no consistent relationship to the local source (NAS, 1986). Thus, the local source of $SO_2$ does not appear to be important at this site.

*Use of Tracers of Opportunity* Knowledge of the chemical composition of aerosol particles and certain HC species combined with knowledge of the sources of these materials provides a means of tracing certain species regardless of the atmospheric transport processes. The method, evolved since the early 1970s and sometimes called *receptor modeling* (Cooper and Watson, 1980), depends on the uniqueness of composition of emitted material, particularly enrichment of tracer species. Tracers of source groupings include such elements as (1) lead (Pb) for automotive exhaust (Pb halides have been used as antiknock fuel additives), (2) sodium chloride (NaCl) for sea salt, (3) silicon (Si) for soil dust, (4) vanadium (V) for residual oil combustion, (5) arsenic (As) or selenium (Se) for coal combustion (or smelting), and (6) potassium (K) for wood combustion. Certain halogenated HCs, such as the Freons, also reflect urban sources over large areas. In an environment such as ENA, the HC tracers are common to many cities. Thus, their use as tracers of opportunity has not been established.

The receptor model technique relies on a material balance of emitted chemicals with that found at a given location. In principle, the argument can be applied both to ambient concentration samples and to wet or dry deposited material. The method, sometimes called *chemical mass balance* (CMB) or *chemical element balance*, requires knowledge of the composition of material emitted from sources as well as measurement of receptor sample composition. The apportionment is formally done within the constraints that

$$C_i = \sum_{j=1}^{p} X_{ij}E_j \quad \text{and} \quad \sum_{j=1}^{p} E_j = 1 \tag{3.5}$$

where $C_i$ is the concentration of the $i$th element in an ambient (or deposition) sample, $X_{ij}$ is the concentration of the $i$th species in the $j$th source, $E_j$ is the strength of the $j$th source (for ambient concentration, $E_j$ is the mass of material from source $j$ per unit mass of aerosol particles in the volume of air sample), and $p$ denotes the number of sources. The apportionment can be done either by straightforward calculation from source and ambient sample composition (e.g., Miller *et al.*, 1972) or by a weighted multiple linear regression with signature elements in the ambient samples as dependent variables and the same elements in the source signatures as independent variables (e.g., Williamson and DuBose, 1983). Weighting is done through an "effective" variance technique (e.g., Clutton-Bock, 1967). In principle, the chemical mass balance technique can be extended to interpret the sources of material in rainwater (e.g., Gatz, 1980). However, the analysis is complicated by the need to account for both soluble and insoluble material in the samples, as well as differences in scavenging efficiency of material (Liljestrand, 1982).

Rahn and his colleagues (e.g., Rahn and Lowenthal, 1984, 1985) have applied the technique of chemical mass balance on aerosol particle samples to infer the contributions to ambient particle concentrations from different regional sources in ENA. Their elemental tracer system is based on the seven elements As, Se, antimony (Sb), zinc (Zn), indium (In), noncrustal manganese (Mn), and (Earth's) noncrustal V present in samples. Noncrustal Mn and noncrustal V refer to the mass present in excess of that expected from the earth's crust.

$$\text{Noncrustal } X_p = \text{total } X_p - Al_p\,(X/Al)_{ref} \tag{3.6}$$

where X and Al refer to the abundances of the signature elements (Mn, V) and the crustal reference element (Al) in the particle (p) and crustal reference material. At most sites, 50% of the Mn and 90% of the V are noncrustal. In addition to the tracer elements, $SO_4^{2-}$ is measured in all samples by barium sulfate ($BaSO_4$) turbidimetry. Each of the seven tracer elements (X) was chosen because it satisfied the criteria for precise determination by instrumental neutron activation, strong pollution origin, and a largely fine-particle (submicrometer) mass distribution in the aerosol particles.

The seven elements are used as six ratios to Se. Ratios of elements are believed to be nearly independent of differential dispersion and removal and are more stable during transport than individual concentrations. Ratios appear to change by less than 25% during transport of 1000 km in North America and 5000–10,000 km from Eurasia to Barrow, Alaska (Rahn *et al.*, 1985). The choice of reference element is not critical, because the final step in the analysis, regional apportionment of aerosol particle concentration, is independent of reference element.

Whenever possible, Rahn determines regional components by taking multiple daily samples at one or more sites within a known or suspected source or receptor area. The sites are chosen to be regionally representative, that is, nonurban with no strong sources nearby. The samples are then analyzed for the tracer elements, and chemical and meteorological techniques are used to select the samples that best represent emissions of that region. The usual chemical technique for sorting samples is to construct frequency distributions of each X/Se ratio and examine them for modes, or maxima, that represent the major types of aerosols present. From meteorological characteristics associated with the modes, it is usually possible to determine whether they represent local or transported aerosol. An alternative approach to isolating local aerosol is to consider samples taken during stagnation. The samples representing local aerosol are then pooled, and the means and standard deviations of their X/Se ratios are calculated. The collection of six X/Se ratios, with standard deviations, constitutes the elemental signature of that region. As a check, a signature generated within a region can be compared with one generated after transport. The CMB computational program (Williamson and DuBose, 1983) is used for the apportionments. Each element is weighted by its effective variance, which includes the variance of the element in both the sample and the signatures. Uncertainties are calculated with standard error-propagation procedures. Average concentrations of signature elements are usually accounted for to 25% or better. Details of the apportionment procedure are given by Rahn *et al.* (1985). The application of the CMB approach to reactive species is difficult and uncertain unless account can be taken of total S or N emitted and lost relative to the inert particles. If the ratio to an inert element (or other tracer) is calculated, change in this ratio between the source and receptor must be accounted for.

Rahn and his colleagues have proposed a method to apportion ambient $SO_4^{2-}$. Because $SO_4^{2-}$ is largely secondary in the atmosphere (formed by oxidation of $SO_2$ in the atmosphere rather than being emitted directly), it cannot be made a part of the tracer system directly. $SO_4^{2-}$ can be apportioned indirectly among a suite of regional signatures, however, by regressing its concentrations in a series of samples from a given season and site against the regional coefficients $S_j$ over the series:

$$(SO_4^{2-})_i = \sum_{j=1}^{p} S_{ij}E_j \tag{3.7}$$

for $p$ sources, where $(SO_4^{2-})_i$ is the concentration of $SO_4^{2-}$ in the $i$th sample, $S_{ij}$ is the regional coefficient of the $j$th source in the $i$th sample, and $E_j$ is a regression coefficient that represents the derived mean $SO_4^{2-}$/Se ratio for the $j$th source, which is termed the effective $SO_4^{2-}$. The effective $SO_4^{2-}$ represents the mean $SO_4^{2-}$ (per unit Se) associated with the $j$th signature over the series, that is, the initial $SO_4^{2-}$ near the source plus that formed from $SO_2$ during transport to the receptor. Because this regression also uses effective-variance weighting, both the set of effective $SO_4^{2-}$ and its uncertainties are functions of the uncertainties in both the measurements of $SO_4^{2-}$ at the receptor and the initial set of regression coefficients $S_{ij}$. The contribution of the *j*th source region to the $SO_4^{2-}$ of a sample is the product of $E_j$ and $S_{ij}$, which is the effective $SO_4^{2-}$ of the signature multiplied by the coefficient of the signature.

This approach to apportioning $SO_4^{2-}$ has been applied to cases in the northeastern United States (Rahn and Lowenthal, 1985). The apportionments of $SO_4^{2-}$ at two sites are consistent both summer and winter, show similar seasonal variations, resemble closely the apportionments of Se (the tracer element most similar chemically to S), and agree (within 10–15%) with results from transport models and directional studies in New York State (Husain *et al.*, 1983). As a further check on the performance of the $SO_4^{2-}$ apportionment procedure (multiple linear regression against regional coefficients), researchers used it to apportion Se, Zn, and As in the 79 samples from Underhill, VT, during summer 1982. The percentages attributed to the Northeast, Midwest, and Ontario smelters generally agreed to within 5% of those determined by the CMB method normally used for the tracer elements. The reason the $SO_4^{2-}$ regression procedure appears to work satisfactorily is that a quasi-reproducible fraction of $SO_2$ has been converted to $SO_4^{2-}$ by the time the various air masses reach the sites. That is, the air masses are well aged with respect to S. Tuncel *et al.* (1985) have found a similar result for Virginia's Shenandoah Valley in summer.

Results using the tracer system suggested that only half of New England's $SO_4^{2-}$ aerosol comes from the Midwest. Most of the other half comes from the Northeast itself. Only a few percent comes from the Canadian smelters, which are large and obvious sources of $SO_2$ in ENA. Daily variations from northeastern and midwestern sources and Canadian smelter components of northeastern aerosol have been shown to agree closely with large-scale meteorological changes. When airflows persistently from the Midwest, a pulse of midwestern aerosol is seen. When the winds are light or from another direction, midwestern aerosol is weak or absent. Northeastern aerosol is always present as a "foreground." Midwestern aerosol appears in pulses of 1 to 4 days, and smelter aerosol comes in short bursts.

Rahn's use of metal ratios, such as the V/Mn ratio, to infer sources of polluted air in North America has attracted considerable attention as a potential means of segregating local and distant sources. However, the method has been criticized for oversimplifying the metal ratio data and for inferring S with tracer metals. Once such criticism and a rebuttal can be found in an exchange between Kneip *et al.* (1983) and Rahn (1983). Husain *et al.* (1983) also criticized the V/Mn hypothesis. By examining this ratio in particle samples at five New York State sites, they found that the V/Mn decreases by a factor of 2 to 3 from west to east. Furthermore, data from weekly or semiweekly particle samples used by Rahn *et al.* would be distorted if pollution episodes involving mixing of low-V (Midwest) air with high-V (East Coast) air occurred in the sampling period. These two factors would invalidate regional air mass source trajectories estimated by V/Mn, as proposed by Rahn and his co-workers.

Any type of technique that uses multiple elements found in a source plume as a tracer and connects them by a statistical regression scheme is subject to certain mathematical constraints (e.g., Henry, 1982). The tracer ratio method has not been analyzed as yet for its stability in estimating source contributions according to these constraints. A mathematical stability analysis should be attempted to help critique the elemental ratio approach for regional-scale source apportionment.

Rahn *et al.* have theorized that the CMB technique can be extended to precipitation samples. Their proposed technique involves collecting precipitation samples in clean polyethylene bags, followed by freeze drying and analysis of the age and contents by neutron activation. $SO_4^{2-}$ is determined by wet chemistry before freezing. The method has been tested on samples taken at Narragansett, RI.

To transfer regional signatures directly from aerosol to precipitation, one must find tracer elements that are transferred from aerosol to precipitation with the same efficiency—that is, are not systematically fractionated. Rahn has proposed a group: As, Sb, Se, and noncrustal V. Other elements do not appear to be usable in precipitation for various reasons. For example, Zn has high sample blanks with the analytical method and is easily contaminated in the laboratory. Noncrustal Mn is masked by the order-of-magnitude enrichment of crustal Mn in precipitation. Precipitation In is masked by the general enrichment of coarse-particle crustal and marine elements relative to aerosol material.

Rahn's reasoning in support of lack of fractionation for As, Sb, Se, and noncrustal V is that once the major known fractionation of all elements as a function of particle size is accounted for, repeated observation reveals no systematic differences within these four elements that cannot be explained by altitude segregation. Evidence for this conclusion comes from comparisons of elemental data for precipitation and particles via plots of washout ratio versus particle size. The *washout ratio* (*W*) (mass weighted) is or the concentration of an element in precipitation divided by its concentration in the associated air mass, normalized to mass of water and air:

$$\xi = \frac{\text{elemental concentration in precipitation (ng/kg)}}{\text{elemental concentration in air (ng/kg)}} \tag{3.8}$$

A scatter diagram of washout ratio versus particle size, or washout ratio plot, is used to see whether a systematic relationship exists between scavenging efficiency for the elements and their particle size.

Washout ratio plots for 30 precipitation events at Narragansett showed that coarse-particle elements are generally scavenged more efficiently than fine-particle elements, by factors of up to 10. The elements and their washout ratios fall into two clusters, with little pattern within each cluster from event to event. Of the elements in the Rahn *et al.* tracer system, As, Sb, Se, and noncrustal V fall in the lower cluster and appear to be usable without correction. Noncrustal V is normally depleted in precipitation by a factor of roughly 2 relative to the particles; this phenomenon results from vertical stratification and precipitation forming in air aloft, which is more midwestern in character than the surface air. $SO_4^{2-}$ is generally enriched in precipitation relative to finely divided particle elements by factors of 2 to 10. This is evidently related to conversion of $SO_4^{2-}$ in acid rain droplets.

Rahn has tested the methodology for apportioning the constituents in precipitation samples taken in winter 1985 at Narragansett. His results suggest that, relative to the suspended particles, the elements in the precipitation shift toward midwestern sources, by amounts ranging from 3% for noncrustal V to 20–25% for Se, Sb, and As. Se in the precipitation was roughly 25% northeastern and 75% midwestern, as opposed to the normal 50–50% split in aerosol particles. Arsenic, which is typically 50–30% to 40–40% Northeast–Midwest in aerosol particles, was more nearly 20–80% in the precipitation. Altogether, this constitutes a significant shift of at least 25% toward the Midwest in precipitation. $SO_4^{2-}$ shifted toward midwestern sources by nearly 40%, but this may be partially an artifact because of the small number of samples with $SO_4^{2-}$ data. If confirmed by subsequent study, the Rahn work suggests that local sources, even on a regional scale, are discriminated against by precipitation because it incorporates most of its contaminants aloft, in air masses of more distant origin than at the surface.

Another form of tracer analysis has been applied to identification of haze constituents in the Shenandoah Valley of Virginia (Stevens *et al.*, 1984). As did Ferman *et al.* (1981), these investigators found that the light extinction coefficient in the Shenandoah area was associated with fine particles. $SO_4^{2-}$ and its related cations accounted for more than 60% of the fine-particle mass concentrations, and a correlation coefficient of 0.89 existed between visual range and $SO_4^{2-}$. Stevens *et al.* concluded that $SO_4^{2-}$ is a major cause of visibility impairment in the Shenandoah Valley. Factor analysis of the aerosol particle chemical composition indicated that S, Se, and $NO_3^-$ were linked together and associated with air mass trajectories traced back to areas of high emission density of coal-fired power plants. From this association and with Se (a tracer of midwestern coal combustion) and on the basis that $SO_2$ emissions in the Virginia region are more than 90% from coal-fired plants, they hypothesized that the $SO_4^{2-}$, Se, and $NO_3^-$ found in the Shenandoah particulate samples came from a common source, notably coal combustion in power plants.

Questions remain about the validity of the CMB approach for segregating sources of aerosol particle or precipitation constituents. Furthermore, the method provides only very coarse separation of regional contributions. Nevertheless, the results represent an innovative approach to developing empirical SRRs. If sources were "tagged" with releases of chemical tracers, the resolution and reliability of the SRRs could be improved dramatically. Such an approach using releases of inert tracers widely over ENA was proposed by Foley and Hilst (1983). Hidy *et al.* (1985) explore the feasibility of an aerometric tracer experiment extensively. Additional aspects of this approach, including analysis of variance, are noted in Chapters 5 and 6.

# 4 AIR CHEMISTRY AND ATMOSPHERIC PROCESSES

The geographic distributions of $SO_4^{2-}$ and $NO_3^-$ broadly correspond to the zones of high $SO_2$ and $NO_x$ emission density, but atmospheric influences spread the distributions out or distort them. For example, the SURE results documented that temporal changes in atmospheric $SO_2$ and $SO_4^{2-}$ concentrations averaged from daily data to monthly periods far exceeded estimated changes in $SO_2$ emissions in ENA. The percentage of stations reporting high daily ambient $SO_4^{2-}$ levels reflects a measure of regionwide change in $SO_4^{2-}$ that parallels changes in emissions. The estimated fractional change in daily emissions was about 15% with a regular oscillation associated with reduced weekend emissions. The $SO_4^{2-}$ pattern did not show a weekend change, and the amplitude of regional change is much larger, a factor of 2 or more. Similar results were found by comparing within-year or year-to-year estimates of emissions changes and variations in precipitation $SO_4^{2-}$ over the same periods. These generalizations qualitatively support the dominance in exposure patterns of atmospheric processes over emission distributions.

Information about the rates of change expected for different processes suggests that mixing and air mass transport should dominate the distribution of $SO_2$- and $NO_x$-related constituents over ENA. Researchers hypothesize that dry- or fair-weather regional pollution is most heavily influenced by the distribution of sources and by atmospheric transport phenomena. But the significance of deposition varies with surface conditions and the occurrence of precipitation. Nevertheless, much of the research on chemical behavior has been predicated on the notion that transport-dominated patterns established by winds and mixing in the planetary boundary layer are modified only in a minor way by chemical transformations and surface exchange. This assumption basically underlies the theoretically based estimates of source contributions to regional pollutant distributions (SRRs) reported through the early 1980s, as discussed in Chapter 6. The extent to which the transport dominance hypothesis can be applied to regional pollution behavior remains open to question. Circumstantial evidence for the influence of the different atmospheric processes exists from the study of pollution events. This chapter describes the phenomenological aspects of $SO_4^{2-}$ pollution events that give insight into dynamic factors. The descriptive analysis together with knowledge of atmospheric processes serves as a basis for interpreting persistent patterns of behavior found in the atmospheric chemistry of $SO_2$ and $NO_x$.

An important aspect of the analyses described here is the evidence for classification of regional chemical behavior by large-scale (regional-scale) meteorological factors. Such a classification offers an opportunity for interpreting chemical phenomena and their variability in a relatively simple empirical manner. This capability may ultimately serve as a practical route to SRR analysis by blending measurements with efficient theoretically based calculations.

## DESCRIPTION OF EVENTS

The episodic character of regional pollutant behavior is best documented by data on airborne $SO_4^{2-}$ behavior, the nature of which is well illustrated by data from many nonurban sites in ENA. This feature was "discovered" in the 1974 data analyzed for the SURE design (Hidy *et al.*, 1976).

### Event Days

In the SURE analysis (Mueller and Hidy, 1983) three parameters have been used to interpret regional airborne $SO_4^{2-}$ coverage associated with different types of events: $SO_4^{2-}$ concentration, geographic extent, and duration. An event day can be defined by coupling specific concentration ranges with ranges of spatial coverage. The number of successive event days determines the duration of an event. This approach permits all $SO_4^{2-}$ events over the study region to be categorized and summarized for the 15-month period of the SURE observations.

Figure 3.9 illustrates the episodic nature and geographic extent of $SO_4^{2-}$ events and the classification of events based on six data-intensive sampling months. Researchers used a composite data set taken at all stations used in the SURE; this composite assumes that the nine CI stations can represent conditions for $SO_4^{2-}$ as well as can the combination of all 54 SURE stations. Mueller and Hidy (1983) reported the testing of this assumption by establishing the similarities between event occurrence at both groups of stations in the intensive observation periods. The tests indicated that the two groups of station data were comparable in event segregation, with certain differences that are discussed below.

The SURE investigators used the terms *event* and *nonevent*. An event day is said to occur when at least one reporting station measures a 24-hour $SO_4^{2-}$ concentration of 10 $\mu g/m^3$ or more. Event days were further categorized as *regional*, *subregional*, and *nonregional*. These definitions were made on the basis of the geographic area affected by elevated $SO_4^{2-}$ concentrations; duration of events was not used explicitly at this stage. This approach to defining *regional* is complementary to the coarse spatial and temporal scaling introduced by Hidy *et al.* (1978). The later convention of labeling multiday events was adopted with a geographic scale of roughly 1000 km as regional; subregional events are single-day phenomena over 100 km.

Table 4.1 lists definitions of event groups by intensity and geographic coverage of $SO_4^{2-}$ concentration. Although there is some overlap in the spatial coverage requirements for the regional and subregional groups in the $\geq 10$ $\mu g/m^3$ range, the definitions allow no overlap in the $\geq 15$ and $\geq 20$ $\mu g/m^3$ ranges. Thus, an attempt was made to distinguish between the first three groups on the basis of the percentage of stations reporting concentrations in excess of 15 $\mu g/m^3$. This attempt was satisfactory; only 7 of the 186 intensive measurement days failed to fit these definitions. As might be

TABLE 4.1 **Definitions of Event Days by Intensity and Coverage of $SO_4^{2-}$ Concentration**[a]

| | **Percentage of stations with $SO_4^{2-}$ concentrations** | | |
|---|---|---|---|
| **Event group** | **>10 $\mu g/m^3$** | **>15 $\mu g/m^3$** | **>20 $\mu g/m^3$** |
| 1. Enlarged regional event day | 40–93 | 25–72 | 10–61 |
| 2. Regional event day | 40–70 | 15–25 | 5–10 |
| 3. Subregional event day | 25–50 | 5–15 | <5 |
| 4. Nonregional event day | <35 | <5 | 0 |
| 5. Nonevent day | 0 | 0 | 0 |

[a] From Mueller and Hidy, 1983; reproduced with permission of the Electric Power Research Institute.

expected from the less detailed coverage of the nine CI stations, there were considerably more outlier days that did not fit into the discrete event groups during the nonintensive months. Thus, although only 4% of the intensive days were outliers, 15% of the 271 nonintensive days did not fit the definitions as precisely. Judgments regarding the likely regional representativeness for each CI station's $SO_4^{2-}$ concentration had to be made for the outlier days on the basis of synoptic weather patterns. Generally, the regional situations were distinct from subregional cases, although it was sometimes difficult to separate group 1 and 2 regional-scale events.

*Group 1* represents the most intensive enlarged regional event days, with a widespread, usually homogeneous area affected by $SO_4^{2-}$ concentrations greater than 15 $\mu g/m^3$. These were the days characterized by the highest concentrations covering the largest geographic areas. During the six intensive measurement periods, group 1 events with concentrations greater than 15 $\mu g/m^3$ were reported by 11 to 40 stations, and values exceeding 20 $\mu g/m^3$ were reported by 5 to 33 stations. The upper boundaries of these ranges in station coverage were established on the basis of two days during the large, geographically extensive episode, 19 and 20 July 1978. *Group 2* also represents days with large regional coverage of $SO_4^{2-}$ concentrations greater than 10 $\mu g/m^3$. However, this category was distinguished from group 1 by smaller areas of elevated concentrations, frequently with geographic separation between localized "hot spots" with concentrations above 15 or 20 $\mu g/m^3$.

Subregional event days, *group 3*, were those during which less than 50% of the SURE region was covered by $SO_4^{2-}$ concentrations in excess of 10 $\mu g/m^3$. $SO_4^{2-}$ concentrations above 15 $\mu g/m^3$ were measured at only two to six SURE stations, usually in one specific area within 100 to 200 km of its apparent source region.

Two additional groupings were established to include days lacking a regional or subregional $SO_4^{2-}$ accumulation. The nonregional (*group 4*) events include days of localized behavior on which only one or two stations reported 15 to 20 $\mu g/m^3$ $SO_4^{2-}$. Only a few outlier days in this category occurred, and these differed from the coverage criterion by only a few percent. *Group 5* nonevent days were said to occur when none of the reporting stations measured concentrations as high as 10 $\mu g/m^3$.

### Event Duration

An important aspect of the regional $SO_4^{2-}$ analysis is event duration. Researchers determined only for regional (groups 1 and 2) and subregional (group 3) events. The number of successive days having the same air mass with regional or subregional $SO_4^{2-}$ coverage was the criterion. If at least half of the days during such a period conformed to the definition of group 1 or group 2 days, then the event was characterized as regional in scale; otherwise, it was considered subregional. This convention changed slightly the number of days categorized as regional or subregional from totals. Figure 4.1 shows the number and duration of $SO_4^{2-}$ events for six data-intensive sampling months and during the full 15-month period of the SURE. This pattern is similar to that obtained for the full station network operating in the intensive sampling months. A total of 26 regional and subregional events were counted during the six data-intensive sampling months, but four were eliminated because they either ended on an intensive start date or started on an intensive end date. Thus, 14 subregional and 8 regional events fell completely within the six data-intensive sampling months. If the nine nonintensive months are also considered, 24 subregional and 28 regional events are represented. It should be remembered that proportionately more regional events were counted during the nonintensive months on the basis of a nine-station sampling network and that there were more outlier days for these months.

On the basis of combined data sets for the August 1977–July 1978 period, approximately 40 events occurred during a year. About half cov-

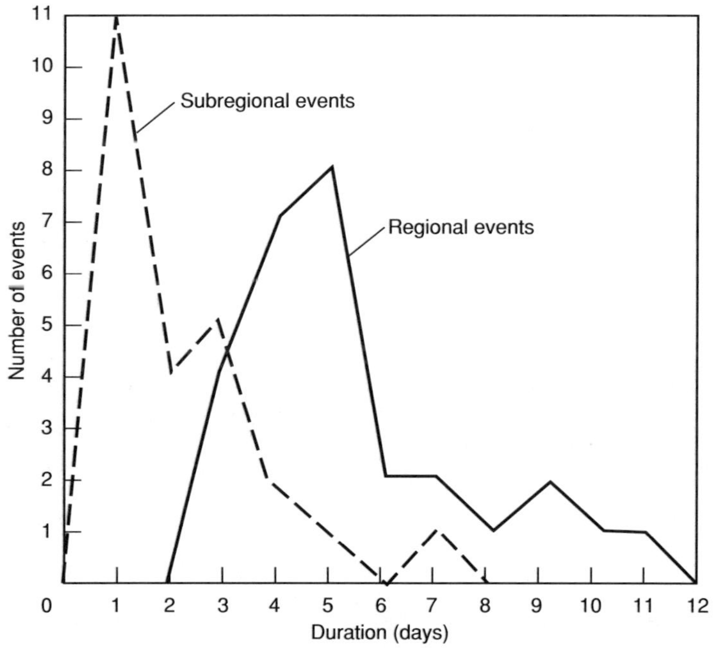


**FIGURE 4.1** Frequency of event duration during full 15-month SURE program. (From Mueller and Hidy, 1983; reproduced with permission of the Electric Power Research Institute.)

ered more than 40% of the SURE region, and about half were more local. Regional events appeared to last from 3 to 11 days with a mean of 5 days; subregional events lasted 1 to 7 days with a mean of 2 days and were more likely to occur in 1 to 3 days. The multiday character of $SO_4^{2-}$ events indicated the importance of the persistence of large-scale meteorological conditions promoting $SO_4^{2-}$ accumulation and, at the same time, minimizing loss of suspended material. According to one survey, such widespread conditions can occur surprisingly often, particularly in summer.

## Representativeness and Seasonal Occurrence

Table 4.2 shows the frequencies of occurrence for the five event groups during the 15-month SURE measurement period and compares the distributions for the six intensive-measurement periods and the full 15-month duration of the experiment. Regional event days (groups 1 and 2) occurred about 30% of the time and subregional events about 15% of the time. The distributions compared favorably; however, a proportionately greater number of regional event days and fewer subregional events occurred during the nonintensive months.

Two reasons are postulated to explain the difference between the intensive and nonintensive months. Either the analysis based on the data from the nine CI stations overestimated the regional $SO_4^{2-}$ coverage, or meteorological conditions conducive to $SO_4^{2-}$ accumulation happened to occur more frequently during the nonintensive months of the experiment.

TABLE 4.2 **Number of Days in Each Event Group during the Period of SURE Observations**[a,b]

| | Event group | | | | | |
|---|---|---|---|---|---|---|
| | **1 Enlarged regional** | **2 Regional** | **3 Subregional** | **4 Nonregional** | **5 Nonevent** | **Total** |
| *Total measurement period* (15 months) | | | | | | |
| Days per group | 86 | 62 | 72 | 142 | 95 | 457 |
| Percent of total days | | 19% | 13% | 15% | 31% | 21% |
| Cumulative percentage | | 19% | 32% | 48% | 79% | 100% |
| *Intensive-measurement periods* (50-station network) | | | | | | |
| Days per group | 23 | 16 | 37 | 77 | 32 | 185 |
| Percent of total days | | 12% | 9% | 20% | 42% | 17% |
| Cumulative percentage | | 12% | 21% | 41% | 83% | 100% |

[a] From Mueller and Hidy, 1983; reproduced with permission of the Electric Power Research Institute.
[b] August 1977, October 1977, January 1978, April 1978, July 1978, and October 1978.

Without further observations of this kind, it is not possible to choose the more likely explanation.

The patterns of seasonal change in $SO_4^{2-}$ concentration are linked closely to the distribution of event occurrences. Regional events in the 1977–1978 period occurred in summer and winter. Subregional events occurred more uniformly from season to season. This pattern might lead one to expect a stronger summer seasonal peak in $SO_4^{2-}$ concentration. However, most sites showed less of a summer maximum in 1977–1978 than was suggested in 1974 data (Hidy *et al.*, 1976).

One reason for this apparent difference is related to the year-to-year variation in meteorological conditions. Because the SURE observations constituted only a little more than a year of data, it was important to establish the representativeness of the set compared with "normal" conditions in the EUS. (*Normal* is taken here to mean the 30-year climatological average.) Using the knowledge gained in the air mass and case study analyses combined with climatological data, researchers attempted to determine the normality of the August 1977 through October 1978 sampling period in terms of the potential for regional $SO_4^{2-}$ events. They compared the monthly climatic average temperature, precipitation, and pressure patterns with the patterns corresponding to weather systems during the SURE period.

To establish an index of normalcy, researchers compared parameter values for each day with normal values based on the previous 30-year average. The results were summarized in terms of departures from monthly temperature and precipitation patterns and characteristic circulation. Considering these data on a month-by-month basis, Mueller and Hidy (1983) reported that August 1977 was nearly normal with respect to temperature, but September was warmer and October was cooler than normal. Precipitation was higher than normal for these 3 months. November 1977 was warmer than normal with excessive precipitation; the remainder of winter and spring 1978 was cooler than normal. December 1977 had above-normal precipitation, but the remainder of the winter was generally drier than normal, except for January 1978 in Ohio and the Northeast. Temperature and precipitation during the summer of 1978 were generally nearly normal, while October 1978 saw cooler, drier conditions than normal. Although it is difficult to generalize from these comparisons, the 13 months appeared to be cooler and drier than normal for the period.

Circulation was characterized in terms of departure from the 30-year normal 70-mb pressure level patterns. The height of the 700-mb surface was found to be higher than normal in the summer of 1977 and generally weaker than normal through the following winter and spring of 1978. The highest occurrence of regional $SO_4^{2-}$ events was in summer with a second maximum in winter. There were considerably fewer subregional events than regional events.

Subjective estimates of the ranges of normal seasonal frequency of occurrence for combined regional and subregional events were based on departures from climatological mean meteorological conditions. Estimates suggest that an anomalous number of winter regional events occurred in 1978. Temperature seems to have had a positive correlation with regional $SO_4^{2-}$ accumulation in all seasons except winter, when low temperatures were associated with near-ground inversions and limited pollutant dispersion potential from ventilation. Anomalies of precipitation (a moisture indicator) had little correlation with $SO_4^{2-}$ production over the region. Comparison of the number of days that departed from the expected seasonal normal ranges of subregional and regional event days showed that the nonintensive study months had more days with meteorology favoring $SO_4^{2-}$ accumulation. Although the seasonal character is apparent, key normalcy parameters also appear to be important for the potential for large-scale $SO_4^{2-}$ events during any season.

## CASE STUDIES

The dynamics of regional-scale $SO_4^{2-}$ events can be described using available regional meteorological observations. Widespread stagnation of air permitting pollution to build up for more than a day was found to be an important factor in all events. Under certain circumstances contaminated air in stagnation evolves into a second stage in which persistent airflow, directed across areas of high $SO_2$ emission density, termed here *ducting* or *channeling*, sustains polluted conditions. Air motion forces the channeling along an advancing, extended frontal system. For purposes of examining different cases, "pure" stagnation buildup is distinguished from development of a channeling condition. The latter is of particular interest because it has not been identified with air pollution before, and it shows straightforward evidence for potential of long-range $SO_x$ transport.

### Fall Stagnation: 27–30 October 1977

Korshover's work (1976) showed that air pollution can take place over large areas under anticyclonic conditions that involve air stagnation. A stagnating anticyclone is defined as a condition in which surface winds do not exceed 3.8 m/s for four or more consecutive days. $SO_4^{2-}$ accumulation should be associated with zones of $SO_2$ precursor accumulation under such conditions; that is, in the absence of dominant transport winds, $SO_4^{2-}$ accumulation should remain closely linked with zones of great $SO_x$ emission density. One case of this type was identified during the period between 27 and 30 October 1977. Maximum $SO_4^{2-}$ accumulation occurred on the group 2 regional event day 28 October, with 24 (60%) of the SURE stations recording concentrations exceeding 10 $\mu g/m^3$. The other 3 days were subregional event days with only 30 to 45% of the stations reporting concentrations in excess of 10 $\mu g/m^3$. These 3 days were one of the longer subregional events documented in the experiment.

Figures 4.2 and 4.3 show the distributions of 24-hour $SO_2$ and $SO_4^{2-}$ concentrations, respectively, for the 4-day period. Contours on these maps are drawn on the assumption that the measurements are representative of regional-scale distributions over the area considered. High concentrations of $SO_2$ are generally associated with areas of highest emission density. The maximum 24-hour average values exceed 3 ppb or about 6.9 $\mu g/m^3$. A broad area of 10 ppb (23 $\mu g/m^3$) covers much of the Ohio River

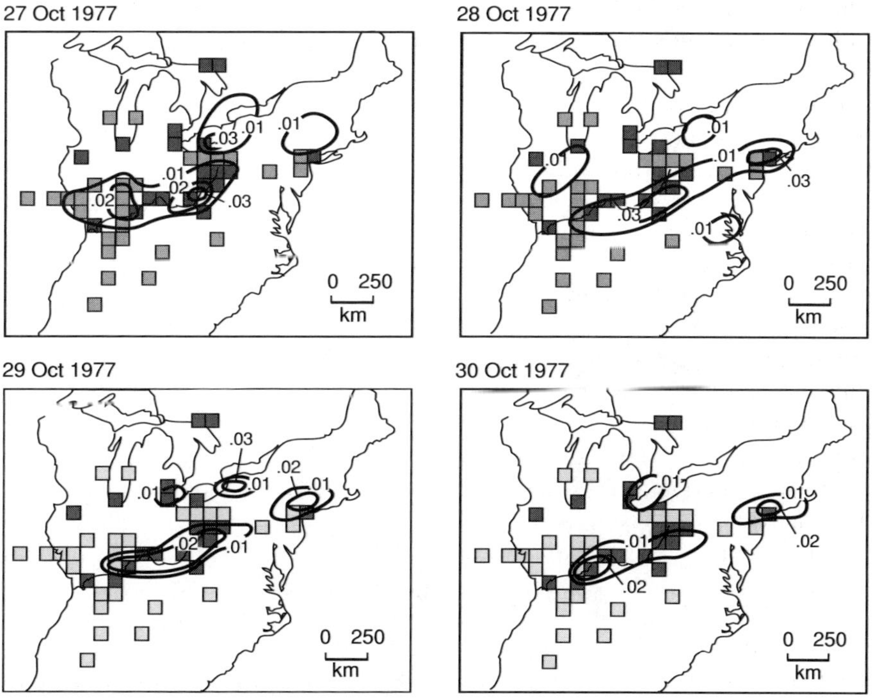


**FIGURE 4.2** Distribution of 24-hours average $SO_2$ concentrations (ppm) during a period of air mass stagnation in October 1977. Shaded areas are blocks of 80 × 80 km areas with fall average $SO_2$ emissions: light shading indicates emissions ≥250 metric tons per day (mT/d), and dark shading represents emissions ≥500 mT/d. (From Mueller and Hidy, 1983; reproduced with permission of the Electric Power Research Institute.)

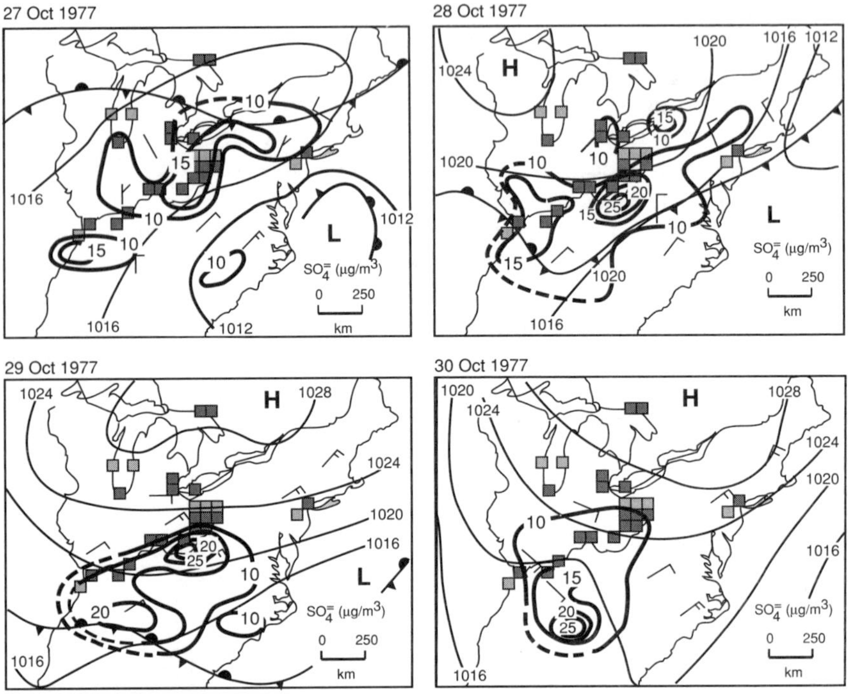


FIGURE 4.3 Distribution of 24-hour average $SO_4^{2-}$ concentrations ($\mu g/m^3$) during the October 1977 stagnation period. Shaded areas are blocks of 80 × 80 km areas with fall average emissions: light shading indicates emissions ≥250 mT/d, and dark shading represents emissions ≥500 mT/d. The 0700 EST surface isobars (mb) and 300-m AGL winds (half barb = 2.5 m/s, whole barb = 5 m/s) are shown. The location of a front is also shown by the barb/half-circle link. High- and low-pressure areas are shown with frontal positions. (From Mueller and Hidy, 1983; reproduced with permission of Electric Power Research Institute.)

Valley and western Pennsylvania. $SO_4^{2-}$ reached levels well in excess of average on 29 October over a broad area of Tennessee, the Ohio River Valley, Pennsylvania, and New York. Except for the Tennessee area, the elevated $SO_4^{2-}$ corresponds to the region of elevated $SO_2$ concentration.

On 27 October, the widespread area of relatively high ground-level $SO_4^{2-}$ concentrations was confined between a coastal storm and a weak stationary front[1] along the Great Lakes. By 28 October, the front advanced southward, while, to the north a high-pressure system moved eastward. $SO_4^{2-}$ continued to accumulate behind the front on the south

[1] Fronts are roughly defined as boundaries between air masses having different thermodynamic characteristics. They are often identified with zones of strong horizontal temperature gradients.

side of this high (pressure) center. The northerly winds were light but effective in restricting the $SO_4^{2-}$ accumulation in a region of converging airflow along the south side of the front. On 29 October, the center of the high appeared farther east. Elevated $SO_4^{2-}$ remained south of the Ohio River Valley over the Tennessee areas of high $SO_2$ emissions and along a convergence region on the north side of the frontal disturbance. During this period, essentially no precipitation was associated with the front, but the relative humidity was high and fog occurred at the surface in the mornings around the region of elevated $SO_4^{2-}$ concentrations. By 30 October, the center of the high remained near its location of the day before, the front had dissipated, and the elevated $SO_4^{2-}$ concentrations began to disappear, except over the sources in Tennessee and the western Ohio River Valley. Everywhere within the 10 $\mu g/m^3$ isopleth on 27 October surface winds were less than 3.8 m/s, making the day one of regional-scale stagnation by Koshover's (1976) definition. The areal extent of the stagnating anticyclone was limited to the southern Ohio River Valley, which was the only area that experienced at least four consecutive days of winds less than 3.8 m/s.

Elevated $SO_4^{2-}$ levels were maintained on the south–southwest side of the stagnating anticyclone; the pattern did not drift with the weak air circulation around the high-pressure center of the air mass. Elevated $SO_4^{2-}$ levels at the ground coincided with $SO_2$ concentrations along the Ohio River Valley. Around the large emission centers in western Tennessee, airborne $SO_4^{2-}$ accumulated at ground level without correspondingly high $SO_2$ concentrations. The elevated $SO_4^{2-}$ zones remained within 100 to 300 km of the zones of high emission density. In general, warm moist air covered the area with considerable cloud cover and morning fog.

Figure 4.4 shows the response of ground-level $SO_4^{2-}$ and $SO_2$ concentrations to restrictions in vertical mixing. The time–height history of potential temperature[2] estimated from rawinsonde data at Huntington, West Virginia, is drawn in the figure, with the mixing height calculated from criteria of Benkley and Schulman (1979). At all three stations, shown in the figure, ground-level $SO_4^{2-}$ accumulated on 28 October with a *reduction* in mixing height and remained relatively high on 29 October even with a daytime *increase* in mixing height. The pattern of 24-hour average $SO_2$ change generally corresponds to the $SO_4^{2-}$ change, as one would expect if the two are linked locally as chemical reactant and product under stagnating conditions.

[2] Potential temperature is often used as an index of the stability of air to vertical mixing. Potential temperature is the temperature a parcel of air would reach if it were brought adiabatically into equilibrium with its surroundings. Air is neutrally stable if its potential temperature is constant with height. If air follows a curve with cooling greater than adiabatic, it is stable; with less cooling, it is unstable to vertical mixing.

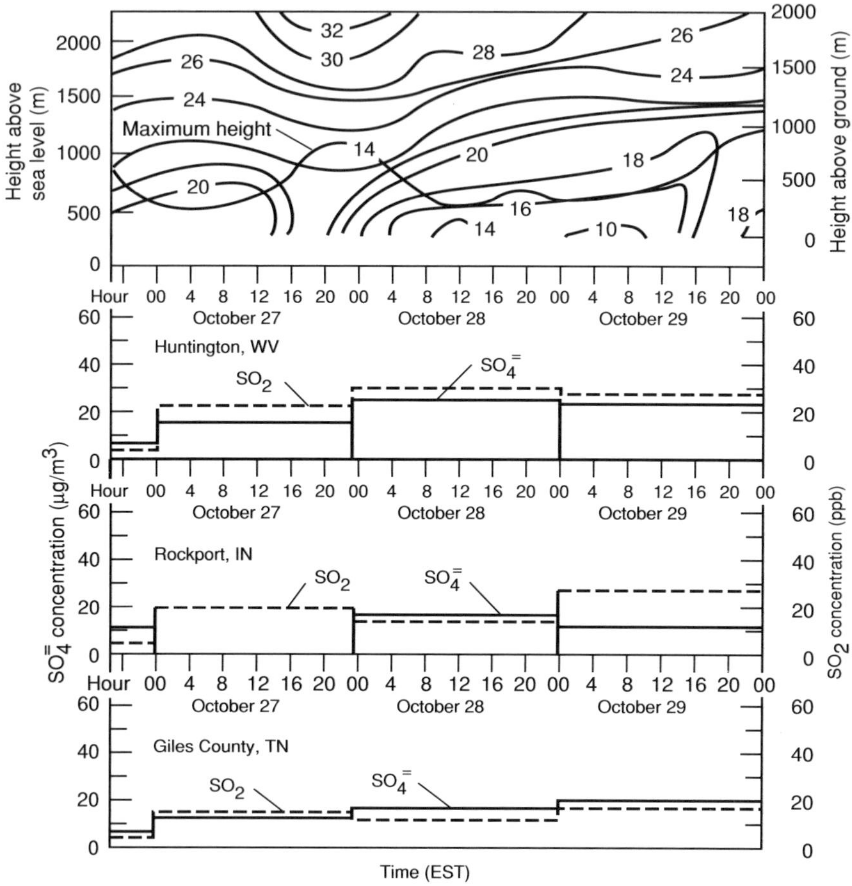


**FIGURE 4.4** Comparison of the time–height distribution of potential temperature (°C) from Huntington, WV; rawinsonde data; and 24-hour average $SO_2$ and $SO_4^{2-}$ measurements during 27–29 October 1977. (From Mueller and Hidy, 1983; reproduced with permission of the Electric Power Research Institute.)

The influence of winds can be deduced from calculated streaklines[3] of surface winds and winds 300 m above ground level (AGL). Streaklines give a qualitative estimate for the movement of air parcels passing over a station at the beginning of a day. Figure 4.5 shows streaklines for 28 October. The winds at a height of 300 m were chosen because that level is representative of transport for most $SO_2$ emissions in the area. Note that

[3] For definition of streaklines, see Prandtl and Tietjens (1934); in steady flow streaklines are the same as particle trajectory paths.

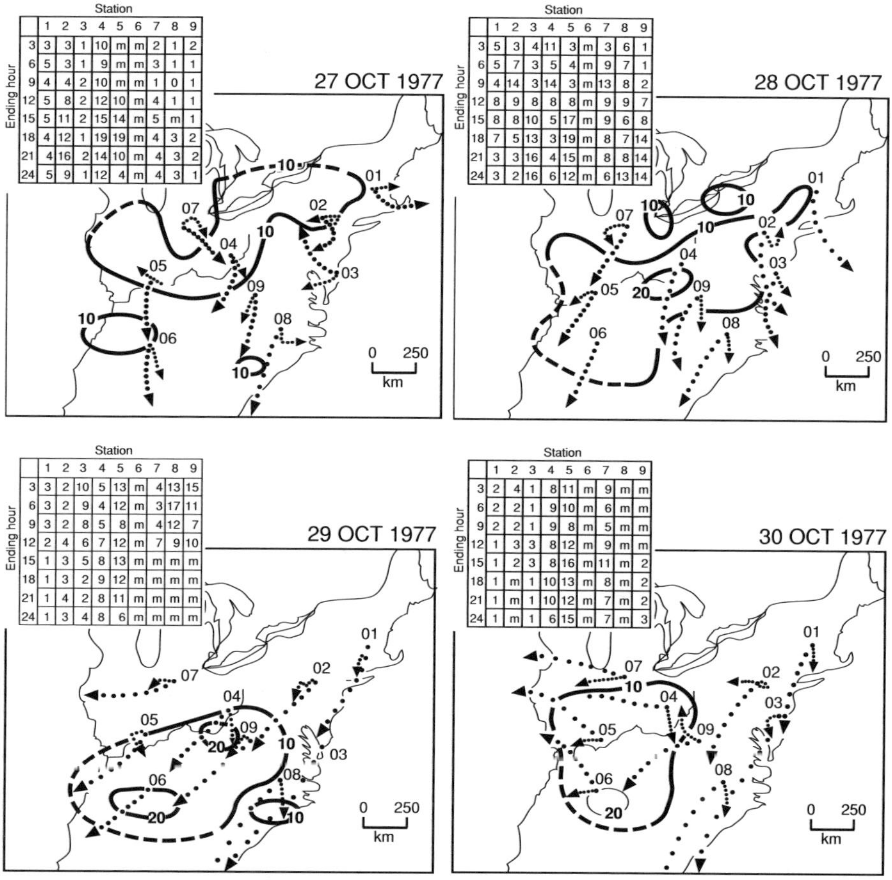


27 OCT 1977

| Ending hour | 1 | 2 | 3 | 4 | 5 | 6 | 7 | 8 | 9 |
|---|---|---|---|---|---|---|---|---|---|
| 3 | 3 | 3 | 1 | 10 | m | m | 2 | 1 | 2 |
| 6 | 5 | 3 | 1 | 9 | m | m | 3 | 1 | 1 |
| 9 | 4 | 4 | 2 | 10 | m | m | 1 | 0 | 1 |
| 12 | 5 | 8 | 2 | 12 | 10 | m | 4 | 1 | 1 |
| 15 | 5 | 11 | 2 | 15 | 14 | m | 5 | m | 1 |
| 18 | 4 | 12 | 1 | 19 | 19 | m | 4 | 3 | 2 |
| 21 | 4 | 16 | 2 | 14 | 10 | m | 4 | 3 | 2 |
| 24 | 5 | 9 | 1 | 12 | 4 | m | 4 | 3 | 1 |

28 OCT 1977

| Ending hour | 1 | 2 | 3 | 4 | 5 | 6 | 7 | 8 | 9 |
|---|---|---|---|---|---|---|---|---|---|
| 3 | 5 | 3 | 4 | 11 | 3 | m | 3 | 6 | 1 |
| 6 | 5 | 7 | 3 | 5 | 4 | m | 9 | 7 | 1 |
| 9 | 4 | 14 | 3 | 14 | 3 | m | 13 | 8 | 2 |
| 12 | 8 | 9 | 8 | 8 | 8 | m | 9 | 9 | 7 |
| 15 | 8 | 8 | 10 | 5 | 17 | m | 9 | 6 | 8 |
| 18 | 7 | 5 | 13 | 3 | 19 | m | 8 | 7 | 14 |
| 21 | 3 | 3 | 16 | 4 | 15 | m | 8 | 8 | 14 |
| 24 | 3 | 2 | 16 | 6 | 12 | m | 6 | 13 | 14 |

29 OCT 1977

| Ending hour | 1 | 2 | 3 | 4 | 5 | 6 | 7 | 8 | 9 |
|---|---|---|---|---|---|---|---|---|---|
| 3 | 3 | 2 | 10 | 5 | 13 | m | 4 | 13 | 15 |
| 6 | 3 | 2 | 9 | 4 | 12 | m | 3 | 17 | 11 |
| 9 | 3 | 2 | 8 | 5 | 8 | m | 4 | 12 | 7 |
| 12 | 2 | 4 | 6 | 7 | 12 | m | 7 | 9 | 10 |
| 15 | 1 | 3 | 5 | 8 | 13 | m | m | m | m |
| 18 | 1 | 3 | 2 | 9 | 12 | m | m | m | m |
| 21 | 1 | 4 | 2 | 8 | 11 | m | m | m | m |
| 24 | 1 | 3 | 4 | 8 | 6 | m | m | m | m |

30 OCT 1977

| Ending hour | 1 | 2 | 3 | 4 | 5 | 6 | 7 | 8 | 9 |
|---|---|---|---|---|---|---|---|---|---|
| 3 | 2 | 4 | 1 | 8 | 11 | m | 9 | m | m |
| 6 | 2 | 2 | 1 | 9 | 10 | m | 6 | m | m |
| 9 | 2 | 2 | 1 | 9 | 8 | m | 5 | m | m |
| 12 | 1 | 3 | 3 | 8 | 12 | m | 9 | m | m |
| 15 | 1 | 2 | 3 | 8 | 16 | m | 11 | m | 2 |
| 18 | 1 | m | 1 | 10 | 13 | m | 8 | m | 2 |
| 21 | 1 | m | 1 | 10 | 12 | m | 7 | m | 2 |
| 24 | 1 | m | 1 | 6 | 15 | m | 7 | m | 3 |

**FIGURE 4.5** Streaklines estimated from space and time interpolation of rawinsonde data at the surface (short dotted-line arrows) and 300 m AGL (long dotted arrows) for 27–30 October 1977. Each segment represents 3-hour surface winds. Streaklines originate at the numbered class I stations at 00 EST and end at 24 EST. Three-hour average $SO_4^{2-}$ concentrations for each segment are included in the inset (m denotes missing data). Isopleths of 10 and 20 $\mu g/m^3$ $SO_4^{2-}$ concentration (24-hour average) are superimposed on daily maps. (From Mueller and Hidy, 1983; reproduced with permission of the Electric Power Research Institute.)

one-half of the $SO_2$ in the SURE region is emitted between 100 and 300 m with 20% above and 30% below. The 3-hour average $SO_4^{2-}$ data in the figure follow the segments of the streaklines with the first tabulated value corresponding to the tip of the streakline arrow. The $SO_4^{2-}$ values ideally would represent material transported during the day if the ground-level values were representative of air at 300 m AGL. In other words, the value of $SO_4^{2-}$ at the beginning of a streakline is the concentration expected to move with the wind 24 hours after passing the station where it was first observed, assuming that no losses or gains occurred and that transport was the only factor. Comparison of these values with the 24-hour average geographic distribution estimated from the 54-station network gives an impression of the role of transport as opposed to other processes influencing $SO_4^{2-}$ behavior.

The streaklines at the surface show the light wind conditions with stagnation, particularly in the southwest sector of the high on 28 and 29 October. Pollution accumulated at the ground over the zones of high emission density and low-level injection. The estimated winds at 300 m AGL were also weak, but sufficiently strong to create conditions for a southeasterly-to-southerly drift that would transport $SO_x$ 300 km or so from source regions. This condition suggests, but does not prove, that emissions of $SO_2$ (and $SO_4^{2-}$ produced aloft) during this case could have traveled southward.

The numbers in insets of the figure are the available 3-hour average fine-particle $SO_4^{2-}$ ($<2.5$ $\mu$m diameter) at ground level. The concentrations would be those for material "transported" away from the site each 3 hours if the atmosphere were well mixed to a height of 300 m. Qualitatively, these data suggest that transported pollution at 300 m would generally have moved relatively low $SO_4^{2-}$ concentrations southward away from the Ohio River Valley, leaving higher concentrations over the high source-density zones. At Scranton, PA, and Duncan Falls, OH, southward transport of air containing elevated $SO_4^{2-}$ concentrations of about 10 $\mu g/m^3$ appears to reinforce the zone of concentrations 10 $\mu g/m^3$ and greater in West Virginia and Virginia and might have contributed to enhancement of the zone of maximum sulfate over Huntington, WV.

## Winter Regional Stagnation: 21–24 January 1978

This multiday event is of interest because $SO_4^{2-}$ concentrations above 20 $\mu g/m^3$ were observed in the Midwest on 23 and 24 January 1978, showing that winter $SO_4^{2-}$ levels can approach those realized at other times of the year with dry air aloft and very low temperature conditions. A subregional accumulation on 21 and 22 January achieved regional coverage and intensity on 23 and 24 January. The high $SO_4^{2-}$ concentrations in winter were unexpected on the basis of earlier results suggesting that maximum $SO_4^{2-}$ concentrations tend to occur mainly during summer. Weather conditions at this time consisted of generally light surface winds with fair, very cold weather between major storm systems. One storm had swept the East Coast 2 weeks earlier, and a second had left a record amount of snow in the Midwest a week before. The air over the East was abnormally cold (10°C below normal) but moist, with temperatures below 0°C and morning ground fogs in the Ohio River Valley.

A general warming trend took place between 21 and 24 January because 22 and 23 January were the sunniest days of the month. Surface winds indicated that stagnation days occurred within the 10 $\mu g/m^3$ $SO_4^{2-}$ isopleth on 22 and 23 January and persisted within the 15 $\mu g/m^3$ isopleth on 24 January over Lake Erie.

The event began on 21 January with localized $SO_4^{2-}$ accumulation to the south of the highest $SO_4^{2-}$ concentrations. On this day, the airflow was northwesterly, forced by a center of low pressure offshore of New England and an advancing high-pressure region from the West. On 22 January, the $SO_4^{2-}$ and $SO_2$ levels accumulated under the high-pressure cell advancing eastward. The area of this accumulation coincided with the region of high emission density in the Ohio River Valley and south of the Great Lakes. A zone of $SO_4^{2-}$ concentrations reaching 10 $\mu g/m^3$ southeast of the Appalachian Mountains was evidently not associated with elevated $SO_2$ concentrations. By the next day, the zone of elevated $SO_4^{2-}$ split into separate areas centered over eastern Kentucky and over the upper Ohio River Valley and Detroit, MI. The high-pressure center was near Huntington, WV. $SO_2$ maxima on this day were concentrated just north of an area in eastern Kentucky and over the upper Ohio River Valley where elevated $SO_4^{2-}$ concentrations were found. The 24-hour average $SO_2$ levels reached a maximum of 60 ppb on 23 January, with $SO_4^{2-}$ concentrations in excess of 20 $\mu g/m^3$. The area of 10 $\mu g/m^3$ $SO_4^{2-}$ concentration also remained southeast of the Appalachians. By 24 January, the event had begun to dissipate, except over Lake Erie, where $SO_4^{2-}$ levels remained elevated and $SO_2$ concentrations increased. The winds shifted from a northwesterly flow to a southwesterly flow between 22 and 24 January. By 24 January, the center of the high-pressure zone had advanced over the middle Atlantic Ocean, and a cold front appearing from the northwest resulted in 0.5 to 1 cm of precipitation starting in the afternoon over the entire SURE region. During this event, the highest $SO_4^{2-}$ and $SO_2$ concentrations tended to be confined to the zones of highest emission density, except for the area southeast of the Appalachians. This situation probably resulted from light and variable winds (less than 2.2 m/s at the surface) from 22 to 24 January in the lower Ohio River Valley. Under these conditions, the zone of elevated $SO_4^{2-}$ levels remained within 300 km of the areas of high emission density. Interestingly, the accumulation of $SO_4^{2-}$ over the lower Ohio River Valley and Tennessee on 22 and 23 January was generally south of the zone of highest $SO_2$ levels. Temperature and humidity increased between 22 and 24 January. Warm air–induced convection with clear skies on 23 January enhanced vertical mixing, which was capped at 1000 m height (relative to mean sea level, MSL), because there was subsidence of cold air aloft. Both $SO_2$ and $SO_4^{2-}$ realized their highest short-term concentration for the event during midday of 23 January under the center of the anticyclone.

Vertical profiles of aerometric parameters measured by aircraft during this event over Duncan Falls, OH, and Lewisburg, WV, suggested accumulation of pollution well below an afternoon inversion base of 1200 m MSL, with the concentration of afternoon $SO_2$ aloft over Duncan Falls many times greater than that found over Lewisburg. The morning profile indicated very high surface concentrations of $SO_2$ below 500 m MSL. Afternoon mixing heights calculated from the Pittsburgh, PA soundings were 650 m MSL on 21 and 24 January, 800 m on 22 January, and 1000 m on 23 January. Temperatures and dew points were very low in these flights, generally below 0°C. Ozone levels between 40 and 50 ppb were measured aloft, which would be typical of clean midcontinent conditions.

During the flights, samples were collected at intervals, providing average $SO_4^{2-}$ and total mass concentrations for RSP at two height ranges, 150–1500 m and 1500–3000 m. These data indicated that the $SO_4^{2-}$ and mass concentrations were highest below 1500 m height as expected, except for Lewisburg, 22 January. Consistent with the ground-level distributions, concentrations of particles aloft were higher at Duncan Falls than at Lewisburg. The concentrations of $SO_4^{2-}$ and RSP mass from the aircraft were similar to, but somewhat lower than, those observed over approximately the same time period at the ground. The averages for the early morning and midday inversion, $SO_2$, particulate, and $SO_4^{2-}$ concentrations increased by midday except on 24 January at Lewisburg.

Without considering the specific wind patterns, the $SO_x$ distributions would suggest a stagnation event similar to the October case. During the January event, the surface winds were light and variable. However, the streakline analysis for 300-m winds revealed a complex localized pattern in which both transport and stagnation may have played a role in distributing $SO_4^{2-}$ over the region. At the beginning of this event, on 21 January, $SO_4^{2-}$ concentrations began to build locally with possible subregional accumulation east and west of the Appalachians. Air generally moved toward the east or the southeast from New England. On 22 January, air stagnation over Tennessee and the lower Ohio River Valley prevailed south and west of the region where the zonal flow persisted. The possibility for transport away from the southwest was of interest for interaction of subregional-scale activity and regional-scale activity.

$SO_2$ increased under the stagnation area in the Ohio River Valley, and $SO_4^{2-}$ concentration increased. Some easterly transport may also have taken place, enhancing local contributions to air over Pennsylvania. By 23 January, the area of stagnation had moved eastward with the center of high pressure. $SO_4^{2-}$ concentrations above 15 $\mu g/m^3$ continued to persist over the southern Ohio River Valley in association with high emission densities in that area, but southerly winds on the lee side of the high-pressure zone appear to have transported $SO_4^{2-}$ throughout the area east of the Mississippi. Despite a persistence of easterly transport to the east of the upper Ohio River Valley, elevated $SO_4^{2-}$ continued to remain there, with the highest concentrations corresponding to the high ambient $SO_2$ and high emission density areas.

The 24 January streakline pattern showed a shift in air movement to a southerly pattern over the entire region, paralleling the frontal disturbance in the West and following the circulation west of the high. A substantial east–west oriented zone of high $SO_2$ concentration remained, but the elevated concentrations had begun to dissipate to the west. The high $SO_4^{2-}$ zone along the east appeared to be localized around metropolitan areas but was also dispersed along the direction of air motion. There was also evidence of northeastward transport east of the Appalachians, perhaps linked with the easterly flow to the West.

On 21 January, the diurnal differences in 3-hour $SO_4^{2-}$ levels was small, and the values were approximately those found over 24 hours in comparable geographic areas. Air transport seemed to account for an apparent widening of $SO_4^{2-}$ distributions around zones of high emission density. A similar pattern appeared on 22 January in the southern part of the area. There may have been a widespread gain in $SO_4^{2-}$ over the Midwest (exemplified by Fort Wayne and Duncan Falls) resulting from $SO_4^{2-}$ production during transport. A comparison of 3-hour–transported $SO_4^{2-}$ with $SO_2$ concentration indicates that on 23 January there appeared to be a net *loss* of $SO_4^{2-}$ compared with the amount ideally transported from Rockport and Fort Wayne but a *gain* (or production) to the east at Scranton. Such a comparison suggested that on 23 January part of the persistent high $SO_4^{2-}$ level near Detroit may have been imported from the South. There were anomalously high $SO_4^{2-}$ concentrations associated with transport, compared with the 24-hour average concentrations in New England. The $SO_2$ concentrations were highest in the areas of net gain, suggesting an imbalance between gain and loss associated with the "local" presence and conversion of precursor gas, $SO_2$.

From the assessment of this January event, Mueller and Hidy (1983) concluded that elevated $SO_4^{2-}$ at levels comparable with those in summer can occur in winter. This event could not be explained by a simple accumulation of pollution during stagnation. The event development evidently involved stagnant air under low inversion conditions, but it was complicated by air mass transport of pollution during its 4-day lifetime. It appears that stagnation played a role in initiating the event in the sense that accumulation of $SO_4^{2-}$ above 10 $\mu g/m^3$ levels began from a broad concentrated baseline at or near this level, and then further accumulation occurred in areas where $SO_2$ emissions were added to polluted air moving into the area. The areas of elevated $SO_2$ concentrations were significantly higher regionally in this winter event than in the October 1977 case discussed previously. The zones of high $SO_4^{2-}$ generally were closely identified with high $SO_2$ regions in both this case and the October event. The presence of cold, relatively dry air and bright sunlight gives an unclear picture of chemical processes taking place. $SO_2$ oxidation seemed to be taking place at rates comparable with summer conditions, but low oxidant concentrations were observed, suggesting weak photochemistry. Since *relative* humidities were high but absolute humidity was low, cold fog and clouds may have been a factor in the chemistry.

### Group 2 Regional Event: 18–23 July 1978

An important example of a summer channeling event was documented in the July 1978 SURE observations. This was the largest extended regional event encountered during the observational periods of the SURE and was one of the largest and most intense observed in the last several years. It represents an excellent example of the group 2 category of regional coverage. Here, the general pattern of a frontal disturbance moving slowly south and eastward with a large high-pressure region dominating the area along the Atlantic Coast is similar to another well-documented case in August 1977. On 17 July, a high-pressure region began to build over Indiana and Ohio behind an advancing cold front (Figure 4.6). A broad area of $SO_4^{2-}$ concentration greater than 10 $\mu g/m^3$ extended northward along the Mississippi River, but the levels recorded in the Ohio River Valley were 8 $\mu g/m^3$ or less as indicated in Figure 4.7. Under the west side of the slowly advancing high-pressure center, both $SO_2$ and $SO_4^{2-}$ accumulated on 18 July (Figures 4.6 and 4.7). This event evidently began with a typical regional situation of air mass stagnation. As the front to the north advanced on 19 July, the winds aloft began to organize into the channeling pattern, forced by the front and the Appalachian Mountains and by the position of the high-pressure center and a frontal disturbance southeast over the Atlantic Ocean. The channeling conditions drew moist warm air from the south across the areas of high emission density and persisted for 5 days, from 19 July to 23 July, but dissipated with the southward advance of a cold front on 24 July.

The accumulation of $SO_2$ persisted with a maximum centered over the upper Ohio River Valley for 3 days, 19–21 July. The maximum 24-hour average concentrations reached 140 ppb at ground level. By 22 July, most of the high $SO_2$ concentrations decreased to levels below 20 ppb, but a zone of high $SO_2$ remained over Pennsylvania, the upper Ohio River Valley, and the mid-Atlantic Coast through 23 July, centered just over the east of the high emission density areas of the upper valley.

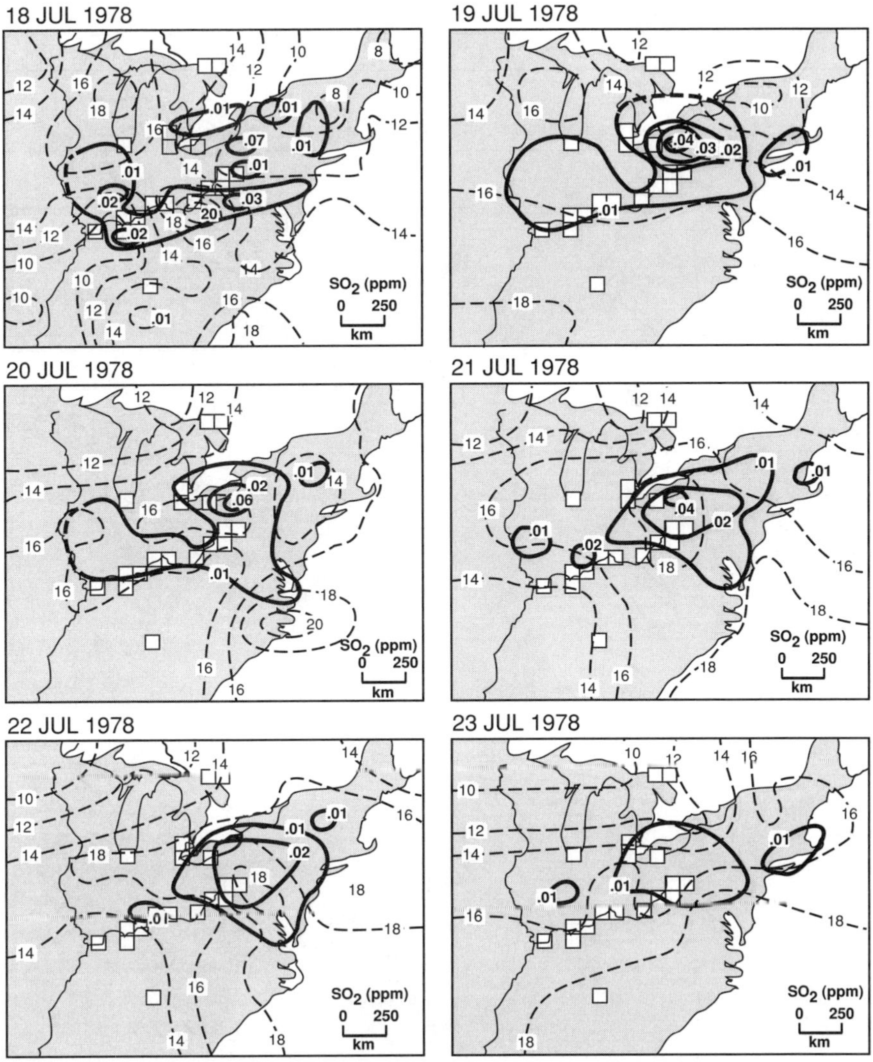


**FIGURE 4.6** Distribution of 24-hour average $SO_2$ concentrations (ppm) for the 18–23 July 1978 period. Light areas are blocks of 80 × 80 km areas with summer average $SO_2$ emissions indicating emissions greater than or equal to 500 mT/d. The dashed lines correspond to lines of constant absolute humidity (g/m$^3$). (From Mueller and Hidy, 1983; reproduced with permission of the Electric Power Research Institute.)

By 19 July, the zone of elevated $SO_4^{2-}$ had changed to a broad ellipse with its major axis oriented east and west. The area of maximum $SO_4^{2-}$ concentrations lay over the region covered by elevated $SO_2$ concentrations. The 20 July $SO_4^{2-}$ distribution was similar, with the ellipse of high

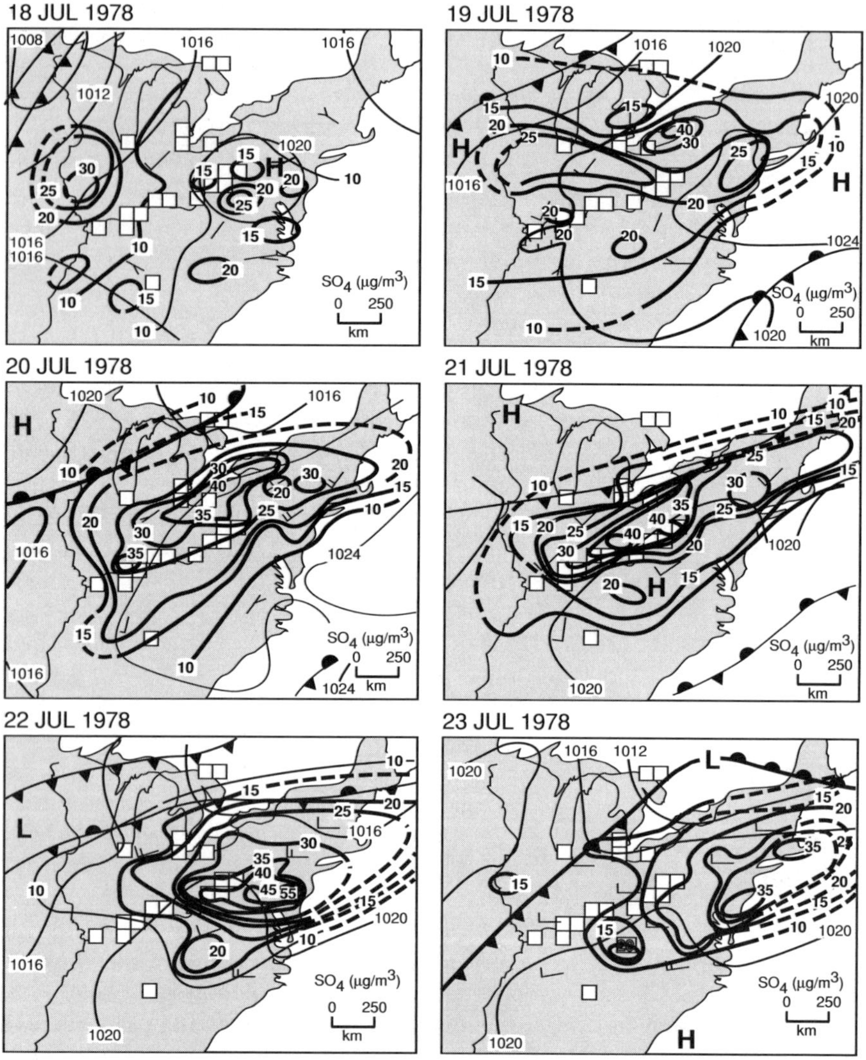


**FIGURE 4.7** Distribution of 24-hour average $SO_4^{2-}$ concentrations ($\mu g/m^3$) for the 18–23 July 1978 period. Light areas are blocks of 80 × 80 km areas with Summer average $SO_2$ average emissions indicating emissions greater than or equal to 500 mT/d. Isobars of surface pressure (mb) are shown in thin lines. High- and low-pressure areas and frontal positions are also indicated. Winds of 300 m are depicted by flags. (From Mueller and Hidy, 1983; Reproduced with permission of the Electric Power Research Institute.)

concentration rotated counterclockwise along the south side of the front. Concentrations of $SO_4^{2-}$ reached the maximum level over a wide area of the upper Ohio River Valley on this day. On 21 July the zone of maximum elevated $SO_4^{2-}$ levels persisted over the zone of high $SO_2$ concentration. At the same time it stretched northeast over New England at levels exceeding 20 $\mu g/m^3$. This high-$SO_4^{2-}$ zone shifted southeastward, centering over Chesapeake Bay on the downwind side of the area of elevated $SO_2$ concentration. The extension of elevated particulate levels over the Atlantic Ocean (where no major $SO_2$ sources exist), as suggested by $SO_4^{2-}$ concentration patterns, was verified in Geostationary Operational Environmental Satellite (GOES) photographs. Analysis of satellite imagery on 22 July (1130 GMT) illustrates clearly the area of dense haze on that day. The zone of haze from the satellite photographs corresponds reasonably well with the $SO_4^{2-}$ isopleth pattern in the SURE. Dense haze was linked with areas of high emission density in the center area eastward to the New York City–Washington, DC strip. The dense haze clearly also extended into the ocean region far to the east. By 23 July, the zone of maximum $SO_4^{2-}$ had advanced further, concentrating along the Atlantic Coast and eastward over the ocean. On 24 July, showery weather associated with the advance of the front dramatically reduced $SO_4^{2-}$ levels to below 10 $\mu g/m^3$ except in the southwest part of Pennsylvania, where they reached 12 $\mu g/m^3$.

The absolute humidity patterns during this event are of interest because they show behavior somewhat distinct from simple moist Gulf air penetration up the Mississippi River Valley and Ohio River Valley. In previous studies reviewed by Hidy *et al.* (1978), such a penetration was documented. However, in the present case, the air was actually drier in the southwestern sector on 18 July, but it was moistened on 19 July by easterly flow from the Atlantic around the high-pressure zone just offshore. On 21 and 22 July, midway through the development of the event, the air to the southwest again dried out, but a very moist region was centered over the upper Ohio River Valley and Pennsylvania.

A time–height history of potential temperature from rawinsonde data at Pittsburgh readily showed the diurnal changes in estimated mixing height, which ranged from nighttime levels as low as 500 m to daytime levels above 1500 m (Figure 4.8). High absolute humidity extended well above the identified mixing height. On 2 days, 20 July and 22 July, a maximum in $SO_4^{2-}$ concentration was observed in the afternoons with daily maximum mixing height. The maximum $SO_4^{2-}$ concentrations on 22 July were associated with low $SO_2$ concentrations. However, maximum $SO_2$ levels at the ground were found early in the event with minimum mixing height. They were highest on 21 July in the presence of the influence of the high-pressure regime advancing eastward. At Duncan Falls,

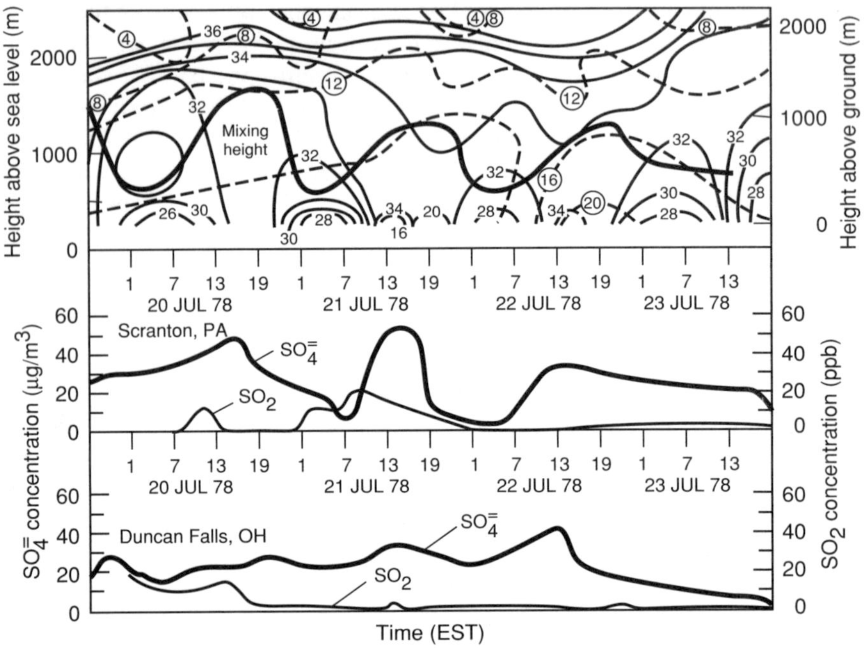


**FIGURE 4.8** Comparison of time–height history of potential temperature (°C in solid lines), absolute humidity (g/m in dashed lines), and mixing height above Pittsburgh, PA and $SO_2$ and $SO_4^{2-}$ at Scranton, PA and Duncan Falls, OH for 20–23 July 1978. Time histories of ground level $SO_2$ (ppb) and $SO_4^{2-}$ ($\mu g/m^3$) concentrations at Scranton and Duncan Falls are shown in the lowest panels. (From Mueller and Hidy, 1983; with permission from Electric Power Research Institute.)

the maximum ground-level $SO_4^{2-}$ concentration was observed with good mixing at midday. The high $SO_2$ levels were recorded only on 20 July, consistent with the stagnation buildup associated with the high-pressure center position.

It is also interesting to note the large difference in absolute humidity aloft between this summer event and the January event. The difference ranges from a factor of 2 or 3 to a factor of 10. If cloud processes were most important in $SO_4^{2-}$ production, relative humidity (liquid water content) would be most important. Absolute humidity differences would be linked more to gas-phase chemistry in terms of $H_2O_2$ production and OH, $HO_2$ linkage (see also Appendix A). Relative humidity was much higher in the January case than in July even though the absolute humidity was low in winter. This would suggest a strong effect of condensed-water chemistry in the January episode relative to gas-phase processes. Such a conclusion is confounded by the possible suppression of oxidation reactions in the presence of ice in winter and the reduced levels of oxidant.

Condensed-phase oxidation to $SO_4^{2-}$ is probably of major importance in summer if cloud formation and following evaporation are significant processors of $SO_2$ and oxidants. Studies of aqueous $SO_2$ oxidation by absorbed $H_2O_2$ indicate a high rate in clouds (Penkett *et al.*, 1979; Hegg and Hobbs, 1982). Studies at Whiteface Mountain have shown dramatic differences in $H_2O_2$ concentration in cloud water between summer and winter, with $H_2O_2$ levels being much lower in winter (Kadlecek *et al.*, 1983). If the peroxide mechanism is seasonally modulated by $H_2O_2$ (or other oxidant production), one could expect substantial differences between summer and winter. This question remains unresolved at present, even with the benefit of experiments superseding the SURE in the 1980s.

Morning and afternoon vertical profiles of aerometric parameters were recorded by the aircraft during this case study at Rockport, Duncan Falls, and Scranton. The profiles indicated that warm moist air was present up to approximately 1700 m, with a drier layer of air above. Appreciable $SO_2$ concentrations, with substantial quantities of $O_3$, were recorded with height over all three stations. The profiles of $b_{sp}$ indicate high particulate concentrations over Scranton, with lower levels to the west over Duncan Falls and Rockport.

The aircraft traverses on 20 and 21 July between 535 and 1270 m (MSL) elevation along paths through the heavily polluted air oriented with the prevailing wind direction are shown in Figure 4.9. The flight patterns were oriented with the prevailing southwesterly wind over the Midwest and the corresponding cross-wind pattern over Pennsylvania and New York. The data provide an important picture of the heterogeneous character of the polluted air aloft in a large regional event. It is interesting to note that high $SO_2$ concentrations persisted in the Louisville, KY, area. On both days, the temperature–dew point difference was about 10°C on the Rockport–Duncan Falls traverse, indicating that moist air was confined near the surface. The Scranton–Galloo Island, NY cross-wind traverse indicated more saturation of air aloft than farther west (Figure 4.10). The dew points and temperature difference on the Scranton–Montague, MA traverse indicated moisture conditions similar to those in the western sector.

The homogeneity of the air mass is indicated by the uniformity in values of temperature and dew point for the flights on 20 and 21 July. Comparison of these and the air quality measures shows that the pollution in the air mass is quite heterogeneous. Yet the measures of secondary pollutant production, $O_3$ and $b_{sp}$, are more uniform than primary emissions of $SO_2$.

$O_3$ concentrations generally ranged from 100 to 150 ppb and were somewhat higher in the afternoon than in the morning on both days, morning and afternoon. $O_3$ levels were particularly high in the afternoon near Louisville and Cincinnati, OH and over eastern Lake Ontario on 20 July. High $O_3$ levels were found near Cincinnati and westward on 21 July;

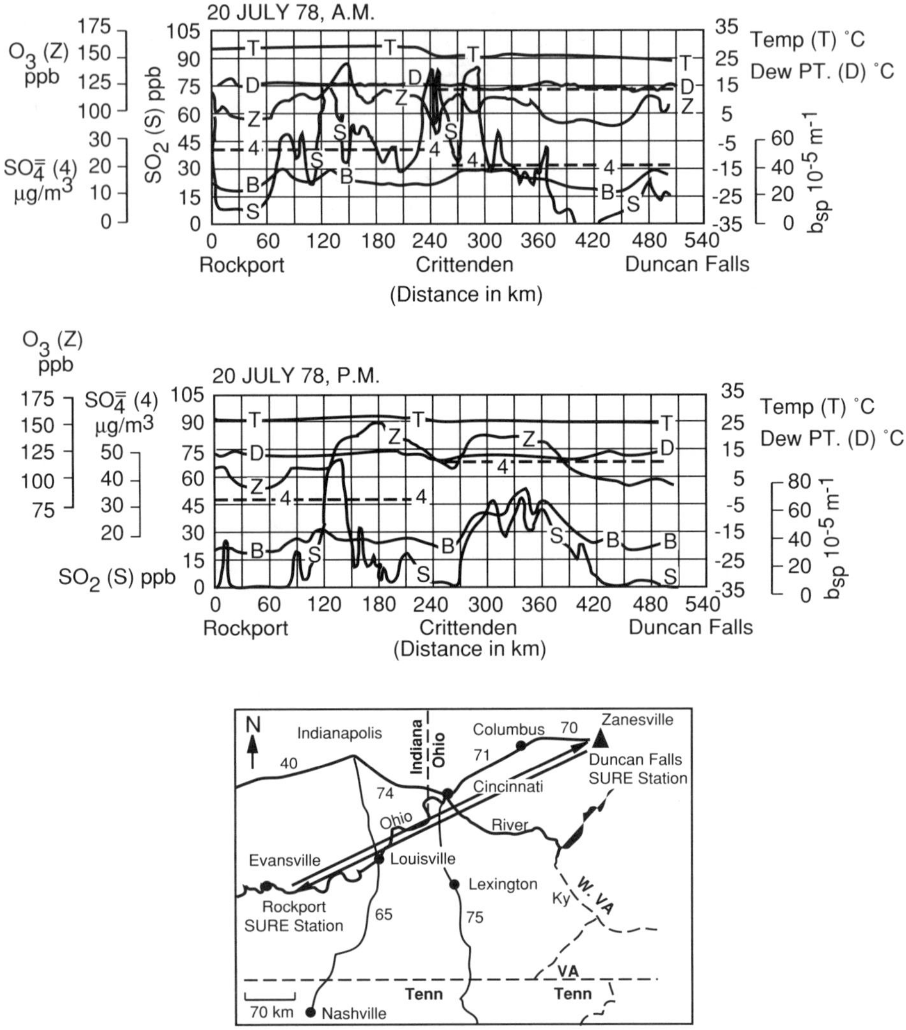


FIGURE 4.9 Aircraft traverses upwind along major axis of elliptical zone of pollution on 20 July 1978. The morning traverse was flown at a mean altitude of 535 m (MSL), the afternoon traverse at a mean altitude of 835 m (MSL). (From Mueller and Hidy, 1983; with permission from the Electric Power Research Institute.)

downwind, $O_3$ tended to decrease northeastward from middle Pennsylvania. The $b_{sp}$ levels tended to be relatively uniform at 30–50 $\times$ $10^{-5}$ $m^{-1}$ in the morning on both days. On 21 July, the afternoon pass showed a systematic decrease in $b_{sp}$ values with a traverse across the downwind zone of the polluted air. Similarly, the afternoon traverse from Scranton

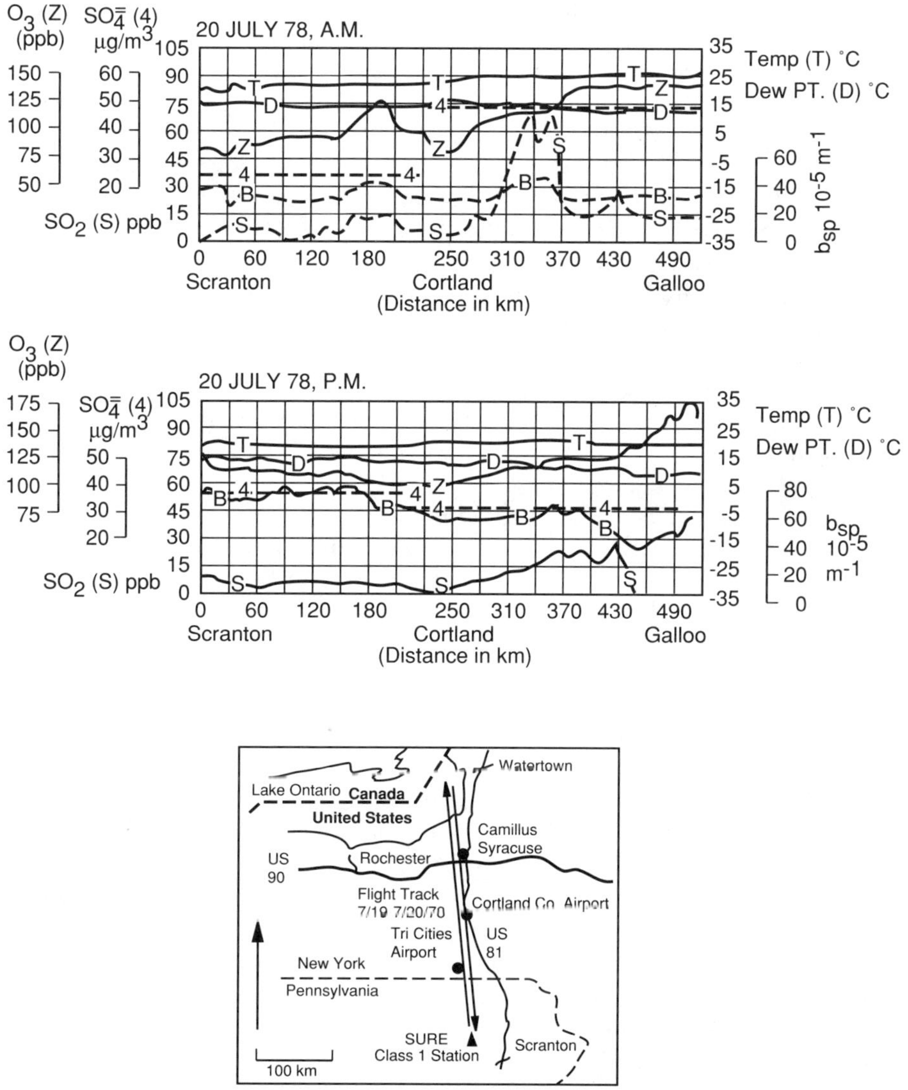


FIGURE 4.10 Aircraft traverses along minor axis of elliptical zone of pollution on 20 July. Flight pattern was between Scranton, PA and Galoo Island, NY at a mean morning altitude of 840 m (MSL) and a mean afternoon altitude of 1150 m (MSL). (From Mueller and Hidy, 1983; reproduced with permission from the Electric Power Research Institute.)

to Galloo Island showed a marked decrease across the zone of maximum $SO_4^{2-}$ to lower $SO_4^{2-}$ levels at the ground. The Rockport–Duncan Falls $b_{sp}$ data suggested a somewhat higher fine particulate level, with higher $b_{sp}$ in the central Ohio River Valley than on the western side. This trend was not as pronounced as the differences farther east.

Although the horizontal resolution was poor, $SO_4^{2-}$ concentrations observed in the aircraft traverses were of interest. First, the concentrations aloft were uniformly high, exceeding 25 $\mu g/m^3$, which compared well with ground-level concentrations on both days. As in the Rockport–Duncan Falls traverses, the change from morning to afternoon $SO_4^{2-}$ levels differed on the 2 days. There was an afternoon increase over the eastern leg on 20 July but an afternoon decrease on 21 July. In contrast, on 20 July, there was an *increase* in $SO_4^{2-}$ from morning to afternoon on the Scranton side but a decrease on the Lake Ontario side, suggesting an accumulation in the highest $SO_4^{2-}$ region. An analogous effect was found in the Rockport–Duncan Falls traverse, on 21 July.

The apparent transport of $SO_4^{2-}$ in polluted air was observed in the 21 July traverse between Scranton and Montague. In this case, the $SO_2$ concentrations were uniformly low. However, the $SO_4^{2-}$ concentrations remained uniformly high. With the eastward movement of the polluted air mass and production by chemical reaction, $SO_4^{2-}$ levels increased significantly between Scranton and Montague in the afternoon, while concentrations in the western part of these traverses remained approximately constant. Here mixing and deposition processes evidently balanced production.

The observations from the aircraft traverses showed that the pollutant concentration distributions aloft are complex during the large regional-scale events. The data suggested a relatively uniform distribution of $SO_4^{2-}$ particles aloft. $O_3$ and fine particulates, as indicated by $b_{sp}$, had a uniformly high baseline, but perturbation of pollutant maxima was clearly identifiable, particularly over zones enriched by pollutant emissions. However, the data indicated regionwide changes in concentration distributions aloft, which are qualitatively consistent with the ground-level measurements.

The mixing height distribution is a distinctive feature of a channeling event as indicated in Figure 4.11. Estimates were made of the morning and afternoon mixing height values for 21 July along the major axis of the ellipse of high pollution along the prevailing wind direction. A profile was also taken approximately along the minor axis of the ellipse (across the dominant wind direction). Cross sections were made of potential temperature, absolute humidity, calculated mixing height, and ground-level $SO_4^{2-}$. Comparison between the morning and afternoon showed a rise in mixing height by late afternoon in the west, with a corresponding dilution of $SO_4^{2-}$ concentrations at the ground. Perhaps more striking was the cross-wind

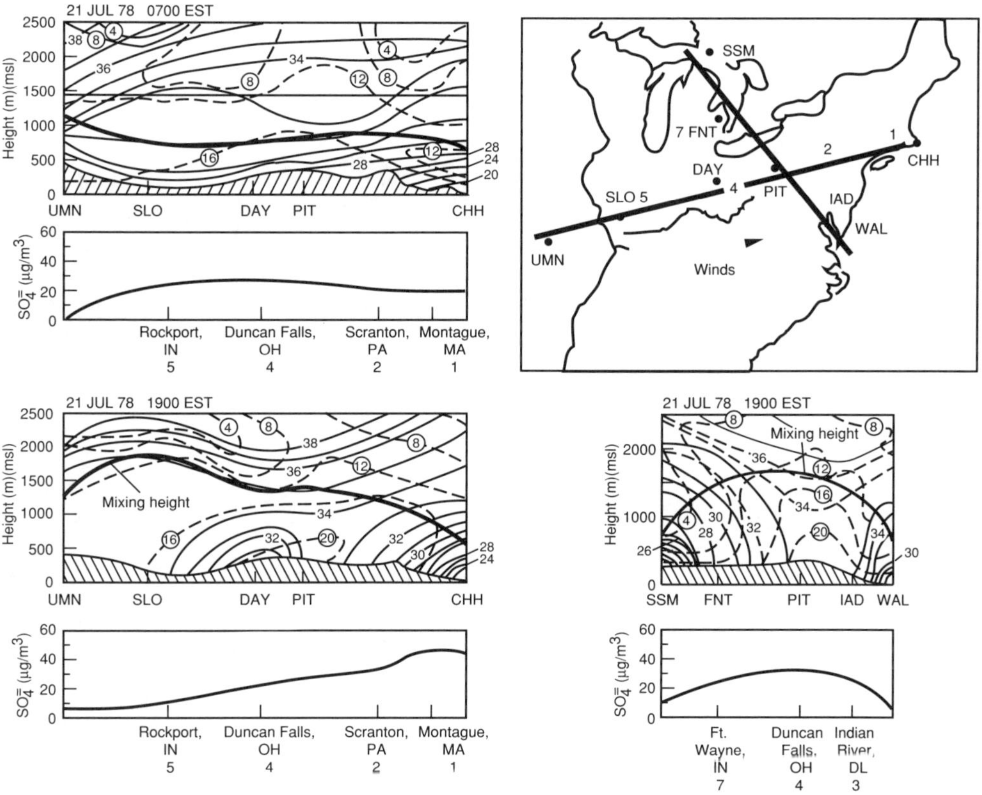


FIGURE 4.11 Potential temperature (°C in solid lines), absolute humidity (g/m³ in dashed lines), estimated mixing height, and observed sulfate distributions for cross sections in the morning and afternoon, 21 July 1978. Data are taken from rawinsonde stations noted in the map in the upper right panel. (From Mueller and Hidy, 1983, with permission of the Electric Power Research Institute.)

cross section (Figure 4.11), which showed the bulge in the mixing height, with a late afternoon maximum in the center of the polluted zone or roughly along the center of the channel of air flowing eastward. This feature, which had not been observed before, is an important one in regional pollution events. The channeling effect was well defined on the morning of 21 July but became less pronounced with the increase in height by afternoon. On the morning of 22 July the mixing height had decreased on the east and west, but it rose again by midday. The channeling structure persisted and was particularly pronounced on the east side of the Appalachians through 22 July.

Completing the picture of this July series are the streaklines suggesting air mass transport (Figure 4.12). They were again calculated for the sur-

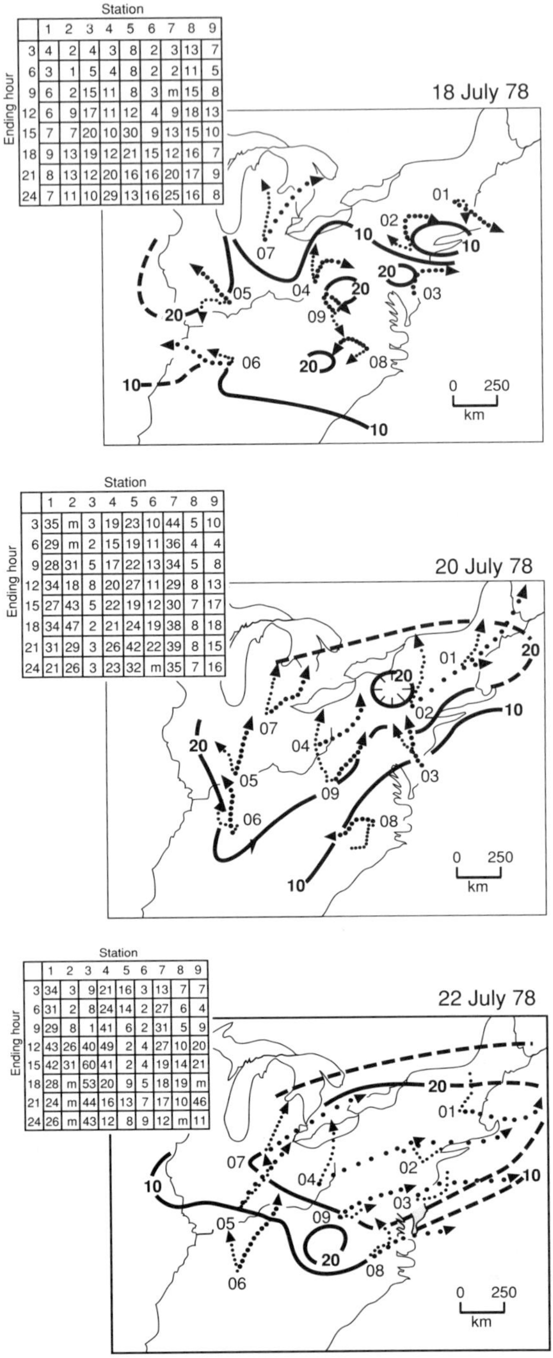


**18 July 78** (Station)

| Ending hour | 1 | 2 | 3 | 4 | 5 | 6 | 7 | 8 | 9 |
|---|---|---|---|---|---|---|---|---|---|
| 3 | 4 | 2 | 4 | 3 | 8 | 2 | 3 | 13 | 7 |
| 6 | 3 | 1 | 5 | 4 | 8 | 2 | 2 | 11 | 5 |
| 9 | 6 | 2 | 15 | 11 | 8 | 3 | m | 15 | 8 |
| 12 | 6 | 9 | 17 | 11 | 12 | 4 | 9 | 18 | 13 |
| 15 | 7 | 7 | 20 | 10 | 30 | 9 | 13 | 15 | 10 |
| 18 | 9 | 13 | 19 | 12 | 21 | 15 | 12 | 16 | 7 |
| 21 | 8 | 13 | 12 | 20 | 16 | 16 | 20 | 17 | 9 |
| 24 | 7 | 11 | 10 | 29 | 13 | 16 | 25 | 16 | 8 |

**20 July 78** (Station)

| Ending hour | 1 | 2 | 3 | 4 | 5 | 6 | 7 | 8 | 9 |
|---|---|---|---|---|---|---|---|---|---|
| 3 | 35 | m | 3 | 19 | 23 | 10 | 44 | 5 | 10 |
| 6 | 29 | m | 2 | 15 | 19 | 11 | 36 | 4 | 4 |
| 9 | 28 | 31 | 5 | 17 | 22 | 13 | 34 | 5 | 8 |
| 12 | 34 | 18 | 8 | 20 | 27 | 11 | 29 | 8 | 13 |
| 15 | 27 | 43 | 5 | 22 | 19 | 12 | 30 | 7 | 17 |
| 18 | 34 | 47 | 2 | 21 | 24 | 19 | 38 | 8 | 18 |
| 21 | 31 | 29 | 3 | 26 | 42 | 22 | 39 | 8 | 15 |
| 24 | 21 | 26 | 3 | 23 | 32 | m | 35 | 7 | 16 |

**22 July 78** (Station)

| Ending hour | 1 | 2 | 3 | 4 | 5 | 6 | 7 | 8 | 9 |
|---|---|---|---|---|---|---|---|---|---|
| 3 | 34 | 3 | 9 | 21 | 16 | 3 | 13 | 7 | 7 |
| 6 | 31 | 2 | 8 | 24 | 14 | 2 | 27 | 6 | 4 |
| 9 | 29 | 8 | 1 | 41 | 6 | 2 | 31 | 5 | 9 |
| 12 | 43 | 26 | 40 | 49 | 2 | 4 | 27 | 10 | 20 |
| 15 | 42 | 31 | 60 | 41 | 2 | 4 | 19 | 14 | 21 |
| 18 | 28 | m | 53 | 20 | 9 | 5 | 18 | 19 | m |
| 21 | 24 | m | 44 | 16 | 13 | 7 | 17 | 10 | 46 |
| 24 | 26 | m | 43 | 12 | 8 | 9 | 12 | m | 11 |

FIGURE 4.12 Surface (short arrow) and 300 m AGL (long arrow) streaklines originating at each class I station for the 18–23 July 1978 period. Isopleths correspond to 24-hour average $SO_4^{2-}$ concentrations of 10 and 20 $\mu g/m^3$. Three-hour average $SO_4^{2-}$ observations by hour-station are listed in the inset (m = missing data). (From Mueller and Hidy, 1983; reproduced with permission of the Electric Power Research Institute.)

Station

| Ending hour | 1 | 2 | 3 | 4 | 5 | 6 | 7 | 8 | 9 |
|---|---|---|---|---|---|---|---|---|---|
| 3 | 7 | 11 | 8 | 22 | 11 | 12 | 15 | 12 | 7 |
| 6 | 7 | 7 | 5 | 25 | 9 | 11 | 13 | 8 | 8 |
| 9 | 8 | 14 | 11 | 20 | 9 | 11 | 18 | 9 | 11 |
| 12 | 8 | 26 | 17 | 18 | 12 | 11 | 23 | 12 | 16 |
| 15 | 7 | 23 | 17 | 17 | 20 | 19 | 43 | 14 | 22 |
| 18 | 8 | 24 | 12 | 13 | 23 | 18 | 44 | 14 | 18 |
| 21 | 14 | 24 | 5 | 13 | 53 | 12 | 36 | 13 | 16 |
| 24 | 28 | 27 | 4 | 28 | 42 | 10 | 45 | 10 | 15 |

19 July 78

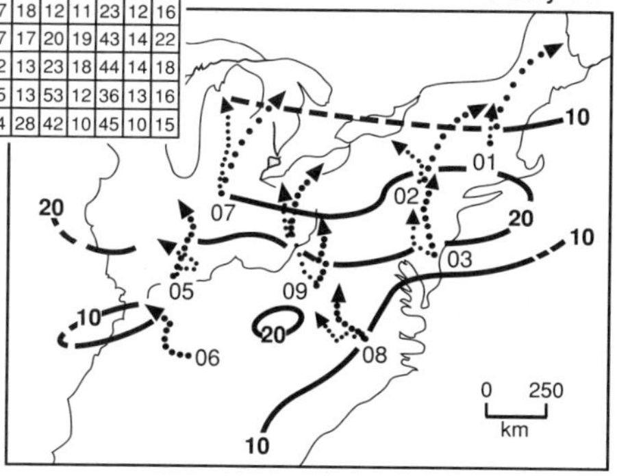


Station

| Ending hour | 1 | 2 | 3 | 4 | 5 | 6 | 7 | 8 | 9 |
|---|---|---|---|---|---|---|---|---|---|
| 3 | 16 | 19 | 2 | 21 | 31 | 16 | 25 | 4 | 9 |
| 6 | 14 | 18 | 2 | 19 | 26 | 11 | 14 | 3 | 6 |
| 9 | 17 | 4 | 3 | 19 | 21 | 12 | 28 | 6 | 11 |
| 12 | 30 | 33 | 7 | 27 | 15 | 6 | 31 | 8 | 18 |
| 15 | 37 | 34 | 10 | 33 | 24 | 6 | 28 | 5 | 22 |
| 18 | 41 | 50 | 8 | 31 | 24 | 8 | 14 | 6 | 15 |
| 21 | 43 | 11 | 8 | 27 | 22 | 5 | 7 | 4 | 22 |
| 24 | 39 | 7 | 11 | 22 | 18 | 4 | 8 | m | 20 |

21 July 78

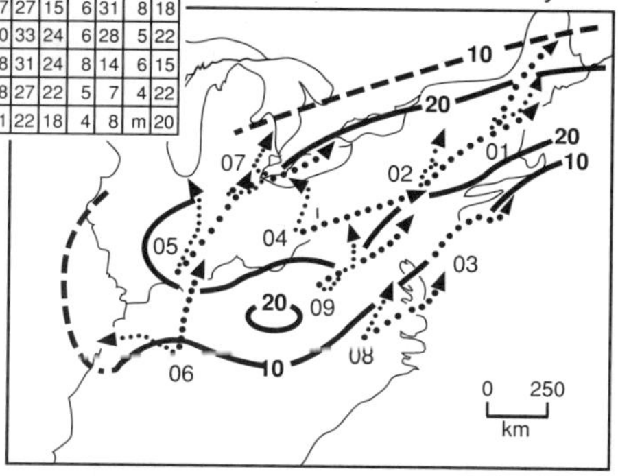


Station

| Ending hour | 1 | 2 | 3 | 4 | 5 | 6 | 7 | 8 | 9 |
|---|---|---|---|---|---|---|---|---|---|
| 3 | 25 | 24 | 30 | 11 | 8 | 7 | 6 | 5 | 3 |
| 6 | 23 | m | 17 | 9 | 6 | 6 | 4 | 5 | 2 |
| 9 | 34 | 93 | 31 | 8 | 4 | 4 | 5 | 4 | 8 |
| 12 | 39 | 23 | 42 | 7 | 4 | 3 | 6 | 5 | 12 |
| 15 | 24 | 11 | 42 | 6 | 6 | 5 | 3 | 9 | 9 |
| 18 | 23 | 18 | 35 | 7 | 6 | 3 | 3 | 10 | 9 |
| 21 | 21 | 11 | 34 | 3 | 6 | 3 | 5 | 7 | m |
| 24 | 15 | 9 | 28 | 4 | 5 | 3 | 4 | 6 | m |

23 July 78

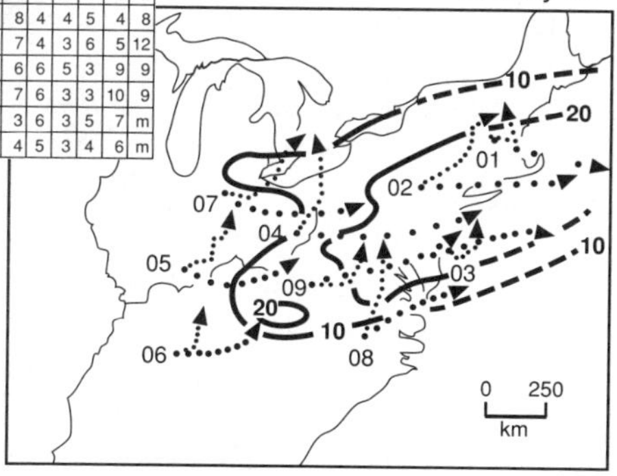


FIGURE 4.12 (*continued*)

face winds and winds at 300 m AGL (Mueller and Hidy, 1983). The series began on 18 July with general stagnation conditions dominated by the anticyclone. The next day, the air movement was weak in the south part of the sector but strengthened with southerly flow toward the front in the north. July 20 was similar to 19 July, but the transport became better defined toward the northeast throughout the Ohio River Valley, Pennsylvania, and New England. This pattern was reinforced and strengthened on 21 July and then broadened by 22 July to encompass the southern areas. By 22 July, the flow veered to become zonal east of the Ohio River. By 23 July, the transport was essentially westerly as the front to the north advanced.

Comparison of the 3-hour $SO_4^{2-}$ levels along these streaklines suggested that the concentrations were consistent with localized accumulation over a 24-hour period. On 19 July, there appeared to be net depletion along the northern edge of the zone of $SO_4^{2-}$ maximum, with maintenance of the high-$SO_4^{2-}$ zone by weak transport from the area of high emission density. The area of high $SO_4^{2-}$ along the east coast over New York appeared to be of local origin rather than the result of transport. Later, on 20 July, the trends for transport of high $SO_4^{2-}$ concentrations were consistent with the extension of the isopleths northeastward well beyond the influence of the areas of high emission density. Transport evidently supported the maintenance of elevated concentrations of $SO_4^{2-}$ in the greater Northeast on 21 July; the 3-hour $SO_4^{2-}$ concentrations agreed reasonably well with the zone of 24-hour maximum levels. On 22 and 23 July, $SO_4^{2-}$ behavior retained its transport-dominated features. By 23 July, transport was sweeping cleaner air eastward, while the dirty air was transported offshore and diluted. Of all the cases studied, this, the largest regional event, offers the strongest evidence for (1) initiation of the event by accumulation under stagnation conditions, (2) regionwide spreading with airflow channeling with long-range transport over $>1000$ km a major feature, (3) net $SO_4^{2-}$ production overcoming dilution and depositional losses, and (4) relatively high $SO_2$ levels lasting at ground level and aloft for 3 to 4 days in the initial stages of the episode.

### Summary

The examples discussed indicate the importance of meteorological conditions in the development of regional events with 24-hour average $SO_4^{2-}$ concentrations greater than 10 $\mu g/m^3$. An operational definition for regional $SO_4^{2-}$ events is proposed that applies to $SO_4^{2-}$ behavior in the greater northeastern United States. Regional events have large, relatively homogeneous areas of $SO_4^{2-}$ greater than 15 and 20 $\mu g/m^3$, which last from about 3 days to more than a week and take place during approximately one-fifth of the year, usually in summer and winter months. Smaller sub-

TABLE 4.3 **Meteorological Regimes Conducive to Sulfate Accumulation and Transport**

| Type | Synoptic pattern | Seasonal preference | Transport winds (300 m) | Afternoon mixing heights | Thermodynamics |
|---|---|---|---|---|---|
| Stagnation | Dominant continental high-pressure cell (surface and aloft) over the Appalachians | Any season | Light (<5 m/s) | Low (<1000 m) | Any temperature, general moist |
| Channeling or ducting | Bermuda High<br>Stationary or slowly advancing frontal system over the Great Lakes eastward to the Atlantic Coast | Summer | Moderate SW winds (7–12 m/s) | High (1500–2000 m) | Warm, moist |

regional events with more localized elevated concentrations are shorter in duration and occur throughout the year about 15% of the days.

There are two extremes of events involving elevation of airborne $SO_4^{2-}$ concentration; one is an outgrowth of the other. The events are described in terms of regionwide stagnation of air masses and channeling by a frontal system exemplifying long-range transport of polluted air across ENA. The former involves widespread accumulation over thousands of square kilometers under a stagnating anticyclone supported by high barometric pressure. The latter begins with a stagnation event and then evolves into channeling of moist, warm air along a frontal disturbance that advances into the stagnated air over a zone of high emission density in the Midwest and passes eastward toward the Atlantic Coast. The channeling cases have readily distinguishable features of long-range transport extending up to a 1000 km away from zones of high emission density, especially along the Ohio River area of industrialization.

''Idealized'' characteristics of the two classes of regional events are summarized in Table 4.3. The regional-scale event involving widespread accumulation of $SO_4^{2-}$ in stagnant air with light winds and low mixing heights represents a ''textbook'' picture of pollution meteorology. In contrast, the channeling situation is quite distinct, involving the maintenance of regionally high $SO_4^{2-}$ levels with moderate wind speeds, particularly from the southwest and west, and relatively high mixing heights. Channeling with persistent winds parallel to frontal systems over distances exceeding 500 km appears to occur mainly in summer or early fall when the cold fronts are oriented from east to west rather than from north to south in winter, with a deep trough. The presence of warm, moist conditions in air flowing northeastward appears to be conducive to accelerating $SO_2$ oxidation. Thus, channeling can maintain elevated $SO_4^{2-}$ concentrations at relatively low $SO_2$ concentrations by enhanced chemical reactivity of the air. The presence of high $O_3$ concentrations with $SO_4^{2-}$ in summer events serves as a surrogate for high chemical reactivity. The descriptive analysis has not distinguished between the significance of homogeneous gas-phase reactions and that of aqueous reactions for summer conditions involving clouds. Reduced photochemical activity in winter suggests that aqueous reactions in clouds or heterogeneous-particle reactions may be dominant in winter cases. But this may not be consistent with reduced $SO_2$ absorption and oxidation on ice particles during winter conditions.

## PREVAILING CLIMATOLOGICAL FEATURES LINK TO ANTICYCLONIC SYSTEMS

Analyses of $SO_4^{2-}$ events documented through the SURE results have been expanded by other studies, including $O_3$ event analyses (Wolff and Lioy, 1980; Vukovich, 1979). Throughout their investigations, these researchers found evidence for distinguishing between patterns of chemically reactive precursor emissions and end-product behavior. These distinctions and the persistent link of spatial patterns with regional meteorological features suggest that a descriptive analysis of prevailing chemical conditions is possible.

Vukovich and Fishman (1986) reported such a climatological assessment of $SO_2$ and $O_3$ behavior. They compared nonurban $SO_2$ and $O_3$ data for several summers in ENA and discussed the data in terms of regional-scale systems with high barometric pressure.

The data indicate that in the summer, areal regimes of high $O_3$ (secondary pollutant) are significantly more variable from year to year with respect to their location than $SO_2$ (a primary pollutant) distributions. Centers of high $O_3$ concentration are seldom found in the same locations. However, the centers of high $SO_2$ concentration were usually found in the region from Illinois eastward to the coast and in the latitude belt of 35°N to 40°N (Ohio River Valley). This region is characterized by high-density $SO_2$ sources. The $SO_2$ concentration pattern tends to be consistent with the observed distribution of $SO_4^{2-}$ concentrations in air and in rainwater during the same period, except that $SO_4^{2-}$ is more widely and uniformly distributed than $SO_2$ (Hidy *et al.*, 1985). The relationship between the observed $SO_4^{2-}$ distribution in rainwater and the distribution of $SO_2$ sources and its subsequent distribution is not surprising because conversion of $SO_2$ to airborne $SO_4^{2-}$ is on the order of 1 day, and wet deposition should be rapid based on the relatively high scavenging efficiency of $SO_4^{2-}$ (Appendix A).

Secondary centers of high $SO_2$ concentration are found over Iowa, Missouri, Oklahoma, and Kansas. There are also significant sources of $SO_2$ in these areas.

Vukovich and Fishman (1986) pointed out that the area affected by high $O_3$ concentrations (>40 ppb) varied from 2 to 5 × $10^6$ $km^2$, whereas the area affected by $SO_2$ concentrations (>10 ppb) varied from 0.3 to 1.3 × $10^6$ $km^2$. Thus, the high $O_3$ concentrations tend to affect an area about 5 to 10 times greater than that for $SO_2$.

In Vukovich and Fishman's analysis, the summer surface pressure distribution over the period 1977–1982 did not vary substantially from year to year. The major high-pressure center associated with a Bermuda high-pressure system appeared over the southeastern portions of the United States. When a large number of migratory high-pressure systems move across the northern portions of ENA, the monthly average surface pressure analysis showed that the Bermuda high-pressure system penetrated to the northwest. An example of this occurred in the July 1980 case when a secondary high-pressure center developed over Ohio, which resulted from five migratory high-pressure systems that moved across that region in that month.

Analysis of the $O_3$ and $SO_2$ distributions and of the path of the high-pressure systems over the period suggested that the variations in the $O_3$ and, to a lesser extent, $SO_2$ distributions appear to be associated with the path and frequency of the migratory high-pressure systems. This correlation was surprising because earlier research had shown that high concentrations of $O_3$ are directly associated with migratory high-pressure systems. Because $SO_2$ is a primary pollutant and high concentrations of $SO_2$ are usually found near sources, spatial variability similar to that for $O_3$ was not expected. However, variations in the magnitude of the $SO_2$ concentrations over source regions have been noted to be associated with the passage of high-pressure systems. This conclusion is consistent with the observations reported by Mueller and Hidy (1983).

The relationship between the path of high-pressure systems and the mean $O_3$ concentration variations are well represented in the 3-year data set compiled by Vukovich and Fishman (1986). In the summers of 1979 and 1981, most of the migratory high-pressure systems moved across the northern part of the eastern two-thirds of the United Sates and the major region of high $O_3$ concentration was found in the northern portions of the United States, stretching from northern Illinois eastward into New England. However, in both years high $O_3$ was also found in the southern part of the United States, and major high-pressure systems were observed to move into the southern states. The $SO_2$ concentration distributions in July and August of 1979 and 1981 have high centers over the regions where there is a high density of $SO_2$ sources (i.e., east-central United States). The location of the center of high $SO_2$ concentration shifted eastward or westward in that zone over the 2 years but, for the most part, remained within the high emission density.

In the summers of 1977, 1978, and 1980, most of the migratory high-pressure systems moved across the central and southern parts of the eastern two-thirds of the United States. In many cases, the systems became stationary in these areas, weakened, and eventually dissipated. Vukovich and Fishman note that these characteristics had a profound effect on the $O_3$ and the $SO_2$ distribution. Centers of high $O_3$ shifted toward the southern portion of the high-density $SO_2$ source region.

Perhaps the most significant difference between the $O_3$ and $SO_2$ distributions is the fact that the spatial behavior of $O_3$ is not directly tied to the distribution of the precursor emissions of the species most responsible for $O_3$ formation (i.e., HCs and $NO_x$). The distribution of HC emissions is closely related to population density and is high in urban areas, including the "northeast corridor," which extends from Washington, DC through Boston. Although this region is noted for high $O_3$, in the 10 months analyzed, centers of high $O_3$ values were found over this general area only in July 1979 and August 1981. This central region, however, is one of relatively high $O_3$ during the 5-year period. The distribution of $NO_x$ emissions more closely resembles the emission pattern of $SO_2$, which is distributed inland around the upper Ohio River Valley. Centers of high $O_3$ concentrations were found inland in July and August 1978 and 1979. Thus, the distribution of $O_3$ at the surface is not necessarily linked to a major source of $NO_x$ and appears to be dependent on regional meteorological factors other than the emissions of the precursors that lead to the photochemical production of $O_3$. One of the most important of these factors appears to be the travel path of major anticyclones. This conclusion is consistent with the association of large-scale anticyclonic systems with regional-scale phenomena that affect both $O_3$ and $SO_4^{2-}$ distributions over ENA.

## Air Mass Classification of Aerometric Data

One way to delineate differences in synoptic-scale meteorology is to generalize the descriptive analysis of Mueller and Hidy (1983) and Vukovich and Fishman (1986) using air mass categorization. Air mass classification can be organized in terms of conditions associated with regimes around areas of high barometric pressure (anticyclonic circulation). These regimes basically change with the formation of stagnation areas under high-pressure zones in such large-scale meteorological disturbances. Hidy *et al.* (1976) first used this approach at their analysis of data collected for the SURE design. This work resulted in the hypothesis that regional events of high $SO_4^{2-}$ concentration are associated with northeastward maritime tropical air penetration, while low $SO_4^{2-}$ concentrations and clean air are identified with continental polar air masses. Baskett (reported in Mueller and Hidy, 1983) extended the earlier work by classifying each day of year during the SURE program according to air mass to describe the air quality and meteorological conditions occurring during the wide variety of measured regional $SO_4^{2-}$ concentrations.

The use of air mass classification to characterize $SO_4^{2-}$ variability involves certain key associations between types of air masses and the intensity of atmospheric processes. Hidy *et al.* (1976) and Mueller and Hidy (1983) made certain assumptions as the basis for their method, but did not discuss them explicitly. A more elaborate attempt to classify $SO_4^{2-}$ variability in terms of air mass characteristics can now be made from descriptive knowledge developed before and during the SURE.

Five basic processes are believed to influence $SO_2$ and $SO_4^{2-}$ concentrations in the atmosphere.

1. *Local dispersion* by mixing should be minimal for the preservation of geographically widespread, elevated concentrations in $SO_4^{2-}$ for periods of days.

2. *Winds* over large geographic areas at heights in the range of emission injection should be persistent in direction and intensity to provide for transport over distances of several hundred kilometers.
3. *Photochemical oxidation potential* should be high, with sunshine and extended fair weather, to permit an excess of $SO_4^{2-}$ production over losses by S deposition.
4. If the photochemical processes diminish, potential for aqueous-phase *oxidation* processes involving wet suspended particulates should be high for $SO_4^{2-}$ production to exceed losses.
5. The loss of $SO_2$ by *surface deposition* should be minimal over large areas to maintain high $SO_2$ concentrations.

Each air mass classification based on synoptic analysis has one or more of these characteristic features maximized or minimized by weather. Ideally, regional-scale $SO_4^{2-}$ pollution events should occur in atmospheric conditions in which local dispersion is minimal, persistent winds are prevalent for transport, oxidation potential is maximum, but the loss at the ground is weakest. In general, this combination is probably not realized in nature. The properties of behavior used for air mass classification can determine when weather conditions should be conducive to $SO_4^{2-}$ accumulation given the five criteria listed above. Table 4.4 summarizes the strength of activity for each of the five processes. Figure 4.13 also shows them schematically in relation to a high pressure–anticyclonic circulation. These estimates are derived from earlier experience using this approach (Hidy *et al.*, 1976, 1978). As an example, maritime tropical (mT) air condi-

TABLE 4.4 **Hypothesized Relative Strength of Processes Important in Determining $SO_4^{2-}$ Concentrations for the Air Mass Types**[a,b]

| **Air mass type**[c] | **Local dispersion** | **Regional $SO_4^{2-}$ transport** | **Photochemical oxidation potential** | **Heterogeneous oxidation potential** | **$SO_x$ deposition** |
|---|---|---|---|---|---|
| cPk | + | − | 0 | − | − |
| cP2 | − | − | + | 0 | 0 |
| cPw | − | 0 | + | + | 0 |
| Tr | + | − | − | 0 | + |
| mT | + | + | + | +(?) | + |

[a] From Mueller and Hidy, 1983; reproduced with permission of the Electric Power Research Institute.

[b] The + symbol denotes higher than average activity; 0 denotes average activity; − denotes lower than average activity

[c] Air mass definitions are shown schematically in Figure 4.13.

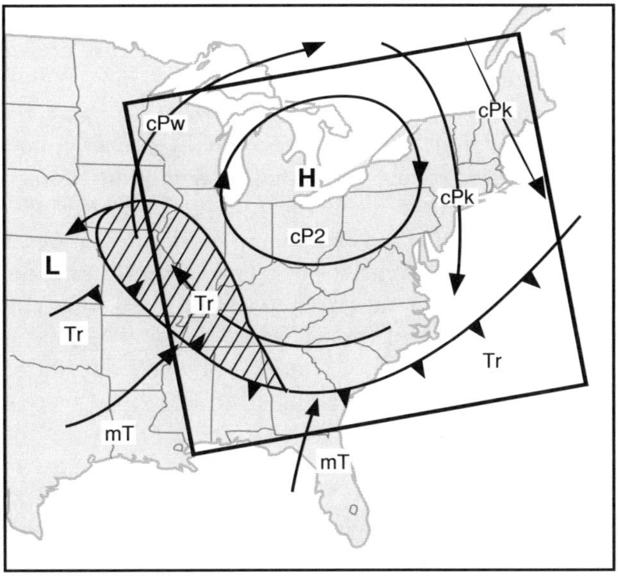


FIGURE 4.13 Air mass categories used for classifying regional air quality. (From Mueller and Hidy, 1983; reproduced with permission of the Electric Power Reserach Institute.)

tions have a mixed dynamic pattern. The continental polar cold (cPk) condition, in another example, is strong in local dispersion by wind action but average or below in other aspects because of cool dry air, deep mixing, and minimal summer occurrence. Its frequent occurrence during winter is associated with low deposition rates over ground with partially dormant vegetation or snow cover. The other continental polar (cP) air mass categories should involve high photochemical potential with fair weather or surface fog and low local dispersion potential. Transitional (Tr) air masses should offer high local dispersion and deposition with deep mixing and precipitation but minimal organized transport and photochemical activity. Of the five classifications given, the cP2 and cPw conditions appear to offer the properties most likely to yield accumulation of $SO_4^{2-}$, but mT conditions are the only ones that have been found to be linked with regionally persistent winds (Hidy *et al.*, 1978).

The link between indexes of synoptic air mass classification and processes affecting regional $SO_4^{2-}$ behavior is shown to have a logical basis. The success of this approach was tested on the SURE data set. The first step was to define the aerometric data base that would be used from each station. The data base consisted of the 24-hour particulate data from the particulate samplers (including TSP, $SO_4^{2-}$, and $NO_3^-$ concentrations) and the 24-hour average gas concentrations of $SO_2$, $NO_x$, and $O_3$. These data were combined with seven meteorological parameters measured at nearby FAA/NWS (Airways) weather stations that were considered most representative of the CI stations. Of the parameters collected on an hourly basis, 24-hour average wind speed, pressure, and total daily precipitation were calculated. The daily maximum and minimum temperatures were selected, and their corresponding dew point depressions were included. From these values, both the relative and absolute humidities were computed at the time of maximum and minimum temperature to give an indication of the diurnal variation in atmospheric moisture content. The data were used to build an aerometric data base for each day of the measurement year for each of the SURE stations. The next step was to categorize the air masses at each station for each day during a 1-year record. The five air mass categories listed in Table 4.4 were used. Each class, except transitional, was intended to have relatively homogeneous temperature and humidity characteristics throughout the area it covered.

The first class (cPk) is generally cool and dry. *Colder* indicates that the air is colder than the land over which it is traveling. This condition signifies the clear air just ahead of, or on the front side of, a high-pressure system. The leading edge of an anticyclone, sometimes distinguished as cP1, is included here in the cPk class. Northerly winds usually dominate with greater than average speeds, due to the upper-level ridge ahead of a high-pressure system. (*Average* is associated with the annual average conditions.) Because polar outbreaks create cPk air masses, one would expect this class to occur most often during winter and at more northerly stations. Occasional instabilities may follow a frontal passage resulting in light snow flurries or showers in cPk air.

Second in progression with the western movement of anticyclones comes the center of a standardized continental high-pressure zone (high, cP2). Although temperature in this class can vary widely depending on season, upper-level support (depth), and latitude of origin, the centers of highs are generally colder than the seasonal average because of radiational cooling under clear skies. Precipitation seldom occurs within the center of a high. The most outstanding feature is poor ventilation associated with low wind speeds and low mixing depths.

Following the center of a high in a normal progression comes the cPw, or continental polar warmed air. As the anticyclone rotates clockwise, the air mass warms as it travels around the southern portion and up the western or back side of the cell. As the stagnating influence of the central high pressure weakens, the winds become more moderate and the mixing depth increases.

A fourth class of air masses, transitional (Tr), used in this study includes the variety of cyclonic systems and mixed air masses with little temperature and moisture homogeneity. Days when more than one air mass moved through an area were also included in this class. Frontal passages, squall lines, subtropical storms, low-pressure systems, and fast-moving systems with residence times of less than 24 hours over the SURE region are all representative of Tr air masses. As one would expect, the characteristic features of these air masses are precipitation, high winds, and deep vertical mixing.

When a continental high-pressure system has moved off the eastern coastline and amalgamated with the Bermuda High, it takes on a maritime tropical, or mT, air mass character. The main difference between this situation and the back side of a continental high is the increase in temperature and water content in the air mass. Because the Bermuda High is a semipermanent feature off the eastern seaboard during the summer months, mT air penetration northeastward generally occurs from May to October. Maritime tropical air can also intrude onto the continent with the western expansion or retrograde motion of airflow under the influence of the Bermuda High. The advance of the cold front from the northwest toward the Appalachians forces a ducting or channeling of warm moist air over the Ohio River Valley region toward the northeast (Figure 4.14). If the mT air condition persists for more than a day, long-range transport of

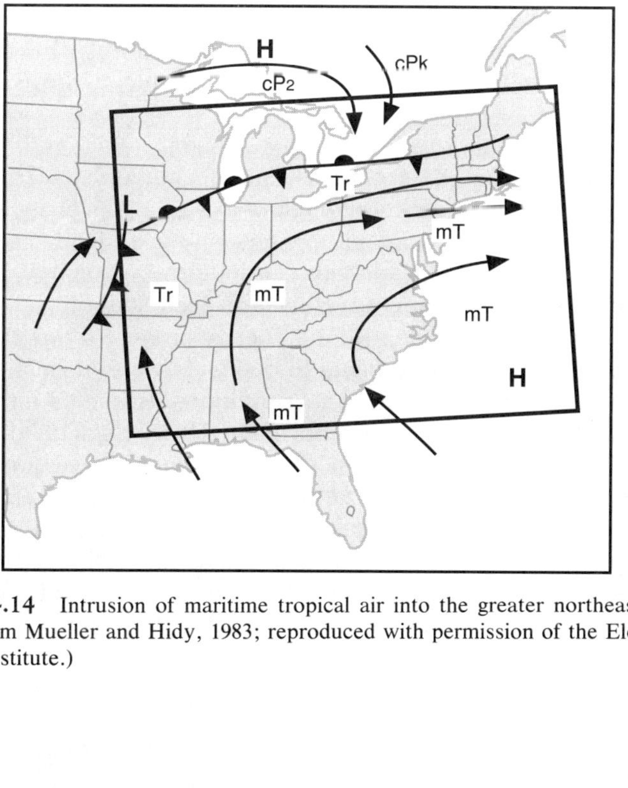


FIGURE 4.14 Intrusion of maritime tropical air into the greater northeastern United States. (From Mueller and Hidy, 1983; reproduced with permission of the Electric Power Research Institute.)

$SO_2$ and $SO_4^{2-}$ across the greater northeastern United States over distances in excess of 1000 km from sources becomes possible.

Baskett's (reported in Mueller and Hidy, 1983) next step was to determine the air mass type at each station for each day of the year. The goal of the air mass classification was to select the air mass that dominated at each CI station for the 24-hour period from midnight to midnight each day during a 1-year record. The year from October 1977 through September 1978 was selected because of very good weather data capture for those 12 months. The SURE stations were grouped in four subregional areas, because inspection of the results showed that these subregions had distinctive differences, possibly aligned with dominance of circulation around highs. Four maps available from the daily weather map service of NOAA were used in the selection process: the 12Z (07 EST) surface and 500-mb maps, the maximum and minimum temperature maps, and the daily total precipitation maps. Interpolation from midnight to midnight was attempted from one map for each day. Generally, the errors of this approach could be attributed to lack of information on temporal variations or incomplete coverage of a subregion by an air mass on any single day.

An average for air mass classifications in each of four seasons and for the full annual period was calculated. The four seasons were selected primarily by the weather and regional $SO_4^{2-}$ patterns during the period rather than by climatological average conditions. The fall season for the annual air mass data base was taken from October and November 1977 and September 1978; winter was December 1977 through February 1978; spring was March through May 1978; and summer was June through August 1978.

*General Results and Discussion* Because parameters were averaged over 24 hours from midnight to midnight and air masses changed at other times at each station, several parameters showed significant standard deviations from an average behavior. Despite the variability in the data base, a self-consistent picture developed at each station. Several rather striking features of the aerometric data behavior were found from analysis. In general, the arithmetic mean $SO_4^{2-}$ levels were highest within mT air masses and lowest in cPk or Tr air conditions. Baskett's findings were consistent with the earlier results reported by Hidy *et al.* (1976). The $SO_4^{2-}$ maximum for mT was greatest in the north coastal and north central subregions, indicating the important influence of stagnation (followed by northeastward air flow) over areas of high emission density in the Midwest. The mT $SO_4^{2-}$ maximum was less prevalent in the central-coastal and south-central subregions where the airflow associated with mT air masses has not passed over extended areas of high $SO_2$ emission density.

High $SO_4^{2-}$ concentrations were associated with high temperatures, high absolute humidity, moderately low pressures and wind speeds, and high $O_3$ concentrations. In this classification method, the possible chemical influences of temperature and absolute humidity could not be segregated from those signatures of air mass character. However, from the theoretical calculations discussed by Mueller and Hidy (1983), it is clear that high humidity, temperature, and $O_3$ are consistent with increased photochemical activity and accelerated oxidation potential for $SO_2$. Twenty-four-hour average $O_3$ levels were similar at most stations, running between 3 and 4 ppb, except for Lewisburg, WV, where levels were generally higher. A secondary maximum in $SO_4^{2-}$ concentration was associated with cPw air masses. This maximum was strong in the north-central and north-coastal subregions and at Giles County, TN, but was not present in central coastal subregions or at Lewisburg. When the cPw $SO_4^{2-}$ maximum was present, $SO_2$ concentrations were also at a maximum. $SO_2$ was generally minimal in mT air masses.

A minimum in $SO_4^{2-}$ concentrations was found in Tr air masses. This situation appeared to be associated with a precipitation maximum. The association led to speculation that the $SO_4^{2-}$ minimum in Tr air may be the result of depletion by wet deposition. Of course, one cannot discount the influence of strong ventilation in Tr conditions, which augments the effect of precipitation in depleting $SO_4^{2-}$ levels.

A principal conclusion regarding $SO_4^{2-}$ and weather systems is that during all seasons, relatively high $SO_4^{2-}$ concentrations consistently occur on the back side of both continental and maritime high-pressure systems. $SO_4^{2-}$ accumulation is frequently initiated during the stagnant conditions under the center of the anticyclone. Photochemically derived conversion mechanisms of $SO_2$ to $SO_4^{2-}$ evidently are also optimal in mT air, as signaled by relatively high $O_3$ concentrations. The conditions under which continental highs take a more southerly track in the summer months and merge with the Bermuda High appear to favor $SO_4^{2-}$ production. Temperature, humidity (relative and absolute), and $O_3$ are also at their highest during the summer, as southwest winds draw the warm, moist air off the Gulf of Mexico on the southwest side of the Bermuda High northeastward access zones of high pollutant emission density.

The results show a major difference between $NO_3^-$ and $SO_4^{2-}$ behavior. With the exception of Duncan Falls and Rockport, $NO_3^-$ was found to exhibit a maximum concentration in cPw air; it was also relatively high in cold, dry cPk air. On the other hand, $NO_3^-$ concentrations were lowest in mT air, in contrast to $SO_4^{2-}$. In the south-central region and at Duncan Falls, $NO_3^-$ was highest in cPk air. $NO_x$ tended to be highest in cPw air, except for the south-central region and at Indian River, DE. In those subregions $NO_x$ was highest in cP2 air. Comparison with the thermodynamic variables indicates relatively high eq $NO_3^-$ concentrations with cool, dry air (low absolute humidity) and lower hourly average precipitation. Interestingly, the high eq $NO_3^-$ levels were also accompanied by low $O_3$ concentrations.

The ground-level $NO_3^-$ behavior[4] throughout the greater northeastern United States differs from that in certain areas of the West, which are heavily influenced by urban or polluted air (for example, Los Angeles, the San Joaquin Valley, CA, and Denver, CO). $NO_3^-$ concentrations were generally much lower in the SURE region than in western urban air. The correlation with low temperatures is similar in both places, but the association with low absolute humidity in the East is not necessarily consistent with the western data (Young *et al.*, 1988).

The role of $NH_3$ in forming particulate $NO_3^-$ is unclear from the SURE results. High $NO_3^-$ concentrations in samples from Los Angeles, the San Joaquin Valley, and Denver appear to be associated with the presence of relatively high $NH_3$ concentrations. However, potentially high $NH_3$ levels in the East, related to cities or agricultural practice (spring and summer fertilization), apparently do not necessarily promote $NO_3^-$ formation. The poorly characterized behavior of $NO_3^-$ (and $NO_y$) in the troposphere remains an important basic issue in atmospheric chemistry. $NO_3^-$ will continue to be of interest to industry because of its contributions to $NO_x$ emissions.

*Seasonal Results* Duncan Falls, OH and Rockport, IN in the Midwest and Montague, MA to the Northeast all exhibited generally higher $SO_4^{2-}$ concentrations on the back sides of highs than in the centers of highs; the highest concentrations occurred in mT air masses in summer. Standard deviations of $SO_4^{2-}$ (not shown) were generally 2 to 4 $\mu g/m^3$ in summer and 3 to 6 $\mu g/m^3$ during other seasons, both in the Midwest and in the Northeast.

Giles (6) and Lewisburg (9) in the south-central region south of the Ohio River Valley did not reflect the same mean pattern, probably because the southerly transport within the lee side of highs carried the major emissions from the valley away from these stations. Central coastal sites, Indian River (3) and Raleigh (8), exhibited generally smaller differences among the three air masses and seasons. Because major source areas do not exist immediately to the south of these stations, high $SO_4^{2-}$ concentra-

[4] These remarks are taken in the context of the known sampling problems with $NO_3^-$ as reported through the early 1980s. Improved sampling methods have been used recently. However, these still have not resolved ambiguities in $NO_y$ concentrations.

tions due to transport on the lee side of highs do not occur. Stations 3, 6, 8, and 9 all exhibited generally large seasonal standard deviations in the 3 to 8 $\mu g/m^3$ range, showing considerable variability in $SO_4^{2-}$ levels relative to stations in the industrialized areas.

At the other extreme, 24-hour averages indicated that the cool dry cPk air on the front side of highs, which occurs primarily between October and March, contained the lowest $SO_4^{2-}$ concentrations (3 to 4 $\mu g/m^3$ throughout the region), as well as the lowest $SO_2$ concentrations, with only average amounts of $O_3$. Aside from anticyclones, the Tr class includes a broad variety of air masses. The stormy or rapidly changing conditions often associated with these weather systems usually produced low ambient concentrations of TSP, $SO_4^{2-}$, and $O_3$. Temperatures in Tr air were typically lower than average, while moisture was higher than average during Tr conditions. Occasionally during the snow or rain showers following passage of a high, however, regional $SO_4^{2-}$ concentrations reached 10 to 20 $\mu g/m^3$ in Tr air masses. As a result, while arithmetic mean concentrations ranged from 4 to 8 $\mu g/m^3$ throughout the region, relatively large standard deviations of 4 to 6 $\mu g/m^3$ were found in Tr air conditions.

Although many of the distinguishing features among seasons are attributable to the seasonal occurrence of certain air masses, several consistent relationships between air masses and aerometric parameters are evident. Daily average surface pressure is a primary parameter that distinguishes the air masses,as they are tied to cyclonic and anticyclonic circulations. Naturally the highest pressure is found under the center of highs; the second highest value usually occurs on the back side of the cell. Seasonally, pressure varies more in winter than in summer because of larger amplitude fluctuations in response to the greater pole-to-equator temperature gradient. Deeper highs (a deep high has greater than average seasonal surface pressure; a shallow high has less) occur in fall and winter, and much shallower anticyclones with less organized support aloft are observed in spring and summer.

The surface pressure can be taken as an indicator of mixing height, because generally the greater the pressure, the lower the mixing height. High pressure gives rise to a more stably stratified boundary layer in fall and winter highs than those in spring and summer, which is reflected in the mean pressures for the cP2 air masses during these seasons. Stations reported mean cP2 pressures in fall and winter from 1020 to 1024 mb, while in spring and summer mean pressures were 1016 to 1018 mb. This pressure–mixing height correlation is consistent with the seasonal patterns of mixing height found by Holzworth (1972). Standard deviations of pressure indicated small variation in spring and summer (3 to 5 mb). The wide variation in fall and winter (5 to 8 mb) allowed the region to "breathe" or ventilate more frequently in the vertical during these seasons.

Wind speed is the other ventilation parameter ascribed to the horizontal dimension of the air mass. Although surface winds (10 m above ground level) included in this analysis did not always relate directly to the amount of flow within the mixed layer, they were a good indication of the occurrence of general stagnation. Based on pressure gradient–derived surface wind speeds, Korshover (1976) concluded that stagnation is most prevalent from August through October, occurring on 19% of the days during those months. This conclusion was based on a 40-year record of surface pressure charts. A stagnation episode was defined as a period during which the surface wind did not exceed 3.8 m/s for a period of four or more consecutive days. A secondary seasonal maximum of stagnation occurs in spring (April through June) on 13% of the days. Stagnation is minimal during the windy periods of late fall and winter, November through March, occurring an average of 5% of the days. A unique July stagnation minimum of 8% of the month was observed for the 1936–1975 record.

The selection of the seasons during the 1977–1978 study year generally coincided with the two maximum stagnation periods. The maximum stagnation months of the year examined in the SURE were June through August. This choice was verified by the fact that the lowest seasonal wind speeds in the center of highs occurred during summer. Generally, the second lowest seasonal average wind speeds took place in fall at each station, and spring and winter were consistently windier within the center of highs. In addition, the lowest variability in wind speed occurred during summer anticyclones, with a mean seasonal standard deviation at all stations of 0.7 m/s. Spring cP2 air masses showed a seasonal standard deviation of 0.9 m/s; fall had a 1.0 m/s seasonal variation and winter had 1.2 m/s. Thus, a consistent pattern is evident in air associated with the central portion of anticyclones associated with highs. These air masses exert strong resistance to the horizontal spread of emissions, confining accumulation of pollution to the core of the anticyclone. In contrast, spatially persistent moderate to high wind speed appears to be the key meteorological condition in mitigating regional $SO_4^{2-}$ accumulation, as exemplified in mT air masses.

The back sides of continental highs generally experience about the same wind speeds during the summer as the centers of the cells, or 0.1 to 0.2 m/s higher. However, other seasons have lower cPw wind speeds within the region. The lee sides of the Bermuda High or an mT air mass, which occurred only during the spring and summer seasons during 1977–1978, were also equal to or 0.1 to 0.2 m/s windier than the center of highs on the average. Lee sides of both continental and Bermuda highs show about the same variation in wind speed as cP2 (about 1.0 m/s). In contrast, both Tr and cPk air masses are considerably more windy and exhibit greater variability; average Tr and cPk winds generally range from 4 to 6 m/s with seasonal standard deviations around 1.5. Winds in these air masses are usually lower in summer and spring and higher in winter and fall.

Pressure and wind speed are the only two parameters that showed the same seasonal pattern as $SO_4^{2-}$ in the air mass analysis. For this reason, the main conclusion to be drawn throughout the region is that within air masses associated with anticyclones, stagnation is most consistently conducive to $SO_4^{2-}$ accumulation. The larger and more persistent the region influenced by the high-pressure zone, the more intensive and widespread the accumulation of ambient $SO_4^{2-}$ concentrations.

## Relation of Event Groupings to Weather Changes

A foundation is now set to relate the five types of $SO_4^{2-}$ events to the general synoptic-scale weather conditions occurring in the SURE region. This procedure provides a synthesis of the meteorological character for each event group. In practice, the meteorological assessment offers opportunities to establish the predictability of each event based on a regional weather forecast. The technique uses the five basic air mass categories to differentiate between the synoptic-scale weather patterns. For this analysis, the dominant air mass affecting the area in which an event took place on each day was chosen to represent the weather conditions affecting that event day.

Table 4.5 lists 12 months from August 1977 through July 1978 dominated by each air mass type during the data-intensive measurement peri-

**TABLE 4.5 Distribution of Each Air Mass Associated with Each Event Group from Table 4.1 (August 1977 through July 1978)**[a]

| | cPk | | cP2 | | cPw | | Tr | | mT | |
|---|---|---|---|---|---|---|---|---|---|---|
| Event group | Days | % | Days | % | Days | % | Days | % | Days | % |
| 1 | — | 0% | 21 | 19% | 19 | 40% | 3 | 2% | 20 | 44% |
| 2 | — | 0% | 23 | 21% | 7 | 15% | 6 | 4% | 12 | 26% |
| 3 | — | 0% | 15 | 14% | 13 | 28% | 14 | 11% | 12 | 26% |
| 4 | 7 | 24% | 37 | 34% | 6 | 13% | 65 | 49% | 2 | 4% |
| 5 | 22 | 76% | 14 | 13% | 2 | 4% | 45 | 34% | 0 | 0% |
| Total days | 29 | | 110 | | 47 | | 133 | | 46 | |

[a] From Mueller and Hidy, 1983; reproduced with permission of the Electric Power Research Institute.

ods of the SURE. The 12-month period provides a picture of event–weather distributions. It reiterates and quantifies the conclusions of the air mass categorization for the region as a whole. The table shows that most all regional $SO_4^{2-}$ events occur under the centers and lee sides of continental highs or in mT air masses. The annual total percentage of events in each air mass for the 12 months is summarized in Table 4.6. The annual distribution of air masses corresponds closely to the distribution of air masses at stations in the center of the SURE region.

Table 4.6 shows that during the 30% of the year when regional events (groups 1 and 2) occur, they are most frequently associated with air mass stagnation under the center of anticyclones; second, with channeling events during mT air masses; and third, with extended stagnation on the back sides of continental highs. On about 40% of the days when the center of a high dominates the SURE region, regional stagnation event days occur. These are the broader and slower-moving stagnating anticyclones that pass over the Ohio River Valley and are most frequently observed in summer over the major sources in the valley. About 55% of the lee sides of those continental highs result in continuation of regional stagnation events. Where the annual total number of days in which the back side of a high-pressure region dominates ENA is less than the number of days under the center of the highs, the warmer air on the lee sides of highs is more conducive to regional events because of enhanced temperature and humidity. This correspondence is even stronger on the lee side of the Bermuda highs, of which 70% result in widespread regional $SO_4^{2-}$ events, most of which can involve long-range transport over distances exceeding

TABLE 4.6 **Annual Percentage of Event Days by Air Mass Category**[a]

| Event group | cPk | cP2 | cPw | Tr | mT | Annual percentage of days in each event group |
|---|---|---|---|---|---|---|
| 1 and 2 | 0 | 12 | 7 | 2 | 9 | 30 |
| 3 | 0 | 4 | 4 | 4 | 3 | 15 |
| 4 | 2 | 10 | 2 | 18 | 0.5 | 32.5 |
| 5 | 6 | 4 | 0.5 | 12 | 0 | 22.5 |
| Annual percentage of days in each air mass | 8 | 30 | 13.5 | 36 | 12.5 | 100 |

[a] From Mueller and Hidy, 1983; reproduced with permission of the Electric Power Research Institute.

500 km. Recall that the air mass analysis performed station by station indicated that the summertime air mass character is the most conducive to the highest $SO_4^{2-}$ concentrations over the largest area of the greater northeastern United States.

Subregional events occur 15% of the year, appearing with equal probability in all air masses except the front side of high-pressure zones. These localized $SO_4^{2-}$ accumulations are frequently associated with localized stagnation near major sources downwind of urban areas. The subregional cases with mT in summer are almost all remnants of extended regional events that have been reduced in size because of frontal disturbances and heavy precipitation. Several subregional events occur in Tr air masses between anticyclones. They are most often associated with light showers and weak frontal activity that follow the lee side of highs; convergence along the frontal zones is probably a significant factor in these kinds of events.

Although the regional events predictably occur under widespread stagnating anticyclones or extensive Bermuda High conditions, the intensity and location of subregional events would be more difficult to predict in a forecasting sense. They most frequently appear under the lee side of high-pressure zones that have a residence time of only about 2 days over the SURE region.

Clearly, an index of air mass stagnation could be developed to relate to the degree of regional $SO_4^{2-}$ accumulation. Korshover's (1976) simplified approach of calculating the surface geostrophic wind is a starting point. However, as the case study analysis shows, transport winds at 300 m above ground level play a major role in determining regional $SO_4^{2-}$ concentrations. Thus, a regional forecasting scheme must also incorporate winds aloft in the planetary boundary layer. Probably winds at the 850-mb level are a good indicator of transport winds, since this region and below are where most of the $SO_x$ emissions are injected. Although the extent and coverage of the stagnation conditions determine the beginning of regional $SO_4^{2-}$ events, the duration of summer events frequently depends on the southwest-only transport condition influenced by the Bermuda High. This transport condition on the lee side of the Bermuda High clearly differs from the stagnation condition during the beginning of the event. Thus, a forecast for regional events should include an iterative upgrading mechanism that considers the succession of days reflecting persistence of synoptic conditions.

Event groupings and weather changes are also reflected in visibility patterns. Zanneti *et al.* (1989) analyzed EUS visual range data between 1979 and 1981 and classified the observations by four different regions in terms of subjective air mass classifications adopted by the SURE. The reports indicate that day-to-day visual range variations are highly sensitive to air mass characteristics in both urban and rural locations. The strongest differentiation is in relative humidity conditions greater or lower than 85%. However, significant regional differentiation also occurs with air mass characteristic. Higher visibility tended to be associated with cP2 or cPk conditions, while the lowest visibilities tended to accompany mT or continental tropical (cT) air mass conditions. The coastal regions generally had higher visual ranges than inland regions but displayed larger variation in different meteorological regimes.

## Air Mass Classifications and Cloud Chemistry

The air mass classification approach has also been used to segregate the MCCP cloud chemistry data (Vong, 1989). This classification scheme involved somewhat different definitions of synoptic conditions than those used for the SURE results.

The data consisted of hourly cloud water samples for the growing season (April–October) and corresponding hourly meteorological data for six sites located at Whiteface, NY (data for 1986–1988) and Whitetop, VA (1986–1988).

Meteorological data consisted of winds, cloud presence, and RH measured at the individual sites, mixed-layer trajectories estimated from a model, and synoptic type evaluated from surface weather maps. Figure 4.13 illustrates the synoptic types; four are associated with the main sectors of a cyclone. The synoptic types are:

1. Pre–warm-frontal clouds associated with low-level easterly winds.
2. Northwest-sector-of-cyclone clouds occurring in direct response to intensifying circulation at low levels (roughly cPk or cP2).
3. Post–cold-frontal clouds resulting from destabilization of the atmosphere due to cold advection (roughly cPk).
4. Warm-sector synoptic-type clouds representing the southeast (SE) sector of the cyclone with characteristics such as high dew points, evening convection, and a potential for nighttime orographic cloud (roughly mT or Tr).
5. Easterly low-level marine flow associated with circulation around high pressure off the coast.
6. Cap cloud of mountain cloudiness in which no important synoptic-scale features were apparent while local airport observations had high cloud bases.
7. Three other synoptic types were pooled into a class (#2) and termed "other synoptic types" (with northwest-sector-of-cyclone clouds). These infrequently occurring types were sta-

tionary-front cloudiness and cutoff-low-aloft (SW to NW flow aloft and S to E flow aloft) types.

For purposes of comparison with the aerometric data classification, a rough correspondence of air mass type is shown in parentheses for cases 2–4.

A site–cloud climatology was developed for a 3-year period (1986–1988) at Whiteface Mountain, NY, Mt. Moosilauke, NH, Whitetop Mountain, VA, Mt. Mitchell, NC, and Shenandoah, VA. Synoptic types for cloudy periods are shown in Table 4.7. At Whiteface, the post–cold-frontal type (3) and warm-sector synoptic type (4) made up almost two-thirds of total cloud hours. The pre–warm-frontal class (1) was the third most frequent, particularly in 1987. At Whitetop and Mt. Mitchell, the warm-sector type (4) was most frequent. This was expected because the normal storm track during summer is through the Great Lakes region. In addition, in the southern United States, long periods of moist southwesterly flow can occur during the summer under the influence of a Bermuda High. The second most abundant (23%) cloud frequency of this type that was observed was a result of easterly flow. This was substantially more than the frequency of this type observed at Whiteface.

Trajectories were obtained from a mixed-layer model that uses rawindsonde measurements to define the transport field (Heffter, 1980; Kahl and Samson, 1986). The trajectory information was used to describe the fraction of hours within each cloud event that the air was located in each of three sectors of potential interest (NW/N, E, SW) over 36 hours and a variable that describes the length of the trajectory path.

The hourly chemical data were aggregated into events and subevents for both precipitating and nonprecipitating clouds to reflect changing meteorological conditions; this definition of subevents was established to reduce autocorrelation in the data (the hourly samples were correlated with immediately previous and succeeding hours by virtue of common meteorology). Removal of this autocorrelation allowed an improved estimate of the true number of "independent samples" of cloud water chemistry for the 2138 hours of sampling.

First, events were defined in the field by an on-site observer. When visibility was greater than 1 km for 15 min, an event was defined to have ended. These events were divided into subevents during data analysis; a new subevent started when the synoptic type changed (e.g., when a front passed the site), a change in trajectory occurred, or 9 hours had passed. Trajectory changes were defined as shifts out of the SW, NW/N, or E sector from the previous hour of cloudiness. The division of events into subevents achieved better resolution of the chemical and trajectory information than could have been achieved by the event definition obtained

TABLE 4.7 **Cloud Synoptic Type Frequency of Occurrence for Mt. Moosilauke, NH; Whiteface Mountain, NY; Shenandoah, VA; Whitetop, VA; and Mt. Mitchell, NC for the April–October 1986–1988 Field Seasons (Year-to-Year Range in Parentheses)**[a]

| | Site | | | | |
|---|---|---|---|---|---|
| **Synoptic type** | **Moosilauke (1000 m)** | **Shenandoah (1014 m)** | **Mitchell (2020 m)** | **Whiteface (1483 m)** | **Whitetop (1689 m)** |
| 1 | 25.5 (23–31) | 6.5 (1–10) | — | 16.5 (9–24) | 0.1 |
| 2[b] | 11.5 (2–18) | 4.5 (1–18) | 2.5 (1–3) | 5.8 (1–9) | 4.2 (1–9) |
| 3 | 11.0 (8–15) | 23.9 (16–35) | 15.2 (13–17) | 32.4 (29–36) | 16.0 (12–20) |
| 4 | 31.8 (25–50) | 11.3 (7–16) | 41.9 (39–43) | 30.7 (21–41) | 45.4 (39–57) |
| 2[c] | 0.1 | 3.2 (0–8) | 2.5 (1–3) | 0.3 | 1.5 (1–2) |
| 5 | 15.9 (1–22) | 43.7 (18–61) | 26.8 (21–34) | 4.0 (0–6) | 23.1 (20–38) |
| 2[d] | 2.0 (1–2) | 1.8 (0–5) | 2.9 (0–4) | 3.4 (0–6) | 5.0 (0–13) |
| 2[e] | 2.2 (0–6) | 5.1 (0–15) | — | 1.4 (0–2) | 0.1 |
| 6 | — | — | 8.2 (1–10) | 5.5 (1–10) | 4.6 (2–6) |

[a] From Vong, 1989.
[b] Northwest sector of cyclone.
[c] Stationary front.
[d] Cut off low-aloft (SW to NW).
[e] Cut off low-aloft (S to E).

from the field observer. Henceforth, we refer to these subevents as cloud events for convenience.

The surface weather types described above were determined by a meteorologist and independently verified by a second meteorologist. These synoptic types are expected to have been precisely estimated relative to the mixed-layer trajectories.

The investigators stratified the data in each of three frequently occurring synoptic types (3, 4, 5) into two classes on the basis of mixed-layer trajectories. This subdivision by trajectory was made under the assumptions that transport direction can vary within a given synoptic type and that this trajectory variation would be useful for describing variation in cloud water chemistry. Trajectory variation could reflect either misclassification of synoptic type, different emission source regions, or errors in the trajectories. Agreement of synoptic type with the expected dominant transport directions lends additional confidence to the use of these variables.

To address the associated uncertainties, two approaches were taken. First, Vong (1989) reportedly analyzed the data with a partial least squares (PLS) model to identify which associations between trajectory and synoptic type are important and to determine how those combinations describe the cloud water chemical composition and its variability. The second approach was to use descriptive statistics to analyze frequently occurring synoptic types.

The nine synoptic-trajectory classes frequently occurring at the MCCP sites were:

1. Pre–warm-frontal, all trajectories.
2. Other synoptic types, i.e., those due to large-scale features but that did not occur frequently (NW sector, occluded front, cutoff high aloft), all trajectories.

3a. Post–cold-frontal synoptic type, NW trajectory.
3b. Post–cold-frontal synoptic type, all trajectories other than 3a.
4c. Warm-sector synoptic type, trajectory from Ohio Valley (defined as SW sector for northern sites but NW sector for southern sites).
4d. Warm-sector synoptic type, all trajectories other than 4c.
5e. Easterly-surface-flow synoptic type, E trajectory.
5f. Easterly-surface-flow synoptic type, all trajectories other than 5e.

6. Cap cloud, all trajectories (summit sites only).

Table 4.8 lists the means for each site, cloud type, and synoptic class. Cap clouds (class 6) and warm-sector (4c and 4d) events had cloud water concentrations significantly higher than concentrations of post–cold-fron-

TABLE 4.8 **Mean values (Large Type) of Cloud Water $SO_4^{2-}$ Concentrations ($\mu$eq/L) and Number of Hours of Synoptic Trajectory Classes and Cloud Type (Small Type) for Synoptic Trajectory Classes and Cloud Types for 1986–1988**[a]

| | Synoptic-trajectory class[b] | | | | | | | | | Mean overall no. hours | Cloud type[c] mean | |
|---|---|---|---|---|---|---|---|---|---|---|---|---|
| **Location** | **wf$^{1}$** | **o$^{2}$** | **cf$^{3,a}$** | **cf$^{3,b}$** | **ws$^{4,c}$** | **ws$^{4,d}$** | **ma$^{5,e}$** | **ma$^{5,f}$** | **cap$^{6}$** | **(NP&P)**[b] | **P clouds** | **NP clouds** |
| Whiteface-1, NY | 110 | 16 | 20 | 76 | 286 | 218 | 60 | 281 | 537 | 128 | 77 | 165 |
| | 175 | 72 | 59 | 217 | 258 | 67 | 24 | 17 | 23 | 912 | 327 | 585 |
| Mossilauke, NH | 146 | 47 | 178 | 45 | 266 | 257 | 94 | 62 | — | 158 | 96 | 249 |
| | 133 | 25 | 2 | 4 | 104 | 11 | 37 | 9 | 0 | 325 | 166 | 159 |
| Whitetop, VA | — | 237 | 147 | 136 | 434 | 226 | 133 | 591 | 1043 | 231 | 134 | 263 |
| | 0 | 95 | 44 | 54 | 47 | 112 | 147 | 44 | 19 | 562 | 119 | 443 |
| Mitchell, NC | — | 679 | 291 | 562 | 964 | 285 | 123 | 302 | 1036 | 281 | 71 | 359 |
| | 0 | 46 | 22 | 20 | 12 | 140 | 144 | 21 | 12 | 417 | 78 | 339 |
| Shenandoah, VA | 84 | 94 | 308 | 140 | — | 57 | 378 | 103 | — | 10 | 69 | 287 |
| | 24 | 25 | 13 | 33 | 0 | 7 | 7 | 12 | 0 | 121 | 75 | 46 |
| All sites P | 121 | 144 | 94 | 105 | 308 | 243 | 121 | 333 | 799 | 178 | NA | NA |
| and NP clouds | 332 | 263 | 140 | 328 | 421 | 357 | 359 | 103 | 54 | 2337 | | |
| All sites only | 93 | 77 | 36 | 48 | 186 | 109 | 59 | 89 | 244 | NA | 87 | NA |
| P clouds | 294 | 96 | 42 | 114 | 128 | 63 | 96 | 21 | 1 | | 765 | |
| All sites only | 178 | 196 | 132 | 149 | 373 | 285 | 152 | 432 | 815 | NA | NA | 239 |
| NP clouds | 128 | 167 | 98 | 214 | 293 | 274 | 263 | 82 | 53 | | | 1572 |

[a] From Vong, 1989.

[b] Synoptic trajectory classes: wf$^{1}$, pre-warm-frontal; o$^{2}$, ''other'' synoptic types; cf$^{3,a}$, post-cold-frontal, NW/N trajectory; cf$^{3,b}$, post-cold-frontal, not NW/N trajectory; ws$^{4,d}$, warm sector, Ohio trajectory; ws$^{4,c}$, warm sector, not Ohio trajectory; ma$^{5,e}$, marine flow, E trajectory; ma$^{5,f}$, marine flow, not E trajectory; cap$^{6}$, cap cloud.

[c] P, precipitating; NP, nonprecipitating.

tal (3a, 3b), pre–warm-frontal (1), easterly (5e), and other (2) clouds. Cap clouds had significantly higher concentrations than any other synoptic type ($P < .01$). Warm-sector events with trajectories from the Ohio River Valley (4c) had higher concentrations than warm-sector events with any other trajectories (4d), but the differences were not significant. The easterly low-level-flow synoptic type (5) displayed a strong dependence on air mass trajectory; easterly mixed-layer trajectories (5e) had relatively low concentrations, but other trajectories had high concentrations (class 5f was not significantly different from 4c and 4d). According to Vong (1989), the easterly flow for class 5f was apparently too shallow to fully characterize pollutant transport.

The southern sites (Mitchell and Whitetop) had the highest concentrations; the northern sites (Whiteface and Moosilauke) had the lowest. The $SO_4^{2-}$ concentrations in cloud water at Mossilauke were within the range defined by Whiteface-1 and Whiteface-2. Shenandoah concentrations were lower than those at the other southern sites but higher than those at the northern sites. Cloud type was also significant, with nonprecipitating clouds having higher concentrations than precipitating clouds.

$H^+$, $NH_4^+$, and $NO_3^-$ concentrations behaved in similar ways to $SO_4^{2-}$ with respect to synoptic type, trajectory, and cloud type.

Vong (1989) compared this classification approach with ones based on (1) cloud event duration and cloud type and (2) mixed-layer trajectory (without synoptic type) and cloud type. Duration described only 20% of the variance in the MCCP data, the trajectory model accounted for 30% of the variance, and the synoptic-trajectory model used here described 43% of the variance. The event duration model described only 11% of the variance in their Mt. Mitchell, NC data if cloud type (precipitating or nonprecipitating) is not used to classify the data. Vong's approach evidently describes the data better than either simple trajectory direction or duration.

Substantial variation remained within each synoptic-trajectory class. This was expected because microscale variation in gas-phase and aqueous-phase chemical processes, scavenging, and meteorology could not be described in the synoptic-trajectory classification scheme. For example, the height of the monitoring site above cloud base would be expected to affect the amount of available liquid water and, therefore, aqueous phase concentrations, but a test of that effect was not possible because cloud base data were not available for all sites. Consistent with a cloud base effect, there were differences between the two Whiteface Mountain collection sites. Nevertheless, the synoptic classification described by Vong characterized several factors that influence the mountain cloud water composition in the eastern United States.

# 5 INHERENT VARIABILITY IN DEPOSITION CHEMISTRY

The previous chapters noted substantial, systematic spatial and temporal changes in mean values of chemical parameters either during sequences of episodes or in terms of diurnal, seasonal, or year-to-year patterns. Such variations or uncertainties are inherent in both emissions observations and atmospheric measurements. They must be taken into account in placing bounds of uncertainty on observed mean conditions and on estimation of SRRs. Indeed, the inherent variability in emissions and atmospheric processes contains information about SRRs, and, in principle, they can be exploited in specially designed experiments.

This chapter summarizes studies of variability in emissions and atmospheric observations. The results are derived from the variety of studies discussed previously. Knowledge of emissions variability largely comes from short-term fuel use and process data focusing on $SO_2$. The long-term records of aerometric data and wet deposition are most useful for describing the nature of these observations. In general, one could segregate variability or uncertainty in data in terms of a measurement or estimation component and a stochastic or random variation associated with atmospheric processes or unpredictable emissions changes. Variations will occur in both spatial and temporal domains. Some information is available on aerometric and deposition measurement uncertainty (Mueller and Hidy, 1983). But little information is available on estimated emissions uncertainty, particularly for time periods of less than a year. This chapter deals mainly with total variability or uncertainty reflected in spatial and temporal changes about average values. Separate characterization of a stochastic component must await additional study of long-term data records in which the sampling and measurement error is determined separately. Although such information was reported in the SURE and NAPAP, an analysis of the data variability was lumped into the analysis of error attributed to measurements.

In this chapter conventional statistical measures of variability are used to characterize temporal and spatial variability, and examples of analysis of variance are noted to indicate how information can be derived from variability in aerometric measurements. A hypothetical experiment is described that would use inherent variability to deduce SRRs.

## EMISSION VARIABILITY

Superimposed on the year-to-year changes in emissions are variations due to operations, uncertainties in emission factors or process throughput, labor strikes, new plant capacity, and other factors. These variations occur in both time and space; they are likely to be random in character after scheduled seasonal or other temporal changes are considered. Little is known about the random variability, but some information has become available from the MATEX feasibility studies (Hidy *et al.*, 1985). The extent to which this variability affects exposure conditions is not known. Some estimate of these influences has been made using SRR model calculations. Evidence from historical records, such as those for the record of Hubbard Brook, New Hampshire, suggest that even cumulative emission changes (in $SO_2$) over a period of years in excess of 2%/yr are difficult to ascertain in ENA deposition observations. Qualitatively, random changes in emissions patterns as large as 20–30% in a specific subregion appear to be lost in the noise of atmospheric processes.

Spatial and temporal variability of $SO_x$ and $NO_x$ emissions in the EUS were obtained from two sources: the emission inventories developed for the region (Klemm and Brennan, 1981; Heisler *et al.*, 1988) and from specialized data provided in Parkhurst *et al.* (1984) for power plants in the Tennessee Valley Authority (TVA) operating region.

### Emission Inventories

The SURE inventory provided a set of appropriate data that shed light on emission variability. They include emissions from small point sources and area sources aggregated into 80 × 80 km grid cells, as well as individual large point sources. Emission estimates for weekdays and weekends by season are provided. The inventory also includes hourly average $SO_2$

emission rates for individual power plants during 5 months (August and October 1977, January, April, and July 1978). In studies of emission variability, $SO_3$ ($H_2SO_4$) emissions are disregarded for regional influence because they are believed to be a small fraction of the total $SO_x$ and would have primarily local influence. No SURE data exist for $NO_x$ emission variability, but it is assumed at best to be at least similar to that of $SO_2$.[1]

Major sources of $SO_2$ are clustered at large urban areas, along the Ohio River, in the TVA region, and just north of Lake Huron. To look at daily and seasonal variability of $SO_2$ emissions, investigators focused on grid squares from which annual average $SO_2$ emissions exceed 500 metric tons per day.

The hourly power plant $SO_2$ emission data from July and October 1977 and January 1978 were combined with the seasonal, weekday, and weekend emission rates for point and area sources to examine temporal emission variations in each source region. The daily emission rate within a region was estimated by adding the power plant emission rates for the day to the weekday or weekend emission rates for the other point sources and the area sources within the region. Weekday–weekend variations during the spring were also calculated, although hourly power plant emission rates for that period were not available for this analysis.

Table 5.1 shows examples of the average emission rate, standard deviation of the daily emissions, and maximum and minimum daily emissions for each season in each region. In the regions with the highest emission rates and for the 3 months of available hourly power plant emission data, the standard deviation of the daily emissions is 5–18% of the average rate. Daily emissions varied from about 70 to 120% of the monthly average. The range of values, expressed as a percentage of the average, was largest in the Detroit region (as much as 52%) but smaller in the upper Ohio River Valley (37%), the lower Ohio River Valley (23%), and northern Ohio–Pennsylvania (28%).

Table 5.2 shows "annual" and "seasonal" average emission rates. The seasonal averages were taken from daily emission rates; with the exception of spring, the seasonal emission rates shown are actually 1-month averages. The annual averages are the averages of the four seasonal averages. In most areas, the lowest emission rates occurred in the summer. The season of highest rates varied inconsistently in different areas. In most areas, the lowest emission rates occurred in the summer.

Although seasonal variations in emission rates were not as large as day-to-day variations within a region, they were significant in some re-

[1] Subsequent data collected in the 1980s through EPRI and EMEFS have not been analyzed for variability, but regional model evaluation has suggested regional difference in $SO_2$ emission inventory quality; $NO_x$ emissions are of lower precision than $SO_x$ (Dennis *et al.*, 1990).

TABLE 5.1 **Example Average $SO_2$ Daily Emission Rates (g/s) and Standard Deviations (%)[a]**

| **Location** | **Spring** | **Summer** | **Fall** | **Winter** |
|---|---|---|---|---|
| Upper Ohio River Valley | | | | |
| Average | 105,055 | 95,210 | 105,443 | 99,502 |
| Standard deviation | 4,312 (4%) | 8,126 (9%) | 9,768 (9%) | 5,933 (6%) |
| Daily minimum | 99,059 (94%) | 78,714 (83%) | 85,451 (81%) | 86,859 (87%) |
| Daily maximum | 108,053 (103%) | 110,969 (117%) | 124,156 (118%) | 112,112 (113%) |
| Northwestern Georgia | | | | |
| Average | 14,757 | 18,928 | 19,559 | 20,679 |
| Standard deviation | 3,695 (25%) | 1,548 (8%) | 2,606 (13%) | 1,907 (9%) |
| Daily minimum | 9,616 (65%) | 16,294 (86%) | 13,143 (67%) | 17,211 (83%) |
| Daily maximum | 17,326 (117%) | 22,008 (116%) | 23,493 (120%) | 24,368 (118%) |
| New York | | | | |
| Average | 13,828 | 8,653 | 12,620 | 24,839 |
| Standard deviation | 1,425 (10%) | 1,347 (16%) | 844 (7%) | 1,667 (7%) |
| Daily minimum | 11,846 (86%) | 6,312 (73%) | 10,701 (85%) | 21,989 (89%) |
| Daily maximum | 14,818 (107%) | 10,223 (118%) | 13,485 (107%) | 26,320 (106%) |
| Sudbury, Ontario | | | | |
| Average | 30,512 | 30,472 | 30,516 | 30,610 |
| Standard deviation | 1 (5%) | 1 (0%) | 1 (0%) | 1 (9%) |
| Daily minimum | 30,511 (100%) | 30,470 (100%) | 30,814 (100%) | 30,609 (100%) |
| Daily maximum | 30,513 (100%) | 30,472 (100%) | 30,517 (100%) | 30,611 (100%) |

| | | | | |
|---|---|---|---|---|
| TVA region | | | | |
| Average | 12,520 | 8,756 | 15,807 | 11,717 |
| Standard deviation | 1,586 (13%) | 1,257 (14%) | 1,623 (10%) | 796 (7%) |
| Daily minimum | 10,314 (82%) | 6,852 (78%) | 11,884 (75%) | 10,522 (90%) |
| Daily maximum | 13,623 (109%) | 11,241 (128%) | 18,569 (117%) | 13,028 (111%) |
| Central Illinois | | | | |
| Average | 10,061 | 10,002 | 8,807 | 7,767 |
| Standard deviation | 786 (8%) | 2,373 (24%) | 2,488 (28%) | 1,713 (22%) |
| Daily minimum | 7,967 (89%) | 4,114 (41%) | 2,368 (27%) | 3,085 (40%) |
| Daily maximum | 10,608 (105%) | 13,634 (136%) | 13,715 (156%) | 9,732 (125%) |
| Southern Indiana | | | | |
| Average | 22,764 | 21,809 | 19,685 | 26,222 |
| Standard deviation | 1,007 (4%) | 2,067 (9%) | 1,797 (9%) | 1,935 (7%) |
| Daily minimum | 21,364 (94%) | 16,160 (74%) | 16,452 (84%) | 22,790 (87%) |
| Daily maximum | 23,464 (103%) | 24,796 (114%) | 22,797 (116%) | 29,899 (114%) |
| Western Michigan | | | | |
| Average | 5,959 | 4,983 | 4,553 | 5,334 |
| Standard deviation | 1,634 (27%) | 1,010 (20%) | 1,021 (22%) | 840 (16%) |
| Daily minimum | 3,687 (62%) | 2,702 (54%) | 2,181 (48%) | 4,230 (79%) |
| Daily maximum | 7,094 (119%) | 7,357 (148%) | 5,967 (131%) | 7,212 (135%) |
| St. Louis | | | | |
| Average | 42,664 | 40,074 | 39,345 | 35,135 |
| Standard deviation | 433 (1%) | 4,075 (10%) | 3,939 (10%) | 2,911 (8%) |
| Daily minimum | 42,062 (99%) | 31,989 (80%) | 28,831 (73%) | 29,977 (85%) |
| Daily maximum | 42,965 (101%) | 48,422 (121%) | 46,482 (118%) | 39,586 (113%) |

[a] From Hidy *et al.*, 1985; reproduced with permission of the Electric Power Research Institute.

TABLE 5.2 **Seasonal Average $SO_2$ Emission Rates (g/s)**[a]

| | Annual | Spring | Summer | Fall | Winter |
|---|---|---|---|---|---|
| Upper Ohio River Valley | 10,302 | 105,055 (104%)[b] | 95,210 (94%) | 105,443 (104%) | 99,502 (98%) |
| Lower Ohio River Valley | 53,272 | 53,185 (100%) | 51,783 (97%) | 50,923 (96%) | 57,195 (107%) |
| Detroit | 38,233 | 42,207 (110%) | 36,050 (94%) | 39,202 (103%) | 35,475 (93%) |
| Northern Ohio and Pennsylvania | 39,584 | 49,835 (101%) | 36,164 (91%) | 38,254 (97%) | 44,082 (111%) |
| St. Louis | 39,304 | 42,664 (109%) | 40,074 (102%) | 39,345 (100%) | 35,135 (89%) |
| Sudbury | 30,527 | 30,512 (100%) | 30,472 (100%) | 30,516 (100%) | 30,610 (100%) |
| Southern Indiana | 22,620 | 22,764 (101%) | 21,809 (96%) | 19,685 (87%) | 26,222 (116%) |
| Northwestern Georgia | 18,480 | 14,757 (80%) | 18,928 (102%) | 19,559 (106%) | 20,679 (112%) |
| New York | 14,985 | 13,828 (92%) | 8,653 (58%) | 12,620 (84%) | 24,839 (166%) |
| TVA region | 12,200 | 12,520 (103%) | 8,756 (72%) | 15,807 (130%) | 11,717 (96%) |
| Chicago | 11,709 | 12,937 (110%) | 10,854 (93%) | 10,597 (91%) | 12,450 (106%) |
| New Jersey | 13,728 | 12,537 (91%) | 11,817 (86%) | 13,108 (95%) | 17,450 (127%) |
| Central Illinois | 9,159 | 10,061 (110%) | 10,002 (109%) | 8,807 (96%) | 7,767 (85%) |
| Northwestern Alabama | 6,328 | 6,284 ( 99%) | 5,424 (86%) | 6,725 (106%) | 6,880 (109%) |
| Western Virginia | 5,167 | 7,643 (148%) | 4,239 (82%) | 4,411 (85%) | 4,374 (85%) |
| Southern Pennsylvania | 6,709 | 7,334 (109%) | 5,278 (79%) | 7,358 (110%) | 6,865 (102%) |
| Western Michigan | 5,207 | 5,959 (114%) | 4,983 (96%) | 3,553 (87%) | 5,334 (102%) |
| Milwaukee | 7,079 | 7,095 (100%) | 6,270 (89%) | 7,616 (108%) | 7,334 (104%) |
| Knoxville | 6,960 | 5,616 (81%) | 7,550 (108%) | 8,311 (119%) | 6,364 (91%) |

[a] From Hidy *et al.*, 1985; reproduced with permission of the Electric Power Research Institute.

gions. Seasonal average rates ranged from 94–104% of the annual average in the upper Ohio River Valley; 97–107% in the lower Ohio River Valley, 94–110% in the Detroit region, and 91–111% in the northern Ohio–Pennsylvania region. Thus, in the regions of highest emission rates the highest and lowest seasonal averages differed by as much as 20% of the annual average.

The data presented do not reflect substantial emissions variations at Sudbury, Ontario; but the large smelter sources there are shut down from time to time by strikes or for maintenance. In the 1970s and early 1980s these shutdowns were equivalent to turning the $SO_2$ emissions from the Detroit source region on and off. Therefore, accurate apportionment of $SO_x$ over ENA requires taking into account such large changes in any given year.

To examine year-to-year variability, investigators relied on estimates of Parkhurst *et al.* (1984) for the TVA region over the period 1978–1983. This case may represent the maximum change expected for a large aggregate of sources, because TVA had not only instituted a systemwide program of emission controls but also experienced a decreased fossil fuel power demand over this period.

The largest annual change at the Cumberland plant was 23% (1981 to 1982). For four plants combined, the largest change, 21%, also occurred between 1981 and 1982. Over the 6-year period, TVA's total $SO_2$ emissions declined 56%, or an average change of 12.8% per year. These data suggest that, at least for the TVA region, between-year changes can be expected to be less than between-season changes in power plant operations.

According to this analysis, day-to-day emissions can vary by more than 20% for large aggregates of sources (and by nearly 100% in the case of the Sudbury region). Because the variability in these emissions estimates is largely controlled by the estimated variability in emissions from power plants—information on variability in other sources not being available—the actual variability in emissions from many source regions may be greater still. Year-to-year variability is such that emissions estimates for one year may be quite inaccurate if applied to a later year.

## TEMPORAL VARIABILITY IN AEROMETRIC AND DEPOSITION DATA

By convention, observations are often stated in terms of mean values over a period of time and a fixed area, with a standard deviation reflecting the degree of uncertainty or variation in that mean.

## Frequency Distributions

Figure 5.1 compares $SO_2$ data from all SURE stations for 6 seasonal months and for 15 months. These results show similar patterns for frequency of occurrence of concentrations. However, the months chosen apparently yielded systematically lower $SO_2$ concentrations in the range near the median value (50th percentile) and below the data for the CI stations taken over 15 months. The measured distribution showed a steep slope change in occurrence below the 60th percentile at approximately 5 ppb $SO_2$. This change is believed to reflect the approximate detection limit of the flame photometric instruments used in the SURE. Extrapolation of the slope of the distribution above the median downward suggests a median $SO_2$ concentration of 3 ppb in the 6 seasonal months and 4.4 ppb for the 15 months, assuming a logarithmic normal frequency distribution. The ranges of concentration levels between the CI stations are shown in Figure 5.1 by the vertical bars. There is a range of as much as an order of magnitude between stations. In any case, this distribution readily suggests that the range of $SO_2$ concentrations observed in ambient air far exceed those expected from $SO_2$ emission variations. The $SO_2$ variability in the air is attributed to meteorological changes influencing ambient concentrations.

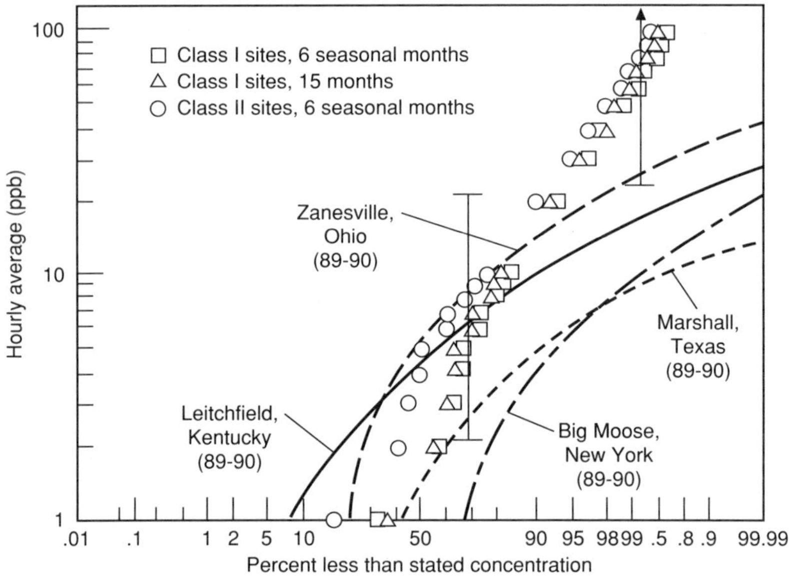


FIGURE 5.1 Sulfur dioxide concentrations observed continuously at SURE sampling sites from August 1977 through October 1978. The vertical bar designates the range of occurrence reported from the SURE CI stations. 1989–1990 EMEFS network data samples are shown by continuous curves. (Data reproduced with permission of the Electric Power Research Institute.)

From site to site the range of differences in the locations near large clusters of sources (Scranton, PA; Duncan Falls, OH; and Rockport, IN) shows systematic occurrence of higher concentrations. The lower concentrations tend to occur at sites at Research Triangle Park, NC, and Montague, MA. The coastal site at Indian River, DL, appeared anomalous with a narrow distribution of occurrence. The frequency curves downward from the range above the 50th percentile suggest that the median $SO_2$ concentrations range from less than 1 ppb at Indian River to 8 ppb at Duncan Falls.

Observations from four EMEFS stations operated about 12 years later are also drawn into Figure 5.1. Unlike the earlier SURE data, the EMEFS data do not follow a logarithmic normal distribution. The EMEFS concentration levels are lower than observed in 1978, especially above the 90th percentile. There is a tendency for higher occurrence of low concentrations and a lower frequency of occurrence of high $SO_2$ concentrations. As in the case of the geographic patterns of mean values, the results appear to reflect generally reduced $SO_2$ emission rates over the EUS.

Figure 5.2 shows the frequency of occurrence of particulate $SO_4^{2-}$ concentrations from the SURE network. The data give a uniform and consis-

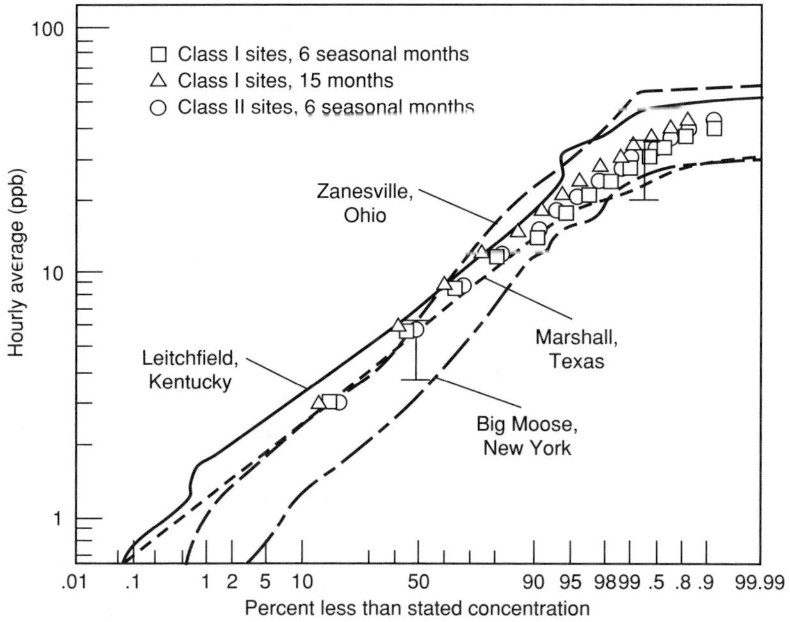


FIGURE 5.2 Sulfate concentrations (24-hour) observed daily at SURE sampling sites from August 1977 through October 1978. Vertical bars show range of SURE data reported for the CI stations. The 1989–1990 EMEFS data for four sites are shown in solid curves. (Data reproduced with permission of the Electric Power Research Institute.)

tent picture of occurrence. The overall median concentration reported at all SURE stations was found to be 6.5 $\mu g/m^3$. From the shallow slope, the distribution of occurrence in $SO_4^{2-}$ concentrations is broader than the $SO_2$ occurrence. The range of station distributions suggests that occurrence of $SO_4^{2-}$ concentrations varies much less with site than that of $SO_2$ concentrations. The lowest median concentration of 4.5 $\mu g/m^3$ was found at Montague; the higher values of about 6.2 $\mu g/m^3$ were uniformly observed at the other stations. The lowest extreme (high) concentrations of 2.5 $\mu g/m^3$ were found at Research Triangle Park and at Lewisburg, WV, whereas highest extremes (40–50 $\mu g/m^3$) were found at Duncan Falls and at Giles, TN. The frequency of occurrence of $NO_x$ and filter-collected $NO_3^-$ shows patterns similar to those of $SO_2$ and $SO_4^{2-}$. They are reported in Mueller and Hidy (1983) but are not included here.

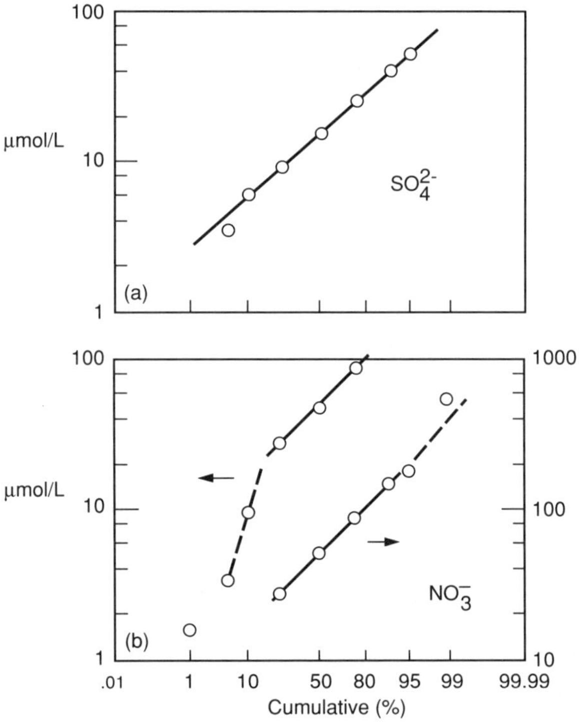


FIGURE 5.3 (a) Cumulative frequency distribution of $SO_4^{2-}$ concentration in precipitation collected at Kejimkujik. Total number of samples was 114. (b) Cumulative frequency distribution of aq $NO_3^-$ concentration in participation collected at Kejimkujik. Total number of samples is 113. (From Hidy *et al.*, 1985; reproduced with permission of the Electric Power Research Institute.)

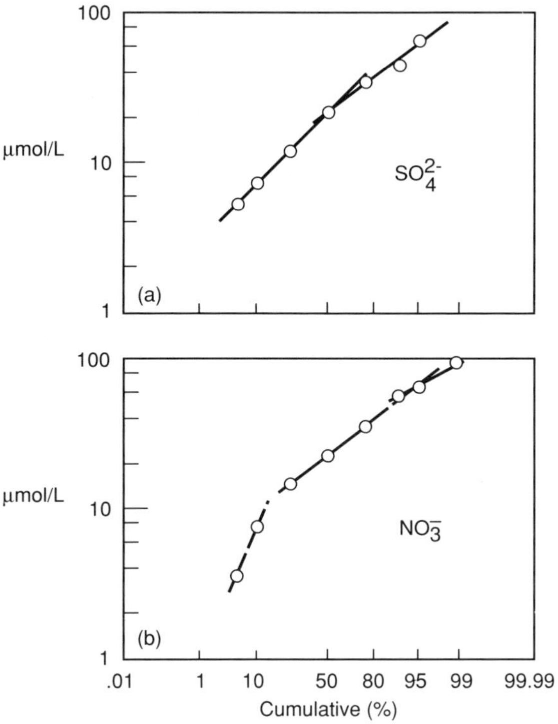


FIGURE 5.4 (a) Cumulative frequency distribution of $SO_4^{2-}$ concentration in precipitation collected at Whiteface Mountain. Total number of samples is 224. (b) Cumulative frequency distribution of aq $NO_3^-$ concentration in precipitation collected at Whiteface Mountain. Total number of samples is 216. (From Hidy *et al.*, 1985; reproduced with permission of the Electric Power Research Institute.)

Sulfate concentrations from four EMEFS stations are also included in Figure 5.2. In this case, these stations suggest a wide range of occurrence of particulate sulfate; there is a tendency for the 1989–1990 data to be similar to the sampling 12 years before.

Figures 5.3 and 5.4 show cumulative frequency plots of precipitation $SO_4^{2-}$ and $NO_3^-$ for two different stations. The data are from the CANSAP and the MAP3S networks. The $SO_4^{2-}$ data, exemplified by Kejimkujik, Nova Scotia, show a straight line on a log probability scale. This indicates a logarithmic normal distribution over the range of observation, except for the lowest concentration. A similar result is found for $NO_3^-$ concentration above approximately 25 $\mu$M. In contrast, the cumulative distribution for $SO_4^{2-}$ and $NO_3^-$ concentration at Whiteface Mountain, NY suggests a blend of two distributions breaking in the midfrequency range (Figure 5.4). The reason for such distinctions is not known. But such differences may reflect the blending to two different modes of occurrence at Whiteface.

Raynor (1984) examined the range of variation in precipitation chemistry by episode or event. His results indicate, as expected, an increasing variability in precipitation chemistry with shortened averaging time. There is substantial variability from event to event or within events. The variability of $SO_4^{2-}$ in these periods may exceed a factor of 5.

The range of temporal variation in precipitation chemistry, seasonally and from year to year, is exemplified in the record (Hinckley, NY) shown in Figure 5.5. The range of values in the monthly data is shown together with the pattern of change in annual mean values for $SO_4^{2-}$ and $NO_3^-$.

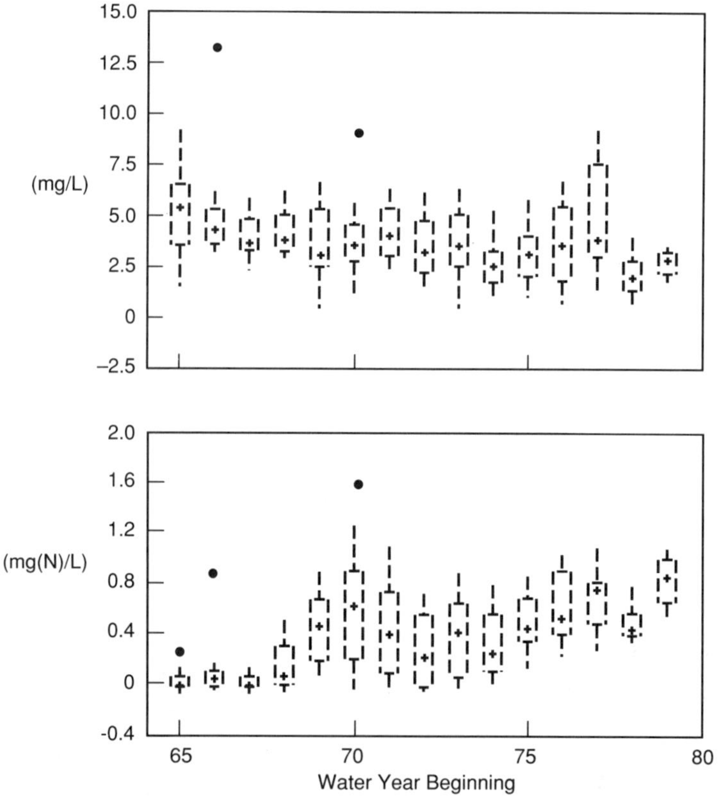


FIGURE 5.5 Box plot of monthly averaged $SO_4^{2-}$ (top) and aq $NO_3^-$ (bottom) observations in bulk deposition sample at Hinckley, New York, between 1965 and 1982. (From Hidy *et al.*, 1984; courtesy of the Air and Waste Management Association.)

These types of plots are useful as exploratory analysis to obtain a display of average behavior and the range of variation in several years of data. The median values are shown with crosses. The standard deviations are indicated by the inner boxed area, the range by the outer vertical lines, and (suspected) outliers by circles. This format is useful for providing a summary of basic statistical parameters displayed in one graph. The results indicate that within-year variation of monthly averaged data can range over more than a factor of 2, with a variance of 40% or more. Year-to-year differences in the medians are a factor of 2 or more during the period record. There is a downward trend of $SO_4^{2-}$ concentration at the site of about 2%/yr, which is similar to that observed at Hubbard Brook, about 160 km away. This change with time has been linked to decreased nearby $SO_2$ emissions (Hidy *et al.*, 1984; National Academy of Sciences, 1986).

## Data Representation by Multivariate Analysis

The phenomenological description of the occurrence and characteristics of regional-scale pollution events provides a framework for interpreting observations. The variations in behavior can also be analyzed for consistency with expectations from the event analysis. In this chapter, the analysis of variance is extended to include such an interpretation.

Three methods are reported in the literature: (1) factor analysis of time-series data, relating aerometric variables; (2) interpretation of factor analysis of spatial covariance; and (3) cluster analysis of precipitation chemistry with aerometric variables. Each provides insight into the significance of atmospheric processes in delineating receptor conditions and the possible pathways for SRRs.

Chapter 4 provided an important perspective on regionally distributed $SO_x$ behavior. The question to be addressed next concerns the extent to which a comprehensive analysis of data supports the interpretation revealed by the detailed examination of a few cases. To research this question, the SURE data set was examined using the statistical techniques of regression on principal components. The method of principal component analysis (PCA) is used to investigate $SO_4^{2-}$ variability over time at individual stations. The terms PCA, natural function analysis, and latent root analysis all refer to the same multivariate linear statistical techniques. References to the statistical method may be found in Harmon (1976). A different approach from the view of chemical applications is found in Malinowski and Howery (1980). Although PCA is actually a special type of factor analysis, the two terms are used interchangeably by some authors.

Scientists have made use of factor analysis, and an increasing number of applications to air quality data have been reported. These approaches have been adopted to interpret the variability in atmospheric $SO_x$ in relation to a large number of aerometric variables (Peterson 1970, 1972; Henry and Hidy, 1979, 1982). Prinz and Stratmann (1968) give a general account of the approach to air quality data analysis that is of interest. Another early application of factor analysis to aerosol behavior is found in the work of Blifford and Meeker (1967). Cleveland and Guarino (1976) used PCA to eliminate the effects of outliers in analysis of $O_3$ data from several sites in New Jersey. Hopke *et al.* (1975), Moyers *et al.* (1977), and Gatz (1982) have applied the methods of factor analysis to aerosol source identification by chemical element analysis. Peterson (1970, 1972) and Padmanabhamurty (1975) have used a form of empirical orthoganal function analysis (EOFA) to relate urban $SO_2$ distributions to emissions and meteorological parameters. PCA is found to be a useful method for highlighting and displaying the complex interrelationships within a large body of data and supporting explanations of ambient $SO_x$ behavior in terms of atmospheric processes.

Although statistical methods for data manipulation are employed, the following analysis is not statistical in spirit. That is, no formal hypotheses are stated explicitly and tested, and no calculations of statistical significance of results are made. The analysis presented is *exploratory* in nature, seeking to find important relationships that appear to be consistent with meteorological and case study descriptions rather than to test preconceived hypotheses. Questions of statistical significance of the results are difficult to answer in a quantitative manner because the variables in the analysis are unlikely to satisfy even approximately the assumptions inherent in rigorous methodologies. Only the main features of the data believed to be intuitively significant have been subjected to interpretation. The issue of formal verification of statistical significance remains open and will require further study.

## Elements of Principal Component Analysis

PCA analysis of data sets addresses several specific questions:

- What, if any, relationship exists between $SO_4^{2-}$ and $SO_2$ levels?
- What other pollutant variables, such as concentrations of $O_3$ and particulates, are related to $SO_4^{2-}$ levels?
- How is $SO_4^{2-}$ concentration related to uncontrollable factors such as weather or temperature and humidity?
- What is the dependence of $SO_4^{2-}$ and the other parameters on seasonal processes or on the diurnal cycles of atmospheric activity?

To address these questions using the factor analysis of the data, several important assumptions were made. Although their applicability on regional scales remains to be tested further, the following assumptions are based on work on urban scales of Henry and Hidy (1979, 1982):

1. The method postulates that a linear relationship exists between variables.
2. The variables of interest are available from the observations taken at a given ground station. Such observations were selected as surrogates of atmospheric processes influencing ambient $SO_4^{2-}$ concentration before the analysis.
3. $SO_4^{2-}$ concentration, taken as the dependent variable, need not be included in the manipulation of the remaining variables, the (linear) combinations of which were expected to yield the mutually independent components affecting $SO_4^{2-}$.
4. Based on accumulated theory and experience of $SO_x$ behavior, the principal components derived from the data are interpretable as indicators of important atmospheric processes.

Linear regression analysis is often used to relate a variable of interest, such as $SO_4^{2-}$ concentrations, to other aerometric parameters. In air quality analysis, these parameters are expected to be related physically to observations, such as wind speed, or are directly or indirectly related to emissions, as, for example, $SO_4^{2-}$ concentrations. The usual goal of multiple regression on observables is to find the best regression, i.e., the equation involving the fewest independent variables that account for a maximum of the variation in the dependent variable. Several alternative techniques for accomplishing this have been developed, for example, stepwise multiple linear regression. However, the usual forms of multiple regression are limited for environmental applications, especially air quality studies, because they do not take into account explicitly the fact that the environmental explanatory variables are usually highly correlated with each other. Thus, regression analysis using measured variables provides empirical relationships between arithmetic properties, but these are not easily interpreted without taking into account knowledge about process mechanisms.

If the variables selected as being "independent" in a regression are not actually independent, two serious difficulties may arise, one related to the mathematics and the other related to interpretation. The mathematical problem is demonstrated graphically by Swindel (1974). He shows that regression coefficients become unstable when the "independent" parameters are correlated. The coefficients are unstable in the sense that small perturbations or errors in the input data are magnified into large changes in the regression coefficients. McDonald and Schwing (1973) showed that regression of mortality statistics on air quality data leads to unstable and therefore inaccurate estimates of the effect of $SO_x$ or $NO_x$ on mortality. Mathematically, the problem is the same as trying to solve a set of linear equations whose determinant is almost zero. This problem has been minimized by others by preselecting for the regression only a few key variables judged to be almost independent. This procedure opens up the possibility of a more subtle but serious error in predetermination of results.

When relating the effects on a dependent variable of intercorrelated "independent" parameters in a regression, there is always uncertainty as to which variables are actually associated with the dependent variable and which have only an apparent association because of their correlation with the truly relevant parameters. However, certain combinations of the observables may be good predictors of the variable of interest, i.e., airborne $SO_4^{2-}$. An example of this kind is found in the work of Henry and Hidy (1979) and its identification of a photochemical smog intensity component in urban air that is a combination of temperature, $O_3$, absolute humidity, and other meteorological variables.

The conventional concept that the best regression equation has the fewest explanatory variables in it is not necessarily applicable to understanding environmental problems, in which complex phenomena represented by combinations of variables rather than one single variable are often of primary importance. Because the basic problems stem from the correlation among the assumed independent variables in the regression, an alternative solution is to replace the original intercorrelated parameters by new uncorrelated variables before the regression analysis is attempted. PCA is used to obtain the new variables of interest. Because the new collections of variables are uncorrelated, numerical instabilities in the calculations are eliminated. The new variables are linear combinations of the original observables that can be interpreted in physical and chemical terms as indications of the underlying atmospheric processes. These key properties are a prime motivation in the choice of PCA as a means of analyzing air quality data.

*Principal Component Representation* The application of PCA stated in mathematical terms requires a discussion of concepts and definitions. The data are assumed to consist of simultaneous observations of $n$ variables for $m$ time periods (e.g., weeks, days, or hours). Then the data set is expressed in matrix form wherein each vertical column contains the observations of $n$ variables for one time period and each horizontal row contains all $m$ observations of one variable. The $j$th observation of the $i$th variable is written $V_{ij}$ and the total data matrix is written as

| | Time period | | | |
|---|---|---|---|---|
| | 1 | 2 | $\cdots$ | $m$ |
| Var. 1 | $V_{11}$ | $V_{12}$ | $\cdots$ | $V_{1m}$ |
| Var. 2 | $V_{21}$ | $V_{22}$ | $\cdots$ | $V_{2m}$ |
| . | . | . | | . |
| . | . | . | | . |
| . | . | . | | . |
| Var. $n$ | $V_{n1}$ | $V_{n2}$ | $\cdots$ | $V_{nm}$ |

It is convenient to normalize the data set so that all the variables have a common average value of zero and a standard deviation of unity. This is accomplished by the transformation

$$Z_{ij} = \frac{V_{ij} - \overline{V}_i}{\sigma_i}$$

where

$Z_{ij}$ = normalized value of $V_{ij}$
$\overline{V}_i$ = average of variable $i = m^{-1} \sum_{k=1}^{m} V_{ik}$
$\sigma_i$ = standard deviation of variable $i$
$[(n - 1)^{-1} \sum_{k=1}^{m}(V_{ik} - \overline{V}_i)^2]^{1/2}$

The mathematical formalism of PCA is to find a set of equations of the following form:

$$Z_{ij} = \sum_{k=1}^{n} a_{ik} P_{kj}, \qquad i = 1, \ldots, n, j = 1, \ldots, m \tag{5.1}$$

where $a_{ik}$ are weighting factors specific to each variable and $P_{kj}$ are the values of the independent principal components that replace the original observables. Figure 5.6 expresses this equation in matrix form and provides a graphic interpretation. The weighting factors are numbers between +1 and −1. The larger the absolute value of the weighting factor, the greater the dependence of the variable on the principal component, and vice versa. The fact that the principal components are independent is expressed by the equation:

$$\sum_{k=1}^{m} P_{ik} P_{jk} = \begin{bmatrix} 0 & \text{if } i \neq j \\ 1 & \text{if } i = j \end{bmatrix} \tag{5.2}$$

From matrix algebra, it can be shown that an infinite number of sets of weighting factors and components will satisfy Eqs. (5.1) and (5.2). Therefore, the definition of a unique principal component deposition of the data requires an additional condition. This condition states that if fewer than

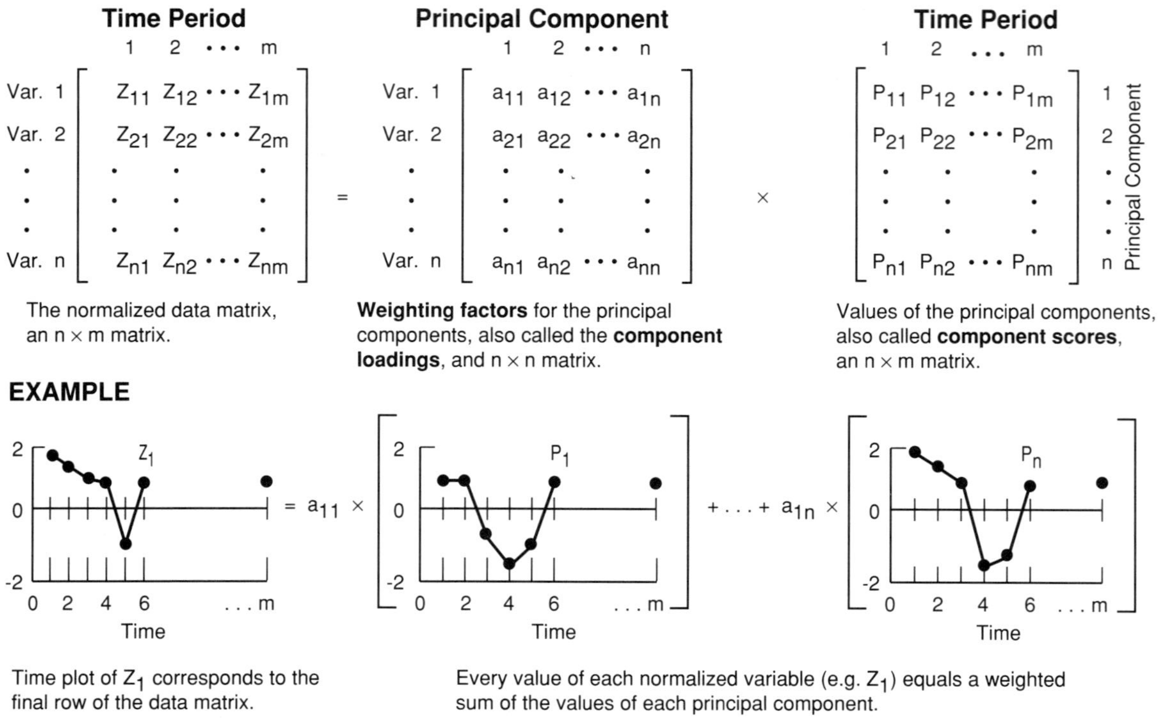


FIGURE 5.6 Schematic description of the prinicipal component of a data matrix. (From Mueller and Hidy, 1983; with permission from the Electric Power Research Institute.)

the total number of principal components is used in approximating the data matrix, then the error in this approximation is minimized in the usual least-squares sense. In symbolic form this is stated as follows:

$$Z'_{ij} = \sum_{k=1}^{Q} a_{ik} P_{kj} \tag{5.3}$$

where the sum is taken over $Q$ principal components. Then the $a_{ik}$ and $P_{kj}$ are chosen to satisfy conditions (5.1) and (5.2) and to minimize the quantity

$$\sum_{i=1}^{n} \sum_{j=1}^{m} (Z_{ij} - Z'_{ij})^2 \tag{5.4}$$

This is equivalent to the condition that each principal component explains a maximum of the variability (sum of squares) of all the variables independent of each of the other principal components. The principal component that fits all the observations best [i.e., minimizes expression (5.4)] is called the first principal component. The second principal component is the best fit to all the data orthogonal (perpendicular) to the first principal components, the $P_{kj}$ in Eq. (5.1) are obtained, and their relation to $SO_4^{2-}$ is determined by linear regression to establish a formula of the form $SO_4^{2-}$

$$[SO_4^{2-}]_j = [\overline{SO_4^{2-}}] + \sum_{k=1}^{n} b_k P_{kj} \tag{5.5}$$

where

$[SO_4^{2-}]_j$ = concentration of $SO_4^{2-}$ for time period $j$ at a station
$[\overline{SO_4^{2-}}]_j$ = average $SO_4^{2-}$ concentration at a station
$P_{kj}$ = value of the $k$th principal component for time $j$
$b_k$ = regression coefficient of component $k$ on $SO_4^{2-}$

Computationally, the regression on the principal components is carried out using the following relationships. The values of the principal components are given by

$$P_{jk} = \sum_{i=1}^{n} (a_{ij}/\lambda_j) V_{ik} \tag{5.6}$$

where $\lambda_j$ is the eigenvalue associated with $P_{jk}$.

The regression of $Y$, the dependent variable, on $P_i$ gives formula (5.5). Because the principal components are uncorrelated, the least-squares estimates of the regression coefficients are given by the formula

$$b_i = \frac{1}{M - 1} \sum_{i=1}^{M} Y_k P_{ik} \tag{5.7}$$

where $M$ is number of cases and $Y_k$ is the $k$th value of the dependent variable $[SO_4^{2-}]$.

***Properties of Data Decomposition*** The application of the PCA can be appreciated by considering certain basic properties of the decomposition of the data set. Five of these properties are noted for their importance in analyzing the aerometric data.

1. *Principal components are uncorrelated.* This property was sought for initially and is of great importance for two reasons. First, it ensures that the multiple regression using major principal components as variables will be stable. Second, if several components are found to have a physical interpretation, then regression on these principal components is a measure of the effect on the dependent variable of the physical or chemical process associated with each principal component.
2. *PCA reduces the number of variables.* PCA of highly intercorrelated data usually produces a small number of major principal components that together account for the observations with less than 10% error. It is common for 20 or 30 intercorrelated variables to be fit to ±10% by only 6 or 7 independent principal components. Furthermore, often only three or four major components can account for 75% of all the variability in the original data set. This organization and condensation of a large collection of observations is useful in itself as a way of efficiently examining the structure of the data as shown by the weighting factors of the major components.
3. *PCA reduces the effects of random error* in the description of the main features of the data set. Mathematically, the effect of random errors in the original data is most heavily borne by the minor principal components, that is, those that explain very little (<10%) of the data variability. Therefore, by restricting the analysis to the major components, much of the random error is suppressed and is identified with the smaller components. Experience has shown that the values of the weighting functions and principal components are quite stable to random errors. Of course, large systematic errors in the observations will seriously affect the PCA results.
4. *PCA uncovers the basic underlying structure of the data.* The intercorrelations in observed variables are defined here to be the structure of the data set. PCA points out explicitly the groups of variables that are highly correlated and the manner in which they are correlated. This is

done by examining the values of the weighting factors in Eq. (5.1). Variables that all have large weighting factors for a major component are highly correlated. If some of the large weighting factors are negative, these variables are negatively correlated with the other variables. Note that principal components are *linear* relationships among the explanatory variables; if these variables include nonlinear functions of the data, then PCA does uncover nonlinear relationships in the data.

5. *Principal components can be interpreted physically*. By examining the variables with large weighting factors for a given component, it is often possible to associate that component with a physical or chemical process in the atmosphere. This is a fundamental hypothesis of PCA interpretation, which has emerged from previous work. For example, a component may have large positive weighting factors for TSP, $SO_2$, and $SO_4^{2-}$ and large negative weighting on mixing height and wind speed. This component is evidently the result of high pollutant concentrations correlated with low inversions and wind speeds. The association of this component with overall dispersion or dilution is straightforward. Other commonly seen components can be identified with photochemistry, humidity, and seasonal variations. When the principal components are identified with physical processes, the principal component regression model relates effects on the dependent variable of each of the processes without the numerical instabilities that exist when regressions are done on the highly intercorrelated original variables.

## Approach to Analysis from Observations

The goal of the PCA was to find relationships between meteorological and air quality variables measured at a specific time and place with $SO_4^{2-}$ levels measured at the same time and place. The crucial assumption that interpretation of the principal components can be related to physically meaningful terms is reasonable only if enough variables representing important processes have been included in the analysis. Thus, it is critical to select measured variables that are indexes of the important processes believed to affect airborne $SO_4^{2-}$. These processes include the geographically distributed emissions of $SO_x$, as well as atmospheric transport, chemical transformation, and surface deposition. Table 5.3 lists the variables chosen for the data sets by Henry (reported in Mueller and Hidy, 1983) and the rationale for the choice. Estimates of variability in $SO_x$ emissions were not in an accessible form at the time the analysis was performed. The PCA analysis does not explicitly account for emission variability at this stage of application. $SO_x$ levels at the stations are assumed to be indicative of emissions in an undesignated functional form.

Before conducting the PCA of the SURE observations, the data set was reviewed carefully to obtain a subset of data that were free of identifi-

TABLE 5.3 **Parameters Selected for PCA Analysis of Data Sets[a]**

| **Parameter (symbol) and units** | **Definition and/or reason for choice** |
|---|---|
| Water-soluble sulfate ($SO_4^{2-}$ in $\mu g/m^3$) | Subject of study |
| Inhalable particulate mass[b] less sulfate (IPMM) in $\mu g/m^3$ | Amount of fine and coarse particulate mass less than 15 $\mu m$ diameter |
| Refined suspended particulate[b] less sulfate (RSPM) in $\mu g/m^3$ | Nonsulfate fine particle mass (<3 $\mu m$ diameter); related to copollutants and possible catalyst of $SO_2$ oxidation |
| 24-hour HIVOL mass concentration (TSP)[c] in $\mu g/m^3$ | Fine and coarse particle loading for less than 30 mm diameter; assumed to be a major covariable with $SO_4^{2-}$ |
| Sulfur dioxide ($SO_2$) in ppm | Major sulfate precursor pollutant; in some manner indicative of $SO_x$ emissions |
| Nitrogen oxides (NO and $NO_x$) in ppm | Indicative of combustion source emissions; important in photochemical oxidation of $SO_2$ |
| Ozone ($O_3$) in ppm | Primary indicator of atmospheric photochemical processes |
| Surface wind speed (SWS) in m/s | Contributor to local dispersion |
| Upper-level wind speed (UWS) in m/s[d] | Representative of regional dispersion |
| Mixing height (MIX) in m[d] | Contributor to dispersion |
| Ventilation (VENT) in $s/m^2$[d] | Derived measure of dispersion = $(MIX)^{-1} \times (UWS)^{-1}$ |
| Surface temperature ($T$) in °C | Related to the speed of chemical processes |
| Surface dew point temperature (DP) in °C | Measure of atmospheric water content; important to photochemical and heterogeneous oxidation chemistry |
| Relative humidity (RH) in % | Derived parameter calculated from $T$ and DP; related to fog formation and heterogeneous oxidation chemistry |
| Absolute humidity (AH) in mb $H_2O$ | Derived parameter calculated from $T$ and DP; potentially important in many chemical oxidation schemes; also a good air mass tracer |
| Station pressure ($P$) in mb | Air mass tracer |
| Cloud cover (CC) in %[d] | Related to available sunlight and photochemistry |
| Precipitation (PPT) in mm/3 h[d] | Related to wet deposition |

[a] From Mueller and Hidy, 1983; reproduced with permission of the Electric Power Research Institute.
[b] In the 3-hour sets RSP values were used; in the 24-hour sets, HIVOL values were used.
[c] Included only in the 3-hour PCA data for intensive months.
[d] Used in 24-hour average data set only, without subtracting $SO_4^{2-}$ concentration.

able errors or outliers that were inconsistent with the majority of related variables at the same time. Data were selected for 3-hour averages for the intensive sampling months and 24-hour averages for a year period between October 1977 and September 1978. For each of nine SURE sites, the 3-hour data subsets were assembled, each consisting of coincident particulate, meteorological, and air quality data from the first five intensive study periods. In addition, the daily 24-hour average measurements taken at each CI site were synthesized into similar comprehensive data sets covering the period October 1977 to September 1978. Considerable effort was made to ensure the quality and reliability of the data. The screening included computer checks for internal consistency and manual scanning of the data files. Because the PCA requires a complete record of all chosen observables, sets of data with missing records had to be eliminated or the missing sections extrapolated. In the case of the 24-hour average data sets, only complete sequences of daily data were retained for the PCA analysis. In general, the analysis relied on the stations with the best data captures to mitigate possible ambiguities introduced by missing data.

*Analysis of the Annual Data Sets* The set of daily average air quality data of the SURE and meteorological data were assembled for the annual period of October 1977 through September 1978. Table 5.4 lists the subset

TABLE 5.4 **Parameters of Annual Class I Data Sets Used in The PCA**[a]

| |
|---|
| Water-soluble $SO_4^{2-}$ from 24-hour high-volume samples taken on Teflon-coated glass fiber filter media ($\mu g/m^3$) |
| Water-soluble $NO_3^-$ from samples as above ($\mu g/m^3$) |
| TSP obtained as above ($\mu g/m^3$) |
| $SO_2$, 24-hour average (ppm) |
| $NO_x$, 24-hour average (ppm) |
| $O_3$, 24-hour average (ppm) |
| Wind speed, average of 24 observations (m/sec) |
| Pressure, station pressure, PRES (mb) |
| PPT in 24-hour period (mm $H_2O$) |
| Maximum temperature in 24-hour period, TMAX (°C) |
| Minimum temperature in 24-hour period, TMIN (°C) |
| Dew point depression at TMAX, DWMA (°C) |
| Dew point depression at TMIN, DWMI (°C) |
| Relative humidity at TMAX, RHMA (%) |
| Relative humidity at TMIN, RHMI (%) |
| Absolute humidity at TMAX, AHMA (mb ($H_2O$) |
| Absolute humidity at TMIN, AHMI (mb $H_2O$) |

[a] From Mueller and Hidy, 1983; reproduced with permission of the Electric Power Research Institute.

of the variables listed in Table 5.3 that were adopted for this computation. This data set is basically the same as that used in the air mass classification analysis described previously. The aerometric data base used in the PCA of the annual data set consisted of the 24-hour particulate data from the high-volume samplers, including TSP, $SO_4^{2-}$, and $NO_3^-$ ($\mu g/m^3$) as well as 24-hour averages of hourly $SO_2$, $NO_x$, and $O_3$ gas concentrations (ppm) from each station. These data were combined with the seven meteorological parameters derived from surface stations. The PCA was performed on the data collection for each of the nine CI sites. To ensure meaningful results representative of the entire year, detailed analysis was restricted to sites with 200 or more days of complete data for all variables selected.

Table 5.5 summarizes the results of the regression of $SO_4^{2-}$ on the principal components for these sites. The number of underlying independent variables in the set of 17 variables ranged from seven to nine, indicating that the observational data are highly intercorrelated. When $SO_4^{2-}$ was regressed on the most significant components, between 39 and 70% of its variability could be accounted for. Typically 60 to 70% of the $SO_4^{2-}$ variance was accounted for by both major and minor principal components of the data. This result was to be expected because $SO_4^{2-}$ is a substantial component of TSP. If the variable nonsulfate particulates (NSP) were used instead of TSP (NSP = TSP − sulfate), then at Rockport the percentage of $SO_4^{2-}$ variance accounted for changes from 58.4 to 64.5% and at Montague the percentage of $SO_4^{2-}$ variance accounted for changes from 59.6 to 73.1%.

To determine what types of variations in the data were most strongly related to $SO_4^{2-}$ change, the analysis was further restricted to the principal components that explained at least 10% of the $SO_4^{2-}$ fluctuations. Table 5.5 shows that of the seven to nine most important uncorrelated principal components, between two and four are thus strongly related to $SO_4^{2-}$. One site, Giles County, had no component meeting this criterion. Therefore, the $SO_4^{2-}$ at this site was not significantly related to the locally measured aerometric variables included in the analysis.

To illustrate the connection of $SO_4^{2-}$ to the actual observables instead of the principal components, the makeup of the principal components was examined in terms of the mix of measured parameters that made up the component. Some of the observables were more heavily weighted in a given component than others. Table 5.6 gives the structure of all the components significantly related to $SO_4^{2-}$ in the annual data sets. A cross (x) means that the component contains a strong contribution from that observable (the weighting factor is $>-0.4$). A minus sign in front of the cross means that the component is negatively related to that observable (the weighting factor is $\leq -0.4$). For a positive relation, $SO_4^{2-}$ is high when the variable is *higher* than average; for a negative correlation, $SO_4^{2-}$ is high when the observable is *lower* than average.

TABLE 5.5 **Summary of Regression of $SO_4^{2-}$ on Principal Components of Annual Sets of 24-Hour-Average Data for SURE Sites with More Than 200 Days of Data**

| | Station 1 (Montague) | 4 (Duncan Falls) | 5 (Rockport) | 6 (Giles County) | 8 (Research Triangle Park) |
|---|---|---|---|---|---|
| Number of components to account for >90% of total variability[a] | 7 | 8 | 8 | 9 | 9 |
| $r^2$ for $SO_4^{2-}$ using these components[b] | 0.66 | 0.70 | 0.59 | 0.39 | 0.62 |
| Number of components explaining >10% of $SO_4^{2-}$ variability | 2 | 4 | 3 | 0 | 2 |
| $r^2$ for $SO_4^{2-}$ using these components | 0.51 | 0.64 | 0.46 | 0.0 | 0.43 |
| Number of days | 279 | 201 | 270 | 201 | 290 |
| $SO_4^{2-}$ mean and standard deviation | 6.0 (5.2) | 12.3 (8.8) | 11.0 (8.0) | 11.3 (7.1) | 8.6 (5.4) |

[a] This is the effective number of degrees of freedom in the set of 17 variables.
[b] $r^2$ = variance.

TABLE 5.6 **Observables Significantly Contributing to Principal Components of Annual Data Sets That Explain More Than 10% of the $SO_4^{2-}$ Variability.**[a,b]

| | **Station 8** | | **Station 1** | | | **Station 4** | | | | **Station 5** | | **Sum** (+) | **Sum** (−) |
|---|---|---|---|---|---|---|---|---|---|---|---|---|---|
| Sulfur dioxide ($SO_2$) | x | x | x | x | x | | −x | x | x | −x | x | 8 | 2 |
| Nitrogen oxides ($NO_x$) | | | x | | | | −x | x | x | −x | | 3 | 2 |
| Ozone ($O_3$) | | | | x | x | x | | | | x | | 4 | |
| Suspended particles (TSP) | | x | x | x | x | x | | | x | x | | 7 | |
| Nitrate ($NO_3^-$) | −x | | x | | | | −x | | x | | | 2 | 2 |
| Wind speed (WS) | | | | x | −x | | | −x | −x | | | 1 | 3 |
| Barometric pressure (PRES) | −x | | x | −x | x | | | | x | | | 3 | 2 |
| Precipitation (PPT) | | | −x | | −x | | | | | | | | 2 |
| Max. temp. (TMAX) | | | | | | | x | | | x | | 2 | |
| Max. dew pt. depression (DWMA) | | | x | | x | | | −x | | x | | 3 | 1 |

| | | | | | | | | | | | | | |
|---|---|---|---|---|---|---|---|---|---|---|---|---|---|
| Min. temp. (TMIN) | | | | | | | x | | | x | | 2 | 0 |
| Min. dew pt. depression (DWMI) | | | | | | | −x | | | x | | 1 | 1 |
| Max. relative humidity (RHMA) | | | −x | | −x | | | x | | −x | | 1 | 3 |
| Max. absolute humidity (AHMA) | | | | | | | x | | | x | | 2 | 0 |
| Min. relative humidity (RHMI) | | | | | | | x | | | | | 1 | 0 |
| Min. absolute humidity (AHM) | | | | | | | x | | | x | | 2 | 0 |
| Assigned physical interpretation (if any) | — | — | High-pressure stagnation | Transport | High-pressure stagnation | — | Seasonality | — | High-pressure stagnation | Seasonality | — | | |
| % $SO_4^{2-}$ variability explained | 23.9 | 18.6 | 25.5 | 25.8 | 24.6 | 15.7 | 12.5 | 11.5 | 25.5 | 10.6 | 10.0 | | |
| $SO_4^{2-}$ coefficient | 2.62 | 2.31 | 2.65 | 2.67 | 4.38 | 3.50 | 3.12 | 3.00 | 4.05 | 2.62 | 2.53 | | |
| % of total variability explained | 7.9 | 6.9 | 14.3 | 10.6 | 17.3 | 8.6 | 34.1 | 11.7 | 14.0 | 33.4 | 7.0 | | |

[a] From Mueller and Hidy, 1983; reproduced with permission of the Electric Power Research Institute.

[b] x = important positive contributor to component; −x = important contributor in a negative or opposed correlation sense.

Table 5.6 contains examples of two types of components that emerged as a common factor in the PCA at CI sites analyzed. These are labeled *seasonality* and *high-pressure stagnation*. In all the cases the majority of the variability of the total data set (not including $SO_4^{2-}$) was explained by a component similar to the seasonal components given in the table for Duncan Falls, OH (station 4). This component is correlated with high temperatures and absolute humidities and with low $SO_2$, $NO_x$, $NO_3^-$, and dew point depression. Temperature and absolute humidity are good indicators of seasonal variation because of their known regular change with season. $SO_2$, $NO_x$, and $NO_3^-$ evidently appeared in this component because they show significant seasonal variations. In this case, $SO_2$, $NO_x$, and $NO_3^-$ tended to be seasonally high when temperature was low (i.e., $SO_2$ peaks in the winter and is at a minimum in the summer).

Because a similar component was common to all the stations, the degree to which $SO_4^{2-}$ was a seasonally related variable indicated how much of its variance is explained by regression on this component. Note that by the nature of PCA this component was uncorrelated with all of the other variability in the data. For Duncan Falls (4) and Rockport, IN (5), just over 10% of the $SO_4^{2-}$ variability was seasonal, with $SO_4^{2-}$ being slightly higher in the summer. For the other components, slightly less than 10% of the $SO_4^{2-}$ was related to seasonal variation. Thus, for the period covered by these data, about 10% of the $SO_4^{2-}$ variability could be related to seasonal differences, which are identified with annual cycles of weather patterns described in the synoptic analysis. In other words, $SO_4^{2-}$ showed a small variation with season according to the PCA. However, there were clear seasonal differences in $SO_4^{2-}$ behavior. $SO_4^{2-}$ tended to show peaks in both the winter and summer. The PCA was not able to distinguish this kind of seasonal behavior. The variables associated with the seasonal principal component, temperature, $O_3$, $NO_2$, and $SO_2$, were seasonal in the sense that they showed a strong maximum in the summer (or winter) with a strong minimum in the winter (or summer).

Table 5.6 contains three examples of components identified with high barometric pressure or anticyclonic stagnation. This component was related to high primary pollutant concentrations ($SO_2$, $NO_x$, TSP), low wind speeds, and high barometric pressure. At Montague, MA (1), Duncan Falls (4), and Rockport (5), fully 25% of the yearly $SO_4^{2-}$ variability was associated with high-pressure stagnation. In fact, sites at Research Triangle Park, NC (8) and Giles County, TN (6) are the only stations of the nine that did not have $SO_4^{2-}$ depending strongly on stagnation. From the meteorological assessment, these sites are probably on the eastern and southwestern periphery of the area dominated by cPw and cP2 air mass situations associated with the eastward advancing continental high-pressure regions. Because the seasonal principal component captured most or all of the seasonal information in the data, the stagnation component took on high values in all seasons, confirming that high-pressure stagnation could occur during any season.

An indication of the dominating influence of specific observables at all stations can be obtained by summing the crosses for each row in Table 5.6. Eight of the eleven most important components influencing $SO_4^{2-}$ variability included *high* $SO_2$ loadings, while two, both seasonal components, have *low* $SO_2$ loadings. Thus, PCA showed that $SO_4^{2-}$ was strongly related to $SO_2$ concentrations measured at the same sites.

TSP was the second observable strongly related to $SO_4^{2-}$ by inspection of Table 5.6. Because $SO_4^{2-}$ is a component of TSP in this particular data configuration, this finding was not surprising. The TSP–$SO_4^{2-}$ relationship may indicate a dependence of $SO_4^{2-}$ on other factors that contributed to high TSP, such as stagnation or the ground-level impact of plumes. $O_3$ was the next most important observable, presumably more for its role as a general surrogate for photochemical oxidation processes than as a direct reactant with $SO_2$ in the gas phase or in clouds.

The effect of meteorological variables was found to be inconsistent. There was a tendency for elevated $SO_4^{2-}$ to be related to high barometric pressure with low relative humidities. The latter observation differs from other analyses of urban data that suggest a positive correlation between relative humidity and sulfate (Henry and Hidy, 1979, 1982). Because relative humidity appeared to be associated mainly with the seasonal or stagnation variables, its negative correlations with $SO_4^{2-}$ may be related to reduced humidity in fair weather. Interestingly, absolute humidity showed little systematic relation to $SO_4^{2-}$ in this data set, as contrasted with urban data sets. Evidently, the absolute humidity was dominated by its seasonal behavior here, and any relation to $SO_4^{2-}$ air chemistry was submerged in the seasonal variation. The seasonal principal component may have contained some nonseasonal information that would make it difficult to assess the importance of a variable such as absolute humidity, which has such large loadings (often greater than 0.9) for the seasonal principal component.

*Analysis of the 3-hour Average Data Sets* To examine the short-term variability of $SO_4^{2-}$ in relation to other observables, a data base of 3-hour averaged air quality and meteorological observations was assembled for SURE CI sites for each intensive period. The parameters included have been listed in Table 5.3. To prepare the data collection, 3-hour precipitable water from local climatological data tabulation, mixing height, wind speed at 300 m, and ventilation coefficient were added. The latter were interpolated from twice-daily rawinsonde observations recorded in the SURE region (Benkley and Bass, 1978).

The station data sets were subjected to PCA, and the $SO_4^{2-}$ was regressed on the resulting principal components, in the same way as the 24-hour averaged annual data. The same general questions were posed for the data analysis during key intensive periods as for the analysis of annual data.

A large amount of information is represented by the PCA of the 3-hour set for the sites in the six intensive months. Only two illustrative data-intensive sampling periods are analyzed here in detail: January and February 1978 and July 1978. The size of the data collections adopted for the PCA is listed by Mueller and Hidy (1983). Generally, only sites with more than half of the maximum possible complete 3-hour data records are discussed. The smaller data sets, such as those available for Lewisburg, could give misleading or biased results and are not discussed here. In one summer case, an exception was taken to this guideline to allow the Duncan Falls station to be included in both the winter and summer analyses for comparison. The analysis of the SURE data-intensive sampling period of August 1977 gave results similar to those presented below for July 1978. The segregation and identification of components for the spring and fall intensive data sets display a pattern corresponding to a mixture of the summer and winter results.

Table 5.7 summarizes the January 1978 results. The PCA showed that of the 16 to 17 variables in the analysis, there are only 8 or 9 truly independent underlying components. The regression of $SO_4^{2-}$ on the first 9 or 10 principal components resulted in 58 to 87% of the $SO_4^{2-}$ variability being explained. This was considered to be a successful regression. The summary tabulation shows that between one and four of the principal components explained more than 10% of the $SO_4^{2-}$ variability at each site and that these components accounted for most of the $SO_4^{2-}$ variance accounted for by the measurements included in the analysis.

Table 5.8 presents the structure of each component, at each site, that explained 10% or more of the $SO_4^{2-}$ variability. As in Table 5.6, only the observables contributing strongly to the components accounting for $SO_4^{2-}$ variability are identified by station. The most important component in accounting for $SO_4^{2-}$ at most stations was associated with high values of primary pollutants (RSPM, IPMM, $SO_2$, $NO_x$), but with low wind speeds (surface and elevated) and low mixing heights. It was reasonable to identify these components as associated with high-pressure stagnation conditions because pollutants from primary emissions should be high during stagnation conditions. Between 20 and 62% of the $SO_4^{2-}$ variability could be accounted for by this type of component. High levels of $SO_4^{2-}$ were also associated with low concentrations of $O_3$ in winter. This suggested that photochemical or oxidant production processes are a minor factor in $SO_4^{2-}$ production in winter.

TABLE 5.7 **Summary of PCA Results for 10 January–4 February, 1978, 3-Hour Sure Data**[a]

| | **Montague 1** | **Duncan Falls 4** | **Rockport 5** | **Giles County 6** | **Fort Wayne 7** | **Research Triangle Park 8** |
|---|---|---|---|---|---|---|
| Number of components to account for ≥90% of total variability[b] | 8 | 3 | 9 | 9 | 8 | 8 |
| $r^2$ for $SO_4^{2-}$ using these components[c] | 0.87 | 0.58 | 0.80 | 0.75 | 0.85 | 0.64 |
| Number of components explaining ≥10% of $SO_4^{2-}$ variability | 1 | 1 | 2 | 4 | 3 | 3 |
| $r^2$ for $SO_4^{2-}$ using these components | 0.62 | 0.40 | 0.51 | 0.69 | 0.71 | 0.53 |
| $SO_4^{2-}$ mean ($\mu g/m^3$) | 5.06 | 9.43 | 10.10 | 9.48 | 8.67 | 6.23 |
| $SO_4^{2-}$ mean and standard deviation ($\mu g/m^3$) | 3.59 | 5.31 | 6.50 | 5.35 | 6.09 | 3.23 |
| Number of 3-hour periods | 156 | 144 | 163 | 139 | 170 | 110 |

[a] From Mueller and Hidy, 1993; reproduced with permission of the Electric Power Research Institute.
[b] This is the effective number of degrees of freedom in the set of 17 variables.
[c] $r^2$ = variance.

TABLE 5.8 **Components of the January Intensive PCA Data Sets That Explain More Than 10% of the $SO_4^{2-}$ Variability**[a]

| | Station 1 | Station 4 | Station 5 | | | Station 6 | | | Station 7 | | | | Station 8 | | Totals (+) | Totals (−) |
|---|---|---|---|---|---|---|---|---|---|---|---|---|---|---|---|---|
| RSPM | x | x | x | | x | x | | x | x | | | x | x | | 9 | |
| IPMM | x | — | x | | x | | | | x | | | x | x | | 6 | |
| $SO_2$ | x | x | x | x | | | x | | x | x | x | | | x | 9 | |
| $NO_x$ | x | x | x | | x | | | | — | — | — | x | | | 5 | |
| $O_3$ | −x | −x | | | −x | | | | −x | x | | −x | | | 1 | 5 |
| Suface wind speed | −x | −x | −x | x | −x | | x | | −x | | | −x | | | 2 | 6 |
| Pressure | x | x | x | | | | | | x | | | | x | | 5 | |
| Temperature | | | −x | | x | −x | | x | | | | x | −x | −x | 3 | 4 |
| Dew point depression | | | x | | | −x | | | −x | −x | | −x | x | −x | 2 | 5 |
| Precipitation | | | −x | | | | | | | | | | | −x | | |
| Cloud cover | | | | | | | | | | | | x | | | 1 | |
| Relative humidity | | −x | | | x | | | x | x | | x | −x | x | x | 5 | 2 |
| Absolute humidity | | | −x | | | | | | | | | x | −x | | 1 | 2 |
| Mixing height | −x | −x | | x | −x | | | | −x | x | | −x | | | 2 | 5 |
| Wind speed at 800 mb | −x | −x | −x | | −x | | | | −x | | | −x | | | | 6 |
| Ventilation$^{-1}$ | | | | | x | | | | x | −x | | | | | 2 | 1 |
| $r^2$ for $SO_4^{2-}$ | 0.62 | 0.40 | 0.24 | 0.27 | 0.27 | 0.13 | 0.18 | 0.11 | 0.40 | 0.18 | 0.13 | 0.20 | 0.10 | 0.24 | | |
| $SO_4^{2-}$ regression coefficient | 2.83 | 3.37 | 3.19 | 3.36 | 2.76 | 1.96 | 2.28 | 1.76 | 3.85 | 2.58 | 2.23 | 1.43 | 1.04 | 1.57 | | |
| % of Total variability explained | 31 | 24 | 26 | 9.2 | 29 | 11 | 8.9 | 5.3 | 31 | 10 | 5.6 | 33 | 16 | 11 | | |
| Assigned physical interpretation | High-pressure stagnation | High-pressure stagnation | High-pressure stagnation | — | Warm stagnation | — | — | — | High-pressure stagnation | — | — | Warm stagnation | — | — | | |

[a] From Mueller and Hidy, 1993; reproduced with permission of the Electric Power Research Institute.

[b] x = important positive condtributor to component; −x = important contributor in a negative or opposed correlation sense.

The role of temperature and moisture in the winter stagnation components differed among sites with no clear pattern emerging. The more southerly sites, Giles County and Research Triangle Park, showed a tendency toward warmer-than-average temperatures and more humid conditions during stagnation periods.

Because relative humidity was considered a measure of liquid water content in particulates, RH may be a surrogate measure of the presence of aqueous–heterogeneous $SO_2$ oxidation rates. The tendency for the positive correlation in winter might identify the significance of heterogeneous chemistry in that season.

Examination of the last two columns of Table 5.8 also shows that refined particulate mass concentrations (less $SO_4^{2-}$) and $SO_2$ concentrations are both strong in the nine major components; they are therefore the dominant variables. Among the meteorological parameters, low surface and upper-level wind speeds (from the rawinsondes) were most often related to high $SO_4^{2-}$ concentrations at the stations. Although each major component was assumed to have a physical meaning, Henry (reported in Mueller and Hidy, 1983) was not able to interpret some of the components in terms of known physical processes.

Analysis of the data from the July 1978 intensive sampling period showed that the patterns of variability exhibited in each of the SURE CI data sets were remarkably similar to one another. In essentially all of the summer data examined, three components were found to explain a majority of the $SO_4^{2-}$ variability as well as the total variability of the data set. The components were identified with major climatological or meteorological variations in the atmosphere: the diurnal temperature cycle, air circulation around high-pressure cells, and warm, moist air mass conditions. This was consistent with the results obtained from the synoptic analysis using air mass classification discussed in Chapter 4.

A summary of the PCA results for the July 1978 data in Table 5.9 shows that 9 or 10 uncorrelated principal components could explain 90% of the variance of the 16 to 17 variables in the data set. Except for the Giles County site, 60 to 80% of the $SO_4^{2-}$ variability could be accounted for by locally measured parameters. Just which variables were the most important and how they related to each other can be seen in Table 5.10, where all of the major contributing observables to the components that account for 10% or more of the $SO_4^{2-}$ are listed in the format adopted for Table 5.6 and Table 5.8.

In each case, the major component, defined as the one that explains most of the variability of the whole data set, was related to the diurnal temperature cycle. In the summer, unlike winter, major changes take place between day and night. During the day, temperatures are high, as are wind speeds, mixing heights, and $O_3$ concentrations. However, rela-

TABLE 5.9 **Summary of PCA Results for July 1978 3-Hour Sure Station Data**[a]

| | Station | | | | | |
|---|---|---|---|---|---|---|
| | Montague | Duncan Falls | Rockport | Giles | Fort Wayne | Research Triangle |
| Number of components to account for ≥90% of total variability[b] | 9 | 9 | 10 | 9 | 10 | 9 |
| $r^2$ for $SO_4^{2-}$ using these components[c] | 0.76 | 0.62 | 0.70 | 0.40 | 0.80 | 0.64 |
| Number of components explaining ≥10% of $SO_4^{2-}$ variability | 2 | 2 | 3 | 1 | 3 | 2 |
| $r^2$ for $SO_4^{2-}$ using these components | 0.69 | 0.53 | 0.54 | 0.22 | 0.72 | 0.49 |
| $SO_4^{2-}$ mean ($\mu g/m^3$) | 12.8 | 13.5 | 12.7 | 8.20 | 12.7 | 8.96 |
| $SO_4^{2-}$ standard deviation ($\mu g/m^3$) | 11.9 | 9.14 | 9.15 | 5.76 | 11.3 | 5.16 |
| Number of 3-hour periods | 129 | 116 | 137 | 127 | 136 | 132 |

[a] From Mueller and Hidy, 1983; reproduced with permission of the Electric Power Research Institute.
[b] This is the effective number of degrees of freedom in the set of 17 variables.
[c] $r^2$ = variance.

tive humidity and sometimes $NO_x$ concentrations peak at night and reach a minimum during the daytime. At three of the sites the diurnal cycle was related to $SO_4^{2-}$ as well. At Montague, Rockport, and Fort Wayne,$SO_4^{2-}$ showed a moderate to weak dependence on the diurnal cycle, peaking in the day. This relationship could have been related to higher mixing heights during the day, which brought $SO_4^{2-}$ down from aloft to the ground or accelerated oxidation rates at midday. $SO_2$ variability was not related to the diurnal cycle at any of the sites during this summer case.

A second major feature derived from PCA at most sites was a component strongly correlated with high absolute humidity and nonsulfate respirable mass concentration and less strongly related to relative humidity contribution, and high mixing heights. The presence of a large absolute humidity contribution together with a strong $O_3$ contribution, did not appear in a single component. Thus, this pair of variables did not have the unique characteristics of a surrogate for chemical reactions that emerged from PCA of urban data (Henry and Hidy, 1979). The reason for this difference is not known.

Because absolute humidity is often identified as a good air mass tracker, this component is called the moist air mass component in Table 5.10. It could explain approximately one-half to one-quarter of the varia-

TABLE 5.10 **Observables Contributing Strongly to Components That Explain More Than 10% of the $SO_4^{2-}$ Variability of the July SURE Data Sets**[a,b]

| | Station 1 | | 4 | | 5 | | | 6 | 7 | | | 8 | | Totals (+) | (−) |
|---|---|---|---|---|---|---|---|---|---|---|---|---|---|---|---|
| Totals | | | | | | | | | | | | | | | |
| RSPM | x | x | x | x | | x | | x | | | x | x | x | 9 | |
| IPMM | x | x | x | x | — | — | — | — | | | x | | x | 7 | |
| $SO_2$ | | | x | | | x | x | x | | | x | | x | 6 | |
| $NO_x$ | | | | | | | x | | −x | −x | x | x | | 3 | 2 |
| NO | | | | | | | x | | −x | | x | | | 2 | 1 |
| $O_3$ | x | | x | | x | x | −x | | x | | | | x | 6 | 1 |
| Surface wind speed | x | | | | x | | | −x | x | | −x | | | 3 | 2 |
| Pressure | | | | | | | | | x | | x | | | 2 | |
| Temperature | x | | | | x | | | | x | | | | | 3 | |
| Dew point depression | x | −x | x | −x | x | | | | x | −x | | x | | 5 | 3 |
| Precipitation | | | | | | | | | | | | | | | |
| Cloud cover | | | | | | | | | | | | −x | | | 1 |
| Relative humidity | −x | x | −x | x | −x | | | | −x | x | | −x | | 3 | 5 |
| Absolute humidity | | x | −x | x | | x | | | | x | | | | 4 | 1 |
| Mixing height | | x | | | x | x | −x | x | x | | | | | 5 | 1 |
| Wind speed at 800 mb | x | | | | | | | | | x | | −x | | 2 | 1 |
| Ventilation$^{-1}$ | −x | | | | | −x | | | | | | | | | |
| $r^2$ for $SO_4^{2-}$ | 0.23 | 0.46 | 0.20 | 0.33 | 0.16 | 0 24 | 0.14 | 0.22 | 0.20 | 0.11 | 0.41 | 0.16 | 0.33 | | |
| $SO_4^{2-}$ regression coefficient | 5.70 | 8.04 | 4.07 | 5.25 | 3.61 | 4 48 | 3.45 | 2.72 | 5.05 | 3.83 | 7.27 | 2.10 | 2.95 | | |
| % of total | 25 | 15 | 14 | 12 | 23 | 15 | 12 | 11 | 22 | 14 | 12 | 15 | 12 | | |
| Variability explained assigned physical interpretation | Diurnal cycle | Moist air mass | — | Moist air mass | Diurnal cycle | Moist air mass | — | High-pressure stagnation | Diurnal cycle | — | High-pressure stagnation | — | — | | |

[a] From Mueller and Hidy, 1983; reproduced with permission of the Electric Power Research Institute.

[b] x = important positive contributor to component; −x = important contributor in a negative or opposed correlation sense.

tion in $SO_4^{2-}$ at some sites. This component was probably related to large-scale penetration of mT air and associated regional $SO_4^{2-}$ events. An extension of regression analysis to include variables that make direct use of regional meteorology (such as the air mass categories) may be useful, but more study is needed to elucidate the role of regional transport in determining $SO_4^{2-}$ concentrations.

The third type of component found by the PCA in the summer data at most sites showed a strong dependence on relatively high barometric pressure, low wind speeds, and concomitant high concentrations of pollutants ($SO_2$, IPMM, and RSPM). This component was taken as an indicator of air mass stagnation, similar to that found in the analysis of the annual data set. At Fort Wayne, 41% of the $SO_4^{2-}$ variance was attributed to high-pressure stagnation conditions, which perhaps appeared to be a reflection of the importance of cPw air mass conditions in this part of the Midwest.

Although $SO_2$ was an important component in the cold winter stagnation components, it was missing from warm summer stagnation. This distinction could be related to differences in $SO_2$ deposition during cold, dry periods compared with warm, moist air periods. Low deposition rates associated with cold, dry winter stagnations would lead to high $SO_2$ levels. More rapid deposition during warm summer stagnation conditions would deplete $SO_2$ in the air.

The significance of a given observable at all SURE stations is provided by the sum of occurrences in the last two columns of Table 5.10. These results suggest that non-$SO_4^{2-}$ refined particulate matter—RSPM ($\leq$2.5 $\mu$m diameter)—was most often found associated with high $SO_4^{2-}$ concentrations at the sites examined. Next were the nonsulfate IPMM ($\leq$15 $\mu$m diameter), and close behind was $SO_2$. The relationship of non-$SO_4^{2-}$ and $SO_4^{2-}$ was interpreted to mean that $SO_4^{2-}$ was a tracer for the remainder of the particulate material greater than 2.5 $\mu$m in diameter. This relationship was partly circumstantial because $SO_4^{2-}$ is also related to the IPMM, which included particles less than 15 $\mu$m; the origins of larger particles are probably not similar to those of $SO_4^{2-}$ in the fine particles.

$O_3$ was nearly as important an observable as $SO_2$ in accounting for $SO_4^{2-}$ variability. The significance of $O_3$ is associated with photochemical oxidation processes that form $O_3$, $H_2O_2$, and other oxidants, as well as $SO_4^{2-}$. As a surrogate of photochemistry, $O_3$ is a consistent indicator of the accelerated oxidation of $SO_2$ to $SO_4^{2-}$. Yet this may also be partly circumstantial, because $O_3$ concentration is often maximum at midday with its highest local production rate, and $SO_4^{2-}$ is mixed downward from aloft with $O_3$ in the afternoon. The two, then, would have appeared related by mixing as well as by chemical processes. Without inclusion of another relevant observable, such ambiguities could not be clarified.

Among the summer weather variables, high mixing heights and low relative humidity emerged most frequently in components related to $SO_4^{2-}$. The mixing height association differed from the winter situation, in which low mixing heights were most often identified with high $SO_4^{2-}$ concentrations. The tendency toward negative $SO_4^{2-}$ correlation with relative humidity is interesting compared with the winter results in Table 5.8. This tendency might be associated, in absence of exchange from clouds, with high midday photochemical production of $SO_4^{2-}$, which would take place with the (diurnally) lowest relative humidities. This effect did not emerge in the analysis of the 24-hour average data over a year because the diurnal resolution was absent.

*Assignment of Physical Meaning to Principal Components* The PCA results from July 1978 were examined for the day-to-day changes in principal component values (loadings) as descriptors for the regional-scale meteorological events. The results support the interpretation of principal components given above as representative of stagnation and transport of warm, moist air. The arguments are illustrated by comparing values of component loading with the meteorological changes and $SO_4^{2-}$ changes from 18 to 23 July. The analysis of two components from the PCA of July 1978 data at three SURE CI sites, one each in Massachusetts, Ohio, and Indiana, serves as an example. Table 5.11 gives the weighting factors (loadings) of observables for these components for each site. The boxed parts of the table enclose the variables that show reasonably consistent behavior for each site. These principal components were highly correlated with $SO_4^{2-}$.

The component called the *moist air mass* or *transport component* was associated with higher than average absolute humidity, relative humidity, and fine-particle mass. $O_3$ concentrations tend to be slightly elevated relative to their average. Of course, high relative humidity requires small dew point depression, as seen in the table. The association of this component with regional transport is made on the basis that absolute humidity is a good air mass tracer. Thus, this component may be related to incursions of warm, moist air, which are often associated with regional $SO_4^{2-}$ events in the summer. The *stagnation component* is most strongly associated with higher than average particulate mass concentration and lower than average wind speed. Pressure and $NO_x$ tend to be elevated.

To examine the reasonableness of this physical interpretation, the numerical value of each of these two principal components was calculated for each 3-hour period from 18 to 23 July 1978. Twenty-four hour averages of the values for each component were calculated for each day and are compared with the 24-hour average regional $SO_4^{2-}$ concentration distribu-

TABLE 5.11 **Contributors[a] of Observables. (For moist air the solid boxes highlight key variables; for stagnation the dotted boxes highlight key variables.)[b]**

| | Site 1 (Massachusetts) | | Site 4 (Ohio) | | Site 7 (Indiana) | |
|---|---|---|---|---|---|---|
| | **Moist air[c]** | **Stagnation[d]** | **Moist air[c]** | **Stagnation[d]** | **Moist air[c]** | **Stagnation[d]** |
| RSPM | 643 | 259 | 644 | 674 | 226 | 774 |
| IPMM | 425 | 263 | 622 | 695 | −151 | 773 |
| $SO_2$ | 331 | −324 | −081 | 391 | 132 | 468 |
| $NO_x$ | 287 | 181 | −003 | 364 | −453 | 397 |
| NO | −135 | 273 | −059 | 270 | −280 | 247 |
| $O_3$ | 272 | 050 | 218 | 399 | 344 | 259 |
| Surface wind speed | −109 | −425 | −114 | −244 | −072 | −438 |
| Pressure | −201 | 543 | −118 | 285 | −017 | 422 |
| Temperature | 048 | 073 | 019 | 052 | −198 | −012 |
| Dew point depression | −593 | 043 | −530 | 447 | −756 | −017 |
| Precipitation | 120 | −321 | −105 | −188 | 156 | 105 |
| Cloud cover | 126 | −221 | 321 | −098 | 182 | |
| Relative humidity | 499 | −261 | 534 | −441 | 761 | 032 |
| Absolute humidity | 729 | 060 | 612 | −410 | 596 | 009 |
| Mixing height | −572 | 189 | −151 | 258 | 096 | −034 |
| Wind speed at 800 mb | −281 | −226 | 387 | −317 | 486 | −087 |
| Ventilation | 104 | 281 | 025 | −022 | −220 | 104 |

[a] Values of component weighting factors (loadings) × 100.
[b] From Mueller and Hidy, 1983; reproduced with permission of the Electric Power Research Institute.
[c] Identified with high relative and absolute humidity.
[d] Identified with high particulate concentrations, high barometric pressure, and light winds.

tion for that day (Figure 5.7). A large value, about +1 or larger, indicates a significant contribution by the component. A value near zero shows little or no relationship one way or the other. A value of 0.1 or less indicates that the component was significantly absent that day.

Using the event analysis from the previous chapter, the period starts on 18 July with widespread regional air mass stagnation, by the $SO_4^{2-}$ concentration for the day. The values of the stagnation component are high for the three widely distributed sites. The transport (moist air mass) component is near zero or negative, indicating its lack of importance. On 19 July, the stagnation deepened, with very high values of the stagnation component. Again, the transport component does not seem to be a significant factor, with values very near zero. The map for 20 July shows an interesting pattern. Much of the Ohio River Valley is under the stagnant air mass of the previous 2 days, but the $SO_4^{2-}$ concentration peak seems to be moving toward the northeast. The stagnation component is high over Indiana with a low transport component; in Ohio, both are equally high; and in Massachusetts, the transport component dominates the stagnation component. This pattern continues for 21 and 22 July. The associated values of the transport component are large positive values that are con-

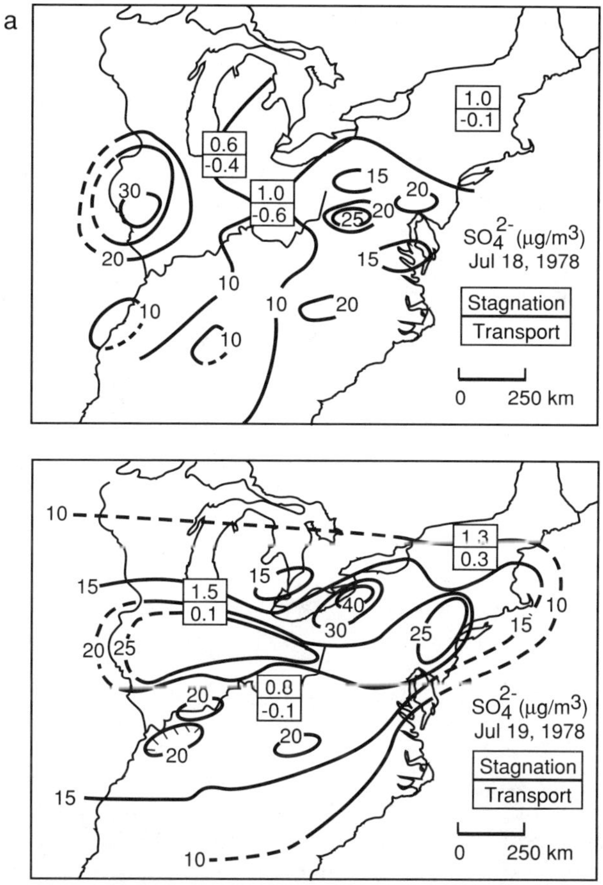


FIGURE 5.7 (a) Twenty-four-hour average values of transport and stagnation components for sites 2, 4, and 7 as calculated from PCA of the 3-hour average data and plotted on the 24-hour average $SO_4^{2-}$ map for 18–19 July 1978. (b) Twenty-four-hour average values of transport and stagnation component for sites, 2, 4, and 7 as calculated form PCA of the 3-hour average data and plotted on the 24-hour average $SO_4^{2-}$ map for 20–21 July. (c) Twenty-four-hour average values of transport and stagnation component for sites, 2, 4, and 7 as calculated form PCA of the 3-hour average data and plotted on the 24-hour average $SO_4^{2-}$ map for 22–23 July. (From Mueller and Hidy, 1983; courtesy of the Electric Power Research Institute.)

b

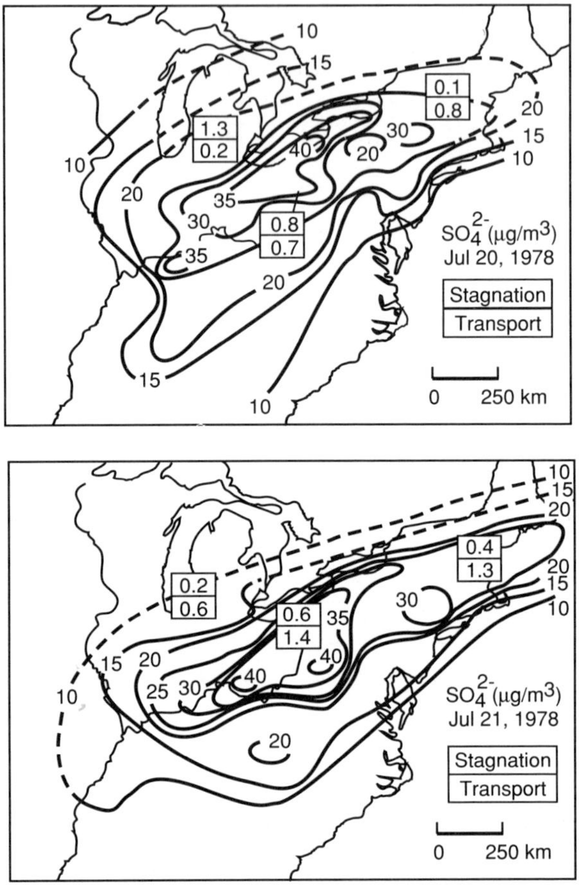


FIGURE 5.7 (*continued*)

sistent with the physical interpretation assigned to this parameter. On 23 July the event ends with the air mass containing high $SO_4^{2-}$ concentrations, apparently still intact, moving over the ocean. The stagnation component becomes strongly negative at Duncan Falls; both the transport and stagnation component are weak but positive at Montague.

*Extended Data Analysis* The analysis of interrelated variables by PCA undertaken for the SURE results could be improved by considering additional variables reflecting regional-scale phenomena. Lioy *et al.* (1982) attempted such an extended analysis for the Lewisburg location. They added variables reflecting upwind trajectory wind speed and upwind mix-c

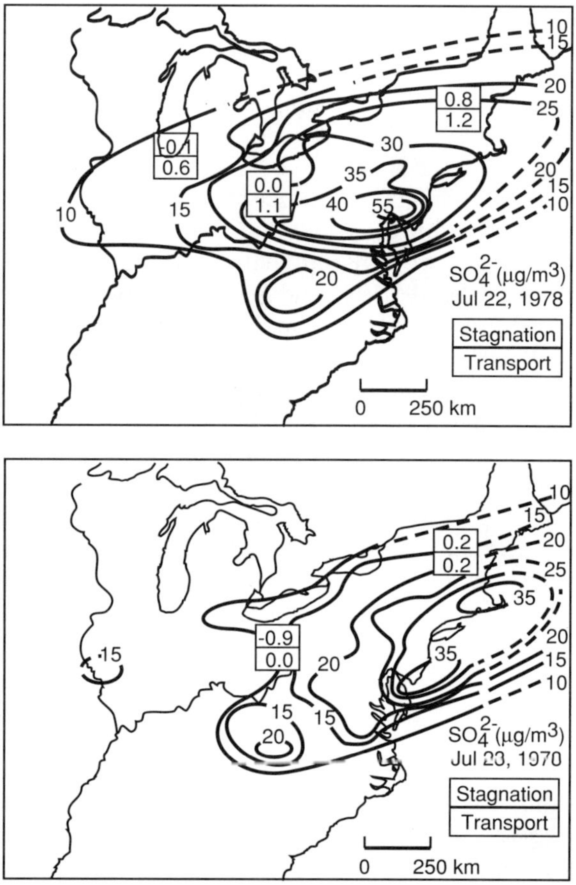


FIGURE 5.7 *(continued)*

ing heights, as well as regional daily maximum $O_3$ concentrations. In addition, a daily $SO_2$ emission parameter was included that reflected emissions accumulation along a 24-hour upwind trajectory. These researchers used data from several sources. The $O_3$ concentrations were obtained from the monitoring networks within the northeast quadrant of the United States (Mississippi River to Atlantic Ocean and Tennessee to Canada), and the values used were the daily 1-hour maximum concentrations. The daily maximum value was used because it has been commonly regarded as representing the $O_3$ concentration available in the mixed layer of an air mass, since local sources have been shown to produce negative interference at other times of the day. The pertinent periods were August 1977,

June 1978, and July 1978. However, the data for the entire 1976 photochemical oxidant season were used to provide a calibration year for selection of the subregional $O_3$ monitoring sites for the model development.

In the Lioy *et al.* (1982) study *subregional* was defined as a specific $O_3$ site that is highly correlated ($r > 0.7$) with surrounding $O_3$ monitoring sites within an 80- to 120-km radius. The protocols used to define the study months, which include a number of pollution episodes, were previously described for $O_3$ by Lioy and Samson (1979).

Climatological data were assembled for Lewisburg, Cincinnati, and Salem, Virginia, for August 1977 and June and July 1978. Measurements, including temperature, dew point, and relative humidity, were obtained from either the (SURE) data base (Lewisburg) or routine local climatological data summaries issued by the NWS. From appropriate data, the inverse of the relative humidity and the absolute humidity were computed. The quantity $(1 - RH/100)^{-1} = IRH$ is theoretically related to the aerosol water content (e.g., a curve through data in Figure 3.14).

Thirty-six-hour backward air parcel trajectories were constructed for Lewisburg by means of Heffter's (1980) mixed-layer model to examine the history of air parcels before they passed through Lewisburg. In addition, the average wind speed and mixing height were tabulated separately for the various 6-hour upwind trajectory segments. The development and limitations of these analyses have been described in previous studies by Samson (1980). Ambient $SO_2$ and water-soluble $SO_4^{2-}$ concentrations measured at Lewisburg were obtained from the SURE. Average upwind $SO_2$ estimates were obtained using the MAP3S emissions grid and interpolating the emissions along discrete segments of each trajectory. Average $SO_2$ emissions, in kg/km$^2$, were obtained for each 6-hour segment. Subsequent 24-hour average upwind $SO_2$ emissions were calculated.

The Lioy *et al.* (1982) PCA was confined to examination of 92 available cases for Lewisburg (August 1977, June and July 1978). Ambient $SO_2$ measurements were initially considered as a variable, but because there were many values below the detection limit, that variable was eliminated.

The interpretation of the components depends on the $a_{ij}$ loadings, which are illustrated in Table 5.12. The identity of each component is determined by the highly significant $a_{ij}$ loading of individual variables on a particular component. For Lewisburg, the first seven components explained a total of 80% of the variance of this variable set. Each component had an eigenvalue $>1$, which is usually considered as the critical level for statistical significance, and the components were considered as new independent variables for the $SO_4^{2-}$ regression analyses. Besides these, two components had a single variable with a high loading ($a_{ij} > 0.86$), but statistically these were only minor components (each had an eigenvalue $<1$). However, because these components could be described physically

TABLE 5.12 **$a_{ij}$ Component Loadings for Lewisburg**[a,b]

| | Component 3 | Component 12 | Component 9 | Component 2 | Component 1 |
|---|---|---|---|---|---|
| TMAXLEW | 0.099 | 0.022 | 0.011 | 0.023 | 0.826 |
| DWLEW | 0.061 | −0.035 | −0.049 | −0.193 | 0.862 |
| RELHUML | −0.078 | −0.082 | 0.012 | −0.051 | 0.208 |
| 03CINLAG | 0.521 | −0.043 | −0.013 | 0.281 | 0.039 |
| SALEMO3 | 0.215 | 0.109 | 0.092 | 0.073 | −0.001 |
| LEWISO3 | 0.178 | 0.266 | 0.065 | 0.833 | 0.153 |
| IRH | −0.051 | −0.105 | 0.031 | −0.189 | 0.023 |
| IRHNL | 0.132 | 0.116 | −0.022 | 0.405 | 0.264 |
| IRHAVL | 0.013 | 0.056 | 0.098 | 0.868 | −0.071 |
| IRHAVC | 0.015 | 0.056 | −0.036 | 0.797 | −0.293 |
| LCMINT | −0.060 | −0.082 | −0.114 | −0.156 | 0.203 |
| IN24WD | 0.831 | 0.089 | 0.027 | 0.071 | −0.013 |
| IN36WD | 0.915 | 0.093 | 0.049 | 0.074 | 0.124 |
| MIX24 | 0.048 | 0.074 | −0.031 | 0.060 | −0.037 |
| MIX36 | −0.030 | 0.103 | −0.203 | −0.058 | 0.221 |
| VENT | 0.717 | −0.019 | 0.023 | −0.004 | 0.007 |
| AHUM | 0.010 | −0.032 | 0.029 | 0.011 | 0.950 |
| NSPLEW | 0.139 | 0.851 | 0.028 | 0.325 | −0.058 |
| WDSPD24 | 0.100 | 0.086 | 0.192 | 0.157 | −0.173 |
| MASO24 | 0.058 | 0.023 | 0.951 | 0.082 | 0.007 |
| Percent of $SO_4^{2-}$ variance[b] | 17.9 | 16.2 | 5.8 | 5.6 | 5.6 |

[a] From Lioy *et al.*, 1982.
[b] O3CINLAG, Cincinnati prev. day $O_3$; SALEMO3, max. daily $O_3$ in Salem; LEWISO3, max. daily $O_3$ in Lewisburg; IRHAVL, avg. RH Lewis.; IRHAVC, avg. RH prev. day Cinc.
[c] This is the percent of $SO_4^{2-}$ variance accounted for by a particulate principal component in the multiple regression equation.

as representing upwind $SO_2$ emissions (MASO24) and the remaining particulate mass (NSPLEW), they were not immediately eliminated from consideration for use in the subsequent regression analyses. Further review of the first seven components found that neither NSPLEW or MASO24 scored highly in any of these components, which suggests that in this case each may be independent of the other variables used in the PCA.

The results obtained for the regression analyses on the principal components indicated that five components yielded significant regression coefficients ($P < .05$), and they produced a multiple $r = 0.72$, accounting for 52% of the variation of the Lewisburg $SO_4^{2-}$. Table 5.12 shows the $a_{ij}$ loadings for these components (in order of significance in the regression) and the percentage of $SO_4^{2-}$ variance accounted for in the regression.

Among the significant components, two were major components and, as previously hypothesized, the other three were the minor PCA components containing MASO24 and NSPLEW.

The components in Table 5.12 describe possible relationships for explaining the occurrence of summer $SO_4^{2-}$. Principal component 3 is related to regional $O_3$ concentration, wind speed (IN24WD or IN36WD), and ventilation, indicating an association with upwind photochemical smog processes. Principal component 12 had a high positive $a_{ij}$ for the TSP mass difference, which suggests an association with particulate matter (secondary or primary) other than $SO_4^{2-}$. Principal component 9 is associated with upwind $SO_2$ emission (3 to 27 hours upwind), which is the precursor of (secondary) $SO_4^{2-}$. Principal component 2 is correlated with local $O_3$ and the IRH, which suggests the influence of local photochemistry and droplet growth. Principal component 1 had high loadings for local temperature and absolute humidity. Temperature can be related to photochemical reaction intensity and air mass character. Absolute humidity is related indirectly to smog processes, although it, too, can be associated with air mass characteristics. Thus, the component could be related to photochemical smog processes, but a more likely explanation is the association of $SO_4^{2-}$ with changes in air mass character.

The Lewisburg analysis showed that PCA methodology can identify the fact that upwind meteorological processes and source emissions are necessary to account for the variation of summertime $SO_4^{2-}$. This result again provides evidence of a long-range transport component in station behavior.

Based on the regression analyses, approximately 50% of the $SO_4^{2-}$ variance could be explained by the five principal components. The addition of all other calculated components increased the variance to 64%; however, this latter analysis may have been an artifact of the repetitive nature of summertime conditions.

In another study, Wolff *et al.* (1984) applied a PCA analysis similar to that of Lioy *et al.* (1982). They included local aerometric variables with mixing height and air mass trajectory data in a set of measurements from the Blue Ridge Mountains in Virginia in July–August 1980. Their analysis generated results similar to those of Henry (Mueller and Hidy, 1983) and Lioy *et al.* (1982). At this site, the single largest component, representing the residence time of an air parcel traveling from a midwestern source area to the Blue Ridge area and photochemistry, accounted for over 70% of the fine particles and $SO_4^{2-}$ concentration variability at the site. This finding suggests that for the 1980 summer period, transported pollutants from the midwestern source area dominated mean $SO_4^{2-}$ and fine-particle levels. These researchers also found a small but detectable association (about 4.8% of fine-particle variability) with organic aerosols on the Blue Ridge Mountains. The organics were presumed to have natural origins.

Evidently, heterogeneous chemical mechanisms were a small factor because of the high photochemical association with airborne $SO_4^{2-}$ behavior.

## SPATIAL VARIABILITY OF DATA

### Ambient Concentrations

*Intersite Correlations* The spatial variability in ambient $SO_2$ and $SO_4^{2-}$ aerosol concentrations has been investigated with 24-hour averaged $SO_2$ and $SO_4^{2-}$ data from the SURE using the intensive sampling periods. Data were screened to ensure that at least 25 days of data were available from each intensive period and that at least 20 days of concurrent data were available from any pair of stations. From the screened data, pairwise intercorrelations have been computed (Henry, reported in Mueller and Hidy, 1983; and Hidy *et al.* 1985).

From the resulting correlation coefficient matrices, selected values have been plotted on maps to:

- Explore correlation behavior with increasing intersite distance
- Discern differences between spatial correlation patterns of $SO_2$ and $SO_4^{2-}$
- Examine possible differences in the correlation behavior of different geographic clusters of stations
- Investigate seasonal differences in correlation behavior

Using the 24-hour averaged $SO_4^{2-}$ data collected at SURE sites during the intensive sampling periods, Henry calculated intersite correlations and paired station covariances and correlations. He constructed maps of isopleths showing distributions of intersite correlations for each period. The regional character of 24-hour $SO_4^{2-}$ concentrations is apparent. Correlations of 0.80 or better extend for 250 km northeast from Rockport, IN and 500 km northeast and 500 km southwest of Warren, PA. Based on the July data, a scale of variation defined by 0.8 correlation is about 250 km in the southern Ohio River Valley and about 500 to 700 km in the area of northwestern Pennsylvania and western New York State.

Intersite correlation maps for the January–February intensive period show that the oval of 0.75 correlation around Rockport extends about 250 km to the southwest, the opposite direction. The most dramatic difference is the intersite correlation around Warren. No intersite correlations of 0.8 or greater appear. Correlations of 0.75 appear in a narrow band across northern Pennsylvania. Evidently, during the January–February sampling period of 1978, there was virtually no regional character to $SO_4^{2-}$ concentration distributions in the area around Warren that was analogous to the summer conditions. The Appalachian Mountains seem to be an important barrier to eastward transport in the winter, because the correlation isopleths of Rockport parallel this range.

*Time-Lagged Correlations* Interpretation of intersite correlations incrementally shifted in time gives an additional perspective on the significance of air mass transport in summer. The case studies of ducting situations in August 1977 and July 1978 show wind conditions conducive to polluted-air transport from the lower Ohio River Valley northeastward. If the long-distance transport of $SO_4^{2-}$ is a dominant factor in the summer months, one would expect a peak in correlation of $SO_4^{2-}$ concentration at a time lag corresponding to the ratio of distance to average wind speed for sites at different distances downwind.

Time-lagged correlations on $SO_4^{2-}$ concentrations between SURE CI sites along the predominant transport path up the Ohio River Valley northeastward were analyzed. Between Rockport and Fort Wayne, 400 km northeast, the correlation decreased steadily from its maximum value of 0.5 at 0 time lag. For Duncan Falls, about 500 km eastward along the Ohio River Valley, there was a peak in the intersite correlations at 12 hours time lag. Farther downwind, at 1000 and 1500 km, the correlations did not rise much above 0.2. The peak at 24 hours in the Rockport versus Montague correlation may be related to transport downwind, but the peak correlation was very weak. The peak of 0.4 in the Rockport versus Duncan Falls curve seemed to be high enough to indicate a physical connection. However, the implied transport speed of 500 km in 12 hours was 42 km/h. This is too high to be consistent with wind at ground level. Winds aloft at 300–600 m are high enough for this transport rate, at least in ducting episodes. In stagnation conditions, such transport rates probably are not realized even in winds aloft.

This analysis is limited by its inherent assumption that most long-range transport occurs in the same direction and at the same speed throughout the month. Alternative analyses remain to be done. One possibility is to use changing wind or transport velocities explicitly. For example, time lag could be measured as a function of wind speed. Wind direction could be incorporated by picking a different site for each day that is most nearly upwind from the fixed site on that day. Another possibility is to restrict the analysis to days on which long-range transport is believed to be occurring based on synoptic meteorology.

The results of this and other intersite correlation analyses discussed by Hidy *et al.* (1985) are summarized as follows:

1. $SO_2$ concentrations during summer or winter show no regular pattern of correlation with increasing intersite distance. Over the typical intersite distances represented here, variations in summer $SO_2$ are mainly out of phase.

2. Winter $SO_2$ concentrations are highly correlated principally at sites in the Northeast.
3. Summer $SO_4^{2-}$ is more highly correlated in the Midwest than the Northeast.
4. $SO_4^{2-}$ is geographically more widely correlated than $SO_2$ during both the summer and winter.
5. Both winter $SO_2$ and $SO_4^{2-}$ have an east–west orientation to the intersite correlations among the northern tier of sites but not among southern sites.
6. Summer $SO_4^{2-}$ is predominantly southwest–northeast to the intersite correlations.

These results are of interest for comparison with the attempts to find emission and precipitation associations (Hilst and Chapman, 1990). If $SO_x$ scavenging consistently involves removal of $SO_4^{2-}$ as cloud nuclei on a broad scale, correlation analysis results 2–4 appear consistent with these investigators' conclusions (see also p. 99).

*Inferred Vertical and Horizontal Variability* The covariance of atmospheric constituents in the vertical (through the planetary boundary layer) and horizontal (over distances of 20 to 60 km) was further explored by analyzing light scattering using the light-scattering coefficient ($b_{sp}$), a surrogate for $SO_4^{2-}$, and $SO_2$ concentration data taken during selected aircraft spirals during SURE intensive measurement periods. The approach was to calculate interlevel and interspiral correlations using data from sets of spirals flown in the vicinity of the Rockport and Duncan Falls stations during August 1977 and January 1978, respectively.

The correlation coefficients ($r$) for $b_{sp}$ were plotted as a function of level separation (Hidy *et al.* 1985). Contrary to expectations based on stable nighttime temperature profiles, the interlevel correlations were larger at night than during the day, indicating that the vertical mixing during the night is better than during the day. Negative daytime correlations for $b_{sp}$ occurred at intervals above 800 m, which suggested that heterogeneity in the mixing layer occurred, and convective mixing is important. Hidy *et al.* (1985) estimated that the mean aerosol mixing height over the locations for the spirals during an August 1977 measurement period was 920 m.

The correlations for both $b_{sp}$ and $SO_2$ shed further light on the chemical distributions aloft (Figures 3.11 and 3.12). The correlations for the daytime Rockport spirals suggested that the degree of mixing of aerosol particles and $SO_2$ was similar above 400 m. However, the fact that the $SO_2$ interlevel correlations did not show both minima and negative values at 800 and 1200 m (as do those for $b_{sp}$) indicated that the vertical gradient of

$SO_2$ between 800 and 1200 m was less than that of $b_{sp}$. The fact that the correlations showed a more rapid initial fall off with height and more "structure" for $SO_2$ than for $b_{sp}$ implied that below 400 m $SO_2$ is less well mixed vertically than were the aerosol particles.

A comparison of the daytime $SO_2$ correlation plots for August and January showed a more rapid decrease in correlation with height during winter. This decrease is greatest above 800 m, consistent with a mixing height between 400 and 800 m, which is lower than that for August 1977.

On the scale defined by the spacing of the aircraft spirals, horizontal variability at three levels (200, 600, and 1200 m) was examined for day, night, and season. The interspiral correlation coefficients were calculated for $b_{sp}$ and $SO_2$. The correlation for $b_{sp}$ between spirals was good at night in the vicinity of Rockport, IN; it was poorer between one set of spirals at 200 and 600 m during the day. During January, $SO_2$ was homogeneously distributed within levels among all spirals above 600 m but was patchy at 200 m.

Interspiral $SO_2$ correlation during August was moderate to poor at all levels for both the day and night flights, presumably because of coarse plume dispersal and convective mixing. Paired layer correlations for $SO_2$ in the night flights ranged from 1.0 at 200 m, to $-1.0$ at 600 m, to 0.0 at 1200 m. There is no explanation for this behavior. Differences between $b_{sp}$ and $SO_2$ correlation patterns suggest distinctive character of horizontal and vertical distributions of $b_{sp}$ as sulfate particles versus a primary pollutant ($SO_2$).

## Wet Deposition

Spatial variability in wet deposition has been examined along with ambient concentration data. Observations from the regional network stations are available to characterize a coarse-scale variability over several years. Data characterizing mesoscale variability are much more limited. A data set covering tens of kilometers was collected on an event basis by investigators from Rensselaer Polytechnic Institute as part of the Integrated Lake Watershed Assessment Study (ILWAS) and reported for the period May 1978 through August 1979. Seven stations were located within a 15-km radius of Eagle Bay in the Adirondack Mountain region of New York. Six of the stations were paired. Two of them were collocated, and the other two pairs were each located within a small watershed. A single wet-only collector was located in the town of Big Moose, New York.

Although precise locations of the collectors were not reported, intersite distances have been estimated. These range from 0 km for the collocated pair to 33 km between the Sagamore Lake and Woods Lake pairs. Johannes *et al.* (1981) calculated intersite correlations for precipitation amount and ionic loadings. The intersite correlations for $SO_4^{2-}$ loadings were plotted as a function of the estimated intersite distances. The intersite correlations were 0.93, and distance had no apparent effect on the degree of correlation. Thus, at least for this area, $SO_4^{2-}$ deposition varied uniformly over distances of about 30 km.

Spatial variability was also examined differently by calculating the coefficient of variation among the seven ILWAS stations for monthly precipitation amount and for monthly volume-weighted $SO_4^{2-}$ concentrations. These data suggest that although the precipitation can vary over more than 160% between stations for any given month, the coefficient of variation for precipitation amount is less than 30% over the network, and for 12 of 16 months the $SO_4^{2-}$ concentration variability is less than that for precipitation amount. This result suggests that the data from any one of the seven stations serve as adequate indicators of precipitation quality within the area covered by the network, even in hilly terrain.

The regional network data are useful for estimating a coarse scale for $SO_4^{2-}$ and $NO_3^-$ variability. Analysis of these data via intersite correlations also gives information on the average anisotropic nature of spatial variability resulting from emissions and persistent atmospheric processes. This interpretation is illustrated using precipitation chemistry data for $SO_4^{2-}$, $NO_3^-$, $H^+$, and $NH_4^+$ from APN, MAP3S, NADP, and UAPSP sites. These data are useful for such analyses as a self-consistent set of weekly volume-weighted averages. Hidy *et al.* (1985) analyzed the weekly averaged data, which were then screened for years 1980, 1981, and 1982, eliminating stations for which there were no values from 10 or more months and for 33 or more weeks. For data available in late 1983, this screening process resulted in data from 50 sites being left from 1980, 62 sites from 1981, and 26 sites from 1982. Correlations were then computed using the weekly averaged data.

To investigate intersite correlations for $SO_4^{2-}$ and $NO_3^-$ concentrations in ENA precipitation, paired correlations for a year were grouped according to intersite distance intervals and averaged within each interval, and the averages were plotted as a function of mean distance within the intervals. The maximum correlations for both species within a given year, which occur at either 0 or 100 km, were essentially equal. There was a general reduction in correlation coefficient with distance—as would be expected—but the decrease with distance was greater with $NO_3^-$, which suggests that the characteristic "scale" for $NO_3^-$ deposition, including residence time, is less than that for $SO_4^{2-}$ deposition. An apparent increase in correlation at distances greater than 2000 km suggested that sites with this separation were generally remote from major source areas, reflecting a uniform, weak "continental" level of $SO_4^{2-}$ expected for background sites.

To investigate whether or not directional anisotropies or geographic distinctions exist in the behavior of intersite correlations with distance, the correlations between a base site and other sites along transects or within subregional site clusters were plotted as a function of distance. The base sites and the other sites associated with each grouping are given by Hidy *et al.* (1985). Because weather systems tend to move into the Northeast from a southwesterly or westerly direction, it is natural that correlations between sites aligned along a southwest-to-northeast direction would differ from those between sites aligned in other directions.

Some of these calculations corresponded to an intuitive understanding of the physical situation related to transport patterns; others did not. For example, precipitation $SO_4^{2-}$ at sites in 1980 to the southwest of Zanesville, OH correlated moderately with that at Zanesville to 700 km; the correlations showed no dramatic decrease with distance. This result would be expected if, on the average, the sites were influenced by the same storm systems. In contrast, in 1981, $SO_4^{2-}$ correlation was significantly higher (0.68 versus 0.48) with a site at 220 km and dropped precipitously to less than 0.1 at 640 km. A detailed meteorological analysis to reconcile these differences has not been performed, but the result indicates the importance of multiyear analysis to determine consistent behavior. Along a northeasterly corridor from Zanesville, $SO_4^{2-}$ correlations are moderate to poor, with no apparent pattern with distance for both 1980 and 1981. A similar situation was found for the $SO_4^{2-}$ correlations along an ENE-to-ESE transect, which suggested a meteorological similarity through the region. Along northwest and southeast (cross-wind) transects from Zanesville, $SO_4^{2-}$ correlations were generally poorer than they are along the more westerly transects (i.e., those to the ENE and ESE). Again, no significant pattern with distance was apparent in the values.

Because none of the intersite correlations within geographic clusters of stations showed any apparent influence of distance, the average and standard deviation of the correlation coefficients were calculated to see how these statistical indicators varied with geographic location. Because the sizes of the selected station clusters were different, only correlations between the base station and stations within 300 km were included in the calculations. From the results of Hidy *et al.* (1985), certain generalizations and specific observations could be made.

1. In any given year, $NO_3^-$ is more poorly correlated spatially than is $SO_4^{2-}$.
2. The Midwest showed the greatest change between years in both $SO_4^{2-}$ and $NO_3^-$.
3. New York–Pennsylvania and southern Appalachia station clusters showed the highest intersite correlations compared with the Midwest and coastal Atlantic data.
4. A Piedmont cluster showed the lowest overall correlation relative to the New York–Pennsylvania region and other observations.

The first observation is consistent with the smaller characteristic scale for $NO_3^-$ than $SO_4^{2-}$ inferred from the aggregated correlation behavior. The second observation suggests that meteorological conditions in the Midwest were quite dissimilar for 1980 and 1981. The third observation suggests that weather patterns in the Northeast and South were more uniform than in the Midwest and Middle Atlantic coastal areas. The fourth observation may indicate the influence of local sources on some of the stations in the New York–Pennsylvania cluster.

Hidy *et al.* (1985) also reported an analysis to see if there were seasonal differences in the patterns of intersite correlations. A more limited data set was used from MAP3S and UAPSP stations. Correlations were calculated for summer (June, July, August $SO_4^{2-}$), winter (November, December, January $SO_4^{2-}$), and summer $NO_3^-$. For summer $SO_4^{2-}$ within a midwestern station group, there is a monotonic reduction in correlation with distance for the four stations within 300 km, but its meaning is obscured with an apparent increase in correlation beyond 300 km. Winter $SO_4^{2-}$ data displayed no clear correlation pattern with distance. Summer $NO_3^-$ correlation behavior was qualitatively similar to that of summer $SO_4^{2-}$. Summer $SO_4^{2-}$ correlations, using Big Moose, NY as the base station, showed a consistent decrease with increasing distance. Stations at 200 ± 50 km showed similar correlations during summer or winter. Whiteface Mountain, only 100 km from Big Moose, was found to be poorly correlated during the winter. This anomaly may reflect an influence of altitude or terrain difference that is unaccounted for.

Test calculations of this kind using conventional correlation analysis reveal complex geographic patterns of precipitation chemistry behavior within regional-scale processes. Further analysis of this kind with data sets covering several years combined with spatial $SO_x$ and $NO_x$ emission distribution data could lead to further elucidation of SRRs. The following gives examples of additional analyses that could eventually assist in characterizing SRRs from monitoring data.

## ANALYSIS OF SPATIAL VARIABILITY

Two methods of analysis have been used to characterize spatial variability as derived from intersite correlations. The first is an objective mapping technique called *Kriging* (David, 1977). The second uses EOFA. Both methods are outlined, but the results are used quite differently. The Kriging analysis, as described, shows an example of data extrapolations for spatial characterization using a historical set of monitoring data on precipitation. The EOFA example illustrates how covariance analysis can give considerable insight into the regional behavior of airborne particles given an existing data set.

## Kriging Representation

A form of empirical spatial analysis has emerged from geostatistics (Bilonik and Nichols, 1983; Bilonik, 1985). For these objective mapping techniques, the determination of the spatial correlation structure of the process takes on fundamental importance. If the data are insufficient to estimate this structure, objective mapping is not possible. Kriging can be used as a design method for networks or for testing the adequacy of existing monitoring networks to characterize spatially representative behavior. In the past, networks have been designed intuitively. Meteorologists have traditionally defined the problem in terms of the spatial correlation function (SCF) or the spatial covariance function. The spatial correlations are used for map construction in two different ways: (1) a theoretical SCF is fitted to the empirical correlations and is used to determine weights or (2) the empirical correlations are used directly to determine the weights. The second approach can lead to problems because not all functions are SCFs. Geostatisticians have relied on the equivalent concept of spatial variance (the inverse of spatial covariance). The spatial variance function is known as the *semivariogram*. A unified, consistent methodology exists for fitting theoretical semivariograms to empirical semivariograms so that data can be extrapolated. The methodology includes a range of frequently applied models and the ability to nest several models so that almost any variance structure, whether isotropic or anisotropic, can be easily accommodated.

Besides producing an optimally estimated map, the SCF, or equivalently the semivariogram, can be used to gauge the corresponding estimation error in the map. The square root of the estimation variance is a measure of the estimation error and can itself be plotted as a map. By suitably using this error map, the addition of the new sampling sites to a network can be made rationally or a new network can be designed rationally from the results of earlier measurements. In addition, once the spatial variance structure has been estimated, error maps can be generated for modified or alternative network designs before new stations are actually installed.

Kriging methods applied to the USGS bulk deposition data in New York and Pennsylvania provide examples of spatial structuring for $H^+$ and $SO_4^{2-}$ in precipitation. In the former, $H^+$ spatial variability was treated approximately as a temporally steady condition. Later, the $SO_4^{2-}$ deposition analysis accounted for year-to-year variability. The data from as many as 22 stations were extrapolated over the New York State area in 80-km$^2$ squares using the method between 1965 and 1978. In winter 1980–

1981, 122 stations were available, extending the network from Illinois and Minnesota to the east coast.

For steady-state conditions geostatistical methods employ the sample semivariogram $y_z^*$ to estimate the spatial variance (Journel and Huijbugts, 1978):

$$y_z^*(h, t) = [2N'(h, t)]^{-1} \sum_{i}^{N'} [Z(x_i + h, t) - Z(x_i, t)]^2 \tag{5.8}$$

where $Z(x_i, t)$ is any regionalized variable, $h$ is a vector that represents both distance and direction, and $N'(h, t)$ is the number of pairs of stations at time $t$ that are separated by $h$. An asterisk is used to distinguish sample estimates from theoretical values. Because $h$ represents both direction and distance, anisotropies can ordinarily be accommodated, if necessary. The semivariogram assumes that the variance between two locations $x$ and $x + h$ at time $t$ is a function only of the relative distance (and possibly direction) $h$ and not the absolute locations.

Including temporal variability is a straightforward extension of the formula in Eq. (5.8). Let $Z(x_i, y_i, t_i)$ represent the $H^+$ or $SO_4^{2-}$ deposition of the $i$th sample at location $(x, y)$—the gnomonic projected coordinates—at time $t$. Then the sample semivariogram for increment $(\Delta x, \Delta y, \Delta t)$ is

$$\begin{aligned} y^*(\Delta x, \Delta y, \Delta t) = {} & [2N'(\Delta x, \Delta y, \Delta t)]^{-1} \\ & \times \sum_{l=1}^{N'} [Z(x_i + \Delta x, y_i + \Delta y, t_i + \Delta t) \\ & - Z(x_i, y_i, t_i)]^2 \end{aligned} \tag{5.9}$$

where $N'(\Delta x, \Delta y, \Delta t)$ is the number of sample pairs separated by the increment $(\Delta x, \Delta y, \Delta t)$. Note that, by definition,

$$y(\Delta x = 0, \Delta y = 0, \Delta t = 0) = 0$$

The units for $y$ are $kg^2/ha^2$-$yr^2$.

It was not possible to make full use of Eq. (5.9) for the 1965–1979 USGS data. Spatial anisotropy cannot be reliably estimated because of the small number of stations. Therefore, a simpler isotropic equation was used:

$$\begin{aligned} y^*(\Delta r, \Delta t) = {} & [2N'(\Delta r, \Delta t)]^{-1} \\ & \times \sum_{i=1}^{N'} [Z(r_i + \Delta r, t_i + \Delta t) - Z(r_i, t_i)]^2 \end{aligned} \tag{5.10}$$

where $r = (x^2 + y^2)^{1/2}$ and $\Delta r = (\Delta x^2 + \Delta y^2)^{1/2}$.

Equation (5.10) handles space and time coordinates simultaneously in this approach, automatically accounting for variability in time and in spatial variance. In one study, Bilonik and Nichols (1983) considered only spatial variability for $H^+$ in precipitation. Equation (5.10) was modified to calculate a long-term average variation.

Bilonik and Nichols' (1983) results for $[H^+]$ show that the yearly 80-$km^2$ block estimates for $[H^+]$ displayed a weak tendency to decrease from west to east. Exceptions were the years 1969, 1976, and 1978. (The years 1965 and 1979 were not estimated; each had far fewer than 12 months of data.) The average $[H^+]$ for 1966 to 1978 was 0.042 meq/L (or 4.38 pH units). The largest difference between blocks within a given year was 0.026 meq/L and occurred in 1974. Across years, the largest difference was 0.037 meq/L (largest maximum in 1974 or 1976 versus the smallest minimum in 1971). The block estimation standard deviations range from 0.005 to 0.015 meq/L, but overall the average was about 0.009 meq/L. A standard error calculation similar to that performed for 1974 indicated that the block maps showed some evidence of spatial detail. The average increase in estimation standard deviations for the point estimates of $[H^+]$ compared with that for blocks was almost 60%. However, the differences between the minimum and maximum point estimates were not significantly larger than for the block estimates. Bilonik and Nichols concluded that point maps based on the USGS data could not be used to support regional differences in $[H^+]$ over New York State for any of the years in the period 1966–1978. An improvement due to a regular sampling grid could result in an average decrease in estimation standard deviations of over 30%.

Bilonik's (1985) extended analysis for $SO_4^{2-}$ indicated that even though the mix of available stations changed from month to month, the estimation variance changed very little from year to year. The Kriged estimates for all blocks from 1965–1979, roughly contained within the borders of New York State, were averaged together each year. The overall deposition average was almost 50 kg $SO_4^{2-}$/ha-yr. For most blocks, the estimation standard deviation was about 15% of the average block $SO_4^{2-}$ deposition. Most block estimation standard deviations were between 6 and 9 kg/ha-yr.

Bilonik (1985) summarized averages for the behavior of $SO_4^{2-}$ deposition over New York State and immediate neighboring areas. No overall trend was apparent between 1966 and 1978. Because these yearly averages were averages of Kriged estimates, the yearly averages were underestimated (because sample points were positively correlated). The semivariogram for the 1980–1981 data exhibted a strong anisotropy in the north–south direction. This was interpreted as a spatial drift in the data; in such cases Kriging may lead to misleading deposition estimates. More precise estimates can be made using a modified procedure called "universal" Kriging.

Two types of information are needed for universal Kriging: the type of drift and an estimate of the semivariogram with the drift removed. Usually, the use of first- or second-degree polynomials in spatial variables is sufficient to handle most types of drift. For the 1980–1981 data, the drift was handled by a first-degree polynomial in the $y$ (north–south) coordinate. The more difficult problem, in general, involves the estimate of the residual semivariograms, i.e., the semivariogram free of the drift. When the drift is confined to one direction, the appropriate residual semivariogram is simply the empirical semivariogram in the direction with the least variability. Therefore, for the 1980–1981 USGS data, this entailed modeling an empirical semivariogram in the east–west direction only.

Bilonik (1985) constructed an $SO_4^{2-}$ deposition map for the region covered by the 1980–1981 USGS network and a corresponding map of estimation standard deviations. Even within the "convex core" of the network (i.e., the region delimited by the smallest convex polygon that completely encompasses the network), the tendency for the $SO_4^{2-}$ deposition to increase toward the south was evident. Typically in the more northern blocks, the $SO_4^{2-}$ deposition in the winter of 1980–1981 was about 5–12 kg/ha. For the more southern blocks, the deposition was about 19–26 kg/ha. From within the interior to about 50 km outside the convex core of the network, the estimation standard deviations ranged from about 2 to 4 kg/ha. At 200–300 km outside the convex core, the estimation standard deviations increased to about 4–6 kg/ha. At 300–400 km, the estimation standard deviations became very large. The map of estimation standard deviations showed that for these data, it was reasonable to extrapolate up to about 200–300 km from the core of the network, subject to certain limitations. In particular, the region near the Great Lakes in Canada could be reliably mapped, even though no samples were actually taken in Canada.

Bilonik (1985) extended the Kriged $SO_4^{2-}$ deposition from the boundaries of the USGS network so that the extent of reasonable extrapolation could be determined and the maximum number of samples could be used to estimate blocks near and beyond the convex core of the network. One of the consequences of doing this was the appearance of some small negative $SO_4^{2-}$ deposition estimates. Most of the negative block estimates were at the extreme northeastern boundary of the map. A few negative block estimates were at the edge of the southwestern region of the map. These presented no inconsistency, since the estimation standard deviations were extremely large (on the order of 100-kg/ha-yr), indicating that reasonably accurate estimates were not possible in these regions. Another artifact of deliberately overextrapolating placed the highest $SO_4^{2-}$ deposition blocks beyond the core of the network. The blocks beyond the core were estimated by fewer samples from one side only. Therefore, they were more subject to the influence of a few high-$SO_4^{2-}$ samples than blocks in the interior of the network. Again, this is not a problem as long as the estimation standard deviations were taken into consideration.

Bilonik (1985) compared an $SO_4^{2-}$ deposition map created by universal Kriging with the subjectively determined map prepared by Peters and Bonelli (1982). Qualitatively, there were similarities between the two. However, it appears that the Peters and Bonelli $SO_4^{2-}$ deposition map was hand-drawn, and no direct quantitative reliability estimates are possible. Although widely used hand-drawn contour maps give an impression of accuracy, in many cases the data are being asked to provide more information than is possible, given the calculated spatial correlations of the network stations.

According to Bilonik, the reliability of the Peters and Bonelli map could only be approximated compared with an objective mapping technique, such as universal Kriging. From this perspective, one could infer that the subjectively drawn map was a reasonable interpolation of the data. The universal Kriging map and the hand-drawn map agreed in broad outline as to the general level of $SO_4^{2-}$ deposition in large regions. In addition, both showed a tendency for $SO_4^{2-}$ deposition to increase from north to south. The differences that did exist between the maps occur for two reasons: different mapping methods and the screening process used in the studies. Because of these differences, some of the highest areas in the handdrawn map did not show up in the Kriged map. Although the screening method removed some of the higher deposition values, some of the smoothing was also caused by the Kriging technique. This is characteristic of all linear estimators.

One major difference between the two maps lies in the extent of extrapolation and interpolation used. Peters and Bonelli (1982) did not extend their estimates over Lake Michigan and part of Lake Superior, even though these regions were completely surrounded by samples. Clearly, even though no monitoring stations were placed in the lakes, the computation of reasonable $SO_4^{2-}$ deposition estimates was possible by Kriging. A more controversial area is the Canadian land between the Great Lakes. This area is not completely surrounded by observations, although there were samples on essentially three sides. The Kriging results, however, showed that the $SO_4^{2-}$ deposition can be estimated reasonably well and hence could be estimated from the data. The same uncertainty occurs for the southern Ohio–western West Virginia region and northern Illinois. Peters and Bonelli were conservative in extrapolating at the boundaries of the network. Their point estimates are almost completely contained within the convex core of the data. Assuming there is reason to believe that the semivariogram is applicable for some distance outside the convex core, the estimation standard deviations showed that block estimates may be reliable several hundred kilometers out from the core.

Empirical semivariograms for both the USGS 1965–1979 data and 1980–1981 winter data exhibit regular patterns that can readily be calculated. Because of the different sizes of area covered by each network, it is not surprising that different models were needed for each data set. Although both data sets were nonstationary in the spatial dimensions, a more extreme indication of the nonstationarity in mean level was observed in the semivariogram for the 1980–1981 winter data. This entailed the use of universal Kriging for the 1980–1981 data, whereas ordinary Kriging could be used for the 1965–1979 data.

The differences in the levels of variability between the data sets seemed to be a reflection of (1) much higher average $SO_4^{2-}$ deposition levels during 1965–1979 in New York State versus much lower levels during the winter of 1980–1981 over the much larger north central–northeastern region and (2) differences in sample intervals. Bilonik (1985) noted that part of the difference in (1) stemmed from the fact that summer $SO_4^{2-}$ deposition levels tended to be higher than winter levels in this region. Since the 1965–1979 data included all 4 seasons whereas the 1980–1981 data included only winter data, more variability would be expected in the 1965–1979 data. However, as the size of the region increases, more variability would also be expected because of changes in mean deposition level. This phenomenon is partly reflected in the anisotropy observed in the 1980–1981 semivariograms, but some of this variability must be connected with the generally higher $SO_4^{2-}$ deposition levels reflected in the 1965–1979 data.

The methods illustrated by Bilonik's calculations represent only a subset of spatial statistical methods that can be applied to environmental problems. The insights gained with these methods have been useful for objectively assessing spatial heteorgcncities, as well as for designing new monitoring networks including the EMEFS, modifying existing networks, and helping to combine information from several overlapping networks. Objective estimation of spatial patterns also verifies subjective handdrawn maps of deposition.

### Empirical Orthogonal Functions and Their Interpretation

The separation and identification of dominant contributing factors to spatial correlation using EOFA have added important insight into regional-scale processes. At this stage, such results have advanced analytical methods employing monitoring data further than Kriging. The EOFA method essentially reorganizes a set of observations from many different stations into two independent data groups using matrix algebra and the singular value decomposition theorem. EOFA seeks relationships between observations of one variable at many different sites. In the examples described below, the variables are either 24-hour average airborne $SO_4^{2-}$ or TSP concentrations taken at the SURE CI and CII stations during the seasonally representative sampling months. The observations are organized in matrix form. In this case, each horizontal row of the data matrix contains measurements from all sites for a given day, and each vertical column contains all the observations from one site. Thus, $S_{ij}$ represents the $i$th day's value of $SO_4^{2-}$ at site $j$. As before, the concentration data are transformed, but this time only by subtracting the average value for each site, thereby centering the data about the average monthly value at the station,

$$Z_{ij} = S_{ij} - \overline{S_j}, \qquad i = 1, \ldots, m, j = 1, \ldots, n \tag{5.11}$$

where

$Z_{ij}$ = centered $SO_4^{2-}$ data for day $i$, site $j$
$S_{ij}$ = uncentered $SO_4^{2-}$ data for day $i$, site $j$
$\overline{S_j}$ = average $SO_4^{2-}$ for site $j$ = $(m^{-1}) \sum_{k=1}^{m} S_{ij}$

Figure 5.8 shows a matrix decomposition based on EOFA. The values of each EOF can be thought of as a map of $SO_4^{2-}$ concentration components; the map of observed $SO_4^{2-}$ for any day then is a weighted sum of the EOF maps.

A model for EOFA relates the ambient $SO_4^{2-}$ concentration at a given station as the sum of a mean value at the station and a series of temporally weighted spatial factors. These factors are derived mathematically using properties of the spatial covariance matrix. Ambient sulfate concentration can be written as

$$SO_4^{2-}(i, j) = \overline{SO_4^{2-}}(j) + \sum_{k=1}^{n} a_{ik} P_{kj}, \qquad i = 1, \ldots, m, j = 1, \ldots, n \tag{5.12}$$

where

$SO_4^{2-}(i, j)$ = 24-hour $SO_4^{2-}$ concentration for day $i$ and site $j$
$\overline{SO_4^{2-}}$ = average of the 24-hour $SO_4^{2-}$ concentrations at site $j$ for the duration of sampling (1 month)
$a_{ik}$ = weighting factor of the $k$th empirical orthogonal function
$P_{kj}$ = value of $k$th empirical orthogonal function (in $\mu g/m^3$) of $SO_4^{2-}$ at site $j$
$n$ = number of sites
$m$ = number of days of data

Note that the values of $P_{kj}$ corresponding to each EOF do not vary with time but vary only with space (i.e., from site to site) in this formulation. The values of $a_{ik}$ vary only with time. Thus, Eq. (5.12) formally describes the separation of $SO_4^{2-}$ variability into EOFs as temporarily constant patterns in space (or maps) and the weighting factors, which are patterns in time. In other words, the EOFs are maps or patterns of incremental $SO_4^{2-}$ concentration that, when added together, can reproduce all the observed distributions of $SO_4^{2-}$ concentration on any given day.

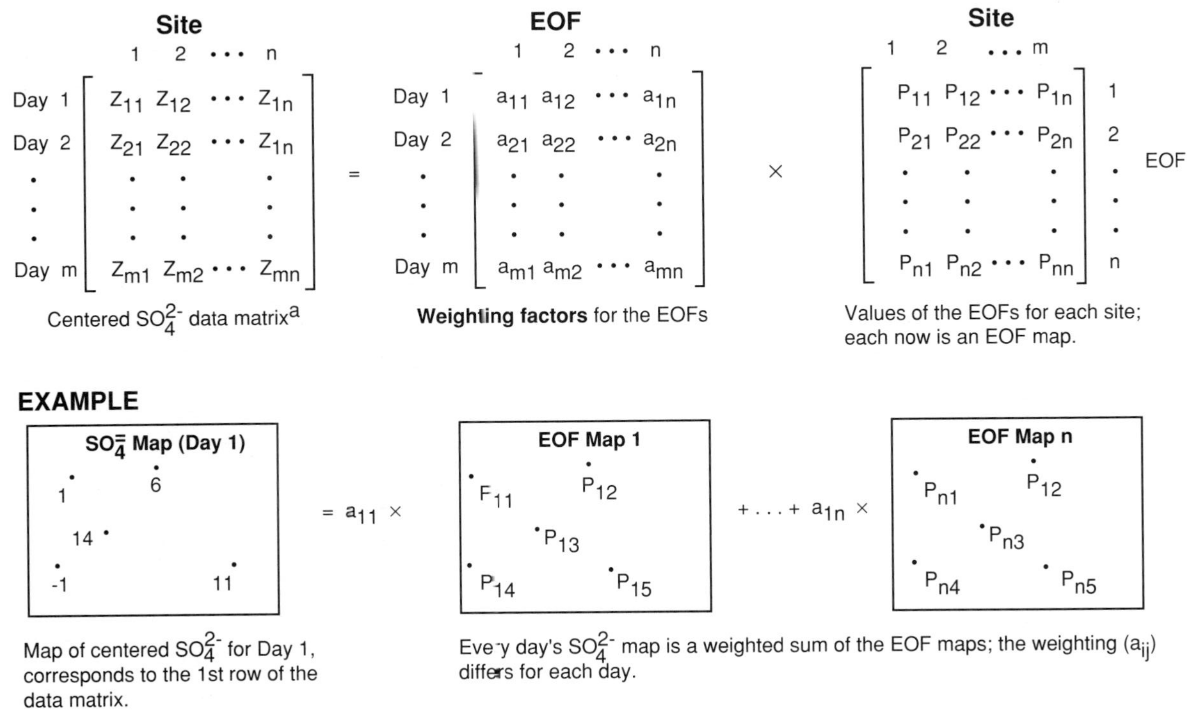


FIGURE 5.8 Illustration of the EOFA for $SO_4^{2-}$ observations, decomposing the matrix of observations into the product of two matrices of derived variables. (From Mueller and Hidy, 1983; reproduced with permission of the Electric Power Research Institute). For each site, centered $SO_4^{2-}$ = observed $SO_4^{2-}$ − average $SO_4^{2-}$ site.

The values of the EOFs, $P_{kj}$, and the weighting factors, $a_{ik}$, are derived from mathematical decomposition of the covariance matrix of paired station data taken on the same day *into* its two component matrices. The EOFs are calculated from characteristic (eigen) functions from the mathematical maximizing property for the eigenvectors of the covariance matrix, using the theory of singular value decomposition as previously described for the PCA.

The $P_{kj}$ in Eq. (5.12) are derived mathematically from the diagonalization of the covariance matrix. The corresponding $a_{ik}$ come from the eigenvectors weighting each $P_{kj}$. The EOFs are spatially orthogonal; the first one explains a maximum of the variability of $SO_4^{2-}$ in terms of covariance between all combinations of site pairs. The second EOF explains a maximum of the variability not explained by the first EOF, and so on.

Experience has shown that the bulk of the spatial variation in a regional air quality observable can often be captured in just three or four EOFs that together explain 70% or more of the variability in the data (e.g., Mueller and Hidy, 1983).

EOFA analysis of August 1977 SURE station $SO_4^{2-}$ data showed two dominant spatial patterns (Figure 5.9a and b) for 67% of the spatial variability for the month (Hidy *et al.*, 1985). In other words, of the more than 30 stations used to calculate the EOF, only two independent variables were needed to explain 74% of the data. The two EOF patterns showed large areas of characteristic spatial scales on the order of 500 to 1000 km. However, there were also areas that showed significant changes over scales of about 100 km. The first EOF had a broad (negative) maximum centered on western New York State extending in a generally east–west direction. The second EOF had a (negative) maximum over Indiana and extended in a north–south direction. Inspection of the daily observations that make up the covariance patterns suggested that the two patterns seem to be indicative of the two major classes of $SO_4^{2-}$ events of summer. These were stagnation on the lee side of a high-pressure cell (the second EOF) and transport by ducting of airflow up the Ohio River Valley (the first EOF).

The $SO_2$ EOF patterns for the August 1977 period, in contrast, showed no evidence of strong regional variation. The first EOF showed that over half the variability in the entire $SO_2$ data set was associated with a single station near Lake Erie. There was no large-scale pattern in any of the EOFs, which suggests that the characteristic spatial scale of $SO_2$ in summer was too small to be captured by the coarse spacing of the SURE stations.

The analysis of $SO_4^{2-}$ by spatial EOFs was also carried out using the October 1977 and January 1978 observations. The first two EOFs for October 1977 were similar in pattern to the summer months, except that

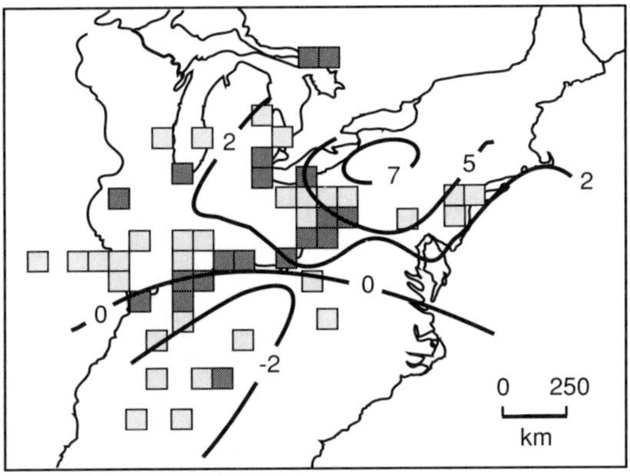


a. First sulfate EOF, explaining 42% of the total covariance.

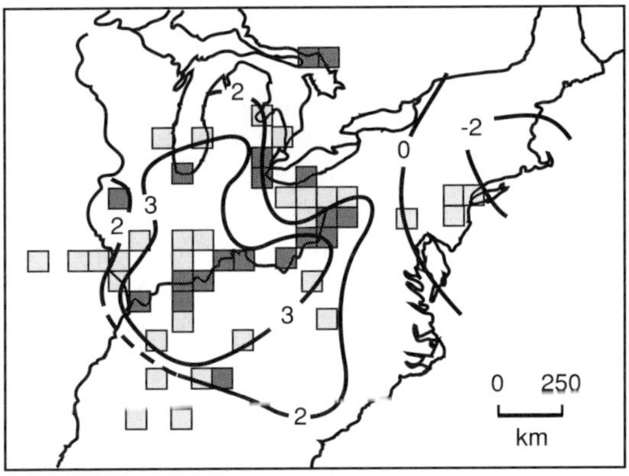


b. Second sulfate EOF, explaining 24% of the total covariance.

**FIGURE 5.9** First and second $SO_4^{2-}$ EOFs for August 1977 intensive. Numbers are $\mu g/m^3$ $SO_4^{2-}$ departure from the mean. Shaded areas are blocks of 80 × 80 km areas with summer average $SO_2$ emissions: light shading indicates emissions greater than or equal to 250 mT/d, and dark shading represents emissions greater than or equal to 500 mT/d. (From Mueller and Hidy, 1983; reproduced with permission of the Electric Power Research Institute.)

the EOF identified with stagnation conditions became more important than the EOF associated with air mass transport conditions. This is consistent with expectations from the detailed meteorological description of specific cases detailed earlier in Chapter 4. The analysis of the January 1978 airborne $SO_4^{2-}$ data produced the first two EOFs, which together explained 68% of the covariance of the $SO_4^{2-}$ at all the sites. The regional patterns differ from those found in summer. The EOF associated with the largest fraction of $SO_4^{2-}$ variability showed a type of east–west regional pattern over the southwestern part of the SURE grid that did not emerge in the results from other months. The second EOF showed a weak north–

south pattern, with almost no regional character signified by large areas without high or low values. The shift southward in the EOFs in the winter reflected an apparent seasonal change observed in the monthly 24-hour $SO_4^{2-}$ concentration averages.

Not all the particulate observables showed regional patterns as clearly as the $SO_4^{2-}$ distribution. For example, the TSP data analyzed for the summer 1977 showed different forms than airborne $SO_x$. Two patterns again dominated TSP variability. The first EOF for TSP was similar to the first $SO_4^{2-}$ EOF, but with a broad east–west maximum over the upper Midwest, not over Pennsylvania and New York as in the case of the $SO_4^{2-}$ EOF. The second EOF for TSP showed no regional character at all. It consists of a number of isolated pockets centered around zones of high emission density.

## Correspondence between Emissions and Ambient Concentrations

The association between the EOFs and emission distributions for both $SO_4^{2-}$ and TSP is readily seen in the EOF distributions. The zones of high emission density qualitatively underlie the patterns. For July 1978 the EOFs were very similar to August 1977. Both show the first EOF associated with transport showed a maximum about 200–300 km to the east from the Lake Erie region and the upper Ohio River Valley area sources. The second EOF showed a maximum just north of the Ohio River area with a southerly pattern reaching into Tennessee. This component was superimposed directly on the high emission density areas in the lower Ohio River area.

The 1977 and 1978 summer TSP EOFs showed a clear distinction from the airborne $SO_4^{2-}$ patterns; they appeared to be identified locally with the heavy band of industrialization through the Great Lakes subregion and with specific areas of high particulate emission density, which coincided with $SO_2$ emission patterns in many areas.

In a sense, the spatial patterns of the EOFs give a definition of a *zone of influence* of sources on receptors. Statistically, the areas of high emission density associate with a systematic $SO_4^{2-}$ variability over two distinct characteristic distances. The transport-related EOF extends in an elongated zone of influence from the upper Ohio River area northeastward with a characteristic distance scale of impact of about 500–700 km. In contrast, the stagnation-related EOFs in both summer and winter suggest a maximum impact centered on the subregional emission density areas usually within a 100-km radius, with an extent not exceeding over $\sim$300 km, corresponding to the EOF values greater than an absolute value of five. Thus, in any given month or year, the zones of maximum source influence may be predictable and should be related to synoptic-scale meteorological behavior.

Additional appreciation for the association between the $SO_2$ emission distributions and the EOF structure can be obtained by considering an analogy to the mass conservation equation for $SO_4^{2-}$ concentration. The relationship governing $SO_4^{2-}$ concentration can be written as

$$\frac{\partial c(x, y, t)}{\partial t} = -\mathbf{v} \cdot \nabla c + r \qquad (5.13)$$

where

$c(x, y, t)$ = airborne $SO_4^{2-}$ concentration
$\mathbf{v}$ = wind vector
$\nabla c$ = spatial gradient of the concentration
$r$ = totality of sources and sinks of $SO_4^{2-}$

Henry (reported in Mueller and Hidy, 1983) has noted that, hypothetically, the EOFs are thought to be analogous to particular steady-state solutions of Eq. (5.13), which, when added together, will describe $SO_4^{2-}$ concentration patterns when the left side of the equation is zero.

When the concentration pattern is in a steady state, i.e., the time derivative is zero, the dot product of the gradient of a dominant EOF and a characteristic wind field should look like a map of net production (or loss) of $SO_4^{2-}$. Henry's hypothesis was tested on the first EOF for 24-hour $SO_4^{2-}$ data during August 1977 (Mueller and Hidy, 1983). For simplicity, the wind field was taken as the unit vector (1,1), which is a uniform flow from the southwest to northeast. This wind field is a simple idealization of the transport winds prevailing in August 1977; it corresponds to a movement of air of one grid cell per day. Such a wind field crudely simulates a dominant airflow pattern aloft at least during ducting episodes, characteristic of significant air mass transport. An inverse-distance-squared interpolation was used to assign a value of the EOF to each grid square, and a standard second-order finite-difference scheme was used to evaluate the gradient. The dot product with the vector (1,1) was taken. Figure 5.10 is a plot of the results. Given these assumptions, the numbered contours are proportional to the apparent *net* $SO_4^{2-}$ production values, less deposition, but including both primary and secondary $SO_4^{2-}$. Because the map of EOF gradients was derived from data for only 30 stations, the geographic agreement of calculated distribution of $SO_4^{2-}$ production corresponds surprisingly well with the distribution of high $SO_2$ emission density. In particular, the areas of heavy $SO_2$ emissions in Ohio, western Pennsylvania, and West Virginia are well defined by this calculation. The isolated maximum over Connecticut may show the downwind production of $SO_4^{2-}$ from the New Jersey and New York City source areas.

In a later analysis, Henry *et al.* (1991) tested the EOFA gradient hypothesis for sulfate behavior in the southwestern United States. This "model" was tested using air mass trajectory calculations for the wind

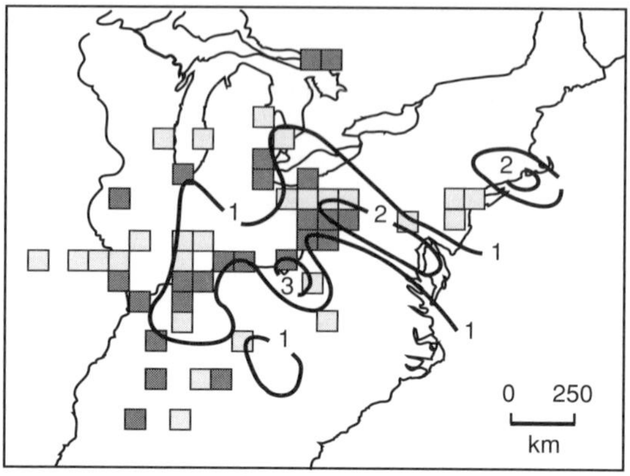


FIGURE 5.10 Comparison between calculated gradient in the first $SO_4^{2-}$ EOF for August 1977 and the distribution of $SO_2$ emissions. Shaded areas are blocks of 80 × 80 km areas with summer average $SO_2$ emissions: light shading indicates emissions greater than or equal to 250 mT/d, and dark shading represents emissions greater than or equal to 500 mT/d. (From Mueller and Hidy, 1983; reproduced with permission of the Electric Power Research Institute.)

field with EOFs calculated from sulfate data taken in the region. The estimated gradients in EOFs identified $SO_x$ source areas and strengths that were consistent with known geographic distribution of sources. Further study has not been attempted in ENA, but the EMEFS data set offers an opportunity in the future to verify this hypothesis.

## AIR MASS CLASSIFICATION BY MULTIVARIATE ANALYSIS

### Cluster Analyses of Meteorological Conditions at MAP3S Stations

Efficient objective interpolation would be assisted with a classification scheme to reduce variance in the data. A key to achieving acceptable levels of uncertainty in acid deposition source apportionment would be the segregation (or clustering) of natural spatial and temporal variability of the atmosphere into climate classes. These are qualitatively distinguishable sets of meteorological conditions that tend to occur together and together exert control over the transport, transformation, and deposition of acid species and their gaseous precursors. It is therefore of interest to see if the meteorological data collected in association with a long-term wet deposition monitoring program tend to fall naturally into distinct climate classes. Preliminary inspection of the UAPSP data by Martin (1984) suggested pronounced distinction between storm types with respect to storm chemistry. Raynor and Hayes (1982) also found evidence of differences associated with synoptic-scale disturbances.

Stratification of meteorological conditions relating to wet deposition was attempted by Samson in the MATEX feasibility studies (see Hidy *et al.*, 1985). The classification analysis differs from the SURE study employing EOFA in that its purpose is to identify clusters of days with similar meteorological conditions that encourage long-range transport accumulation and deposition of acidifying species, irrespective of ambient pollutant concentrations. Stratifying meteorological conditions in this manner helps segregate dominant weather regimes from stochastic elements of variability. The information is useful, ideally, for optimizing locations for sites of observational networks to capture the spatial and temporal characteristics of these dominant weather regimes.

Samson and Moody (1986) applied cluster analysis to four periods: wet days (days with precipitation) in winter, wet days in summer, dry days in winter, and dry days in summer. Table 5.13 lists the meteorological data that were available for clustering.

TABLE 5.13 **Available Meteorological Variables and Those Selected for Clustering of Wet and Dry Cases of MAP3S Data**[a]

| | Selected subsets | |
|---|---|---|
| **Available** | **Wet cases** | **Dry cases** |
| 850-mb level[b] | | |
| Height | | |
| Temperature | x | x |
| Mixing ratio | | x |
| Wind direction | | |
| Wind speed | | |
| *u* component | x | x |
| *v* component | x | x |
| Potential temperature | | |
| Transport level average | | |
| Mixing height | | x |
| Wind velocity shear | | |
| Precipitation | | |
| Amount | x | |
| Intensity | | |
| Duration | x | |
| Trajectory meteorology | | |
| Total upwind precipitation | | |
| Average upwind wind speed | x | x |
| Percent of time over source regions | | |

[a] From Samson and Moody, 1986.

[b] Some data available for 700-mb level.

- The rawinsonde data were obtained from the National Climate Center.
- Values of temperature, mixing ratio, wind velocity, and height of 850- and 700-mb levels, as well as the surface temperature, pressure, and mixing ratio, were extracted for the two rawinsonde stations closest to each MAP3S site.
- Two parameters were calculated from the ensemble vertical wind and temperature data. A transport layer depth was calculated using the technique of Heffter (1980). A weighted mean wind shear parameter (which combines both speed shear and directional shear) was computed for two layers: surface to 850 mb and 850 mb to 700 mb.
- The MAP3S station precipitation and precipitation chemistry data were obtained from the MAP3S data tapes.
- Three years of data (1979–1981) were used in constructing backward trajectories terminating at the eight MAP3S sites. The University of Michigan (UMACID) model was used to generate the trajectories. Average Lagrangian precipitation amounts were interpolated over successive 12-hour segments, 0–12 hours, 12–24 hours, and 24–36 hours upwind from each of the MAP3S sites.
- Wind speeds 12 hours, 24 hours, and 36 hours upwind of the sampling sites were estimated from the lengths of the 3-hour trajectory segments terminating at those times. The average upwind wind speed along the trajectory was also computed. (To remove any bias due to trajectories terminating prematurely, all trajectories were carried out to 96 hours and where necessary extended stochastically using a technique developed by Samson and Small, 1984.)

This particular set of meteorological parameters was chosen for the initial clustering because (1) the data were readily available or could be easily constructed in an objective manner and (2) (intuitively at least) any of the parameters could conceivably be an important key to understanding different regimes of transport, transformation, and deposition.

The MAP3S stations used for the analysis were Whiteface Mountain, NY; College Station, PA; Oxford, OH; Ithaca, NY; Charlottesville, VA; Urbana, IL; and Lewes, DE. At each of the MAP3S sites used, the cluster analyses were done separately for each of the two nearest upper air stations, summer or winter season, and wet days or dry days. Because incorporation of the full data set listed in Table 5.13 was found to be impractical, only subsets of the full set were used. Subsets of variables for wet days and dry days, respectively, were chosen subjectively in an attempt to accentuate air mass characteristics that might be characteristic of long-range transport, transformation, and deposition of S and N species. For cases involving precipitation, several parameters were chosen: precipitation duration, precipitation amount, 850-mb temperature, 850-mb *u*-component, 850-mb *v* component of winds, and average wind speed along the upwind trajectories terminating at the sampling station. For dry cases, the two precipitation parameters were replaced by the 850-mb mixing ratio and the depth of the mixed layer.

Although many other alternatives could have been chosen, the cluster metric used was the average Euclidean spacing between variables (e.g., Ward, 1963). A conventional software package was used to construct dendrograms and to carry out the successive clustering steps. Clustering was discontinued when the dendrogram spacing measure jumped significantly, which tended to occur at about 12–16 groupings. Of course, many of the resulting classes contain only one or a few outlier cases.

Summary descriptions of the dominant climate classes were prepared for each combination of MAP3S station, rawindsonde station, season, and wet or dry conditions. Each class that emerged from the cluster analysis was described by comparing, for each variable, the mean values in the class with the mean values for the period of study.

The results are illustrated by comparing the principal cluster classification results for two particular combinations of MAP3S station (Whiteface) and rawinsonde station (both Albany, NY and Maniwacki, Ontario) for wet and dry cases during summer only. The cluster analyses for summer dry and summer wet days at the Whiteface MAP3S site and rawinsonde data produced 12 cluster classes each. If classes that contain only a few cases are disregarded, four classes dominate the summer dry cases and three classes dominate the summer wet cases. These are described qualitatively in Table 5.14.

According to the clustering algorithm, the dominant cluster class in summer dry conditions (53 out of 156 cases) might be expected to favor physicochemical transformation of $SO_2$ and $NO_x$ (because of higher than average moisture and temperature during fair summer weather) and also to favor stagnation with higher than average ambient concentrations and deposition (because of light wind speed and shallow mixed layer). For this cluster class, the characteristic trajectories terminating at Whiteface appear to arrive from the WSW. This might suggest that for such conditions, source regions in western New York and the upper Ohio River Valley could contribute especially heavily to deposition at Whiteface Mountain as implied from case studies in Chapter 4. In summer wet conditions the dominant cluster class (18 days out of 45) corresponds to air masses originating in the western part of New York and in southern portions of Ontario. One could suppose that for these days the "signatures" of the large powerplants in north-central and western New York and the smelter

TABLE 5.14 **Summary of Dominant Cluster Classes (Whiteface MAP3S Site and Albany Rawinsonde)**

| | Description of dominant cluster |
|---|---|
| | Summer dry days 1979–1981 |
| Number of days in cluster class[b] | |
| 53 | Above average moisture; light wind speed; just above average temperature; shallow mixed layer; and WSW winds at 850 mb |
| 39 | Below average moisture; below average wind speed; average temperature; close to average mixing depth; and light variable NNW winds at 850 mb |
| 28 | NW winds at 850 mb; below average temperature; below average moisture; below average wind speed; average mixing depth |
| 9 | Very deep mixed layer; above average moisture; moderate wind speed; average temperatures; and NW wind at 850 mb. |
| | Summer wet days 1979–1981 |
| Number of days in cluster class[c] | |
| 18 | Shorter than average precipitation events; below average precipitation amount; below average wind speed; average temperature; and WNW wind at 850 mb |
| 9 | Lower than average precipitation duration; above average amount; above average wind speed; above average temperature; strong WSW winds at 850 mb |
| 5 | Average duration; below average amount; above average wind speed; below average temperature; strong NW winds at 850 mb |

[a] From Samson and Moody, 1986; reproduced with permission of the Electric Power Research Institute.
[b] Total number of summer dry days = 156
[c] Total number of summer wet days = 45

at Sudbury, Ontario, might occur as large pulses in the wet deposition field at Whiteface, as the still relatively coherent plumes from these distant point sources pass intermittently over Whiteface Mountain.

To test the uniqueness of the climate classification obtained at Whiteface with Albany rawinsonde data, the cluster analysis was repeated with rawinsonde data from Maniwacki, Ontario. Table 5.15 shows the resulting dominant cluster classes for summer dry and summer wet days. Comparison of the two tables shows that the largest summer dry cluster class in the Albany data set (53 cases) compares well with the largest summer dry cluster class in the Maniwacki set (69 cases), in terms of their descriptive air mass characteristics. However, the smaller cluster classes in each set show little or no correspondence. It seems premature therefore to interpret these classes in terms of conventional air mass descriptors (e.g., continental polar or maritime tropical). Similar side-by-side cluster comparisons were made for the other MAP3S stations locations, each with two alternative choices of rawinsonde data.

TABLE 5.15 **Summary of Dominant Cluster Classes (Whiteface MAP3S Site and Maniwacki Rawinsonde) Analysis**[a]

| | Description of dominant cluster |
|---|---|
| | Summer dry days 1979–1981 |
| Number of days in cluster class[a] | |
| 69 | Below average wind speed; above average temperature; above average mixing ratio; WSW winds at 850 mb; shallow mixed layer |
| 30 | Above average wind speed; slightly below average temperature; below average mixing ratio; NW winds at 850 mb; below average mixed layer |
| 16 | Light winds; average temeprature; below average mixing ratio; fairly strong SW winds at 850 mb; mixing depth twice as deep as average |
| 11 | Below average temperature; vey low mixing ratio; N winds at 850 mb; close to average mixing depth; average wind speed |
| | Summer wet days 1979–1981 |
| Number of days in cluster class[b] | |
| 17 | Shorter than average precipitation duration; above average precipitation amount; average wind speed; temperature close to average; fairly light WNW wind at 850 mb |
| 9 | Average precipitation duration; below average precipitation amount; below average wind speed; below average temperature; N winds at 850 mb |
| 7 | Average precipitation duration; below average precipitation amount; high wind speed; above average temperature; ENE wind at 850 mb with large variability |
| 4 | Short precipitation duration; variable precipitation amount; low upwind wind speed; above average temperature; NE wind at 850 mb |

[a] From Samson and Moody, 1986; reproduced with permission of the Electric Power Research Institute.
[b] Total number of summer dry days = 155
[c] Total number of summer wet days = 50

Cluster analysis has been extended to examine precipitation chemistry in terms of meteorological variables. For example, Moody and Samson (1986) have reported an analysis for 1 year of data at a UAPSP station at Rockport, IN. The data set was composed of 112 precipitation samples collected in 1982. The precipitation gauge from Rockport was used to determine the actual hours of each precipitation event, and the average precipitation intensity and maximum precipitation intensity were calculated. Cloud data were obtained for this site using the time of maximum precipitation intensity. The Rockport data were used to examine the relationships between cloud type and analyze concentrations. Five low-cloud types occurred with sufficient regularity during precipitation events to allow analysis of the associated concentration data. The five cloud types were stratus (layer clouds), cumulus (head clouds), stratocumulus (heaped layers), cumulonimbus (vertically developed cumulus clouds), and stratus and cumulus occurring simultaneously. When cloud type was obscured or otherwise not known, the cloud type unknown was reported.

Nonparametric statistical tests were applied to $SO_4^{2-}$, and $NO_3^-$ to determine if distributions were significantly different for the five low-cloud types. Separate tests were made for each season. The cumulative distributions of $SO_4^{2-}$ and $NO_3^-$ concentrations showed some interesting variations with cloud type. Examination of the distributions of warm season $SO_4^{2-}$ for the five low-cloud types illustrates that high $SO_4^{2-}$ concentrations were always associated with the cumuliform cloud types: stratocumulus, cumulus, and cumulonimbus. Stratus clouds, on the other hand, and simultaneous stratus and cumulus formations had uniformly low concentrations, with no value greater than 60 $\mu$mol/L. Similarly, the highest $NO_3^-$ concentrations were observed to occur with cumuliform clouds. Stratus clouds had $NO_3^-$ concentrations of less than 50 $\mu$mol/L. These variations of concentration with cloud type were not observed during the cold season.

A cluster analysis was applied to the data (separated by season) from different stations including Rockport. A PCA of the meteorological data was made to determine which variables contribute most to meteorological variability. In the warm season the variability was primarily associated with precipitation intensity, precipitation amount, scavenging potential, and height of the cloud tops; these were used as the cluster variables. The cluster analysis grouped 41 cases into 5 categories based on the values of these meteorological parameters.

Inspection of $SO_4^{2-}$ concentrations versus precipitation amount, intensity, and cloud top height was made, and points were labeled with the number of the cluster into which each event was classified. Precipitation intensity and precipitation amount were highly correlated, showing a strong inverse relationship with the $SO_4^{2-}$ (and $NO_3^-$) concentration.

Cluster 2 had the highest precipitation weighted average $SO_4^{2-}$, $NO_3^-$, and $H^+$ concentrations. It was distinguished by high cloud tops (all over 3.1 km) and had the lowest average precipitation amount. Cluster 4 also had very high cloud tops (90% greater than 2.7 km) but was associated with heavy precipitation and high precipitation intensity. Half of the events in this cluster were associated with severe weather and occurred with cumulonimbus clouds. The $SO_4^{2-}$ concentrations in this cluster were low, producing relatively low weighted average concentrations of $SO_4^{2-}$, $NO_3^-$, and $H^+$. Cluster 3 was characterized by relatively high precipitation intensity and amount but had the lowest cloud tops (<2 km). Weighted concentrations of major ions were very low in cluster 3. The most important factor in determining concentrations might appear to be precipitation amount. However, cluster 1, which had uniformly low precipitation amount, exhibited a large range in $SO_4^{2-}$ (and $NO_3^-$) concentrations. Although the highest $SO_4^{2-}$, $NO_3^-$, and $H^+$ concentrations were all associated with low precipitation amount, low precipitation amount was not always associated solely with the highest $SO_4^{2-}$ concentrations. Cluster 5 was small, containing only three events, and the data were insufficient to produce a statistically significant result.

Clusters representing conditions that consistently produced low precipitation weighted average concentrations of $SO_4^{2-}$, $NO_3^-$, and $H^+$ have been defined, and they appeared to be defined by relatively great precipitation intensity and precipitation amount. This suggested that dilution is responsible for these lower concentrations. Clusters 1 and 2, which contained high weighted average concentrations, were associated with low precipitation amount and intensity; however, there was great variability in the chemical composition of these clusters.

Because of the seasonality observed in some of the analyte concentrations, the data were separated into cold season (November–March) and warm season (April–September) sets. The cold season was subdivided further into snow and rain events. At northern sites, $SO_4^{2-}$ concentrations measured during the cold season were consistently higher in rain than in snow. Differences in $NO_3^-$ concentrations between winter rain and snow were less apparent, but for some sites $NO_3^-$ was observed to be more concentrated in snow than in rain.

The wind direction associated with each precipitation sample at all the eastern UAPSP sites was examined to determine what influence it might have had on $SO_4^{2-}$ concentrations. The 850-mb (approximately 1500 m above sea level) wind direction was used as the direction of air being scavenged by the precipitating clouds. An average $SO_4^{2-}$ concentration associated with each 45° wind sector was determined. This analysis did not identify any consistent site-to-site pattern of $SO_4^{2-}$ transport, and there were large variations in concentration around the average values. Wind direction alone did not appear sufficient for determining $SO_4^{2-}$ transport patterns.

To see whether seasonal differences in precipitation chemistry could be related to seasonal differences in the occurrence of various cloud types, the chemical characteristics of generic cloud type categories were summarized. Because the mean concentrations of analytes were not markedly different between cloud types, other factors such as transport, precipitation rate, and precipitation type may have a more controlling influence on analyte concentrations.

The relationship between upwind emission strength and the $SO_4^{2-}$ and $NO_3^-$ concentrations in precipitation was investigated for selected UAPSP sites. The amount of emissions upwind of a site was estimated using the 850-mb wind direction and wind speed and emissions data from the SURE. These data were used to calculate the emissions strength within 50 km or 12 hours of transport (whichever came first) of the UAPSP sites. No significant correlation was found between the estimated $SO_2$ emissions into the atmosphere upwind of the site and the observed $SO_4^{2-}$ concentrations in the precipitation at the site. Samson and Moody (1986) speculated that the lack of correlations might be related to (1) incorrect emission strength calculations, (2) contributions from distances greater than 500 km, (3) removal of the analyte and its precursor before storms' arrival at sampling locations, or (4) the nonrepresentativeness of 850-mb wind flow extrapolation for estimating upwind emissions input.

The cluster analysis was used to define periods of similar meteorology for three UAPSP sites: Tunkhannock, PA; Gaylord, MI; and Rockport, IN. Tunkhannock cold season events were categorized into similar meteorological periods using conventional air mass characteristics of temperature, moisture content, and wind direction. This demonstrated that the technique could identify different synoptic weather situations successfully. Although the mean chemical compositions of precipitation between groups were in general statistically different, the variability within groups overlapped that among groups. The cluster analysis showed two sets of weather patterns that produced relatively low $SO_4^{2-}$ concentrations but included $NO_3^-$ concentrations that correlated with acidity. These weather patterns were characterized by dry air, subzero temperatures, and winds from either the northwest or south.

Calculated air mass trajectories served to further segregate the data. They were clustered into significantly different transport regimes for the cold season at Rockport and Gaylord. The differences in chemical composition for clusters defined by transport were significant for both sites but were greater for Gaylord events. At the Rockport site, the highest cold season mean precipitation-weighted concentrations of $SO_4^{2-}$, $NO_3^-$, and $NH_4^+$ were associated with relatively low wind speed transport from west of the Rockport site. The highest mean acidity, however, was associated with low wind speed flow from south of the site. The lowest mean concentrations of $SO_4^{2-}$, $NO_3^-$, and acidity were associated with flow from the southeast. At Gaylord, the highest cold season mean concentrations were associated with flow from the south. The lowest concentrations occurred with relatively fast flow from west to northwest of the site.

Mixed-layer air mass trajectories, temperature, and precipitation intensity were used to define periods of similar conditions for Rockport warm season events. Significant chemical differences were influenced by stagnation upwind and precipitation upwind (along the trajectory). The highest mean concentrations were associated with two clusters, both representing relatively low wind speed situations with flow from east and west of the site. The lowest mean concentrations of $SO_4^{2-}$, $NO_3^-$, and acidity occurred with flow from the south along the Mississippi River valley.

Generally, Samson and Moody (1986) found it easier to define periods of uniformly low concentrations rather than uniformly high concentrations. Specifically characterizing a set of meteorological conditions that always led to high concentrations was not possible with the UAPSP data.

An exploratory comparison of the clusters between seasons for one site indicated that similar transport situations occurred in both the warm and cold seasons. At the Rockport site, the mean concentrations of analytes during relatively high wind speeds in both seasons were found to be similar. Large differences in mean seasonal $SO_4^{2-}$ and $NH_4^+$ concentrations occurred during relatively low wind speed situations. Thus $SO_4^{2-}$ concentrations at Rockport were not uniformly higher during the warmer months for all transport situations. These findings suggest that certain transport patterns (i.e., with high wind speeds) yield characteristic $SO_4^{2-}$ and $NH_4^+$ concentrations regardless of season.

The Samson and Moody (1986) results demonstrated that it is possible to define quantitatively periods of meteorological similarity and to do so in an objective manner, in contrast to the subjective synoptic classification in Chapter 4. If the results can be extrapolated, they suggest that the meteorological factors that have the greatest influence on precipitation chemistry include precipitation type, transport patterns, upwind stagnation, precipitation en route to the receptor, and precipitation amount and intensity (dilution). These results are consistent with expectations of the receptor and aerometric analyses in Chapters 3 and 4.

Samson and Moody were unable to resolve the relatively large variability in chemical concentrations of precipitation observed within meteorological clusters. Parallel chemical measurements in clouds and on the ground would be needed to characterize the causes of this variability. For improved results, future measurement networks for atmospheric deposition also should be collocated, to the degree feasible, with concurrent observations of wind flow between the surface and the base of the precipitating clouds as well as with observations of cloud parameters. Some of the difficulties encountered by the authors were associated with the changing meteorological conditions over the course of the 24-hour sampling periods. It was difficult to determine which meteorological conditions existed when ionic deposition was greatest.

Fernau and Samson (1990) extended the analysis of Moody and Samson (1986). The later work was conducted for 10 sites in ENA selected from the 28 UAPSP and APIOS locations (Montague, MA; Underhill,

VT; Longwoods, ONT.; Gaylord, MI; Zanesville, OH; Raleigh, NC; Rockport, IN; Lancaster, KS; Clinton, MS; and Uvalda, GA). Calculated backward air mass trajectories between 1979 and 1983 from the ATAD model were used for the cluster analysis, along with the UAPSP and APIOS precipitation event data for the same period. Air mass trajectories were calculated for representative 72-hour periods.

The resulting clusters represented groups where mean air mass transport were statistically different from one another and correspond to types of high- and low-pressure patterns on daily weather maps. Ward's (1963) method of calculation produced clusters with roughly equal numbers of occurrence, with recurring patterns from different sets. The clusters were found to be relatively insensitive to changes in numbers of sites or years used and the number of hours upwind used to define distance vectors.

The clustering by air mass trajectory was found to be successful in defining wet, polluted, or nonpolluted sulfate deposition clusters in mean deposition patterns. Some clusters were found to be in relatively dry conditions over much of ENA, while others were associated with high precipitation. Some of the clusters had high sulfate deposition over much of the region and/or had a ratio of percent sulfate deposition to percent water deposition greater than 1.0. The highest sulfate deposition over the widest areas was associated with mean transport patterns with slow air mass movement over large areas of high $SO_x$ emissions, and these were frequently persistent over several periods or followed persistent clusters. Some of the cluster deposition distributions are significantly different from each other at the 10 sites, although the spatial patterns were not statistically compared. Within-cluster variability was large and there was overlap among clusters. Some of this may have been due to changes in deposition because of frontal passages, something not discriminated by the 3-day deposition period. It was not possible to define clusters that were almost always heavily polluted. This was especially true because some clusters with very polluted precipitation also included many dry periods. Within clusters, sulfate deposition was generally higher in the warmer months. Overall, cluster anlaysis of air mass transport vectors has identified the weather patterns which are likely to either lead to large deposition events or contribute large amounts of deposition over the course of a year.

Although the separation of meteorological cases that lead to distinctive deposition patterns is facilitated by clustering, there remain ambiguities that are qualitatively analogous to those in the subjective analyses. Nevertheless, the results obtained by Samson and colleagues have been used as a means of compositing characteristic conditions that can represent an annual average for model calculations (e.g., Dennis *et al.* 1991). (See also Chapter 6.)

## VARIABILITY AND SOURCE–RECEPTOR RELATIONSHIPS

### Application of Covariance Analysis

Application of objective analysis[2] provides a possible means of extracting SRR information from spatial variability in aerometric or deposition observations. This possibility was suggested by Eddy *et al.* (1984) as a scheme for designing an empirical method to estimate SRRs. Furthermore, objective analysis schemes for *a priori* network design and data analysis are available and could be applied to experimental designs, exemplified by the Massive Aerometric Tracer Experiment (MATEX). With an objective analysis method, a formal means of organizing measurements and their error and uncertainty, including interpolations, could be set up. This approach provides a means of optimizing ground network and aircraft sampling to provide data for a stated SRR estimation scheme.

The application of objective analysis treats the SRR estimation problem by specifying (1) a desired confidence level, (2) a means of determining a space–time-varying observable (deposition), (3) criteria of space–time "volume" of estimate on interpolation from measurement, and (4) an optimal scheme for estimating apportionment among several possible source contributors, with desired confidence intervals stated within the SRR estimates. The analysis provides a complete design sampling network and data analysis scheme for meeting the five design elements, minimizing variances, experiment duration, and station number density.

By suitably stratifying a field of observations (e.g., into meteorological classes or EOFs), a statistical model for total variance may be used to further distinguish within-class and between-class parts and noise variance (residual not otherwise explained).

The basic set of data that drives the subsequent analyses is the structure of the covariance field for the observation set. In the absence of suitable field data, covariance structure functions are postulated for design purposes, from *a priori* knowledge of the signal generation process. Covariance *structure functions* can also be composed from long-term data records simulated from theoretical calculations. The error variance analysis for design can be strengthened considerably if the observable covariance fields are decomposed into their principal eigen (EOF) patterns, grouped within distinguishable meteorological classes (or weather regimes) and assigned misclassification probabilities. The weather regimes would be identified with cluster analyses or air mass segregation.

[2] This term refers to a technique for interpolation or extrapolation of data from a few stations to characterize spatial and temporal conditions over a region (Brady, 1977).

An analysis of spatial covariance fields then drives the data interpretation and empirical SRR determination scheme, such as the MATEX. The analysis is organized to include the following three points:

1. Optimal, objective multivariate regressions to link predictors (such as $SO_2$ emissions and meteorological and aerometric variables) to predictands (such as wet and dry deposition of total S) within distinct climate classes.
2. Optimal objective analysis interpolation methods to generate unbiased, efficient estimates of point-average or area-average grid point values of observation fields for different averaging times and spatial domains. These techniques can also be used to make the explicit connections between confidence intervals of estimates and the spatial and temporal resolution of the network.
3. Design of optimal sampling networks to preferentially capture "signal" variance and select against "noise" variance. Nonlinear programming techniques can be used to optimize the sampling network design for a wide range of relative values, costs, and practical constraints or restrictions on sampling protocols.

The mathematical aspects of the method are summarized below. For additional details, see Hidy *et al.* (1985) and Eddy and Samson (1986).

Suppose a time history for an observational variable $Y^\circ$ at a sampling location. The pattern is considered continuous but could be a set of discrete point values. Assume that the particular climate class (or other data grouping) scheme to which this portion of the $Y^\circ$ time history belongs is known. Then $Y^\circ$ can be represented by the following components:

$$Y^\circ = \mu + Y_c + Y_s + \varepsilon_s \tag{5.14}$$

where

$Y^\circ$ = the actual observation (e.g., a wet deposition value)
$\mu$ = the mean background value (the population mean of $Y^\circ$ over the entire ensemble of observations)
$Y_c$ = the mean value of the particular climate class, expressed as a deviation from $m$
$Y_s$ = the source signal identified, expressed as a deviation from the within-class mean
$\varepsilon_s$ = the noise in the observation—small-scale fluctuations, instrument error, and anything else that cannot be accounted for

Because perfect knowledge about the population is unachievable, $\mu$ and $Y_c$ are estimated as

$\overline{\overline{Y}}$ = the sample estimate of $\mu$
$\overline{Y}_c$ = the sample estimate of $Y_c$

and

$$Z = Y^\circ - \overline{\overline{Y}} - \overline{Y}_c \tag{5.15}$$

is the predicted residual signal portion at an observed point.

Adopting matrix notation, assuming that $Z$ is an $n \times m$ matrix representing $N$ sets of observations at $M$ points, an optimal objective analysis of the signal is set up. The standard linear regression model is then

$$Y_s^\circ = Z\beta + \varepsilon \tag{5.16}$$

where

$Y_s^\circ$ = the predicted (source) residual signal + noise ($Y^\circ - \overline{\overline{Y}}{}^\circ - \overline{Y}_c$) analysis point
$Z\beta$ = the "ability" of the signal embedded in the observation points to estimate the signal at the analysis point
$\beta$ = the set of population regression coefficients
$\varepsilon$ = the set of errors in the signal estimates (assumed to have constant variance across the set of realizations)

The analysis estimate of $Y_s^\circ$ becomes

$$\tilde{Y}_s = Z\tilde{\beta}$$

$$\tilde{\beta} = [Z^TZ]^{-1}[Z^TY_s^\circ]$$

The analysis error based on $Y_s$ is just

$$\eta = Y_s^\circ - Y_s$$

and the *variance* of the residual error is given by

$$\sigma_\eta^2 = Y_s^{\circ T}Y_s^\circ - Y_s^{\circ T}Z[Z^TZ]^{-1}Z^TY_s^\circ \tag{5.17}$$

The first term on the right is the total variance of the signal + noise at the analysis point. The second term is that part of the variance explained by the predictors $Z$. The fraction of variance unexplained is then

$$1 - \frac{[Y_s^{\circ T}Z][Z^TZ]^{-1}[Z^TY_s^\circ]}{Y_s^{\circ T}Y_s^\circ} \tag{5.18}$$

Note that the covariance matrices $Z^T$ and $Z^TY_s^\circ$ control the fractional error variance. These matrices can be specified in different ways, depending on the application:

1. If an *optimal objective analysis* of a set of observations is desired, then $Y_s^\circ$ (a set of predictands) and $Z$ (a set of predictors) are both known, and the regression shows the fraction of variance in $\sigma^2(Y_s^\circ)$ explained by $Z$.
2. If the intent is *optimal* interpolation to $Y_s^\circ$ from a set of observations $Z$, the $Z^TZ$ is known, but not the covariance matrices $Y_s^{\circ T}Z$

or $Y_{\mathrm{s}}^{\circ\mathrm{T}} Y_{\mathrm{s}}^{\circ}$. These must be specified analytically or constructed from theoretical model results.

The usefulness of the regression $Y_{\mathrm{s}}^{\circ}$ upon $Z$ to explain results would hinge on the relative magnitudes of the different estimates of the total variance of $Y^{\circ}$, as based on (1) the population mean, (2) the between-group means, and (3) the within-group means. That is

$$\sigma_{Y^{\circ}}^{2} = \sigma_{\bar{\bar{Y}}}^{2} + \sigma_{\bar{Y}_{\mathrm{c}}}^{2} + \sigma_{Y_{\mathrm{s}}}^{2} + \varepsilon_{\mathrm{s}}^{2} \tag{5.19}$$

For an objective analysis (or for a network sampling design) to perform well in its mission to reinforce signal and suppress noise, a good "background" model to help specify $\sigma_{\bar{\bar{Y}}}^{2}$ is needed, as well as a good climate classification (or cluster analysis model) to help specify $\sigma_{\bar{Y}_{\mathrm{c}}}^{2}$. The value of subdividing the total predicted data set into groups depends on the extent to which the total variance as estimated by the between-group means is larger than the total variance as estimated by the within group means ($\sigma_{Y_{\mathrm{s}}}^{2}$). In other words, the ability to distinguish the signal from the noise in an observation depends on how well one can associate variable behavior with a predictable component that can be tied to changes in meteorological and other conditions.

*Role of Covariance Function* However derived—whether synthesized from model data, chosen analytically, or constructed from actual field observations—the spatial covariance function is absolutely fundamental to applying linear regression, objective interpolation, and source apportionment in optimal ways. The covariance function must supply the elements for $Z^{\mathrm{T}}Z$, $Y_{s}^{\circ\mathrm{T}}Z$, and $Y_{s}^{\circ\mathrm{T}}Y_{s}^{\circ}$, i.e., for all possible pairs of observations points and analysis (or interpolation) points. Moreover, partitioning $Z^{\mathrm{T}}Z$ should use a statistical technique that allows one to apply what is known about the various characteristic patterns associated with different climate classes to filter out optimally the predictively useful signal structure from the random noise structure in the covariance field. It is of considerable importance in the experimental design to obtain such signal structure from available observations wherever possible.

The eigenvector approach to variance partitioning leads directly to a promising candidate method for source apportionment (Eddy *et al.*, 1984). One begins by assuming that the spatial covariance patterns of individual sources can be inferred by direct measurements of source signatures with suitable tracers or can be estimated from model simulations made over different climate classes. For example, a large set of source-region-specific covariance functions can be constructed from theoretical model results. From this perspective, the problem of source apportionment becomes an exercise in spatial variance partitioning. The signal estimate of Eq. (5.15) can be rewritten for the case of $L$ sources:

$$\tilde{Y}^{\circ}_{gk} = \overline{\overline{Y}} + \overline{Y}_{gk} + \sum_{\ell=1}^{L} \tilde{Y}_{gk\ell} \tag{5.20}$$

where

$\tilde{Y}^{\circ}_{gk}$ = the estimated total signal at the observation point $k$, for class/group $g$
$\overline{\overline{Y}}$ = the background at point $k$
$\tilde{Y}_{gk\ell}$ = the contribution from source $\ell$ to the signal at point $k$, for class/group $g$
$\overline{Y}_{gk}$ = the class/mean for climate class/group $g$, point $k$.

The signal total variance becomes

$$\sigma^2_{\tilde{T}} = \sigma^2_{\tilde{g}} + \sum_{\ell=1}^{L} \sigma^2_{\tilde{g}k\ell} \tag{5.21}$$

where

$\sigma^2_{\tilde{g}}$ = the between-class signal variance
$\sigma^2_{\tilde{g}k\ell}$ = the within-class signal variance for source $\ell$

and the components of noise variance become

$$\sigma^2_{\tilde{e}} = \sigma^2_{\overline{\overline{Y}}} + \sigma^2_{\overline{Y}_g} + \sum_{\ell=1}^{L} \sigma^2_{\eta\ell} + \sigma^2_{ss\ell} \tag{5.22}$$

where

$\sigma^2_{\overline{\overline{Y}}}$ = the uncertainty in the estimate of background $\overline{\overline{Y}}$
$\sigma^2_{\overline{Y}_g}$ = the uncertainty in the estimate of within class mean $\overline{Y}_g$
$\sigma^2_{\eta\ell}$ = the analysis error associated with the objective analysis $\tilde{Y}_{gk\ell}$
$\sigma^2_{ss\ell}$ = the remaining error (small-scale, instrumentation) associated with $\tilde{Y}_{gk\ell}$

Source apportionment is then equivalent to the problem of finding an optimal minimum-residual variance solution to

$$\tilde{Y}_{gk\ell} = \gamma_{gk\ell} Y_{ogk} \; (\ell = 1, \ldots L) \tag{5.23}$$

and

$$\sum_{\ell} \gamma_{gk\ell} = 1$$

where

$Y^{\circ}_{gk}$ = the total observation (signal + noise) at $k$
$\gamma_{gk\ell}$ = the apportioning parameter for source $\ell$

***Candidate Apportionment Procedure*** Let the long-term (expected value) of the mean contribution at observation point $k$ from source $l$ in class $g$ be

$$\overline{Y}_{\mathrm{T}kg\ell}$$

with signal variance

$$\sigma^2_{gk\ell} = E(Y_{\mathrm{T}gk\ell} - \overline{Y}_{\mathrm{T}gk\ell})^2$$

and total noise variance:

$$\sigma^2_{\mathrm{e}gk\ell}$$

The zeroth-order estimate of source attribution within class $g$ could be

$$\tilde{Y}_{kg\ell} = \gamma_{gk\ell} Y^{\circ}_{kg} \tag{5.24}$$

where

$$\gamma_{gk\ell} = \overline{Y}_{\mathrm{T}gk\ell} \Big/ \sum_{\ell=1}^{L} \overline{Y}_{\mathrm{T}gk\ell}$$

and an error variance given by

$$\sigma^2_{\tilde{Y}_{gk\ell}} = \sigma^2_{gk\ell} + \sigma^2_{egk\ell}$$

This is simply apportioning by relative expected long-term signal means. An improvement on this can be made if the estimate of apportionment is adjusted by an increment $\delta_{gk\ell}$ that reduces the variance, $\sigma^2_{gk\ell}$. Here $\delta_{gk\ell}$ is deduced by some logical partitioning of this signal variance. This can be done with empirical orthogonal functions (eigenvector) methods derived from spatial covariance functions.

Earlier it was noted that an observation field could be split up into a set of weighted space patterns. Furthermore, a time series of these observation fields could be expressed as a product of weights that are functions of time only and space patterns that are time invariant. If the field of $\sigma^2_{gk\ell}$ over all points $k$ is captured largely by one such eigenpattern, then this pattern can be gainfully used in the calculation of the partitioned component, $\delta_{gk\ell}$.

If one further denotes

$\overline{Y}_{gk\ell}$ = class mean value from source $\ell$ at point $k$ (for unit emission)

one can then define a residual of the observation at point k as

$$Y^{\circ\prime}_{gk} = Y^{\circ}_{gk} - \left( \sum_{\ell=1}^{L} \overline{Y}_{gk\ell} \right) \tag{5.25}$$

The first eigenpattern from each of the $L$ sources is used to estimate this field of $k$ residuals. Let $E_{gikl}$ mean the component at location $k$ of the $i$th eigenpattern for source $\ell$ and class $g$. Then the residuals can be expanded out in the eigencoefficient $E_{gik\ell}$ as:

$$\begin{aligned} Y_{g1}^{o'} &= \alpha_{g1}E_{g111} + \alpha_{g2}E_{g112} + \cdots + \alpha_{gL}E_{g11L} \\ &\quad + \varepsilon_{g11} \\ Y_{g2}^{o'} &= \alpha_{g1}E_{g121} + \alpha_{g2}E_{g112} + \cdots + \alpha_{gL}E_{g12L} \\ &\quad + \varepsilon_{g12} \\ &\ \ \vdots \\ Y_{gk}^{o'} &= \sum_{\ell=1}^{L} \alpha_{g\ell}E_{g1k\ell} + \varepsilon_{gk\ell} \\ &\ \ \vdots \end{aligned} \tag{5.26}$$

For convenience,

$$W_{g\ell}^{\mathrm{T}} = (E_{g11\ell}, E_{g12\ell}, \ldots, E_{g1k\ell}, \ldots, E_{g1k\ell})$$

where

$$W_g = (W_{g1}, \ldots, W_{g\ell}, \ldots, W_{gL})$$

Then a first estimate of $\alpha_g$ is obtained by a regression technique:

$$\tilde{\alpha}_g = [W_g^{\mathrm{T}} W_g]^{-1}[W_g^{\mathrm{T}} Y_g^{o'}]$$

$$\tilde{Y}_g' = W_g \tilde{\alpha}_g$$

leaves too much freedom in the estimation process; that is, the relative probabilities of "observing" the $L$ eigenpatterns for a given set of source emissions on a given day do not enter in the selection of $\alpha_g$ in this type of solution. Instead, the residual variance is minimized using a standard iterative procedure where $K$ = number of observation points, $L$ = number of sources, and OF = an objective function to be minimized:

$$\mathrm{OF} = \sum_{k=1}^{K} \left( Y_{gk}^{o'} - \sum_{\ell=1}^{L} \tilde{\alpha}_{g\ell} E_{g1k\ell} \right)^2 \tag{5.27}$$

subject to the constraint that

$$P_1(\alpha_{g1} < \tilde{\alpha}_{g1}) = P_2(\alpha_{g2} < \tilde{\alpha}_{g2}) = \cdots = P_\ell(\alpha_{g\ell} < \tilde{\alpha}_{g\ell})$$

Here $P_i(\alpha_{gi} < \tilde{\alpha}_{gi})$ means the probability that $\alpha_{gi}$ is no larger than $\tilde{\alpha}_{gi}$.

The subtlety of the method enters in the way the $\tilde{\alpha}_{g\ell}$ are determined from the probability distributions of the respective eigenpatterns. The condition imposed is that, when scaled by relative source strength, the principal eigenpattern from each source $\ell$ has equal probability of contributing to observation $k$. The values $P_\ell$ come from the time-dependent set of principal components associated with the within-class eigenpatterns.

To achieve equiprobable selection of the principal eigenpatterns for the $\ell$ sources requires that the coefficient $\tilde{\alpha}_{g\ell}$ be proportional to the normalized source strength, or

$$\tilde{\alpha}_{g\ell} = S_\ell \tilde{h}_{g\ell} \tag{5.28}$$

where the source strength $S_\ell$ are normalized so that $\Sigma\, S_\ell = 1$. For an iterative solution, proceed as follows:

1. Beginning with source $\ell$, select $\tilde{\alpha}_{g\ell}$ (using a logical search to decrease the objective function).
2. Calculate the $h_{g\ell}$ from Eq. (5.28).
3. Find the frequency $[P_\ell(h_{g\ell} < \tilde{h}_{g\ell})]$ from the cumulative frequency distribution for source $\ell$.
4. Use the $P\ell$ on the $L - 1$ other plots to find the other $L - 1$ values of $\tilde{\alpha}_{g\ell}$.
5. Put these equiprobable values of $\tilde{\alpha}_{g\ell}$ into Eq. (5.27).
6. Repeat steps 1–5 until Eq. (5.27) becomes a minimum.

This procedure would lead to an adjustment $\delta_{g1k\ell}$ to the source apportionment. That is, a revised contribution of the $\ell$th source in class $g$ is:

$$\tilde{\tilde{Y}}_{gk\ell} = \tilde{Y}_{gk\ell} + \delta_{g1k\ell} \tag{5.29}$$

where

$$\delta_{g1k\ell} = \left(\tilde{\alpha}_{g\ell}E_{g1k\ell} \Big/ \sum_{\ell=1}^{L} \tilde{\alpha}_{gk\ell}E_{gk\ell}\right) Y_{gk}^{\circ\prime}.$$

Ideally, to verify this approach one would (1) calculate eigenpatterns for each source with preselected models; (2) do multiple experiments with releases of different distinguishable tracers, one from both sources; and (3) aggregate the different tracers at a point (i.e., pretend that the tracers are indistinguishable) and use the method outlined to apportion the source contributions.

The actual partitioning for each experiment series could be determined from the unaggregated daily tracer data in combination with sensor data giving the sum of all $\ell$ contributions. Over many realizations, the method should yield a reliable portioning. The method was proposed as a basis for a station network for the MATEX (Hidy *et al.*, 1985) but has not been implemented.

## An Optimal Sampling Network Design

On practical grounds, the design of sampling networks must seek optimal trade-offs between desired minimization of observational noise variance on the one hand and operational, logistical, and cost feasibility on the other. The analyses of covariance methods already discussed are directly useful here as well. They can be applied to the closely coupled problems of relating a desired source contribution or signal-to-noise ratio of an observation to the size and space–time resolution of the sampling network and of seeking sampling designs that provide the most variance reduction (or other benefit) for a given level of cost.

The process of optimization begins by recognizing that the network residual variance, given by Eq. (5.17), is dictated primarily by the intrinsic covariance structure function that characterizes the observational field. But because the structure function must be estimated from observations over a discrete set of stations, estimates of the matrixes $Z^{T}Z$ and $Z^{T}Y$ are strongly dependent on the space–time separations between observations. Therefore, the reduction of observational variance that can be achieved by a given network design will depend critically on

- how realistically the covariance structure function can be approximated *a priori* (whether analytically or as estimated from modeled or actual observational data), and
- how accurately the candidate network design can resolve the space–time structure of the covariance field for a given cost or resource commitment.

Success in approximating the covariance function will be determined largely by whether an adequate and efficient climate classification scheme is constructed with which to group observational fields (or their modeled surrogates) under different conditions. The space–time behavior of a field (such as wet deposition) is quite different for different climate classes; therefore a network optimized for one class is not likely to be optimal for another. Compromise is therefore necessary, and an optimal network that emerges will depend on both the relative probability of occurrence and the relative importance of each class to the overall (e.g., long-term average) space–time behavior of the field of interest.

Analysis appears to identify suitable climate classifications (or other aggregations of meteorological variables) by which to stratify the observational fields of tracer, ambient reactive species, and deposition (see Chapter 4). A method also exists for decomposing these (already stratified) fields into their principal (e.g., signal-carrying) eigenstructures. Suppose that both the stratification and the signal/noise partitioning implied have been successfully achieved. An objective scheme for optimal (cost-effective) placement of a set of sampling stations can now be developed given the following:

1. The observational signal-to-noise ratios required to meet experimental end product objectives
2. The covariance structure functions and principal components for the "important" classes
3. The relative probabilities of occurrence and misclassification of these classes
4. The relative values of information from different sampling stations (or sampling platforms)
5. The operational and other practical constraints that may limit the scope or resolution of a sampling technique (e.g., restrictions on aircraft flight profiles)
6. The relative costs associated with the numbers, locations, and kinds of samplers

Eddy *et al.* (1984) offered a means by which one can proceed for optimal network placement. To seek solutions that *minimize an objective function* that combines all the factors enumerated above,

$$\mathrm{OF} = \Sigma_{\ell}\, \Sigma_{n_g}\, C_{\mathrm{ng}}\, \Sigma_i\, \Sigma_j\, \mu_{ij} P_{ij} \sigma^2_{\mathrm{RES}}(i, j)$$

$\ell$ = the number of sources

$n_g$ = the number of analysis points (grid points)

$C_{\mathrm{ng}}$ = the cost of error at analysis point $n_g$

$i, j$ = indices that range over the climate classes: $i$ = correct, $j$ = incorrect

$P_{ij}[\mu_{ij}]$ = the probability (the cost) of misclassification; that is, of assigning class $j$ when class $i$ is true

$\sigma^2_{\mathrm{RES}}(i, j)$ = the network residual variance defined by a particular configuration of samplers for set $i, j$ of classes

This last term requires some elaboration. In general, the network residual variance is given by

$$\begin{aligned} \nu\sigma^2_{\mathrm{RES}} &= [Y^{\circ\mathrm{T}} Y^{\circ}] = [Y^{\mathrm{T}} Z][Z^{\mathrm{T}} Z]^{-1}[Z^{\mathrm{T}} Y] \\ &= [Y^{\circ\mathrm{T}} Y^{\circ}] - \beta^{\mathrm{T}}[Z^{\mathrm{T}} Z]^{-1}\beta \end{aligned} \tag{5.30}$$

where $\nu$ is the number of degrees of observational freedom. Now if class $j$ is selected mistakenly when class $i$ is the correct choice, the network residual variance $\sigma_{\mathrm{RES}}$ is given by

$$\nu\sigma^2_{\mathrm{RES}}(i, j) = [Y^{\circ\mathrm{T}} Y^{\circ}]_i - 2[Y^{\mathrm{T}} Z]_i \beta_j + \beta_j [Z^{\mathrm{T}} Z]_i \beta_j \tag{5.31}$$

in which both the correct terms $[\ldots]_i$ and the incorrect terms $[\ldots]_j$ enter.

The network optimization problem then reduces to a problem of minimizing the objective function under the specified set of constraints. The minimization is accomplished iteratively by a conventional but complicated nonlinear programming (NLP) procedure (pp. 254–256). In brief, the solution scheme follows:

1. Select the set of grid analysis points at which the objective function is to be evaluated. Note that the locations specified for the analysis points exert considerable influence on the optimal sensor locations ultimately derived.
2. Choose a reasonable starting vector, that is, a tentative set of sensor locations upon which to iterate. Because for computational efficiency the initial distribution should be matched at least approximately to the expected structure(s) of the principal covariance patterns, some preliminary experimentation will likely be required.
3. Calculate the objective function from Eq. (5.27).
4. Use the NLP method to "move" the sensor locations. (First derivatives of the objective function may be required.)
5. Recalculate the objective function, and calculate whether any improvement has been gained; return to step 4 and iterate until the improvement is marginal.
6. Add sensor locations and recalculate their optimal placement considering the entire set together.
7. Stop when the objective function is a minimum and the marginal improvement gained by adding more sensor locations is not worth the incremental cost of doing so.

Alternatively, the procedure could begin from the other end, i.e., starting with the maximum conceivable number of receptor locations and then removing them one at a time. At each iteration the location(s) that contributes the least improvement in overall OF per unit receptor cost would be discarded.

The NLP objective network sampling procedure is not restricted to use of fixed ground stations. In the MATEX investigation Eddy *et al.* (1984) made some sample calculations to show how, once an optimal surface network was chosen, one could construct optimal aircraft flight paths to achieve further overall reductions in network noise variance. At the time these laborious calculations were performed, the covariance structure functions were not available from observational data, so Eddy *et al.* (1984) adopted arbitrary (but generally reasonable) analytical forms as surrogates.

To show how the network design approach might contribute to optimal placement of samplers for a deposition experiment, Eddy *et al.* (1984) carried out a generic network analysis for an upper Ohio River source.

Beginning with the tracer-pollutant distributions generated from a theoretical model, they calculated the *principal eigenvectors* of each of the following six covariance fields:

1. Winter/passive tracer release distribution
2. Summer/passive tracer release distribution
3. Winter/ambient $SO_2$
4. Summer/ambient $SO_2$
5. Winter/total S wet deposition
6. Summer/total S wet deposition

In the case studied, the first eigenpattern explained most of the variance.

When these patterns were plotted, they readily suggested appropriate "starting vectors" for network optimization. The patterns themselves appeared to be representable by functions whose defining parameters require only 8–16 effective degrees of freedom, which implied that they could be resolved by networks with comparable numbers of sampler locations.

Accordingly, network optimization experiments were made beginning with a 6-station network and also beginning with a 14-station network (to explore both extremes implied by the above degrees-of-freedom argu-

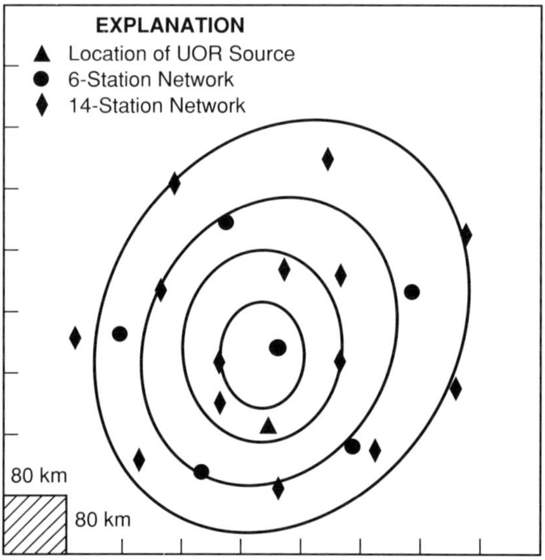


FIGURE 5.11 "Optimal" 6-station and 14-station networks for upper Ohio River source. Grid point spacing is 80 km. Covariance structure function is typical of wet deposition footprints. (From Hidy *et al.*, 1985; reproduced with permission of the Electric Power Research Institute.)

ment). The "first-guess" sensor placements were chosen based on the covariance patterns. The NLP-generated "optimal" patterns vary with constituent sampled but resemble those shown schematically in Figure 5.11. Note that about half of the stations in the 14-station network are quite tightly packed—in effect, to have the best chance of resolving the concentration gradients in the near vicinity of the source. This sort of behavior was subsequently confirmed with more detailed calculations reported in the MATEX study (Hidy *et al.*, 1985).

# 6 THEORETICAL INFERENCE OF SOURCE–RECEPTOR RELATIONSHIPS

## OVERVIEW

By their nature, observations give only a very limited picture of atmospheric or deposition conditions in time and space. Fixed station measurements are limited in their areal representativeness; the degree of their regional representativeness is not known. The uncertainty in representativeness of ground measurements becomes apparent when one considers the station spacing adapted in the SURE, for example, in which 54 stations were used to characterize a region of $\sim 10^7$ km$^2$. In a regional grid of 80 km $\times$ 80 km a single station may represent a nominal minimum area of $10^4$ km$^2$. During a day, a station typically samples air at a rate of 2 m$^3$/h or about 50 m$^3$/d. This is taken to represent observed changes in an air volume of $10^{10}$ m$^2$ $\times$ 10 m height, or $10^{11}$ m$^3$ in a day. The station thus uses only a $5 \times 10^{-8}$% sample to represent such a large volume; such widely spaced stations hardly give confidence in representation.

Because an aircraft flies at 200 km/h, it samples air that resides over much larger spatial areas than are characterized by a fixed station. Thus, airborne observations are difficult to interpret as representations of average spatial distributions over an hour or more. Uncertainties in inferring conditions aloft compared with those on the ground are also due to lack of data spanning several hours to days. Both ground and aircraft observations give only circumstantial evidence of SRRs because no specific link is established by means of tracers. In other words, the contributions of sources to total exposure generally can be inferred only indirectly and without confident prediction.

With the limitations of interpreting observations, workers have turned to mathematical descriptions of atmospheric behavior to describe (and predict) atmospheric conditions. The use of mathematical models complements well the knowledge inferred from observations. Models incorporate or synthesize knowledge of physical and chemical processes to describe atmospheric characteristics using selected spatial and temporal averaging schemes. The "accuracy" of a model to describe atmospheric behavior depends on the fidelity of description of observations given constraints of process knowledge and mathematical parameterization. Basically, a model resolves the representational problem posed by a finite, fixed sampling network and sporadic aircraft sampling. But it leaves unresolved other problems that derive from attempting to describe deterministically "chaotic" natural phenomena. Most of the models assume mathematically linear forms relating concentration and sources that greatly simplify analysis. Linearity extends to an additional property that the total exposure of a receptor can be determined from the sum of all contributing sources. In principle, one can calculate the geographic spread of pollution from each source, then total these contributions at given receptor locations to represent exposure. Atmospheric $SO_x$ and $NO_x$ chemistry is known to be nonlinear in character (Appendix A), as is wet deposition and perhaps dry deposition. The effect of nonlinearities in the system for source effects on receptors remains under debate but has been explored using theoretical models.

In general, models are tested against measurements for their accuracy in describing natural phenomena. Given the limitations of regional-scale observations, a well-defined verification of performance is difficult to establish. Researchers who rely on measurements and those who use model output to characterize air chemistry have debated this issue for some time. Regardless of the viewpoint, there is no doubt that scientific convention requires ideally that the model output be a well-defined, quantitative representation of the observations. The difficulty lies in specifying *a priori* the tolerance specifications for model error, given uncertainties in the measurements.

A further problem lies in model output. In general, the output is so voluminous that the computations are usually summarized. The summarized output is readily comparable with available observations, but it often submerges key information on source contributions to receptors. This problem has been dealt with in various ways, none of which lays the groundwork for satisfactory quantitative, interactive analysis of the calculations with observations. However, considerable progress on at least a conceptual codification has been reported as a part of EMEFS (e.g., Dennis 1990).

Because of their importance to SRR analysis, more than 50 regional-scale models have been produced that rely on various theoretically based approaches. Regardless of the details, all the models represent a solution for an equation governing conservation of mass of chemical species ($i$), written for the rate of change in average mass concentration per unit time in a turbulent medium:

$$\frac{d\bar{c}_i}{dt} = \frac{\partial \bar{c}_i}{\partial t} + \mathbf{V} \cdot \nabla \bar{c}_i = \nabla \cdot \nabla (K\bar{c}_i) + S_i + L_i$$

(Rate of change in time moving with air mass) = Rate of change in fixed reference system) + (Advective or transport component) = (Turbulent diffusion) + (Sources) + (Losses) (6.1)

with suitable initial and boundary conditions (including specification of sources at the surface). The vector **V** is the wind field ($U$ is the velocity in the downwind direction), $V$ the cross-wind direction, and $W$ the vertical direction. The wind velocity vector can be divided into a mean **V** plus a fluctuation $\mathbf{v}'$. The boundary conditions would include source emissions and surface loss by dry or wet deposition. The averaging is assumed to be over a time duration or an ensemble of realizations of atmospheric conditions, where the instantaneous concentration field, $c_i = \bar{c} + c_i'$ is made up of its average value and a fluctuating component $c_i'$ accounting for the stochastic or turbulent character of the system. Equation (6.1) is written with certain assumptions about averages of fluctuating quantities in the fluid. In particular, by definition the average $\overline{(c')^2} = 0$, and correlations of wind velocity fluctuation $\mathbf{v}'$ and $c'$, e.g., $\overline{u'c'}$, are expressed in terms of $K\ \nabla\bar{c}_i$.

Equation (6.1) is a mathematical statement that the rate of concentration change in time following a coordinate frame of reference in a moving air mass is equal to its rate of change, in a framework fixed to the earth's surface, plus an advective (horizontal wind) component. Vertical motion or convection is frequently small compared with horizontal air transport so that $\overline{\mathbf{V}}$ is approximated by the horizontal vector components. The rate of change in concentration is also equal to the sum of a turbulent diffusion component and the sources and sinks of material in an air volume. $K$ is the eddy diffusivity, assuming that diffusion can be written in terms of the product of $K$ and the local gradient in concentration, $\nabla\bar{c}_i$. Specification of $K$ can range from simple empirical constants to functions of space and time of day. Sources ($S_i$) are designated in the air volume and include chemical reactions *in situ*, both in the gas phase and in the condensed phase contained in clouds. Sinks or losses ($L_i$) of species ($i$) as in Eq. (6.1) can include chemical reactions converting species from ($i$) to ($j$) as well as scavenging processes in or below clouds that move more material from one layer to another. Multilayer models may include sources in layers above the ground as well as at the ground, depending on the height of injection.

The temporal resolution of most models depends to a large extent on the resolution of the input emissions and meteorological data. In most models, the meteorological data are the critical input for calculating spatial distributions. Models can be categorized as short-term or episodic (time scales on the order of a few hours to a few days) and long-term (monthly/seasonal/annual). Similarly, models can be classified as mesoscale (space scales on the order of hundreds to thousands of kilometers) or regional–synoptic (scales on the order of hundreds to thousands of kilometers). These classifications are based on the assumptions inherent in the model (e.g., constant versus diurnal variation, instantaneous versus diffusion-limited mixing), time steps, and grid sizes used in the temporal and spatial description. However, these delineations are not rigid for some models. In principle, a user could modify a model to achieve considerable flexibility in its application, although the modifications are not always straightforward. For example, several models that track pollutants in parcels of air (air mass trajectory models) obtain long-term averages by sequential simulation of short-term periods, so they could be used for short-term predictions. They are, however, relatively poor performers in this regime in the sense that calculation of long averages tends to smooth out the effects of short-term phenomena.

The spatial resolution of regional models is $O(10^4)$ $km^2$ and larger. The same schemes can be used to calculate atmospheric processes on the subregional (meso-) scale of $10^2$ $km^2$, to account efficiently for fine structure of concentration patterns near sources in the coarser regional calculations. Care must be taken in such calculations to ensure that the approximations adopted for the regional scale are also adaptable to the scale of hundreds of kilometers or less.

## INVENTORY OF MODELS

Essentially no model exists that can simulate in complete detail the transport (advection and dispersion), chemistry (gas phase, aqueous phase, heterogeneous), and removal (dry and wet deposition) processes because of incomplete understanding of the processes, computational constraints, and the extensive data requirements of comprehensive models. Most practical models rely on simplified parameters of the physical and chemical processes believed to be important for S and N behavior. The relative significance of processes is rationalized intuitively or subjectively. The level of detail varies significantly from one model to another. Table 6.1 summarizes the basic features and properties of several transport/deposition models applicable to regional air quality behavior. These models range from simple to comprehensive schemes that produce complicated three-dimensional grid calculations. Most have been applied only to the study of the transport and deposition of sulfur compounds, and almost all have been applied under conditions of relatively smooth terrain and uniform precipitation fields.

Models traditionally have been classified according to their mathematical framework. Those with reference frames fixed at the surface of the earth are called Eulerian; those with frames fixed on moving parcels or puffs of pollutants are called Lagrangian. Models can also be classified as numerical or analytical, depending on the techniques used to solve the mass conservation equations. It is convenient to subdivide models into four categories: analytical, Lagrangian, statistical, and Eulerian.

## Analytical Models

Analytical models are based on the closed-form solutions of the mass conservation equations. In general, an analytical solution of the complete three-dimensional unsteady-state equation (6.1) is not possible. Thus, analytical models must make use of several simplifying assumptions (e.g., steady-state uniform winds, linear chemistry, simple eddy diffusivity profiles) to reduce the equations to a tractable form. These assumptions limit the applicability of analytical models to highly idealized situations.

## Lagrangian Models

In Lagrangian models, volumes of air are followed as they are transported by the wind. The models keep track of the pollutant content of the moving parcels. Two general types of Lagrangian models are puff models and parcel models. They are sometimes referred to as source-oriented and receptor-oriented, respectively, depending on whether the air motion is traced "forward" from sources or "backward" from receptors. These definitions are ambiguous because most Lagrangian models can compute backward as well as forward trajectories and operate in both source- and receptor-oriented modes.

In a puff model the injection of emissions into the air is simulated incrementally with the time-stepping process as a puff whose size is scaled to emission rate. The puffs expand with dispersion along the direction of the wind. The positions of puffs are traced as a function of time. At any instant, the concentration field is given as the sum of concentrations due to each puff. The computational costs for such models are directly

TABLE 6.1 **Transport/Deposition Models for Regional Application**

| Model | Type | Temporal and spatial scales | Dispersion | Advection | Transport winds | Mixing height | Chemistry | Removal processes—deposition: Dry | Removal processes—deposition: Wet | References |
|---|---|---|---|---|---|---|---|---|---|---|
| STATMOD (Canada/U.S.) | Statistical | Regional, long term | H: Uniform with directional weighting factors<br>V: Uniform (or instantaneous) | H: Constant wind used in solution<br>V: None | Seasonal or annual 600 m wind rose | Constant | S: Linear, constant<br>N: Linear, constant | Linear, constant $V_d$ | Linear, related to precip. rate | Venkatram *et al.* (1982), Venkatram (1984), Venkatram and Pleim (1985a), Venkatram and Pleim (1985b) |
| Fisher (U.K.) | Statistical | Regional, long term | H: Uniform with directional weighting factors<br>V: Eddy diff., analytic solution for primary, uniform for secondary | H: Constant wind used in solution<br>V: None | Statistical distributions for wind speed and directions | Statistical average | S: Linear, constant<br>N: Linear, constant | Linear, constant $V_d$ | Linear, constant | Fisher (1978), Fisher (1982), Fisher and Clark (1985), Wright *et al.* (1984) |
| TVA (MODEL T) | Statistical | Regional (ENA), long term | Uniform with directional weighting | 360° spread weighted by direction and diluted by wind speed | Statistical distribution for wind speed and direction | Spatial constant, vary seasonally | S: Linear Seasonal vary | Linear, seasonal, vary | Linear, seasonal, vary | Clark *et al.* (1987) |
| MIT (U.S.) | 2-D analytical | Regional, long term | H: Eddy diff. (analytical solution)<br>V: Uniform | H: Constant wind used in analytic solution<br>V: None | Seasonal average | Constant/ seasonal variation | S: Linear, constant/ seasonal variation<br>N: Linear constant | Linear, constant/ seasonal $V_d$ | Linear, related to precip. rate | Fay and Rosenzweig (1980), Fay *et al.* (1985), Golomb *et al.* (1985) |

| | | | | | | | | | | |
|---|---|---|---|---|---|---|---|---|---|---|
| CEGB-SS (U.K.) | 3-D analytical, steady state | Mesoscale, short term | H: Gaussian<br>V: Eddy diff., analytical solution for $SO_2$, uniform for $SO_4$ | H: Analytic solution<br>V: None | Uniform wind, derived from observations | Constant | S: Linear, constant<br>N: Not treated | Linear, constant $V_d$ for $SO_2$, $SO_4$ not deposited | Linear, related to precip. rate for both $SO_2$ and $SO_4$ | Maul (1980) |
| ARL-ATAD (U.S.) | 1-Layer Lagrangian | Regional, short term | H: Gaussian<br>V: Uniform within transport layer | H: Transport layer trajectory, simple Euler integration<br>V: None | Analysis of upper-air data | Diurnal variation in TLD—transport layer depth | None (can incorporate linear chemistry) | Linear, constant $V_d$ | Linear, related to precip. rate | Heffter (1981), Heffter (1980) |
| ARL-MS (U.S.) | Multilayer Lagrangian | Mesoscale, short term | H: Gaussian<br>V: Eddy diff. (finite-difference solution) | H: Transport layer trajectory for all layers, simple Euler integration<br>V: None | Analysis of upper-air data and adjustment of surface winds | Transport layer estimated from extent of vertical diffusion | None (can incorporate linear chemistry) | Linear, constant $V_d$ | Linear, related to precip. rate | Draxler (1977), Draxler (1979), Draxler (1980) |
| ARL-MLA-TAD/BAT (U.S.) | Multilayer Lagrangian | Regional, short term | H: Uniform, width growing with time<br>V: Uniform within each layer, no mixing among layers at night | H: Winds within each layer used to determine trajectories, simple Euler integration<br>V: None | Nearest data used | Diurnal variation | None (can incorporate linear chemistry) | None (can incorporate linear removal) | None (can incorporate linear removal) | Draxler (1982a), Draxler (1982b), Draxler and Taylor (1982), Draxler *et al.* (1982), Heffter (1983) |
| BNL-AIRSOX (U.S.) | Multilayer Lagrangian | Regional, long term/short term | H: Gaussian<br>V: Eddy diff., Crank–Nicholson solution | H: Mixing layer trajectory for all layers, simple Euler integration<br>V: None | Analysis of upper-air data | Seasonal/ diurnal variation | S: Linear, seasonal/ diurnal variation<br>N: Not treated | Linear, seasonal/ diurnal variation in $V_d$ | Linear, related to precip. rate. | Ball (1981), Kleinman *et al.* (1979), Kleinman (1981) |

*(continued)*

TABLE 6.1 *(continued)*

| Model | Type | Temporal and spatial scales | Dispersion | Advection | Transport winds | Mixing height | Chemistry | Removal processes—deposition: Dry | Removal processes—deposition: Wet | References |
|---|---|---|---|---|---|---|---|---|---|---|
| PNL-RAPT/ MILRAPT (U.S.) | 1-Layer or multilayer Lagrangian | Regional, long term/short term | H: Uniform<br>V: Gaussian until puff growth to layer depth, uniform after | H: 100–1000 m. trajectory, simple Euler integration<br>V: None; MLRAPT interlayer trans. from potential temp. | Analysis of upper-air data | Diurnal variation | S: Linear, diurnal variation<br>N: Not treated | Linear, constant $V_d$ | Linear, related to precip. rate. | Ball (1981), McNaughton (1980), McNaughton (1981), McNaughton *et al.* (1981), Davis (1984) |
| Eliassen and Saltbones (Norway) expt. model | 1-Layer Lagrangian | Regional, long term | H: Uniform<br>V: Uniform | H: 850-mb trajectory, constant acceleration integration<br>V: Transport between boundary layer and free troposphere | Analysis of 850-mb winds | Daily variation | S: Linear, constant<br>N: Not treated; photochemical $SO_x/NO_x/HC$ mechanism, solved by quasi-steady-state approx. | Variable aerodynamic resistance for $SO_2$; constant $V_d$ for $SO_4$; constant $V_d$ for $NO_2$, PAN, $SO_2$, $HNO_3$ and $SO_4$. Diurnal, latitudinal var. for $O_3$ over land | Linear related to precip. rate for both $SO_2$ and $SO_4^{2-}$<br>Linear, constant; applied when RH exceeds 90% | Eliassen *et al.* (1982a), Eliassen *et al.* (1982b), Eliassen and Saltbones (1983), Eliassen and Saltbones (1975), Eliassen (1978) |
| Eliassen and Saltbones (Norway) routine model | 1-Layer Lagrangian | Regional, long term | H: Uniform<br>V: Uniform | H: 850-mb trajectory, constant acceleration integration<br>V: None | Analysis of 850-mb winds | Constant | S: Linear, constant<br>N: Not treated | Liner, constant $V_d$ for $SO_2$, overall decay for $SO_4$ | Linear, constant for $SO_2$<br>Empirical for $SO_4$ | Eliassen and Saltbones (1975), Eliassen (1978), Eliassen and Saltbones (1983) |

| | | | | | | | | | | |
|---|---|---|---|---|---|---|---|---|---|---|
| Prahm and Christensen (Denmark) | 2-D Eulerian | Regional, short term | H: Eddy diff.; pseudospectral technique, third-order Adams Bashforth integration<br>V: Uniform | H: User-specified transport layer winds; pseudospectral technique; third-order Adams Bashforth integration<br>V: None | Analysis of 850-mb and surface winds | Constant | S: Linear, constant<br>N: Not treated | Linear, constant $V_d$ | Not treated (can incorporate linear removal) | Prahm and Christensen (1977) |
| PHENIX (U.S.) | 2-D Eulerian | Regional, short term | As above | As above | Analysis of upper-air data | Constant | Treats one primary–secondary pair with 1st-order chemistry, constant rate | Linear, constant total (dry + wet) deposition velocity) | See dry deposition | Murphy (1981) |
| ACID (U.S.) | 1-Layer Lagrangian | Regional, long term/short term | H: Gaussian<br>V: Uniform within transport layer | H: Mixing-layer trajectory, simple Euler integration<br>V: None | Analysis of upper-air data | Monthly variation | S: Linear, diurnal, and monthly variation, higher rates in wet conditions; Option: rate a function of $HC/NO_x$ ratio.<br>N: Not treated | Linear, diurnal and monthly variation in $V_d$ | Linear, related to precip. rate | Kahl *et al.* (1985), Moody *et al.* (1983), Samson *et al.* (1983), Samson and Small (1984), Samson *et al.* (1986) |
| AES-LRT (Canada) | 1-Layer Lagrangian. Concentrations determined by averaging 3 trajectories at diff. levels | Regional, long term/short term | H: Uniform<br>V: Uniform | H: Trajectories at 1000, 925, and 850 mb, constant acceleration integration<br>V: None | Objective analysis of observations | Diurnal and seasonal variation | S: Linear, constant<br>N: Not treated | Linear, constant $V_d$ | Linear, related to precip. rate; $SO_2$ washout is temp., pH dependent | Olson and Oikawa (1983), Voldner *et al.* (1981), Bottenheim *et al.* (1984) |

(*continued*)

TABLE 6.1 *(continued)*

| Model | Type | Temporal and spatial scales | Dispersion | Advection | Transport winds | Mixing height | Chemistry | Removal processes—deposition: Dry | Removal processes—deposition: Wet | References |
|---|---|---|---|---|---|---|---|---|---|---|
| RIVAD (U.S.) | 1-Layer Lagrangian | Regional, long term/short term | H: Uniform, width increasing with time<br>V: Uniform within puff, puff grows with time until limited by surace below and mixed layer above | H: Transport layer trajectory, simple Euler integration<br>V: None | Analysis of upper-air and surface data or obtained from numerical wind model | Diurnal variation | S and N: Linear parameterized reaction mechanism; enhanced $SO_2$ oxidation rates in urban areas; inst. $SO_2$ oxidation by ambient $H_2O_2$ within cloud fixed nighttime $NO_2$ oxidation rate const.; $NH_4NO_3$/$NH_3$/$NHO_3$ chem. equilibrium | Linear, constant $V_d$ determined by surface type | Linear, related to precip. rate and type; $SO_2$ removal determined by equilibrium scavenging assumption | SAI (1985) |
| TRANS (Canada) | 4-Layer Lagrangian | Regional/mesoscale, long term/short term | H: Gaussian<br>V: Uniform within mixing layer | H: Transport-layer winds used to advect all layers, constant acceleration integration<br>V: None | Adjusted surface geostrophic winds | Diurnal and seasonal variation | S: Linear, seasonal, and diurnal variation<br>N: Linear, seasonal, and diurnal variation | Linear, seasonal, and diurnal variation in $V_d$ for all species | Linear, related to precip. rate; $SO_2$ removal dependent also on pH and temp. | Scholtz and Weisman (1985), Weisman (1980), Weisman (1983) |

| | | | | | | | | | | |
|---|---|---|---|---|---|---|---|---|---|---|
| PUFF (Netherlands) | 3-Layer Lagrangian. Layers: surface, mixing, reservoir | Mesoscale, short term | H: Gaussian<br>V: Gaussian in mixing layer, uniform in reservoir layer | H: Reservoir and mixing-layer trajectories determined using winds in layer with larger pollutant mass, simple Euler integration<br>V: None | Derived from hourly measurements and power law profile | Diurnal variation | S: Linear, solar radiation dependent<br>N: Not terated | Linear, resistance analogy for $V_d$ | Not treated (can incorporate linear removal) | van Egmond and Kesseboom (1983a), van Egmond and Kesseboom (1983b), Dumont *et al.* (1985) |
| CEGB-TD (U.K.) | 1-Layer Lagrangian, time dependent | Regional, short term | H: Uniform, width increasing with time<br>V: Eddy diff., analytical solution for $SO_2$ uniform for $SO_4$ | H: Mixing-layer trajectory, Runge–Kutta integration<br>V: Estimated wind speeds used to adjust analytic profiles | Adjusted surface geostrophic winds | Diurnal variation | S: Linear, constant<br>N: Not treated | Linear, constant $V_d$ for $SO_2$, $SO_4$ not deposited | Linear, related to precip. rate for both $SO_2$ and $SO_4$ | Maul (1980), Maul (1982) |
| MESOS (U.K.) | 2-Layer Lagrangian | Regional, long term/short term | H: Uniform, width increasing with time<br>V: Uniform depth increasing with time unit puff grows to layer size | H: Separate trajectories above and below mixing layer, with representative wind fields, simple Euler integration<br>V: None | Adjusted surface geostrophic winds | Diurnal variation | 1st-order radioactive decay (for radionuclides). Can treat 1st-order chemistry similarly | Linear, variable aerodynamic resistance | Linear, related to precip. rate | Apsimon *et al.* (1985a), Apsimon *et al.* (1985b) |
| MOE-PUFF (Canada) | 1-Layer Lagrangian | Regional, long term/short term | H: Gaussian for short term, uniform for long term<br>V: Uniform | H: Mixing-layer trajectories, constant acceleration integration<br>V: None | Adjusted surface geostrophic winds | Seasonal variation | S: Linear, seasonal variation<br>N: Not treated | Linear, seasonal variation in $V_d$ | Linear, related to precip. rate | Ellenton *et al.* (1983), Ellenton *et al.* (1985) |

(continued)

TABLE 6.1 *(continued)*

| Model | Type | Temporal and spatial scales | Dispersion | Advection | Transport winds | Mixing height | Chemistry | Removal processes—deposition: Dry | Removal processes—deposition: Wet | References |
|---|---|---|---|---|---|---|---|---|---|---|
| CAPITA-MCARLO (U.S.) | 1-Layer Lagrangian | Regional long term/short term | H: Determined by random perturbation of individual trajectories<br>V: Uniform | H: Mixed-layer trajectory, simple Euler integration<br>V: None | Modified surface winds | Constant | S: Linear, constant<br>N: Not treated | Linear, constant total (dry and wet) removal rate | See dry dep. | Patterson *et al.* (1981) |
| MESOPUFF-II (U.S.) | 3-Layer Lagrangian | Mesoscale, short term | H: Gaussian<br>V: Initially Gaussian, then uniform | H: Separate trajectories in mixed layers, determined by a two-step bilinear interpolation<br>V: None | Analysis of surface and upper-air data | Diurnal variation | S and N: Linear parameterized reaction mechanism; $NH_4NO_3$/ $NH_3$/$HNO_3$ chem. equlibrium | Linear, resistance analogy | Linear, function of precip. rate and type | Scire *et al.* (1983a), Scire *et al.* (1983b), Scire and Lurmann (1983) |
| EURMAP-Series (Europe/U.S.) | 1-Layer Lagrangian; multilayer | Regional, short term, long term | H: Uniform, width growing with time<br>V: Uniform; puff growth with time; eddy diffusion layers | H: Adjusted 850-mb winds used to advect puffs, weighting with puff height, simple Euler integration; weighting in layers<br>V: None | Analysis of 850-mb winds, or combined surface and 850-mb winds; terrain corrections | Seasonal or diurnal variation; Diurnal variation | S: Linear, constant<br>N: Not treated; 5 species chem added | Linear, constant $V_d$; rate inverse prop. to puff height | Linear related to precip. rate | Bhumralkar *et al.* (1980), Johnson *et al.* (1978), Bhumralkar *et al.* (1981), Bhumralkar *et al.* (1983), Bhumralkar *et al.* (1984), Endlich *et al.* (1983, 1984), Brodzinsky *et al.* (1984) |

| | | | | | | | | | | |
|---|---|---|---|---|---|---|---|---|---|---|
| ENAMAP-2 (U.S.) | 3-Layer Lagrangian | Regional, long term/short term | H: Uniform, constant expansion rate<br>V: Eddy diff. among sublayers Crank–Nicholson solution | H: Sublayer representative winds used in determining trajectories, simple Euler integration<br>V: None | Adjusted, terrain-corrected surface and 850-mb winds (logarithmic profile) | Diurnal variation | S: Linear, const.<br>N: 5 species parameterized linear chemistry, diurnal variation | Linear, diurnal variation in $V_d$ | Linear, related to precip. rate and type | Bhumralkar *et al.* (1984), Brodzinsky *et al.* (1984), Endlich *et al.* (1983) |
| RELMAP (U.S.) | 3-Layer Lagrangian; puff | Regional long term (ENA) | Uniform by day; no nocturnal mixing | Constant at 339 $km^2/h$ | 1000-mb winds in lowest layers; 850-mb winds upper layer; weighted in middle layer | Monthly max/min | S: Linear diurnal; Seasonal | Linear, seasonal and land use | Linear seasonal varying | Dennis *et al.* (1990) |
| ASTRAP (U.S.) | Multilayer Lagrangian | Regional, long term | H: Guassian, parameters determined from long-term trajectory statistics<br>V: Eddy diff., solved by Gaussian moment conservation scheme solution | H: Single (transport) layer trajectory; simple Euler integration<br>V: Parameterization leakage to free troposphere | Analysis of upper-air data | Seasonal variation in transport layer | S: Linear, diurnal, and seasonal variation; enhanced rate to account for oxidation in nonprecip. clouds<br>N: Linear, diurnal, seasonal variation | Linear, diurnal, seasonal variation in $V_d$ | Linear, related to precip. rate | Shannon (1981), Shannon (1985a), Shannon (1985b), Streets *et al.* (1985) |
| ARIAM (U.S.) | 1-Layer Lagrangian | Regional, long term | H: Gaussian<br>V: Uniform | H: Transport layer trajectory, simple Euler integration<br>V: None | Analysis of upper-air data | Diurnal variation | S: and N: Linear, constant/seasonal variation | Linear, constant/seasonal $V_d$ | Linear, related to precip. rate and type | Pleim *et al.* (1983), Yamartino *et al.* (1985) |

*(continued)*

TABLE 6.1 (*continued*)

| Model | Type | Temporal and spatial scales | Dispersion | Advection | Transport winds | Mixing height | Chemistry | Removal processes—deposition: Dry | Removal processes—deposition: Wet | References |
|---|---|---|---|---|---|---|---|---|---|---|
| EDRAB (U.S.) | 1-Layer Lagrangian | Regional, short term | H: Gaussian<br>V: Uniform | H: Mixed-layer or isentropic trajectories depending on stability<br>V: Implicit in isentropic mode; none in isobaric mode | Analysis of upper-air data | Diurnal variation | None (can incorporate linear chemistry) | None (can incorporate linear removal) | None (can incorporate linear removal) | Dittenhoefer and Ferullo (1983) |
| PLMSTAR (U.S.) | Multilayer, multicolumn Lagrangian | Mesoscale, short term | H and V: Eddy diff., solved by hybrid Crank–Nicholson/ fully implicit | H: User-specified level winds used to transport all layers, simple Euler integration<br>V: None | Analysis of surface and upper-air data | Diurnal variation | $HC/NO_x/SO_x$ photochemical scheme, solved by predictor–corrector method, added aerosols and top/cloud chem. | Linear, resistance analogy | None (can be incorporated); param. complex of chem. | Godden and Lurmann (1983), Lurmann *et al.* (1985), Seigneur *et al.* (1984), Seigneur *et al.* (1985), Seigneur *et al.* (1986) |
| KNMI (Netherlands) | 3-D Eulerian | Mesoscale, short term | H: Uniform<br>V: Eddy diff., (Crank–Nicholson solution) | H: Pseudo-spectral technique, 4th-order Runge–Kutta scheme<br>V: None | Analysis of surface and 850-mb data | Diurnal variation | S: Linear, constant<br>N: Not treated | Linear, resistance analogy for $V_d$ | Linear, related to precip. rate | van Dop and De Haan (1983a), van Dop and De Haan (1983b) |
| GRID (Netherlands) | 3-Layer Eulerian; layers: surface, mixing, reservoir | Mesoscale, short term | H: Eddy diff.; pseudospectral technique, leapfrog integration<br>V: Uniform | H: Numerical advection; pseudospectral technique, leapfrog integration<br>V: None | Derived from hourly measurements and power law profile | Diurnal variation | S: Linear, solar radiation dependent<br>N: Not treated | Linear, resistance analogy for $V_d$ | Not treated (can incorporate linear removal) | van Egmond and Kesseboom (1983a), van Egmond and Kesseboom (1983b) |

| | | | | | | | | | | |
|---|---|---|---|---|---|---|---|---|---|---|
| SURADS (U.S.) | 3-D Eulerian | Regional, short term | H and V Eddy diffusion; finite element | H and V: Finite element using layer-averaged winds | Analysis of surface and upper-level winds | From meteorol. boundary-layer model | S: chemihomog. gas phase no clouds or aqueous reactions | Linear $V_d$ by land use | None | Mueller and Hidy (1982) |
| RTM-II (U.S.) | 2-Layer hybrid Lagrangian–Eulerian | Regional, short term | H: Eddy diff.; solved by SHASTA algorithm<br>V: Uniform<br>Subgrid scale:<br>H: Gaussian<br>V: Gaussian | H: Solved by ShASTA algorithm using layer-averaged winds<br>V: Entrainment/detrainment across mixed layer | Geostrophic or analysis of upper-air data | Diurnal variation | S: Linear, diurnal and latitudinal variation<br>N: Not treated | Linear, resistance analogy | Linear, related to precip. rate | Liu *et al.* (1982), Stewart *et al.* (1983a) |
| RTM-LT (U.S.) | 2-Layer Eulerian | Regional, long term | H: Eddy diff.; solved by SHASTA algorithm<br>V: Uniform | As above | As above | As above | As above | As above | As above | Stewart *et al.* (1983b) |
| STEM (U.S.) | 3-D Eulrian | Regional, short term | H and V: Eddy diff.; finite element; Crank–Nicholson Galerkin solution | H and V: Finite element; Crank–Nicholson Galerkin solution | 3-D non-divergent (from upper-air data or met. model) | Diurnal variation | S: Linearized photochemical oxidation scheme with constant heterogeneous rate<br>N: $SO_x/NO_x/$ HC (chem added; e.g., chem/clouds added, aerosols added by Euler method. | Linear, resistance analogy | None (can be incorporated); added parameterized, coupled with aq. chem and cloud phys. | Carmichael and Peters (1980), Carmichael and Peters (1984a), Carmichael and Peters (1984b), Charmichael *et al.* (1986) |

*(continued)*

TABLE 6.1 (*continued*)

| Model | Type | Temporal and spatial scales | Dispersion | Advection | Transport winds | Mixing height | Chemistry | Removal processes—deposition: Dry | Removal processes—deposition: Wet | References |
|---|---|---|---|---|---|---|---|---|---|---|
| ADOM (Canada/West Germany/U.S.) comprehensive model | 3-D Eulerian | Regional, short term | H and V: Eddy diff.; finite difference with Euler integration V: Eddy diff.; finite differencing of the flux formulation of the advection–diffusion equation; hybrid Crank–Nicholson/fully implicit scheme | H: Blackman cubic spline interpolation with Euler time marching V: See treatment of vertical dispersion | From meterological and boundary-layer model | From meterological and boundary-layer model | $SO_x/NO_x/HC$ homogeneous gas-phase chemical mechanism; cloud physics module coupled with aqueous-phase equilibrium/chemistry; solved by Boris–Young predictor corrector method | Linear, resistance analogy | Parameterized, coupled with aqueous-phase chemistry and cloud physics | ERT (1984a), ERT (1984b), ERT (1985) |
| RADM (NAPAP SRR model) | 3-D Eulerian | Regional, short term long term by simulation from characteristic days | H and V: Eddy diffusion; finite difference similar to ADOM | H: From MM4 model V: Mass continuity in pressure coordinates | From MM4 meso-scale model | From meterological boundary-layer model | Complex non-linear gas-phase chemistry in $SO_x/NO_x/HC$, cloud phys/aqueous chem. residue with cloud predictor | Linear resistance analogy tied to boundary layer model | Parameterized, coupled with aqueous chemistry and cloud physics | NCAR (1985), Chang *et al.* (1990) |

related to the number of sources treated. The superposition of individual puffs to determine concentrations requires that such models treat chemical and removal processes linearly.

In the parcel-type model the pollutant content of an air parcel is followed until the parcel arrives at a selected receptor point. Precomputed back trajectories are generally used for this purpose. During its travel, the parcel receives emitted material from the various sources it passes over and undergoes transformation and removal processes. As opposed to puff models, parcel models can include nonlinear chemical processes.

*Statistical Models* A subset of analytical or Lagrangian models makes use of smoothing of simple analytical results through calculation and analyzing a large number of realizations. Statistical models are based on the premise that long-term (seasonal or annual) averages of pollutant concentration and deposition can be estimated using information about the statistical properties of meteorological factors influencing deposition and deposition patterns themselves. The statistical models assume that the details of these processes are not important in determining long-term averages and can be simulated by coarse statistical approximations. Simple horizontal and vertical pollutant profiles are assumed, and chemistry and scavenging are treated as linear processes with appropriate long-term averaged parameters.

Straight-line or simple curvilinear air mass trajectories, with mean wind speed and direction determined from seasonal or annual wind roses, are assumed for estimating the transport of pollutants from source to receptor. The mathematical framework of statistical models is Lagrangian. They are semiempirical in the sense that they rely extensively on parameters that are often manipulated to provide a good fit between observations and model predictions. The theoretical simplicity and empiricism of these linear models limit their applicability, particularly for emission scenario evaluations. However, their ease of application and small computational requirements make them an attractive tool for qualitative estimation of geographic patterns for long-term averages of ambient concentration and deposition.

## Eulerian Models

Eulerian models represent the group in which the air volume of interest is divided into a two- or three-dimensional grid of cells, and Eq. (6.1) is solved numerically to yield concentration and deposition values at every grid point. Because of their ability to treat unsteady physical processes in three dimensions and handle nonlinear chemistry, Eulerian models are physically more realistic than the other types of models discussed thus far. However, their complexity translates into extensive input data and computational requirements, which has limited their use to episodic or short-term simulations. Recently, long-term simulations have been tried using a scheme to construct an annual or seasonal average from composite runs made up of "prevailing" meteorological conditions determined from cluster analysis (e.g., Chang *et al.*, 1990).

Another disadvantage of Eulerian models is the artificial diffusion associated with a numerical solution generated for the advection equation [Eq. (6.1)]. This "pseudodiffusion" can be reduced by using sophisticated numerical integration schemes. However, pseudodiffusion effects remain a problem generating uncertainty in results, which may be particularly acute for simulations of large point sources under stable atmospheric conditions.

Another disadvantage of Eulerian schemes is the problem of sub-grid-scale plume mixing. For practical reasons, Eulerian models involve calculations with spatial resolution of about 100 km × 100 km. Within these grids, pollution from point sources is assumed to mix instantaneously in the grid volume. This "initiation" problem has presented a dilemma for model designers. One pragmatic solution involves a hybrid approach. The RTM-II model (Liu *et al.*, 1982) is a hybrid Langrangian–Eulerian model in which major point sources are treated on the subgrid scale as puffs advected along a trajectory and undergoing Gaussian horizontal and vertical dispersion until they expand to a point where their dimensions approach grid cell sizes.

Regional-scale Eulerian model development began in the mid-1970s and has culminated in the 1980s with two complicated codes, reflecting the best available scientific description compatible with large-scale computing. The two models are the (U.S.) regional and deposition model (RADM) and the Canadian/German acid deposition and oxidation model (ADOM) (e.g., Chang *et al.*, 1990; Environmental Research and Technology, 1985). Both codes are designed to simulate simultaneously the behavior of $SO_x$, $NO_x$, $NO_y$, VOC or HC, and anthropogenic chemical constituents over the greater ENA. The principal advances required over earlier models have been (1) ability to assimilate complex emission inventories of $SO_x$, $NO_x$, HC, and cation species; (2) addition of sophisticated methods to estimate meteorological conditions, including cloud location, liquid water content, and precipitation rate; (3) aqueous cloud and fog chemistry; and (4) spatially and temporally heterogeneous dry and wet deposition rate. One of the models, RADM, is coupled with a prognostic meteorological model that provides an extended-duration estimate of the state of the atmosphere. The two models, with STEM II, basically are capable of investigating known, theoretical, nonlinear relationships in the meteorological, chemical, and physical system described.

## PROCESS DESCRIPTION

### Transport and Advection

Transport is distinguished from dispersion in modeling by time and space. Transport is identified with the horizontal and vertical motion of an air parcel's center of mass by an average wind velocity, which can be calculated from standard meteorological data and interpolative schemes. Dispersion refers to the expanded or increased volume of the air parcel caused by wind fluctuation (or turbulence) over temporal and spatial scales smaller than the averaging time or space interval used in the model.

Models vary considerably in how they simulate the movement of pollutants with winds, how they generate a wind field, and how they choose a suitable transport layer. Although most of the models listed in Table 6.1 are equipped with their own wind field generation or interpolation schemes, they are designed to use wind fields derived from other models with simple modifications as input to their computations.

Analytical models and some statistical models treat horizontal transport in the simplest way by assuming a uniform wind speed between the source and receptor. Next in hierarchy of complexity are the single-transport-layer Lagrangian models that assume a constant wind velocity for the duration of each trajectory segment (simple Euler integration). An unresolved problem in the use of single-layer average winds is the choice of a suitable transport layer (this is also true for the simpler statistical and analytical models). Because wind speed and direction generally change with height, the choice of a wind for the long-range transport of pollutants is considered an important factor.

On the other hand, A. Venkatram, a developer of one of the important statistical models, believes otherwise. He suggests that correlation of a calculated trajectory has little analogy to atmospheric measurements. He considers that a trajectory is correct if its calculated wind field produces a "correct" picture of the average concentration fields.

Most models differ in their choice of a transport layer. For example, some models use mixed-layer trajectories; others use winds averaged in a 0.1- to 10-km layer. One model uses a constant barometric pressure height (850 or 925 mb) isobaric trajectory; others use adjusted surface geostrophic winds. The selection of alternate, effective transport layers can drastically change the path of a trajectory when strong wind shears are present (e.g., in a frontal storm) (Michael and Raynor, 1983; Michael and Hales, 1984). Eliassen (1984) compares surface geostrophic trajectories and isobaric 850-mb trajectories for transport distances out to 1000 km. The mean distance between the two trajectories for 179 simulations was 246 km and the median distance was 140 km. Thus, the determination of a true trajectory involves considerable uncertainty. A large part of this uncertainty also arises from the limited amount of meteorological information available for trajectory models and the length of the trajectory time step (to a certain extent, these two factors are interdependent). For example, the ATAD model uses upper-air data that are typically available at a spatial resolution of approximately 400 km and a temporal resolution of 12 hours; the MS model supplements the upper-air observations with hourly surface data at a spatial resolution of approximately 100 km. The trajectory time steps for the ATAD and MS models are 3 hours and 1 hour, respectively. The transport layers in the two models are determined by different methods. Rao *et al.* (1983) have compared 24-hour trajectories, originating at Buffalo, NY, in December 1978, for the two models. They found that, on the average, the linear distance from the source to the trajectory end point for the ATAD model was 62% greater than that for the MS model. This difference was attributed to the bias of the ATAD model toward the upper-level winds. They also found a consistent directional discrepancy between the trajectories. The largest differences were found when the wind direction was changing rapidly because of the passage of a storm system through the area of interest.

Other multiple-layer Lagrangian models solve the problem of defining a single-layer average wind by advecting the sublayers with different horizontal winds under stably stratified conditions. However, this approach has the disadvantage that vertical diffusion among the sublayers cannot be treated satisfactorily.

The choice of a suitable transport layer and the resolution of the meteorological input data limit the accuracy of the computed trajectory. However, the numerical algorithm for calculating trajectories may also be an important consideration under certain situations. The assumption of persistence of wind velocity along a trajectory segment is not always justified, particularly for large integration time steps. Several models use the so-called constant acceleration scheme (Walmsley and Mailhot, 1983) in which the value of the mean advecting wind for a time step is determined as the average of the winds at the beginning and end of the trajectory segment. Because the end point of a segment is determined from the wind used for that segment, this process is an iterative one. Other schemes used to compute trajectories are the modified Euler method, a Runge–Kutta integration scheme, and a bilinear interpolation method.

The most elaborate advection algorithms are used in the two- and three-dimensional Eulerian models. Because traditional upstream finite-differencing methods can give rise to large numerical diffusion error, a variety of sophisticated numerical schemes have been developed to alleviate this problem. Examples are the pseudospectral method, the sharp and smooth transport algorithm (SHASTA) method, and the use of finite elements, cubic splines, or antidiffusive velocities.

The Eulerian grid models have the most extensive wind field requirements, the statistical and analytical models have the least, and the requirements of Lagrangian trajectory models vary from model to model. Precise estimation of pollutant transport is controlled by the accuracy of the input wind fields.

Models that explicitly simulate vertical motion are the three-dimensional Eulerian models. The vertical component of motion is disregarded in the statistical and analytical models, most of the Lagrangian trajectory models, the two-dimensional Eulerian models, and the Dutch three-dimensional Eulerian model. Vertical motion is probably important over long transport distances and can be especially important in the regions of strong updrafts and downdrafts associated with convective storm elements, in the vicinity of frontal systems characterized by the slow and steady ascent of warm air over a cold air mass, and in orographic clouds formed by the lifting of air over gross terrain features. These latter processes provide an effective mechanism for the transport of boundary-layer air and associated pollutants to the free troposphere.

Layer-averaged or isobaric constant-pressure surface Lagrangian trajectory models cannot simulate vertical motion. However, vertical motion can be implicitly treated in isentropic (constant entropy surface) trajectory models under certain conditions (e.g., Eliassen and Saltbones, 1975). In an isentropic trajectory model for dry air, parcels are followed on a surface of constant potential temperature. Thus, vertical motions in dry air are intrinsically built into the method. However, the inability of an isentropic trajectory model to account for latent heat exchange processes limits its applicability to a noncondensing, nonprecipitating atmosphere (Takacs and Starheim, 1983). In a moist isentropic trajectory model, on the other hand, parcels are followed on a surface of equivalent potential temperature.[1] Thus, latent heat effects and the associated enhanced vertical motions are implicitly taken into account. Both the dry and moist isentropic calculations are inapplicable under convectively unstable atmospheric conditions that can cause the potential temperature surface to become nearly vertical. In addition, isentropic trajectory models are not able to handle situations in which the isentropic surfaces may intersect terrain (Lazaro, 1985).

Artz *et al.* (1985) compared trajectories from the ATAD model and an isentropic trajectory model. Three-day back trajectories, terminating at Charlottesville, VA, were computed for the two models. In the absence of frontal movement, the two trajectories agreed. On the other hand, trajectories computed through frontal precipitation systems frequently dis-

[1] Potential temperature is corrected for latent heat release by adding the term $\lambda\ q/c_p$, where $\lambda$ is the specific heat of water vaporization, q is the specific humidity, and $c_p$ is the specific heat at constant pressure.

agreed. To address this question, the EDRAB model (Dittenhoefer and Ferullo, 1983) incorporates both the moist isentropic and mixed-layer trajectory calculations. This hybrid model uses the isentropic approach during stable saturated or unsaturated conditions and switches to the mixed-layer method of ATAD during convective conditions associated with the daytime boundary layer and strong cold fronts. An extension of the isentropic approach is also used in the multilayer Lagrangian model (MLRAPT), in which vertical motions of the pollutant are simulated through the use of layer-averaged potential temperatures. Whenever the difference between parcel temperature and the environmental potential temperature exceeds a certain value, the parcel is moved vertically to a new layer that is closer to its potential temperature. The problem associated with this approach (and with all isentropic approaches in general) is that potential temperature does not always have a monotonic variation with height; thus, trajectory computations for such conditions do not converge to a unique solution (Dittenhoefer and Ferullo, 1983).

Vertical motion under all atmospheric conditions can be realistically simulated only in a three-dimensional Eulerian model. Under certain conditions, the isentropic trajectory approach appears to be physically more realistic than the constant-level trajectory approach. However, the isentropic trajectory approach has not been used widely in long-range transport and deposition models because of its various disadvantages, including the problems of obtaining precise potential temperature profiles from the limited amount of data available and the increased computational expense of computing isentropic trajectories.

### Dispersion

The representation of horizontal and vertical dispersion processes varies in complexity from model to model, ranging from the simplest treatment in statistical models to the most complex treatment in Lagrangian and Eulerian grid models. Although the horizontal and vertical dispersion processes in the atmosphere are correlated, particularly over the short term, most models treat the two processes independently.

Statistical models, which are based on the assumption that trajectories originating from a source are approximately straight over the distance of interest, use the simplest treatment of the horizontal dispersion of pollutants. They assume that pollutants are distributed uniformly around a source region. To account for the distribution of wind directions, the uniform distribution function is modified by a directional weighting factor equal to the probability of the wind blowing from the source to the receptor divided by the probability of the wind blowing with equal frequency from all directions. These weighting factors can be derived from long-term wind roses. Alternatively, following the work of Bolin and Persson (1975), a Gaussian distribution for the pollutants can be assumed, with the dispersion parameters estimated from long-term trajectory statistics. An early version of STATMOD used this approach, which predicted annually averaged wet sulfur deposition measured in 1977 at sites in eastern North America (Venkatram *et al.*, 1986). Analytical models are also forced to rely on a simplified treatment of the dispersion process, either by assuming a constant eddy diffusivity or a Gaussian profile.

Several Lagrangian models treat the horizontal dispersion problem by assuming horizontal uniformity within a puff or plume of pollutant, with the width of the plume remaining constant in time. The reasoning behind this approach is that horizontal diffusion is considered a function of synoptic-scale wind variation only. Thus, horizontal diffusion can be simulated by the variation of a large number of trajectories over a long averaging period. Because these models ignore the smaller-scale horizontal diffusion around the plume centerline, they are suitable for predicting long-term averages only, although many have also been used for episodic simulations (e.g., Olson and Oikawa, 1983). To include small-scale effects, this approach is extended by allowing the width of the plume to increase with time, maintaining horizontal uniformity within the plume. Further detail may be introduced by allowing horizontal concentration variations in the form of a Gaussian distribution, with parameters determined as a simple function of travel time, from stability considerations, or from an analysis of long-term trajectory statistics. A similar approach is taken in the CAPITA-MCARLO model, which simulates the horizontal dispersion by a Monte Carlo technique, using repeated random perturbations of the movement of pollutants about the mean flow. The Ontario Ministry of Environment (MOE) PUFF scheme uses a uniform distribution for annual simulations and a Gaussian distribution for applications less than a year (season/month/day). Models that use temporally and spatially varying eddy diffusivities to simulate achieved the most realistic treatment. The eddy diffusivity concept is also used in most of the complex Eulerian grid models.

The effect of wind shear (variation in wind speed and direction with height) on plume dispersion is explicitly treated in three-dimensional Eulerian models. Strong vertical gradients in frontal systems or stable conditions are most important for the longitudinal dispersion of the plume at long travel times and is also an important consideration in the selection of a layer-averaged wind to transport the pollutants in single-layer models. Some multilayer Lagrangian models approach this situation by simulating advection of stably stratified layers by a different horizontal wind. However, most of these models do not address the problem of vertical mixing between the sublayers. For the simulation of long-term concentrations, however, wind shear effects are often ignored, since the dispersion on this time scale is dominated by the large-scale variations in the wind field (Eliassen, 1984).

As in the case of horizontal diffusion, the treatment of vertical dispersion varies among the models. The most common approach is to assume that pollutants disperse instantaneously within a mixed or transport layer. Some multilayer Lagrangian models, such as ATAD, simulate the effects of wind shear by transporting sublayers with different horizontal winds under stable conditions (at night), allowing complete mixing within the layer.

Vertical dispersion is associated with atmospheric turbulence whose intensity is characterized as a function of the ratio of thermal to frictional motion (the Richardson number). Some models use analytical expressions for the vertical profiles, derived by using constant eddy diffusivities or simple eddy diffusivity profiles. Some Lagrangian models use a Gaussian vertical profile, with dispersion parameters determined from stability considerations. The most realistic treatment, using spatially and temporally varying eddy diffusivities, is achieved by the three-dimensional Eulerian models and some multilayer Lagrangian models (with single-layer transport winds). Another advantage of these models is their ability to divide emissions among the sublayers by calculating the plume rise either explicitly or subjectively.

### Mixing Height and Dispersion

Vertical mixing or dispersion works to dilute injected material in the air. The volume available for dilution is estimated in terms of a mixing height, which in turn is specified by the vertical profile of eddy diffusivity. Not only does the mixing height define the vertical extent of the region (the air volume) in which pollutants are assumed to be instantaneously mixed, but also it specifies the height-averaged transport wind. Thus, a model's treatment of the mixing height is important in determining its applicability to a particular situation.

The mixing height varies diurnally with low values during the night and high values during the day. For predicting short-term concentrations in air, both diurnal and spatial variations must be considered. Models that use constant or spatially uniform climatological values are best suited for the prediction of long-term averages, although several have also been used for predicting daily concentrations.

A modified version of the Eliassen and Saltbones model (Eliassen and Saltbones, 1983) allows daily and spatial variation in mixing height, using 12 GMT (Greenwich mean time) radiosonde data to determine the base of the inversion. Some models use a climatological value of mixing height for long-term applications (Kleinman *et al.*, 1979; Ball, 1981) and specify diurnal variations (with no spatial variation) for short-term applications (Kleinman, 1981). Other models incorporate a diurnal cycle between specified nighttime minimums and daytime maximums with no spatial variations. Several models use twice-daily temperature soundings to determine mixing heights. However, the ATAD model also uses a Gaussian technique with an average nighttime coefficient for initial releases during the night. The RTM-II and RTM-LT two-dimensional Eulerian models allow a diurnal cycle between daytime and nighttime mixing heights obtained from temperature soundings. The MS model, which uses a box diffusion model to simulate vertical dispersion, estimates the transport layer (for advection purposes) as the layer containing 90% of the mass of the pollutant. Some models carry out a more detailed treatment in which the early diffusion profile and the mixing height are determined by stability and micrometeorological considerations. STEM-II, for example, uses mixing-layer heights derived from meteorological models or from an objective analysis of observations. The ADOM Eulerian model receives its input from a diagnostic module that combines information from a large-scale numerical weather prediction model with that from a high-resolution atmospheric boundary-layer model; RADM uses the output from a predictive mesoscale meteorological model as the basis for its mixing calculations.

## Chemical Transformations

The transformation of $SO_2$ and $NO_x$ to $SO_4^{2-}$ and $NO_3^-$, respectively, is controlled by a variety of chemical and meteorological factors, such as the concentration of other pollutants, humidity, photochemical effects, and the presence of clouds. A realistic treatment of these transformation processes entails the use of elaborate gas- and aqueous-phase chemical mechanisms that incorporate the appropriate diurnal variations in reaction rate constants. This approach is limited by computational constraints and uncertainties in knowledge of several chemical processes. Thus, most models assume simple first-order (linear) chemistry with invariant rate constants to represent the oxidation processes. Trajectory models do not incorporate chemical processes, but they can be easily modified to incorporate linear transformation processes. The ARL-ATAD model in particular has been widely used in acid deposition modeling.

For various reasons, most of the attention in past long-range transport and deposition modeling has been focused on sulfur species. The $NO_x$ conversion processes have been neglected in most models. However, $NO_3^-$ is an important contributor to acidic deposition, particularly in the western United States, where analysis of precipitation chemistry data in cities has shown $NO_3^-$ concentrations in precipitation comparable to or greater than $SO_4^{2-}$ concentrations (Zeldin and Ellis, 1984).

Oxidation of SO2 to $SO_4^{2-}$ can occur homogeneously in both the gas and aqueous phases. In addition, heterogeneous reactions between gases and liquids adsorbed on solid surfaces may also occur, although these are generally considered less important in the troposphere (see also Appendix A).

The primary pathway for the oxidation of $SO_2$ in the gas phase involves the hydroxyl (OH) radical:

$$SO_2 + OH \rightarrow HSO_3 \xrightarrow{O_2,\ H_2O} H_2SO_4 + HO_2$$

The rate of conversion of $SO_2$ in the gas phase depends on the OH concentration, which depends on the level of $NO_x$ and HCs as well as the sunlight intensity. However, both theory and experiments suggest that the $SO_2$ oxidation reaction is not accompanied by a net consumption of the OH radical because the overall reaction yields the peroxy radical, $HO_2$, which leads to regeneration of OH (Stockwell and Calvert, 1983). Therefore, the effective gas-phase oxidation rate may be considered approximately linear with respect to $SO_2$, although the OH concentration itself may vary depending on other factors, which provides some justification for the linear transformation rates used in most models. Several of these linear models, however, also employ seasonally varying rates (lower values in winter to reflect the lower OH concentrations resulting from reduced photochemical activity) rather than fixed rate constants. The linear models do not account for the diurnal and latitudinal variability in photochemical activity and are suitable only for long-term predictions. However, a diurnal variation for short-term applications is incorporated by specifying different daytime and nighttime oxidation rates (Kleinman, 1981). Specified diurnal and seasonal cycles are also used in some linear Lagrangian models. A modified version of (UM) ACID parameterizes the effect of the $HC/NO_x$ ratio by scaling baseline reaction rates with a reactivity function, which is determined using a simple regression analysis on results from box model simulations of a photochemical mechanism with diurnally averaged rate constants (Moody *et al.*, 1983; Samson *et al.*, 1986). The PUFF models use a first-order reaction rate constant that is dependent on measured solar radiation. The Eulerian models RTM-II and RTM-LT allow a diurnal and latitudinal variation in the reaction rate constant. The MESOPUFF-II model uses a parameterized rate constant for daytime gas-phase oxidation, which is a function of total solar radiation, background $O_3$ concentrations, and atmospheric stability. The parameters were developed from an analysis of nonlinear chemical mechanisms for $SO_2$ and $NO_x$ oxidation. Nighttime oxidation rates are assumed to be constant. One form of the Eulerian model, STEM, uses a linear photochemical oxidation rate, which depends on OH and $HO_2$ radical concentrations. These concentrations are estimated from steady-state considerations and estimated or specified concentrations of other species, such as NO, $NO_2$, $H_2O$, and CO. A linear version of the ADOM model treats gas-phase $SO_2$ transformation as a first-order process with a specified diurnal variation in the rate constant.

The most realistic treatment of the homogeneous gas-phase chemical processes is carried out in the comprehensive Eulerian models such as ADOM and RADM and some Lagrangian parcel models (Eliassen *et al.*, 1982a, 1982b). These models employ condensed gas-phase photochemical mechanisms that typically include 100 reactions and 20 to 40 species, with reaction rates dependent on several parameters such as temperature, solar radiation, and latitude. Photochemical effects and the effect of other atmospheric constituents, such as HC and $NO_x$, in $SO_2$ oxidation rates are explicitly taken into account.

$SO_2$ also undergoes rapid oxidation in the aqueous phase if oxidants are present. These oxidation rates are often of the order of 100% per hour—much higher than typical gas-phase oxidation rates (a few percent per hour under summer daytime conditions). However, the lifetime of individual clouds is short. Long-term average oxidation rates in cloudy air are probably comparable with those in dry air. Two primary oxidizing agents for typical atmospheric conditions are $H_2O_2$ and $O_3$, although other pathways exist (e.g., trace metal–catalyzed oxidation by molecular oxygen). The $O_3$ oxidation route becomes less important as the acidity of the cloud water increases, while the effective $H_2O_2$ oxidation rate is almost independent of acidity. Oxidation rates in solution depend on a number of factors, such as the oxidant concentrations; the presence of buffering agents, such as $NH_4^+$ and carbonate (from soil dust) that depress acidity levels and thus enhance the $SO_2$ rates by $O_3$; and the presence of catalytic agents, such as iron and manganese, that promote oxidation by dissolved molecular oxygen. Moreover, the assumption of linear chemistry may not be valid, because aqueous-phase oxidation is often limited by the availability of oxidants like $H_2O_2$.

Few models explicitly take the aqueous-phase oxidation of $SO_2$ into account. Most models implicitly incorporate the oxidation effect in the wet deposition calculations using $SO_2$ wet removal rates that are higher than would be the case were $SO_2$ an inert component in the aqueous phase (since $SO_2$ by itself is not very soluble). This treatment assumes that there is an ample supply of oxidant available and the rate of oxidation is controlled only by the $SO_2$ concentration (linear transformation). Some models introduce an additional first-order term to represent aqueous-phase chemistry. For example, the ASTRAP model uses higher gas-phase oxidation rate constants than most other models to reflect aqueous-phase oxidation in nonprecipitating clouds; the ACID model enhances the gas-phase rate whenever a trajectory encounters precipitation. The ENAMAP-2S and STEM-I models add a constant term to their gas-phase oxidation rate constants to represent aqueous-phase effects. The MESOPUFF-II model adds the rate constant to the gas-phase rate constant as a function of relative humidity.

Only the more elaborate Eulerian models, RADM and ADOM, can account for dispersion of material and production of particulate $SO_4^{2-}$ by evaporation of clouds. Most clouds form in the atmosphere and evaporate without precipitation. $SO_4^{2-}$ produced in cloud particles is left as an aerosol residue after droplet evaporation. Thus in some circumstances clouds are believed to accelerate $SO_4^{2-}$ particle formation. Furthermore, the enhanced turbulence in clouds disperses material vertically over that expected in dry air. Current models do not include this calculation explicitly.

Explicit, detailed treatment of aqueous-phase chemistry is achieved primarily in the comprehensive three-dimensional Eulerian models. These models also use available meteorological information either to model cloud and precipitation water contents and other relevant cloud variables or to estimate them using tables generated from cloud model simulations. One version of the PLMSTAR model is an example of a Lagrangian model that explicitly simulates aqueous-phase chemical processes in stratus clouds and fog (Seigneur *et al.*, 1984, 1985, 1986).

Aqueous-phase oxidation of $NO_x$ to $NO_3^-$ is generally accepted as being relatively slow compared with gas-phase chemical processes under typical atmospheric conditions. Therefore, this oxidation route is not included even in models in which aqueous-phase chemistry is explicitly treated, except in a version of the PLMSTAR model, where it is included for completeness (Seigneur *et al.*, 1984, 1985, 1986).

As in the case of $SO_2$ gas-phase oxidation, the formation of strong acid in the $NO_x$ oxidation cycle involves primarily the OH radical:

$$NO_2 + OH \rightarrow HNO_3$$

where the rate constant is a factor of 8 higher than that for the $SO_2$ + OH reaction. Unlike $SO_2$ oxidation, however, this reaction proceeds with net consumption of the OH radical, although it is only one of several HO-$HO_2$ radical chain termination reactions that occur in the troposphere. The use of effective first-order kinetics to describe $NO_x$ oxidation is difficult to justify.

Most of the models to date have been developed primarily to describe the transport and deposition of S compounds. Very few have been used to estimate the deposition of $NO_3^-$. However, most of the linear S chemistry models can be easily extended to include N chemistry. First-order oxidation rates with constant or seasonally varying rate constants have already been incorporated into recent versions of STATMOD (Venkatram, 1984) and Fisher's statistical model (Wright *et al.*, 1984). The AES-LRT model has similarly been modified (Bottenheim *et al.*, 1984). This model was further extended to include rudimentary PAN chemistry with monthly and diurnal variability incorporated in several parameters. Linear oxidation rates are also assumed in the ASTRAP model, with a specified diurnal and seasonal variation in the rate constants. The MESOPUFF-II model parameterizes the first-order rate constant for $NO_x$ oxidation to inorganic ($HNO_3$, $NO_3^-$) and organic (PAN) nitrates as a function of atmospheric stability, background $O_3$ concentrations, and the $NO_x$ concentration itself. The parameters are determined in a fashion similar to that of $SO_2$ oxidation. The rate constant is a function of the $NO_x$ concentrations at the previous time step. The model also explicitly treats the chemical equilibrium of the $HNO_3$, ammonia ($NH_3$), and $NH_4^+/NO_3^-$ system. The ENAMAP-2N model has been developed specifically to model $NO_x$ chemistry. The model accounts for five N species: NO, $NO_2$, PAN, $HNO_3$, and $NO_3^-$. NO and $NO_2$ are assumed to be in equilibrium. $NO_2$ is oxidized to PAN, $HNO_3$, and $NO_3^-$ in a specified molar ratio. PAN acts as a temporary $NO_x$ reservoir, decomposing slowly at night to form $NO_2$. All rates and equilibria vary diurnally to reflect photochemical effects. A version of a prototype ADOM model with linear chemistry has also been applied in the EUS (Hidy *et al.*, 1985). $NO_x$ is oxidized to $HNO_3$ and organic nitrates in a specified molar ratio, with a specified diurnal variation in the oxidation rate constant. However, the role of HCs is disregarded in all of these simple parameterizations of the $NO_x$ oxidation process.

As in the case of S chemistry, a realistic representation of $NO_y$ chemistry can be achieved only by simulating the coupled HC/$NO_x$/$SO_x$ system. This approach is followed in comprehensive Eulerian models and in Lagrangian models.

## Removal Processes

The removal of a pollutant from the atmosphere occurs by dry and wet deposition. Dry deposition occurs continuously; wet deposition occurs intermittently. Both dry and wet removal processes contribute to the deposition of acidic species. Strictly speaking, the chemical transformation of one species to another constitutes yet another removal mechanism for the original species.

The removal of gaseous and particulate pollutants by dry deposition occurs at the surface of the earth. It is preceded by the transfer of the airborne material from just above the surface to the surface (see also Appendix A). The entire process can be viewed as consisting of three distinct steps, as described in Chapter 3:

1. Transport of material through a surface or turbulent air layer to the immediate vicinity of the surface. This step is controlled by turbulent diffusion in the surface layer and is a function of wind speed, atmospheric stability, and surface roughness. It is independent of the species being transferred. This step is often referred to as the aerodynamic (or atmospheric) component of the dry deposition process.
2. Diffusion of material through a thin layer in laminar or quasi-laminar flow (laminar sublayer) or a deposition layer, just adjacent to the receptor surface. This air layer is typically a few millimeters thick, and the transport through this layer is controlled by molecular processes. This process depends on the molecular properties of the pollutants and air and on micrometeorological parameters. This step is the deposition-layer component of the dry deposition process.
3. Uptake at the surface. This process depends on several pollutant and surface characteristics, such as the solubility and reactivity of the depositing species and the physical structure, chemical composition, and biological properties of the receptor surface. This step constitutes the surface (or canopy) component of the dry deposition process.

Dry deposition of a pollutant, therefore, involves several interacting and complex individual processes, some of which are not well understood. The simplest representation of the dry deposition process is a definition of a deposition velocity that, when multiplied by the concentration of the material at some height just above the source, gives the downward vertical flux of the species to the surface. If each of the three steps in the dry deposition process can be visualized as offering a resistance to the removal of pollutant from the atmosphere, then, by analogy to electrical flow through a series of resistors, the deposition velocity can be estimated as the inverse of the sum of the individual resistances—the aerodynamic resistance, $r_a$; the deposition layer resistance, $r_b$; and the surface resistance, $r_s$. For the dry deposition of particulate matter, the surface resistance is usually ignored because the particle is considered to deposit immediately on encountering the surface. In addition, the settling velocity of the particle, if appreciable, contributes to its deposition rate. An adequate treatment of dry deposition requires a realistic parameterization of these processes, several of which are only partly understood. For example, the uptake of gaseous pollutants at the surface, which is often the rate-limiting step, is particularly uncertain for many species and surface/plant types.

Several models use constant or seasonal values of the deposition velocity. These models do not account for the diurnal variability associated with atmospheric stability; the variation in plant stomatal activity, which is a function of solar radiation, or the spatial variability associated with land use. The ASTRAP, ACID, and ENAMAP-2N models specify diurnal and seasonal variability. The ENAMAP-2S model uses tables to estimate deposition velocities as a function of season, atmospheric stability, and land-use type and also allows for reduced uptake of $SO_2$ by plant surfaces at night by using lower deposition velocities under nighttime conditions. The Lagrangian model of Eliassen *et al.* (1982a, 1982b), which simulates relatively detailed atmospheric chemistry (and was developed primarily to predict $O_3$ concentrations), specifies constant deposition velocities for all depositing species except $O_3$, for which a diurnal and latitudinal variation in deposition velocity over land is specified (dry deposition of $O_3$ over water is assumed to be negligible).

The most complete treatment of the dry deposition process is carried out in models that estimate the individual resistances to pollutant transfer and compute an effective deposition velocity as the inverse of the total resistance. The deposition flux to the ground is calculated as the product of the deposition velocity and a representative surface concentration. This removal can be applied naturally as a lower boundary condition in models that solve the vertical diffusion equation analytically or numerically. One-layer models, which have no vertical structure, remove material throughout the depth of the puff or the layer. The first-order removal rate constant is given by the deposition velocity divided by a characteristic length scale (usually the mixed-layer height). This also is known as the source depletion technique and may result in excessively high dry deposition rates in the near field and low dry deposition rates farther downwind. A more realistic approach is taken in some multilayer models that do not treat vertical dispersion processes explicitly. These models allow for removal at the surface (surface depletion technique) by defining a surface layer, which is typically a few tens of meters thick, in which a concentration gradient is maintained by the dry deposition process.

The removal of pollutant by precipitation (rain or snow) can occur by in-cloud scavenging (rainout) and below-cloud scavenging (washout). In the rainout process, the dominant mode of removal of airborne material is by its absorption into cloud droplets followed by its transfer to the precipitating phase at a rate determined by the microphysical processes occurring within the cloud. The absorbed pollutant may undergo reaction within the cloud to form products that are also removed by the precipitation. In the washout process, the pollutant is scavenged by precipitation falling below the cloud base. Because the precipitating hydrometeors fall quickly, there is much less time for the absorbed pollutants to react to form secondary species in the washout process.

As opposed to the dry deposition process, which occurs only near the surface, wet removal processes take place throughout the depth of clouds overhead, or the rain-bearing zone. Thus, the wet flux to the surface must be computed as the sum of wet removal from all volume elements aloft. In general, the wet removal rate depends on several parameters, including height above the surface, precipitation rate and type, cloud characteristics, and the ambient concentrations of the removed pollutant and other pollutants with which it may react or that may affect its solubility. Clearly, wet removal processes can be treated in a physically realistic manner only in models that simulate cloud physics and aqueous-phase chemical and mass-transfer processes. Other models use more simplified and parameterized approaches. These parameterizations usually do not distinguish between in-cloud and below-cloud scavenging and treat the entire wet removal process by means of bulk removal parameters. However, these models vary considerably in the parameterizations they adopt.

Wet deposition is commonly parameterized by two approaches. The first uses the scavenging coefficient, which is a rate constant for the linear removal of a pollutant. The scavenging coefficient is generally a function of the precipitation rate. The wet removal rate of a pollutant is given by the product of the scavenging coefficient and the local (gas-phase) concentration of the pollutant. The wet flux to the surface is found by integrating the wet removal rate with respect to height throughout the depth of the precipitation zone. Models generally ignore the height dependence of the scavenging coefficient. They simply estimate the wet flux to the surface as the product of an average scavenging coefficient and the vertically integrated concentration.

The second approach uses the scavenging (rainout or synonomously washout) ratio, which is defined as the ratio of the concentration of the pollutant in surface-level precipitation to its concentration in surface-level air. The wet deposition at the surface can then be estimated as the product of the washout ratio, the concentration of the airborne material, and the precipitation rate. Unlike the scavenging coefficient, the scavenging ratio does not require a vertical integration of concentration to obtain the deposition rate (although this distinction is irrelevant for models that assume vertically uniform concentration profiles). In a manner similar to the scavenging coefficient, however, the scavenging ratio approach assumes a first-order removal process. The first-order rate constant (scavenging coefficient) is estimated as the product of the scavenging ratio and the precipitation rate divided by a length scale that is either the mixed-layer height or the depth of the rain-bearing layer.

The representation of wet removal as a first-order process is assumed to be adequate when the scavenging is irreversible, as is the case for aerosol particles and highly soluble gases, because the rate of removal does not depend on the amount of material already scavenged. For gases with limited solubility, such as $SO_2$, the amount scavenged is determined by equilibrium considerations, which, in turn, are determined by solution properties such as temperature and acidity. Thus, the assumption of irreversible washout may not apply to $SO_2$. However, by assuming that $SO_2$ is oxidized rapidly to $SO_4^{2-}$, which is then efficiently washed out, an irreversible scavenging coefficient for $SO_2$ can be defined that is of the same order of magnitude as that for $SO_4^{2-}$ removal, typically $10^{-4}$ $s^{-1}$ (Fisher, 1978, 1983). This value is usually much higher than what would be obtained by calculating the solubility-limited removal of $SO_2$ under typical atmospheric conditions (Fisher, 1978). Several studies (e.g., Fisher and Clark, 1985; Venkatram and Pleim, 1985a, 1985b) have shown that it is necessary to assume efficient removal of $SO_2$ during wet conditions to explain observed patterns of wet S deposition.

The parameters actually used in most models tend to be empirical in the sense that they are estimated by fitting model predictions to observations or are obtained from semiempirical models (e.g., Scott, 1978). The simplest treatment is to assume a constant scavenging coefficient. The most common parameterization is to assume some dependence of the scavenging coefficient on the precipitation rate. A linear dependence is usually assumed. Models that use the concept of the washout ratio also effectively assume a linear dependence between the scavenging coefficient and the precipitation rate. Some models incorporate a washout ratio as a function of solution pH, whereas the removal of other species is treated using a scavenging coefficient proportional to precipitation rate.

The AES-LRT models use a temperature- and pH-dependent washout ratio for $SO_2$ and a constant washout ratio for other pollutants. The RTM-II and RTM-LT models employ fairly detailed parameterizations for the washout ratios of both $SO_2$ and $SO_4^{2-}$ aerosols. The washout ratio for $SO_2$ is based on the equilibrium of $SO_2$ in rain with $SO_2$ in ambient air and includes the effects of temperature and pH. For $SO_4^{2-}$, the washout ratio is based on Scott's parameterization (Scott, 1978) and is a function of rainfall rate, subcloud $SO_4^{2-}$ concentrations, type of precipitation (liquid or frozen), and cloud type and temperature. In addition to using a scavenging coefficient proportional to the rainfall rate, the MESOPUFF-II model distinguished between liquid and frozen precipitation by assuming lower scavenging efficiencies for snow.

The ACID model also assumes scavenging coefficients directly proportional to the precipitation rate for $SO_2$. However, the scavenging coefficient for $SO_4^{2-}$ is proportional to the precipitation rate raised to a fractional power. This approach follows Scott's parameterization for clouds in which the initiation of rain depends on the melting of snow at higher elevations. The ENAMAP-2S and ENAMAP-2N models also follow Scott's approach for developing empirical parameterizations for the scavenging coefficients of the form $W = aP^b$, where $W$ is the scavenging coefficient, $P$ is the precipitation rate, and $a$ and $b$ are empirical parameters that depend on the season (summer, fall/spring, winter). In the ASTRAP model, the scavenging coefficient is proportional to the square root of the precipitation rate up to a certain value of the precipitation rate and is constant thereafter. The reason is that heavier rain contains a higher proportion of large drops, which are less efficient than small drops in washing out material. Similarly, in the prototype ADOM model, wet removal processes are parameterized using a simple empirical scavenging coefficient approach in which scavenging coefficients are proportional to the precipitation rate raised to a fractional power (Hidy *et al.*, 1985).

Although models such as AES-LRT and RTM-II incorporate some realism in their treatment of $SO_2$ scavenging by including the effects of acidity on $SO_2$ solubility, they do not include the transport and deposition of other species necessary to accurately estimate the pH. Examples of such species are $NH_3$ and the cations and anions associated with soil dust.

As pointed out earlier, wet removal of pollutants can be treated realistically only in a model that explicitly simulates the relevant processes of gas- and aqueous-phase chemistry, cloud physics, and the associated mass transfer among the various phases. This approach is taken in three of the comprehensive Eulerian models, STEM II, ADOM, and RADM. Even these models, in their current form, cannot simulate all the processes in detail and must rely on some sort of parameterization.

Several models do not treat wet deposition processes at present. However, they could be modified to incorporate linear removal using some of the parameterizations previously discussed. In addition, Lagrangian models that use a vertical structure can incorporate simple cloud physics and aqueous-phase chemistry to treat wet removal explicitly. These capabilities are already present to a certain extent in a version of PLMSTAR that simulates the aqueous-phase chemistry of stratus clouds and fogs.

## CALCULATION OF METEOROLOGICAL CONDITIONS

### The Wind Field

The ability of a model to estimate accurately the deposition distribution of pollutants depends critically on the accuracy of the estimated transport wind fields. These wind fields must be available at the spatial and temporal resolution of the transport models, and they must be physically realistic. An accurate specification of the wind field is even more critical for flow over complex terrain. Significant mesoscale variations in large-scale flow patterns can be induced by the heterogeneity of the terrain, such as lifting and channeling of flow near obstacles, and the variations caused by differential heating or cooling of the earth's surface.

Limited resolution of observational data necessitates the use of accurate wind field generation schemes that can provide the input required to drive the transport models. Ideally, these schemes should account for variations in topography and produce three-dimensional winds that satisfy physical laws, such as conservation of mass. The most realistic and sophisticated approach is taken in the three-dimensional prognostic calculations (also known as *primitive equation* models), in which the time-dependent equations of conservation of mass, momentum, energy, and water substance are integrated forward in time from a prescribed set of initial conditions to yield gridded three-dimensional wind fields (and other meteorological parameters) at each time step. However, the computational requirements of these schemes are large, and their use in air quality modeling is limited to short time frames. Their use is justified only when the transport model is sufficiently complex (e.g., a three-dimensional, multilayered Eulerian model) to take full advantage of the detailed outputs from a prognostic model. For further details, the reader is referred to Pielke's (1984) review.

Most other applications require more economical wind-field generation schemes. These simple models are often lumped into the category of diagnostic models. However, diagnostic models can be further subdivided, based on their fundamental approach: empirical interpolation, objective analysis, and simplified physics models (Ramsdell, 1981).

***Empirical Interpolation Schemes*** The *empirical interpolation* (EI) models interpolate wind data spatially to drive horizontal wind fields. Goodin *et al.* (1979) describe several commonly used interpolation schemes. Sheih (1983) gives a more recent evaluation of wind-field interpolation schemes. The most common approach is to assume that the grid value is some weighted average of the surrounding data values, where the weighting parameter is a function of the distance between the grid point and the measuring station. Sheih tested several schemes using an analytical wind field and found that best results were obtained when the weighting parameter was inversely proportional to the square of the distance between the grid point and the measuring station.

In addition to spatial interpolation, temporal interpolation is required, because most upper-air observations are available at intervals of 12 hours.

This interpolation is usually accomplished either by assuming persistence of winds for a period centered around the observation time or by linear interpolation between observation times.

Surface data, available every hour, are sometimes used to supplement the upper-air data. These models do not account for terrain effects and are not mass consistent. Their value lies in their ability to compute smooth wind fields from available data in a relatively economical manner.

The ATAD (Heffter, 1981) is one of the most widely used air mass trajectory generation and transport models in LRTAP modeling. The ATAD trajectory model does not actually generate gridded wind fields. Instead, 3-hour trajectory segments are determined using the weighted average of transport-layer winds at each rawinsonde station within a specified radius (scan radius) of a segment origin. The transport-layer depth is calculated in the model, and the average wind in a layer at a rawinsonde station is computed from winds linearly weighted with respect to the depth between mid-reporting levels.

The meteorological processing component of the MESOPUFF-II model computes time- and space-interpolated fields of meteorological variables including the horizontal transport winds (Scire, *et al.*, 1983a, 1983b). Twice-daily rawinsonde data are supplemented with hourly data from the denser network of surface stations to construct wind fields at two user-selected levels: a lower level to represent boundary-layer flow and an upper level to represent flow above the boundary layer. The method used to determine the mixed-layer wind at each grid point is adapted from work of Draxler (1977, 1979, 1980). The mixed-layer-averaged winds are computed first using a representative rawinsonde sounding that is selected based on the thermal stability class at the nearest surface station to the grid point and the time of day. Then the directional change and ratio of wind speeds between the layer-averaged and surface winds at the rawinsonde station are determined. The hourly surface data are spatially interpolated to each grid point by a procedure similar to that used in ATAD.

*Objective Analysis Schemes* The next level of sophistication in the modeling of wind fields is the application of physical constraints to adjust the interpolated wind vectors at each grid point. Objective analysis schemes are an extension of empirical interpolation schemes. However, they are physically more realistic, because they reduce the divergence of the interpolated wind fields (i.e., they satisfy the mass continuity equation). Several objective analysis schemes can also treat simple terrain effects, using appropriate grid geometry and boundary conditions or terrain-following coordinates. In addition, some of the more recent models can also simulate other terrain effects, such as flow over and around obstacles during stably stratified conditions. The choice of the procedure used to reduce the divergence in the interpolated wind fields often controls the computational costs of these schemes.

The MASCON model, for example, was developed to provide wind fields to air quality models for the San Francisco Bay area, which is topographically and meteorologically complex (Dickerson, 1978). The model produces gridded, mixing-layer-averaged, horizontal wind fields as well as a vertical velocity at the inversion base to allow transport of material between the mixed layer and the inversion above. These winds are derived from initial wind fields by variational calculus. The model requires the specification of two empirical parameters that are used to weight the adjustments to the horizontal and vertical components. Flowthrough boundary conditions are specified around the outer boundary; no-flowthrough conditions are specified at the interior boundaries wherever topographic features protrude through the base of the inversion. This simulates wind channeling effect and diversion of flow around obstacles during stable conditions.

MATHEW is a three-dimensional version of MASCON that produces gridded fields of the horizontal and vertical wind components (Sherman, 1978). The model domain is a rectangular box set on the earth's surface with the bottom of the box located at the lowest topographic point in the area of interest. At each horizontal point a solid boundary is specified that is determined by the topography of the application area. Mass-consistent wind fields are derived from initial wind fields using a variational calculus technique with the SOR method as used in the MASCON version.

The ATMOS1 and VARMET wind field models are virtually identical. In fact, the ATMOS1 model relies heavily on work done by the developers of the VARMET code (Davis *et al.*, 1984). Three-dimensional nondivergent wind fields are generated in a terrain-following coordinate system, with the top of the computational domain held constant. An expanded vertical grid system ensures resolution near the surface. Initial winds are adjusted to generate mass-consistent wind fields using variational calculus techniques and the SOR method.

MINERVE-I is yet another objective analysis scheme that is based on the concepts of variational calculus (Caneill *et al.*, 1985). Terrain-following coordinates are used, as in ATMOS1 and VARMET, but the top of the domain is allowed to vary spatially. The adjustment of the initial winds to generate mass-consistent, three-dimensional models is essentially similar to that used in the other objective analysis models discussed so far. However, the generation of the initial horizontal wind fields differs. Surface measurements are spatially interpolated by a polynomial fit in which the domain is divided into triangular subregions by connecting the data points. Goodin *et al.* (1979) compared several interpolation schemes using test data sets of concentration and velocity fields and found that their triangle scheme provided a good compromise between accuracy and computational cost. The scheme determines initial horizontal winds above the surface layer by interpolating available radiosonde data using an inverse distance squared weighting.

MINERVE-II, described by Caneill *et al.* (1985), is similar in many aspects to MINERVE-I; the major difference is that MINERVE-II relies on the use of finite elements instead of finite differences in the integration of the differential equations. The main advantage of finite-element methods lies in their ability to resolve topography accurately.

The wind field generator in the PLMSTAR model conceptually follows a scheme proposed by Goodin *et al.* (1980). The PLMSTAR model extrapolates surface measurements vertically for stations. The extrapolated surface winds supplement the sparse upper-air measurements. The rest of the wind field generation scheme follows the procedure described by Goodin *et al* (1980).

The COMPLEX model (Endlich *et al.*, 1982) is distinguished from other objective analysis schemes by its incorporation of simplified physics, such as the concept of the dividing streamline to simulate flow around terrain features under stably stratified conditions. However, the COMPLEX model applies all these concepts in a distinctly different manner. The model produces mass-consistent horizontal winds in a three-dimensional grid, using a flow-following coordinate system. This coordinate system is similar to a terrain-following coordinate system in many respects. However, under stable conditions the coordinate surfaces may intersect terrain in areas where flow does not pass over an obstacle. Based on this concept, the model defines a critical height, which is the highest height to which the air at a given level can be propelled by the local wind speed against the local potential temperature gradient. COMPLEX uses these critical heights to determine the heights of the flow surfaces. Details of the procedure are provided in Ludwig, *et al.* (1986).

## Simplified Physically Based Models

Both the empirical interpolation and objective analysis schemes rely on the availability of a relatively large amount of input data. Such data are seldom available for roughened or complex terrain. Simplified physically based models, which solve a subset of the equations for airflow using a limited amount of input data, offer an alternative to predicting reasonably realistic wind fields in complex terrain. These advantages usually require increased computation. Some of these models solve the time-dependent equations of momentum and energy until a steady-state solution is achieved. However, they are still referred to as *diagnostic calculations* because they are not used to forecast by integrating the conservation relations (Pielke, 1984).

The complex terrain wind model (CTWM) of Liu *et al.* (1982) is a multilayer diagnostic model based on the fluid continuity equation. The CTWM treats the overall wind convergence (or divergence) within each layer as the sum of several components, each representing a physical process weighted by empirically determined coefficients. Parameterizations are developed for (1) lifting and diversion of flow by topographic effects, (2) vertical variation of horizontal winds from surface friction effects, (3) convergence of the flow from thermally induced circulation, and (4) mountain valley winds.

Danard's (1977) model is a diagnostic one-level model for computing mesoscale influences of terrain, friction, and heating on surface winds, given large-scale data weather charts. The model integrates a finite-difference form of the time-dependent equations for surface pressure, potential temperature, and surface winds, starting from given initial values. The equations are integrated until steady-state fields of these variables evolve that represent the modifying influences of topography. Fluid mass conservation is not used as a constraint. The model simulates phenomena, such as topographic channeling of flow, effects caused by changes in atmospheric stability, land and sea breezes, and katabatic winds. Because the model has no vertical resolution, it does not address lifting and diversion of flow near obstacles.

Another approach is Mass and Dempsey's (1985) modified version of the Danard model, in which the time-dependent equations for the conservation of momentum and energy are solved in terrain-following coordinates. Unlike the Danard (1977) model, this modification does not integrate the surface pressure equation. Instead, it expresses the surface pressure as a function of surface temperature by vertically integrating the hydrostatic equation from a reference level (850 mb) to the surface. Like the Danard model, the scheme is not mass consistent and has no vertical structure. Other differences from the Danard model include an improved numerical scheme and initialization procedure. Mass and Dempsey's scheme is computationally stable and reaches steady state quickly.

Perhaps the most elaborate method for calculating meteorological conditions comes from the project to develop RADM (Chang *et al.*, 1990). This model employs a dynamic meteorological submodel capable of performing four-dimensional data assimilation (FDDA). The meteorological code is based on a miniscale forecasting scheme called MM-4 (Anthes *et al.*, 1987).

MM-4 calculates airflow, pressure, temperature, and water concentration (including clouds and precipitation). The MM-4 model employs the equations of air motion for horizontal wind components, the continuity equation written for pressure, the energy balance equation, and the hydrostatic equation. The fluxes of heat and momentum in the planetary boundary layer are calculated explicitly. The calculation of water concentration is incorporated and uses a balance for water in vapor and condensed form. The initial conditions are specified by observations or by a planetary-scale general circulation model.

The output of MM-4 specifies the meteorological fields every 3 hours. The spatial resolution for regional-scale application is approximately 80 km. However, the model can be applied to spatial scales as small as 5 km. At such scales, MM-4 can simulate atmospheric processes like land–sea breezes and mountain–valley winds.

The MM-4 as FDDA is used in a "pseudodiagnostic" mode when the calculated meteorological fields are forced by observations. FDDA combines the advantages of the self-consistency of a mathematical model with the diagnostic analysis of observations. The use of FDDA is particularly useful for multiday simulations because the pure prediction approach tends to decline in performance as the length of simulation increases.

FDDA is conducted using a Newtonian relaxation method in which the MM-4 model prediction is relaxed toward an observed state by adding to one or more fluid dynamic equations, artificial "nudging" terms that are based on a difference between the model calculation and observations. The nudging terms are proportional to the difference between the model simulation and an analysis of observations calculated at each grid point. As long as the differences are small, the nudging does not dominate the estimated conditions.

The advantages of the FDDA system seem to be based on the ability of the MM-4 to assimilate observations after the initial time of each simulation step, while retaining the numerical information framework of the submodel to estimate mesoscale details, vertical motion, and intervariable consistency that are not contained in analyzed observations. The MM-4 is also capable of simulating nonlinear interactions that cannot be represented by direct interpolation between analyses based on synoptic observations separated by 12 hours. Thus, FDDA suppresses accumulation of large-scale forecast errors without serious loss of the mesoscale flow detail constructed by numeral integration.

## Clouds and Precipitation

Cloud fields are generally input from observations as precipitation. They are treated with different degrees of sophistication. Perhaps most elaborate is that found in RADM and ADOM. Clouds in this model are determined from hourly precipitation data, temperature, and relative humidity soundings.

Three types of clouds are included: (1) precipitating cumulus clouds, (2) precipitating stratus clouds, and (3) fair-weather cumulus clouds. The base of all cloud types is calculated to be the condensation level in rising air. The top of clouds corresponds to the level where water vapor concentration is unsaturated.

The first two cloud types are selected when the meteorological model predicts that precipitation occurs for a given hour on given grid cells. If the vertical temperature profile suggests unstable conditions, cumulus clouds are assumed to be present. If the vertical temperature profile is stable, stratus clouds are predicted where supersaturation of water vapor occurs.

Cloud cover from raising cumulus clouds is calculated for grid cell–averaged precipitation rate, cloud lifetime, and amount of excess water produced by the clouds. The amount of excess water generated by a cloud is the amount of water in the cloud which exceeds the water vapor concentration at a corresponding ambient temperature and pressure. It is calculated from a mass balance equation for water and an energy balance equation according to a model of Walcek and Taylor (1986).

For stratiform clouds, cloud cover is assumed to be 100% in a grid cell. The vertical extent of clouds is estimated from a calculated profile if water vapor supersaturation. Cloudiness and fair-weather conditions is parameterized as a function of relative humidity. Fair-weather clouds are assumed to form when relative humidity is above 70%. Cloud cover is parameterized to increase with relative humidity but is constrained less than 20%.

Cloud fields are calculated for each hour in the simulation. Movement of clouds is not provided for between grid cells, and information on cloud fields is not extended from one hour of the simulation to the next hour. This is a potentially important limitation for calculations with high spatial and temporal resolution.

The amount of precipitation is calculated by the meteorological model on an hourly basis and averaged over each grid cell. The chemical composition of the rain is assumed to be equivalent to that of the cloud. No below-cloud scavenging of pollutants is considered.

In-cloud scavenging is parameterized in terms of equilibrium solubility coefficients, empirical relationships using scavenging coefficients, or aqueous chemical reactions. Particle scavenging can be estimated knowing liquid water content and assuming a droplet size distribution (Appendix A).

Precipitation by snowfall is treated as precipitation by rainfall and is determined according to ambient temperature.

Complexities involve cloud collection on mountain slopes and deposition by dew or settling of fog.

## Reliability of Airflow Estimation

As in the case of the transport/deposition models, the sparsity of observational data, particularly in complex terrain, handicaps a thorough validation of wind field generation schemes. Goodin *et al.*, (1979) evaluated the performance of several empirical interpolation schemes for winds using three data sets. One data set was an idealized concentration distribution to which the exact solution was known; the second data set was a field corresponding to a velocity potential; the third data set consisted of surface $O_3$ concentrations measured in the Los Angeles basin on a particular day. The results of the study indicated that a second-degree polynomial fitting procedure, with an inverse distance-squared weighting scheme, provided a practical compromise between computational cost and desired accuracy.

Sheih (1983) compared the predictive capabilities of four interpolation schemes using a theoretical cyclonic wind field superimposed on a uniform field. Although none of the schemes compared well with the theoretical solution, the inverse distance-squared weighted interpolation had the most accurate wind fields and total divergence. Empirical interpolation schemes are an efficient and economical approach to generating wind fields and have been used extensively in transport models. However, their success depends on the density of meteorological stations, because all the information necessary to describe the variation of the interpolated quantities must be contained in the observational data. The sensitivity to the number of observation stations and their geographic distribution is amplified for complex-terrain applications.

Objective analysis schemes force realism by adjusting the empirically interpolated wind fields using mass-consistency constraints. Models based on variational calculus techniques, such as MATTHEW and ATMOS1, have been applied with air quality models to simulate the transport and dispersion of tracers in complex terrain (Porch, 1980; Porch and Lange, 1983). Such models have been found to agree with measured tracer concentrations. Comparisons with observed winds show that the adjustment procedure improves the prediction of winds in complex terrain. High-quality results are obtained when more data are used to generate the wind fields. The variational calculus techniques can be computationally demanding, depending on the degree of divergence of the initial interpolated wind field. In addition, the final form of the flow field depends on empirically chosen constants (Sherman, 1978).

In comparison, grid-point adjustment schemes (e.g., Goodin *et al.*, 1980; Lurmann *et al.*, 1985) generate mass-consistent wind fields efficiently and have been successfully tested with field measurements and hypothetical flow fields (Goodin *et al.*, 1980). The computational burden can be further reduced for the generation of wind fields over a long period of time (e.g., 1 year) by using an eigenvector method (Ludwig and Byrd, 1980). Objective analysis schemes, such as COMPLEX and MELSAR (Allwine and Whiteman, 1985a,b), have not yet been adequately tested against observations. Objective analysis schemes, like empirical interpolation schemes, require a relatively dense network of meteorological observations to generate reasonably accurate wind fields.

Simplified physically based models, on the other hand, require a limited amount of input data. The CTWM model has been applied to several complex terrain settings where extensive wind monitoring data were available (Yocke, 1981). In general, the model performed reasonably well, particularly for high-wind-speed days, and appeared to treat adequately the effects of topographic blocking, surface friction, and differential heating. The performance during low-wind-speed days was not as good. The CTWM model did not exhibit any significant tendency toward over- or underprediction. A major disadvantage of the model is that it depends on the empirically determined coefficients to weight the contributions of the various processes responsible for the flow field divergence. The Mass and Dempsey model (1985) diagnosed many details of mesoscale flow in complex terrain for a variety of flow directions and diabatic conditions. Simplified physical models incorporate a reasonable amount of realism in their formulation and appear to reproduce expected mesoscale effects using a limited amount of input data. However, compared with the empirical interpolation and objective analysis schemes, simplified physical models require more computation, particularly the multilevel models that must be integrated several times before steady state is reached. Only the prognostic wind models incorporate the most realistic representation of the physical processes that govern local and mesoscale flow patterns. These models can be considered an analog of the comprehensive Eulerian transport/deposition models described above, in which the equations of air motion are solved numerically. Their extensive computational requirements generally limit their use to research applications.

The MM-4/FDDA system was evaluated using both objective verification methods on 12-hour segments of nine 5-day simulations and a subjective verification of synoptic-scale and mesoscale amplitudes, position, and patterns associated with meteorological observations. The objective statistical scheme was used for comparing observations with calculated sea level pressure, winds, temperature, mixing ratio, and precipitation. The performance characteristics are reported by Chang *et al.* (1990). The FDDA system consistently and substantially improved the accuracy of the MM-4 model simulation. The wind and sea level pressure errors that affect pollution transport were reduced to levels expected from objective analyses of the observations. And 12-hour rainfall totals affecting wet deposition contained little bias, while showing significant improvements in rainfall patterns and amounts compared with results from MM-4 control runs. The rainfall estimates contained a 0–7 mm bias with mean arithmetic error of 0.27 cm. The latter is within the estimated uncertainty in rainfall observed with an 80-km scale.

Examination of the simulations showed that the MM-4/FDDA system satisfactorily reproduced mesoscale features not resolved by the rawinsonde observations, such as boundary layer dynamics, strong wind currents, frontal systems, squall lines, or rain bands.

## RESULTS OF MODEL SIMULATIONS

Air quality models provide their output in a variety of forms. Perhaps most common are concentration maps and deposition patterns. For example, Figure 6.1 shows the annual accumulation of total S deposition from the ASTRAP model, which tracked the major $SO_x$ sources in ENA by repeated calculation of air mass trajectories. The ASTRAP model simulates deposition patterns accounting for transport, linear scavenging in clouds, and dry deposition at the surface by first-order rate terms in ambient concentration. This highly parameterized calculation simulates at least qualitatively the expected long-term deposition pattern based on wet deposition measurements.

Similar maps of $SO_x$ concentrations and distributions can be generated on shorter time scales (Figure 6.2). Calculations for an 8-day event (17–24 July 1978) in which precipitation was negligible indicate patterns that are qualitatively consistent with observations. These calculations were performed by the ADOM model, a multilayer Eulerian scheme with great diurnal detail in wind field, (nonlinear) homogeneous gas-phase chemistry, in-cloud processes, and dry and wet deposition. The calculations in Figure 6.2 were for fair-weather conditions without precipitation, which was consistent with observed conditions. The calculated event sequence indicates dry deposition with strongest local areal impact in western Pennsylvania and the lower Ohio River region, as expected from the observed patterns of pollutant concentrations.

Eulerian models can also calculate diurnal patterns. SURADS, for example, produced data for 3-hour averages (Figure 6.3). The calculations correspond to 80 × 80 km grid cell averages in this case. The figure also shows the SURE data from stations in the appropriate grid cell. The calculations generally do not correspond well to observations at individual stations, which raises unresolved questions about the representativeness of single sampling station observations for short-term behavior over

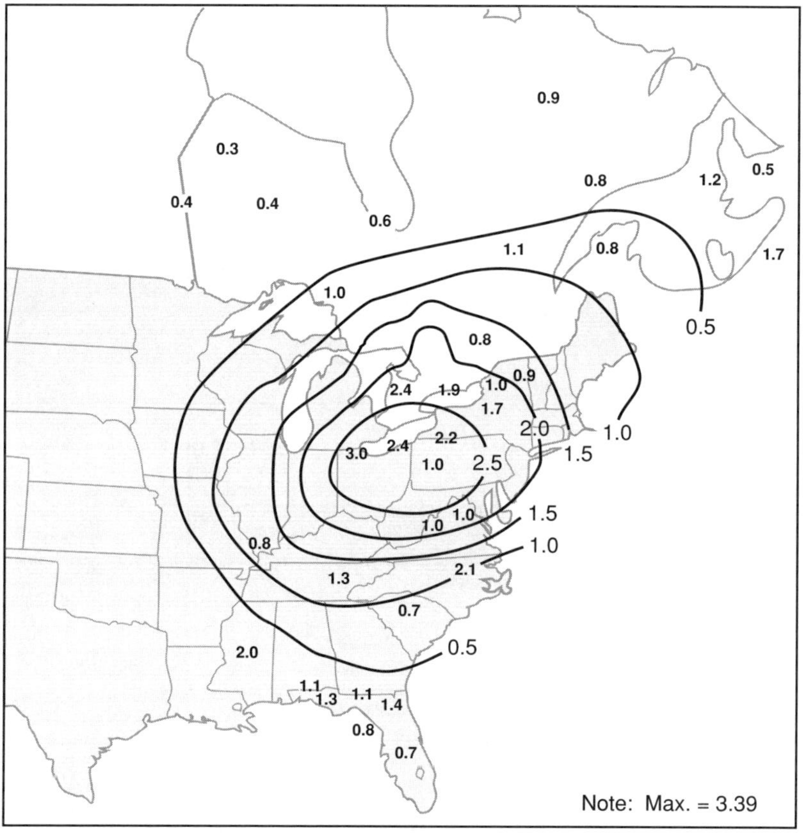


**FIGURE 6.1** Cumulative $SO_4^{2-}$ in rain, expressed as g of S/m$^2$ for 1977 with ASTRAP simulations (isopleths).

large areas. Furthermore, such tests raise important questions about the reliability of the calculations for short-term local behavior of $SO_x$.

## Source–Receptor Relationships

Model calculations are generally well suited to investigate SRRs, particularly Lagrangian calculations that link specific sources with a field of receptors. One important approach, adapted years ago, linked source areas with receptor areas. The method is sometimes termed the *transport matrix concept*. In the method ENA is subdivided into large subregions (or into separate states). A key assumption is that the average pollutant

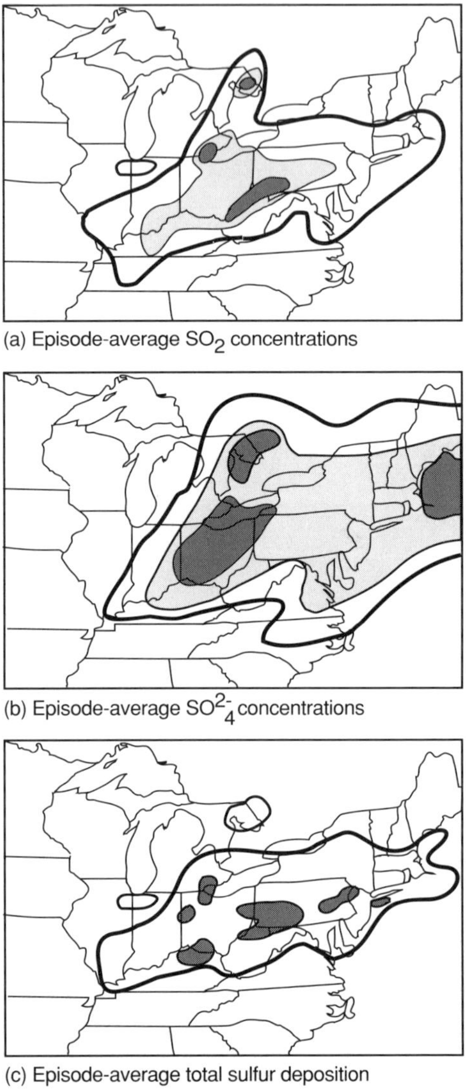


**FIGURE 6.2** Computed episode (8-day) average fields of ambient $SO_2$, $SO_4^{2-}$, and total S deposition for period 17–24 July 1978. $SO_2$ concentrations are expressed as ppb. (Light shading = greater than 10 ppb; medium shading = greater than 15 ppb; dark shading = greater than 25 ppb). $SO_4^{2-}$ concentrations are expressed as $\mu g/m^3$. (Light shading = greater than 15 $\mu g/m^3$; medium sharing = greater than 20 $\mu g/m^3$; dark shading = greater than 25 $\mu g/m^3$.) Total S deposition is expressed as $\mu eq/m^2$-d. (Light shading = greater than 600 $\mu eq/m^2$-d; medium shading = greater than 1,000 $\mu eq/m^2$-d.) (From Hidy *et al.*, 1985, with permission of the Electric Power Research Institute.)

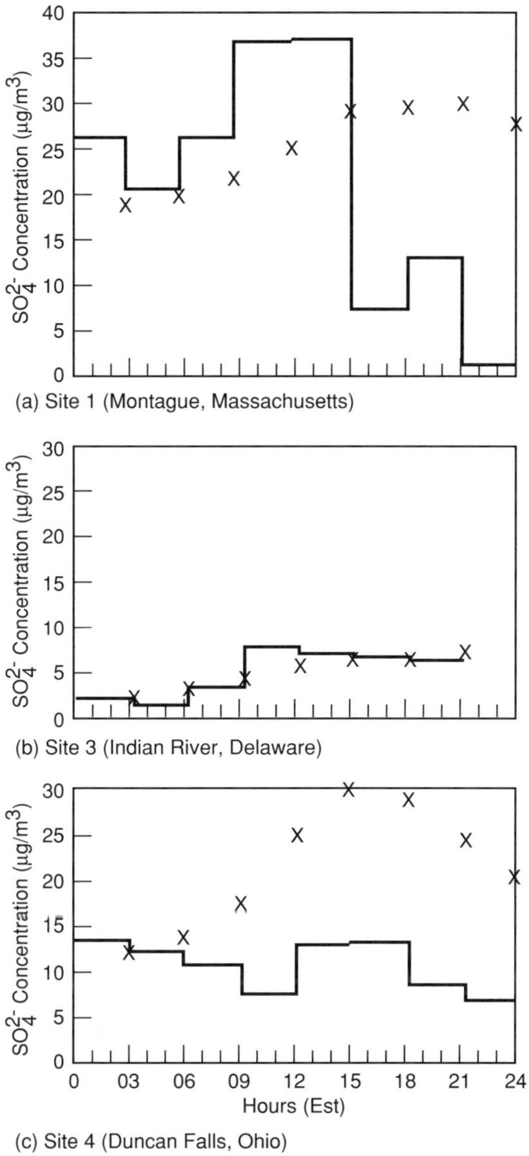


**FIGURE 6.3** Comparisons between observed 3-hour average $SO_4^{2-}$ concentrations (heavy line) and model calculations (X) for the gride cells in which SURE Class I stations were located. Simulation of 5 August 1977. (From Mueller and Hidy, 1983; reproduced with permission of the Electric Power Research Institute.)

concentration (or deposition), $c_i$, in any subregion is a linear combination of emissions of its precursor pollutant in every region and the models are linear in describing dynamical processes. For any region ($i$)

$$c_i = \sum T_{ij} E_j \tag{6.2}$$

where $E_j$ is the emission rate from subregion $j$ and $T_{ij}$ is the transfer coefficient between subregion $i$ and $j$. The method requires a two-dimensional array, a matrix, to determine the concentration (or deposition) values in all subregions in terms of a set of emissions from the subregions. Models have been used to calculate the values in the transfer matrix (Bhumralkar *et al.*, 1980). Results indicate that, fractionally, the bulk of deposition comes from within a particular receptor subregion, but minor contributions come from neighboring subregions.

The variability in percentage of $SO_4^{2-}$ deposited at a given receptor by a source subregion depends on the model used to estimate $T_{ij}$. The values of $T_{ij}$ vary from model to model given the model itself and input assumptions. Table 6.2 illustrates the estimates of fractional receptor exposure by major source area (state) from eight different models for the Adirondacks as the receptors. The range of estimates of fractional deposition is substantial for this receptor area. Differences of this magnitude

TABLE 6.2 **Comparison of Model Predictions for U.S. Sources $SO_x$ and $NO_x$ Contributions to an Adirondack Receptor**[a]

| | % of $SO_4^{2-}$ from source area[b] | | % of $NO_3^-$ from source area[c] |
|---|---|---|---|
| | Mean | Range | |
| Pennsylvania | 18.3 | 13.2–23.6 | 14.1 |
| Ohio | 18.6 | 14.8–26.7 | 9.4 |
| New York | 14.1 | 9.3–24.2 | 11.8 |
| Indiana | 8.0 | 5.1–9.5 | 2.8 |
| West Virginia | 6.4 | 4.9–8.9 | 5.5 |
| Maryland, Delaware, New Jersey, Washington, DC | 4.1 | 2.9–6.4 | 10.0 |
| Massachusetts, Connecticut, Rhode Island | 3.2 | 1.3–5.8 | 4.6 |
| Michigan | 6.6 | 3.9–9.0 | 3.3 |
| Illinois | 4.5 | 2.3–6.5 | 2.7 |
| Kentucky | 3.2 | 1.7–4.4 | 2.6 |
| Other[d] | 13.0 | 5.5–32.4 | 2.6 |

[a] After Mohnen, 1984; reproduced with permission of McGraw-Hill.
[b] Based on eight model predictions, reported by Mohnen (1984); normalized to listed states.
[c] Predictions based on MIT Mode (Fay *et al.*, 1987); excludes Canada and western and southern U.S.
[d] States on the western border of the Mississippi River.

were considered excessively large and consequently too uncertain to rely on the early models for quantitative guidance on targeted control strategy.

More recent calculations of source contributions using the RADM have been reported by NAPAP (1990b). These are listed by geographic region in Table 6.3. Source attribution is done in these simulations by identifying (tagging) specific source emissions, following them to receptors, and calculating their fraction of the estimated receptor total. Although not directly comparable with estimates in Table 6.2, the RADM results including nonlinear chemistry and sophisticated treatment of meteorology are similar, reflecting greater contribution from nearby sources than distant ones.

SRRs based on transfer matrix estimates also have been calculated for $NO_3^-$ deposition. Table 6.2 also lists results for state contributions to the Adirondacks. These estimates are based on calculations from the MIT model for linear chemical conversion of $NO_x$ to $NO_3^-$ and linear deposition rates. As in the case of $SO_4^{2-}$ deposition, the results show major contributions from the states upwind and immediately surrounding the receptor. The results are biased from the southwest to northeast, as expected from the direction of prevailing winds. No model comparisons for $NO_3^-$ deposition have been reported for early models. However, with the complexities of $NO_y$ chemistry involved and differences in calculations of transport and deposition, one would expect a range of estimates similar to $SO_4^{2-}$.

Another example approach to linking sources and receptor conditions has emerged from work at the University of Michigan (Samson and Small, 1984). Like the calculation of a transfer matrix, their technique uses a Lagrangian model in either a backward or forward trajectory mode of calculation. In the former, calculations produce a history of air mass trajectories that would cross over the receptor from areas upwind. In the

TABLE 6.3 **Percentage Contribution of Tagged Emissions of Total Sulfur Deposition (Wet Plus Dry) at Four Receptor Areas Based on RADM Calculations**[a]

| Source area | Emissions in RADM domain | Adirondacks | Poconos | Central Appalachian | Southern Blue Ridge |
|---|---|---|---|---|---|
| Upper Ohio River Valley | 19 | 19 | 23 | 37 | 8 |
| Lower Ohio River Valley | 13 | 8 | 7 | 10 | 9 |
| South | 10 | 11 | 5 | 2 | 0 |
| Canada | 8 | 2 | 2 | 4 | 21 |
| Other | 50 | 60 | 63 | 47 | 62 |

[a] From NAPAP, 1990b.

latter, air mass movements are calculated that originate over a source and travel downwind for a suitable duration. In either case, the family of trajectories generated represent a distribution of receptors affected by sources. The ACID model adopted for the calculations contains highly parameterized estimates for the relevant atmospheric processes.

## Natural Potentials

The results of ACID modeling normalized by the strength of the emissions field to yield simulated concentrations/deposition per unit of emissions, a *natural potential* for contribution to the simulated concentration/deposition. This field reflects the natural variability in the meteorological and other chemical or physical conditions by which a particular source location is coupled through the atmosphere to a particular receptor. Natural potential fields are effectively a statistical approach; they do not depend on the spatial distribution of sources. Their estimation does, however, assume a linear relationship between the rate of $SO_2$ emissions and the resulting $SO_4^{2-}$ concentration and deposition field.

To estimate the transport path of air leaving a source in the forward mode or arriving at a receptor in the backward mode, trajectories are calculated based on observed upper-air winds. If a trajectory represents the center of mass of a parcel of air moving from a source to a receptor, then the probability of a chemically reactive species arriving at a point $x$ at a time $t$, $A_s(x,t)$, can be expressed as

$$A_s(x,t) = \int_{t-\tau}^{t} \int_{-\infty}^{\infty} \int_{-\infty}^{\infty} Q(x, t/x', t')R(t/t') \times D_s(x', t')\Lambda_s(x', t')\, dx'\, dt' \quad (6.3)$$

where $Q(x, t/x', t')$ is the probability of an air parcel located at $x'$ at time $t'$ arriving at a receptor $x$ at time $t$, $R(t/t')$ is the probability of the tracer not being reacted to another species from time $t'$ to time $t$, $D_s(x', t')$ is the probability that the tracer will not be deposited dry at $(x', t')$, and $\Lambda_s(x', t')$ is the probability that the tracer will not be deposited wet at $(x', t')$. The integration is continued over a period $\tau$, which has been prescribed beforehand (generally from 48 to 72 hours). The product of the last three probability functions can be approximated for $s$, for example, by solving the coupled equations

$$\frac{\partial s}{\partial t} = -k_t s - k_d s - k_w s + E_s$$

$$\frac{\partial p}{\partial t} = +k_t s - k_d p - k_w p + E_p \quad (6.4)$$

where $s$ is the probability of S occurring as $SO_2$, $p$ is the probability of $s$ occurring as $SO_4^{2-}$, $k_t$ is a linear rate of transformation of $s$ from $SO_2$ to $SO_4^{2-}$, and $k_d$ and $k_w$ are rates of dry and wet removal, respectively, for $SO_4^{2-}$. Some fraction, $E_p$, (usually thought to be about 1 to 5%) of the initial probability of emitted $s$ is assumed to be $SO_4^{2-}$; the majority of the initial probability is assumed to be $SO_2$. The chemical transformation rate, $k_t$, and the $SO_2$ and $SO_4^{2-}$ dry deposition rates are assumed to vary diurnally and seasonally. Variation is specified from estimates of rate coefficients based on diagrams from seasonal observations. The wet deposition rates depend on precipitation rate as described by a washout ratio. In all cases the probabilities represent the vertical integral over transport layer. The consistent use of units of probability ($km^{-2}$) allows comparisons that would otherwise be awkward. For example, at a given point one can compare in common units the probability of occurrence of inert tracer versus the probability of deposition for a reactive tracer.

Hidy *et al.* (1985) give an example of the calculations generating natural potentials. The calculations were made to compare the dispersion of an inert tracer released from a source, $SO_2$ release from that source, transported $SO_4^{2-}$ product, and S deposition. Results for the lower Ohio River subregion source region illustrate the kinds of behavior that can be calculated. Figure 6.4 displays the average probabilities of observing an inert tracer and ambient $SO_2$ in the region for 6 years. Calculations accounted for the regional wind and precipitation conditions for this period. Figure 6.4 shows that the inert tracer occurs with highest probability nearest the source, reflecting the difference of trajectories from the release point over the 6-year period. The probability of observing tracer is skewed toward the east because the prevailing winds over this period were from the west. The general symmetry of the calculated field is also noteworthy. Trajectory calculations were limited by availability of wind information; hence, extension of the results beyond the eastern seaboard is unwarranted. Figure 6.4b represents the probability of observing (through vertical integration in the air volume) a reactive, depositing tracer that simulates $SO_2$ behavior. Again, the highest probability of observing the reactive tracer occurs nearest the source. This field shares much of the same basic shape and skewness of the field for inert tracer. The probability of observing an $SO_2$-like tracer is significantly smaller far from the source, presumably because of its loss through deposition and conversion to $SO_4^{2-}$. The gradients of the calculated $SO_2$ field are also more pronounced near the source. The $SO_4^{2-}$ field in Figure 6.4c is similar to the fields for inert tracer and $SO_2$. The rate of $SO_4^{2-}$ production of ~3% $SO_2$/h assumed in the model yields an average maximum in $SO_4^{2-}$ about 24–48 hours downwind of the source. The similarity of $SO_4^{2-}$ and $SO_2$ patterns suggests that the 6-year probability of observing $SO_4^{2-}$ is con-

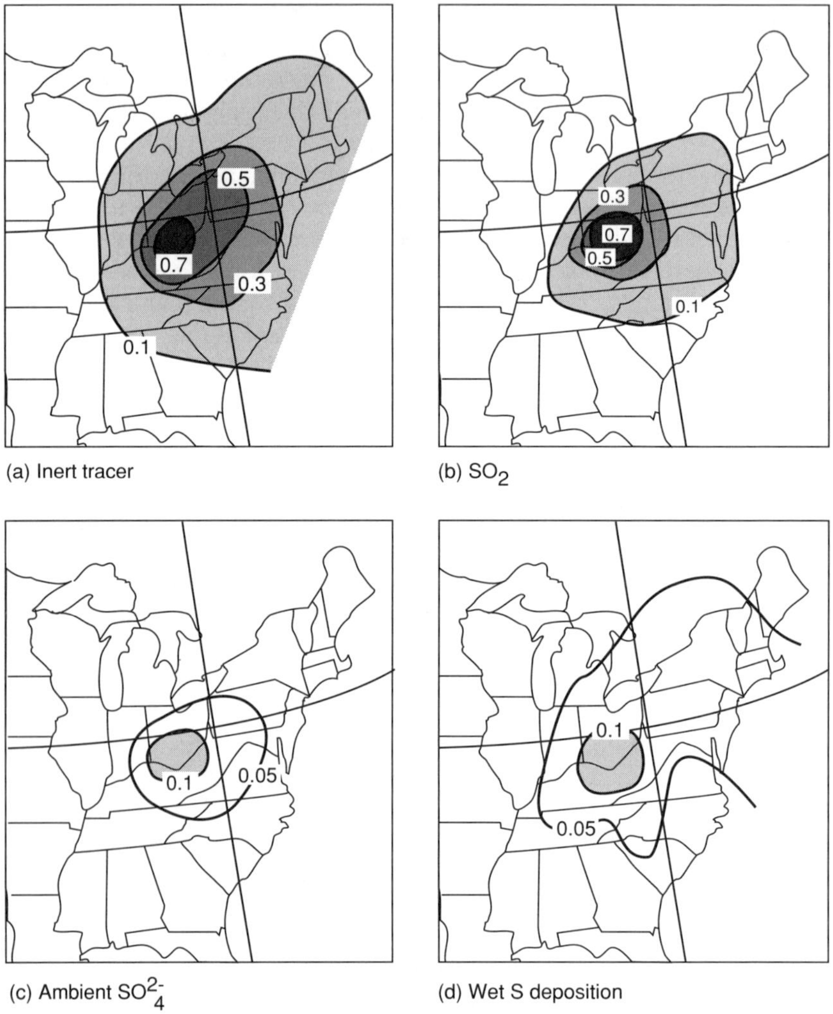


**FIGURE 6.4** Six-year average natural potential functions in fraction of total S emitted calculated from meteorological conditions between 1976 and 1981 for sources in the lower Ohio River subregion. (Courtesy of P. Samson; from Hidy *et al.*, 1985; reproduced with permission of the Electric Power Research Institute.)

trolled more by the diffluence of trajectories than the rate of chemical transformation. By contrast, the probability field for sulfur wet deposition in Figure 6.4d is more skewed toward the northeast than the ambient $SO_4^{2-}$ concentration field. The greater skewness reflects the higher likelihood for southwest winds to encounter precipitation (hence wet deposition) downwind of the source (see also Figure 2.10).

The ACID can also calculate the degree to which an inert tracer released at a source would represent dispersion or deposition of a reactive effluent. This representation is readily achieved by ratioing the probability of occurrence of the reactive species to the inert tracer. For example, the ratios between $SO_2$ inert tracer distributions and S wet deposition versus inert tracer distributions decrease with distance in the Ohio River region (Hidy *et al.*, 1985). The degradation of inert tracer as an $SO_2$ surrogate exceeds 50% after about 500 km from a source area. Calculation of the ratio of inert tracer distribution and the wet deposition pattern in this region suggests that the inert tracer distribution would be of little use in characterizing S wet deposition from the same source.

ACID calculations can show the interannual variability in terms of the coefficient of variation (c.v.) of annual averages in the natural potential distribution. The pattern for transport of passive tracers is similar to that calculated for $SO_2$ distribution (Hidy *et al.*, 1985). The year-to-year variability throughout much of the region is relatively low (c.v. of 0.2 or less for the source examined). The subregions of low c.v. for $SO_2$ tend to be smaller because transformation and deposition processes are less persistent. The calculated variability in wet deposition was found to be much larger between years because of precipitation variability.

ACID calculations yielded additional information on interannual variability of wind direction within the transport layer and the deposition probability with distance from a source. The variability of predominant exposure varied with source and location. The extreme directional variation was found to be at Sudbury, Ontario, especially for wet deposition. The air mass trajectory calculations also showed that the unit transfer matrix decreases exponentially with distance away from a source. Extending in the direction the prevailing wind (southwest), the transfer matrix decreases by 10% of its initial value over a 900-km distance. A source of unit strength will exert an influence of one-tenth its value over this distance along the windward direction. This scale is consistent with other measures, such as spatial covariance and the space–time correlations discussed in Chapter 5. An exponentially decreasing distance of impact, on the average, for annual average conditions reflects the influence of conversion and loss processes in ENA.

## Sensitivity Analysis

Results from model calculations should be sensitive to the atmospheric process description. Observations clearly show the potential for that sensitivity.

From the previous section and case studies, the calculation of the wind trajectory patterns is perhaps the most sensitive component of regional scale modeling. In a further illustration, Figure 6.5 shows 5-day averaged

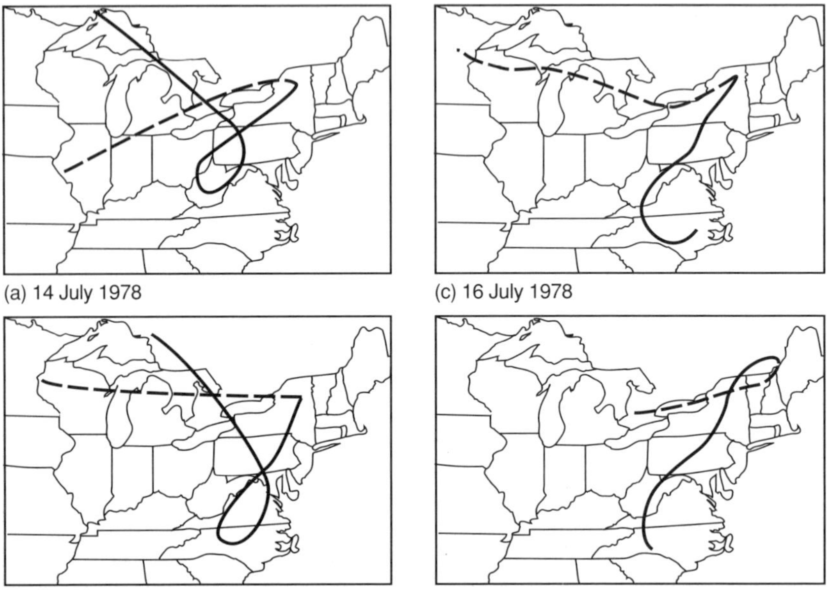

(a) 14 July 1978 (c) 16 July 1978

(b) 15 July 1978 (d) 17 July 1978

FIGURE 6.5 Paths of 5-day layer-average trajectories terminating each day at Whiteface, NY. The solid trajectory corresponds to the 100–400-m transport layer, the dashed trajectory corresponds to the 400–3000-m transport layer. (From Hidy *et al.*, 1985; reproduced with permission of the Electric Power Research Institute.)

trajectories for two different assumed atmospheric layers in which the bulk of transport is assumed to occur. In this example, as in many others, the layer selected makes a major difference in the calculated link between source and receptor (also Figures 4.5 and 4.12). Over an extended averaging period of a season or a month these differences are expected to be smoothed out. Indeed, calculations suggest that this is the case. This wind trajectory average combined with areal averaging of precipitation location and quantity are the strongest factors in accounting for the "success" of predictions made with both simple statistical models and trajectory models. Multiyear average behavior, illustrated by the ACID calculations, should be even more reliable for long-term average SRR estimation if the linear format is valid.

Figure 6.6 illustrates the sensitivity of deposition calculations for (linear) $SO_2$ oxidation rate as a function of travel time along an air mass trajectory. A range between $SO_4^{2-}$ production at a rate of 0.5% and 2% $SO_2$/h with a 1%/h loss rate gives an estimated uncertainty in the deposition flux of ±20% for a factor of ±2 uncertainty in oxidation rate coefficient ($k$). A range in variation of wet deposition over the same range of

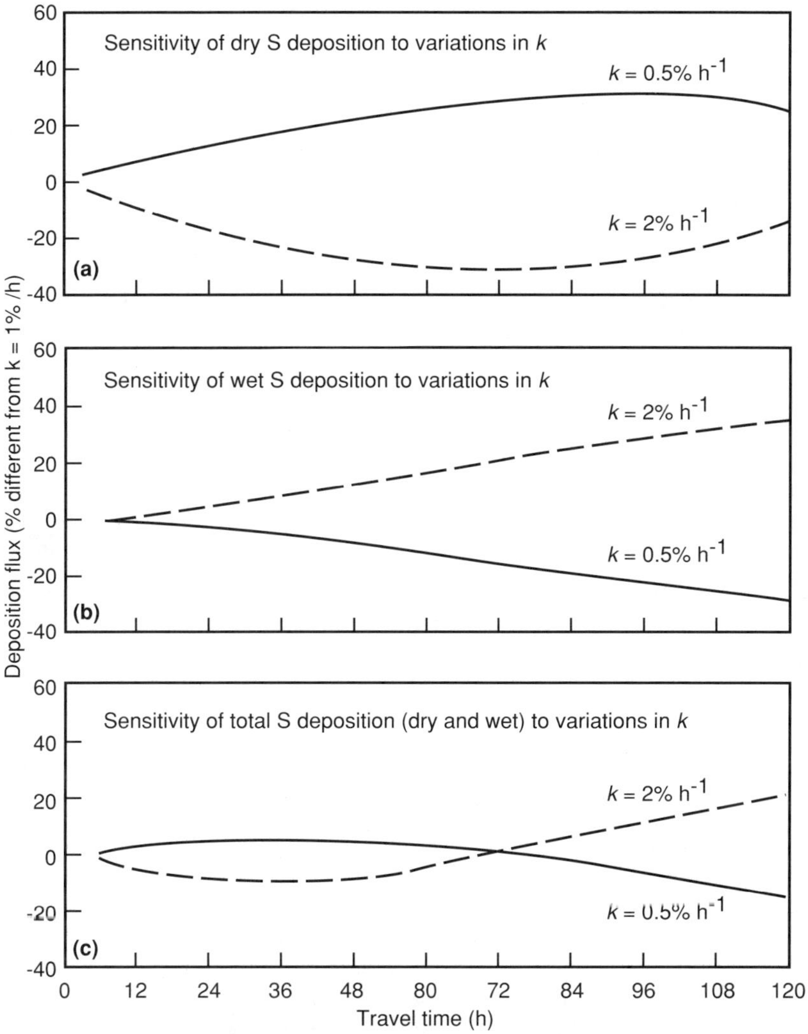


FIGURE 6.6 Sensitivity of S deposition to variation in $k$. The plots shows the difference between model estimate for specified $k$ and $k = 1\%/\mathrm{h}$. (From Venkatram and Pleim, 1985a; reprinted with permission from Pergamon Press.)

uncertainty yields a similar uncertainty in wet deposition flux, but its occurrence is at a travel time downwind. This uncertainty level is reversed in behavior for the rate coefficient compared with dry deposition. Because of this reversal, the uncertainties in deposition rate tend to cancel one another, giving a false sense of confidence in the results for linear chemistry.

*Effect of Nonlinearity in Chemical Transformation Rates* Considerable controversy has developed about the significance of theoretical nonlinearities in model estimates of SRRs. The S and N chemical transformation processes are potentially nonlinear as a result of their link with oxidation chemistry and the HC-$NO_x$ system (Appendix A). As a consequence of intermittent scavenging, oxidation chemistry hydrometeor acidity, and condensed-phase differences, the aqueous system is also thought to be nonlinear. Although dry deposition is theoretically a quasi-linear transfer process at the surface, effective nonlinear behavior can be induced by changes in absorption characteristics of vegetation, as well as the influence of varying surface characteristics.

The potential effects of nonlinearity in rate processes have been explored using the theoretical models. Fair-weather conditions were examined for the July 1978 regional episode characterized by the SURE data. Theoretical nonlinearities in dry chemistry were compared with a linear system using the ADOM model (Hidy *et al.*, 1985). Table 6.4 lists a sample of results for $SO_x$ and $NO_x$ in an 8-day event simulation that summarized calculated differences. The calculations were made using a diurnally varying $SO_2$ oxidation rate, ranging from 0.5% $SO_2$/h to a maximum of 3% $SO_2$/h, compared with a complex, coupled chemistry linking $SO_x$, $NO_x$, and HC emissions with gas-phase chemistry over the region. The results indicate that the nonlinearities in chemical reactions tend to increase ambient $SO_4^{2-}$ concentrations over those expected in a linear mode but decrease the $SO_2$ concentrations. Nonlinear S chemistry does not appreciably influence dry deposition, however, for these calculations. In the case of N chemistry, the nonlinear model tends to increase the presence of $HNO_3$ over $NO_2$ and tends to increase slightly $NO_3^-$ dry deposition because of the difference between deposition rates for $NO_2$ and $HNO_3$.

As important as these distinctions are in detail, perhaps more important are the magnitudes in the differences and the shifts in areal distributions to be expected. The calculations suggest that the differences to be expected between a linear rate expression and the nonlinear format are small for the HC-$NO_x$ emission regime in ENA. The shifts in areal distribution for fair-weather conditions, at least, were also small.

ADOM was also used to simulate the influence of nonlinear chemistry in dry air if a change of 50% in $SO_2$ emissions, or a twofold increase in HC with $NO_x$ emissions constant, were to occur at the beginning of the episode of 18–22 July 1978. This assumes that the $SO_2$ emissions reduction translates directly into ambient $SO_x$ available for reactions. Table 6.4 compares the prediction for diurnally varying $SO_x$ chemistry. The calculations show that ambient $SO_2$ concentrations are generally lower in the nonlinear simulation than in the linear case, which is offset by higher

TABLE 6.4 **Comparison of Linear and Nonlinear Model Results for S and N Species Averages for 15–24 July 1978 Baseline Emissions**[a]

| Station | $SO_2$ (ppb) | | $SO_4^{2-}$ ($\mu g/m^3$) | | S deposition ($\mu eq/m^2/h$) | | $NO_x$ (ppb) | | $HNO_3$ (ppb) | | Deposition ($\mu eq/m^2/h$) | |
|---|---|---|---|---|---|---|---|---|---|---|---|---|
| | Linear | Nonlinear | Linear | Nonlinear | Linear | Nonlinear | Linear | Nonlinear | Linear | Nonlinear | Linear | Nonlinear |
| Sensitive receptors | | | | | | | | | | | | |
| Algonquin Provincial Park, Ontario | 4.5 | 3.4 | 15.5 | 19.9 | 15.3 | 15.6 | 1.00 | 0.58 | 1.85 | 2.15 | 8.8 | 9.2 |
| Parc de Laurentides, Quebec | 1.9 | 1.5 | 7.9 | 9.3 | 6.3 | 6.2 | 1.71 | 1.35 | 1.54 | 1.89 | 4.4 | 4.6 |
| North central Wisconsin | 2.1 | 2.1 | 5.4 | 6.1 | 6.9 | 7.0 | 0.37 | 0.36 | 0.48 | 0.54 | 4.0 | 4.1 |
| Adirondacks, New York | 5.5 | 4.1 | 17.5 | 23.0 | 21.9 | 22.3 | 0.91 | 0.60 | 1.28 | 1.42 | 16.0 | 16.4 |
| Hubbard Brook, New Hampshire | 6.8 | 5.3 | 20.1 | 26.6 | 20.5 | 21.0 | 1.38 | 0.94 | 2.21 | 2.57 | 9.4 | 9.5 |
| Great Smoky Mountains, North Carolina | 3.2 | 2.8 | 6.7 | 9.1 | 9.9 | 10.4 | 1.25 | 0.90 | 1.25 | 1.48 | 13.0 | 14.8 |
| West central Pennsylvania | 15.7 | 14.4 | 20.4 | 30.7 | 38.4 | 40.3 | 4.11 | 2.53 | 3.55 | 4.75 | 22.8 | 27.1 |
| SURE Class I stations | | | | | | | | | | | | |
| Montague, Massachusetts | 8.0 | 6.5 | 19.6 | 26.7 | 24.4 | 25.0 | 2.64 | 1.63 | 3.46 | 4.27 | 16.6 | 18.1 |
| Scranton, Pennsylvania | 10.1 | 8.2 | 21.1 | 30.5 | 36.4 | 38.3 | 1.98 | 1.11 | 2.09 | 2.46 | 29.9 | 32.7 |
| Indian River, Delaware | 8.4 | 7.0 | 18.4 | 26.1 | 22.2 | 23.4 | 2.00 | 1.15 | 4.92 | 5.89 | 15.5 | 16.8 |
| Duncan Falls, Ohio | 17.8 | 16.9 | 19.9 | 30.3 | 40.2 | 42.0 | 3.75 | 2.21 | 3.63 | 4.81 | 20.0 | 23.6 |
| Rockport, Indiana | 18.7 | 17.9 | 18.9 | 29.4 | 37.4 | 39.2 | 3.65 | 2.38 | 3.77 | 5.08 | 14.0 | 16.8 |
| Giles County, Tennessee | 5.1 | 5.0 | 7.2 | 9.7 | 10.9 | 11.4 | 1.05 | 0.74 | 1.63 | 1.91 | 5.5 | 5.9 |
| Fort Wayne, Indiana | 8.9 | 7.7 | 18.9 | 26.3 | 24.1 | 25.2 | 1.90 | 1.15 | 3.13 | 3.82 | 12.6 | 14.1 |
| Research Triangle Park, North Carolina | 7.4 | 6.5 | 14.6 | 20.8 | 19.2 | 20.3 | 2.48 | 1.70 | 3.43 | 4.31 | 11.5 | 12.8 |
| Lewisburg, West Virginia | 7.3 | 6.5 | 12.5 | 18.0 | 9.4 | 18.5 | 1.64 | 1.00 | 2.38 | 2.88 | 9.4 | 10.2 |
| Modeling Region Average | 6.9 | 5.5 | 12.9 | 17.5 | 16.1 | 16.4 | 1.67 | 1.13 | 2.51 | 3.06 | 9.0 | 10.1 |

[a] From Hidy *et al.*, 1985; reproduced with permission of the Electric Power Research Institute.

ambient $SO_4^{2-}$ concentrations in the nonlinear simulation. The differences in ambient concentrations are not large for ambient concentrations between the two cases. For dry S deposition, the differences between the two simulations are very small. These results suggest that nonlinearity in chemical transformations is not a significant factor for S behavior in fair-weather conditions where clouds and wet deposition are not involved.

The ADOM was also used to calculate nonlinear effects in $NO_x$ chemistry for the July 1978 episode (Table 6.4). In this case $SO_2$ emissions were again changed 50%, all else being kept the same. In the case of $NO_x$, the difference between linear and nonlinear chemistry is small at the sites selected; this is also the case for calculated N deposition under dry weather conditions with only changes in $SO_2$ emissions.

Nonlinearities brought about by the intermittency of cloud and precipitation systems have been explored in only a limited way. The ADOM, RADM, and STEM models incorporate these processes in some detail using parameterized one-dimensional treatment of cloud processing. The nonlinearity investigated to date has focused on the aqueous–gas-phase chemistry coupling and not on the intermittency of mixing, condensation, and evaporation. $SO_4^{2-}$ production is believed to be most strongly coupled with aqueous absorption of $H_2O_2$ and $O_3$. The $O_3$ oxidation rate decreases with decreasing pH, while the $H_2O_2$ reaction does not (Appendix A). At cloud droplet concentrations and pH levels expected over ENA, $H_2O_2$ oxidation is believed to be the dominant S chemistry, but some conditions may also lead to $O_3$ oxidation and some contribution by metal-catalyzed processes.

Seigneur *et al.* (1986) have examined the model nonlinearities expected from cloud processes, considering clouds forming and precipitating under conditions characteristic of different locations in ENA. Their calculations are reported in terms of a seasonal linearity index relating ambient $SO_2$ concentration and wet $SO_4^{2-}$ deposition. The calculations indicate that the $SO_2$–wet deposition relationship should be nonlinear in winter months (limited by oxidant concentration). However, the relationship becomes nearly linear during other seasons when oxidants are not limiting but $SO_2$ concentration is. This relation was also found for the east-north-central and Atlantic coastal areas. In the Northeast, linearity was estimated to be stronger in summer and weak in spring and fall. These researchers ascribe the nonlinearity to either the $H_2O_2 + SO_2$ or the $SO_2 + O_3$ reaction. In all seasons but winter, $H_2O_2$ and $O_3$ levels are high, creating oxidant-enriched conditions. In the Northeast the $O_3$ reaction appears to be important, which accounts for the relatively low $SO_2/SO_4^{2-}$ ratio there. In winter the cloud pH is higher because $H_2SO_4$, $HNO_3$, and $H_2O_2$ concentrations are lower; the higher pH and low $H_2O_2$ levels favor the $SO_2 + O_3$ reaction that occurs at a higher rate for high pH levels. Seigneur notes, however, that the $O_3$ reaction may be overemphasized in his calculations because dry air entrainment was not included. This exclusion underestimates the amount of reactants and products present from clear air reactions. Seigneur's calculations suggest, for example, a 32% reduction in total $SO_4^{2-}$ deposition for a 50% reduction in ambient $SO_2$ concentration. Similar changes are found in the areas to the west, but a larger disparity of 20% total $SO_4^{2-}$ reduction and only an 11% wet $SO_4^{2-}$ deposition reduction are estimated in the Northeast.

Both ADOM and RADM have been used to simulate an event in the OSCAR field experiment for 11–15 April 1981. The ADOM calculation incorporates $H_2O_2$ aqueous chemistry with stratus cloud parameterization. The OSCAR conditions included rains with stratus activity in a weak frontal system. The model qualitatively estimated $SO_4^{2-}$ concentrations reasonably well in cells where precipitation was heavy and widespread. The quality of the results compared with observations was best when calculated precipitation matched measurement and worst when precipitation was over- or underestimated (Venkatram *et al.*, 1986). The nonlinearity in chemistry is reflected mainly in the distinction between (in-cloud) oxidant-rich and oxidant-lean conditions in relation to $SO_2$ concentrations. A simulation of estimated change in S deposition with 50% reduction in $SO_2$ emissions showed a linear response in wet deposition only at distances far from sources (>300 km), where $H_2O_2$ exceeds $SO_2$ concentrations. Near sources, where $SO_2$ concentrations are high and the air is $H_2O_2$ lean, wet deposition changes with emissions are much less than a linear projection.

Dennis *et al.* (1990) have reported calculations using a simplified engineering version of RADM. One series involved comparison between a base case and a 50% uniform rollback in $SO_2$ emissions to test whether the rollback would yield a nonproportional (nonlinear) seasonal deposition. The results indicated proportionality held for dry $SO_x$ deposition, consistent with the short-term ADOM calculation discussed above. However, the fractional change in wet deposition was generally less than proportional to emission changes throughout the model region (e.g., Figure 6.9a). The lack of proportional response was worse near large source areas.

Case studies using RADM to simulate the response 50% $SO_x$ emission reduction for three summer and fall events added insight to nonproportional concentration reductions at individual receptors. A linear response to the emission reduction was found in all three cases. In contrast, a nonlinearity was noticeable in a winter case. This was believed to be associated with winter suppression of oxidant with, in turn, decreasing airborne sulfate production.

Dennis *et al.* (1990) have distinguished two inherent types of nonlinearity: (1) "common" nonproportionality and (2) nonproportionality from nonlinear atmospheric processes.

Common nonproportionality can be defined in terms of expectations of a linear relationship between source emissions and receptor conditions. Linear behavior results when source emission reductions yield a corresponding reduction in receptor exposure. Furthermore, a uniform decrease in all source emissions results in a receptor exposure reduction proportional to the sum of the source reductions. In any subregion, the proportion of the deposition field from sources in the reduced-emission subregion would undergo the same percentage reduction as the emissions in that subregion.

Nonproportionality between emission changes and receptor exposure from deposition nonlinearities in atmospheric processes can arise not only from chemical reactions but also from cloud or precipitation scavenging or dry deposition processes. In the models, wet deposition results from a combination of particulate sulfate scavenging, $SO_2$ oxidation in cloud droplets, and in-cloud sulfate production from evaporated droplets. Nonlinear response in either one of these processes involving changes in oxidant concentration can affect wet deposition rates. The Dennis *et al.* analysis of RADM model results suggests that sulfate in wet deposition is linked with roughly a third for each of these contributions. Approximately two-thirds of the sulfate present in precipitation thus could be subject to oxidant limitation effects.

The various analyses have elucidated the nature of nonproportional relationships between emission changes and deposition patterns. However, the results do not show clearly the potential level of uncertainty in model calculations resulting from nonproportional response between emission changes and receptor exposure. Empirically, the calculations suggest that nonlinear behavior is not a severe limitation for estimating dry deposition. However, such behavior could lead to significant errors in estimates of oxidant concentrations that would affect cloud–wet deposition processes. Thus process nonlinearities could produce nonproportional relationships between sources and receptor conditions which will affect emission control strategies.

Concern about adverse environmental effects of deposition of acid $SO_x$ and $NO_y$ species has resulted in a variety of strategies for reducing deposition rates, especially in ecologically sensitive areas. These strategies revert back to emission reductions for $SO_x$ and $NO_x$. Evaluation of the effectiveness of regional emission control strategies requires the use of air quality models. The projected high costs of emission controls and ensuing potential for economic disruption place severe requirements on the reliability of predictions from the models. Thus, in recent years users of model results have required greatly improved standards for measuring the performance of air quality models. Attempts to formalize such criteria have come from workshops, such as the ones reported by Fox (1981) or Schiermeier and Misra (1983). Perhaps the most elaborate study of methods to define performance criteria derives from the recent NAPAP work (Dennis *et al.*, 1990).

Performance can be tested in different ways. The best way relies on comparison with observations of atmospheric conditions, given knowledge of emissions. Additional insight about models can be gained by comparing results from different models that rely on similar or different means of describing atmospheric processes. Performance can be measured in terms of (1) comparison with calculations and observations, where their ratio should average unity, (2) the mean range of difference between calculation and observation and its standard deviation, (3) consideration of the bias of observations and calculation from mean values, and (4) comparison of estimates of model submodule output with field or laboratory experiments. Even today, there are only limited collections of field or laboratory data that are suitable for comprehensive model evaluation. Thus, many of the tests of models are considered only weak measures of reliability.

In addition to internalized uncertainties, a model's predictive capabilities are limited by the reliability of the meteorological and emissions data used to drive the model. It is difficult to objectively assign differences between model predictions and observations to inaccuracies in the field measurements and errors in model formulation. One approach taken to resolve this issue has been to evaluate comparatively several models using standardized input and validation data sets. If the models being compared were equally accurate, they would yield the same results. Even this approach is hampered by the differences in model treatment of input data as well as differences in numerical integration methods for nominally the same mathematical relationships. Comparisons of this kind have been attempted by several groups including Schiermeier and Misra (1983), Ruff *et al.* (1985), Stewart *et al.* (1983), Shannon *et al.* (1982), and Clark *et al.* (1989).

## Measures of Performance

Debates continued through the 1980s about the effectiveness of evaluations. The challenges include (1) developing diagnostic procedures, (2) designing criteria for assessing the scientific significance of levels of agreement in comparisons, and (3) resolving issues of relying on testing over a range of emission conditions. Because none of these challenges are resolved at this time, investigators have retreated to judging the reliability of regional models by a so-called weight-of-evidence approach. This method, as described by Dennis *et al.*, (1990), involves a subjective, iterative process of multiple testing, model and observational intercomparisons, and expert judgment. Rigorous, objective intercomparisons in themselves, as we shall see, have not given sufficient insight into model reliability to assure that they represent quantitatively atmospheric behavior under a range of known conditions.

*Comparison of Model Results* Intercomparison of model results, like the prediction of geographic contributions by source, employing different process parameterizations give information on within-model uncertainty. A typical range for deposition in the Adirondacks was shown in Table 6.2. A further comparison between an Eulerian model (RADM), engineering approximate model (EM), two Langrangian models (RELMAP and ASTRAP), and a statistical calculation (MODEL T) are shown in Figures 6.7 and 6.8. Figure 6.7 indicates the differences in calculation for spatially averaged wet, dry, and total S deposition over ENA. The models have quite different results—the largest for total deposition, 10–16 kg/ha, being associated with ASTRAP and RADM versus the others. Also, there are substantial differences between RELMAP and the other models for wet deposition. Figure 6.8 shows the estimates of percentage contribution of upper Ohio River Valley sources to annual S deposition for four major receptor areas. The models have a range of 3–5% difference between their

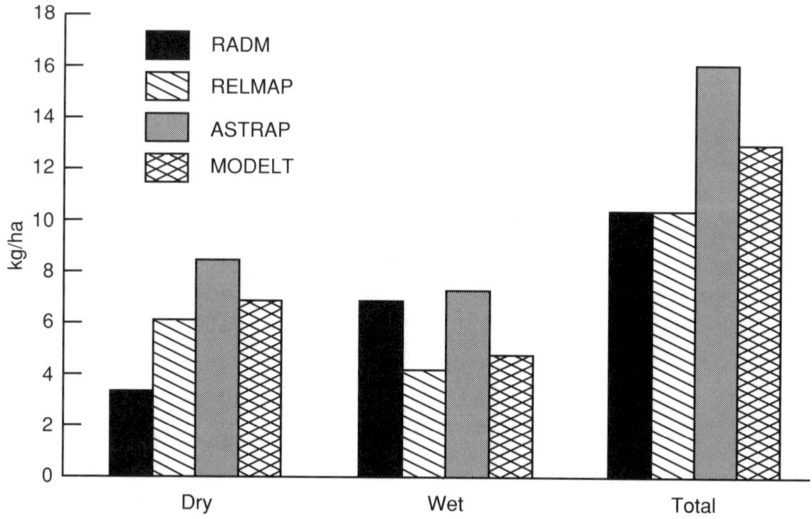


FIGURE 6.7 Comparison of the spatially averaged annual (1985) wet, dry, and total deposition of sulfur for the eastern United States and Canada calculated by the RADM/EM and three simpler models. (From NAPAP, 1990b.)

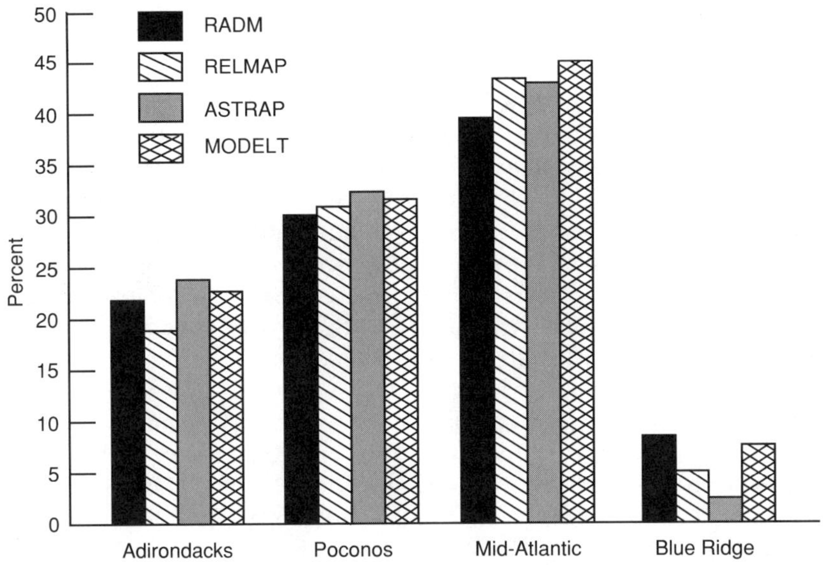


FIGURE 6.8 Comparison of the percentage contribution of upper Ohio River Valley sources to annual total sulfur deposition at four receptor areas calculated by the RADM/DM and three simpler models. (From NAPAP, 1990b).

estimates. These ranges of fractional contribution are generally improved over earlier comparisons. Without detailed diagnostic tests, the reasons for these differences cannot be determined.

***Comparisons with Observations*** More revealing as an empirical test is the comparison of model outputs with observations. This is often called an *operational evaluation*. Direct comparison between model calculations and observations taken at a fixed site is by far the most prevalent test of model performance. However, it is believed to be least rigorous given the inherent incompatibility between regional model estimates over 80 km or larger grids and "point" sampling. Workers also have suggested that models may avoid rejection by this test even though they are flawed. That is, the model may produce a simulation similar to observations for the wrong reasons, thus leading to incorrect guidance for emission control.

A number of model comparisons have been made with observations. In fact, a conventional description of a model includes such a qualitative comparison, either for long-term average data or for short-term episodes. The comparisons are normally reported in terms of correlation or scatter diagrams displaying the average bias of the results, along with data dispersion about mean values.

In general, most models calculating long-term averages perform reasonably well compared with observations (e.g., Eliassen 1980, 1984; also Figure 6.1), because the errors in model calculations from case to case have a tendency to cancel in the long-term sequences. For example, predictions from the Eliassen and Saltbones model, which has been applied extensively in Europe, were found to agree with observations when compared on an annual basis. However, the model did not perform well for shorter-time-scale averages (Eliassen and Saltbones, 1983).

Most of the ENA models performed similarly to the European models. All of the models are able to reproduce the correct order of magnitude of the coarse time- and space-scale features of measured wet $SO_4^{2-}$ deposition patterns, although variations exist among the models in their quantitative estimates.

Formal statistical criteria can be applied to the performance of models. An early comparison of three 24-hour average data sets of $SO_4^{2-}$ was evaluated for the SURADS, as an example (Mueller and Hidy, 1983). The observational data sets were constructed for comparisons. Set A included selected days in January 1977, April 1978, and summer 1978. Set B was the same but excluded two days in summer 1978. Set C excluded two other days in summer 1978. To test the (statistical) hypothesis that the means of the observed and calculated $SO_4^{2-}$ concentrations and the logarithmic concentration are the same, model runs were conducted for the three sets. Comparisons then were made with observations to determine if the variance in the model calculations equaled the observational variance. These results indicated acceptance of the hypothesis for sets B and C. Thus, the SURADS model represented the SURE observations for six independent, selected days, but not for the complete set (A) examined. Formally, one would conclude that SURADS was deficient in its performance for such air quality modeling even though the model simulated qualitatively spatial and temporal patterns of the data.

Regional-scale models have been compared with each other as well as with observational data. Several major studies of this kind have been reported. These are the Memorandum of Intent on Trans-Boundary Air Pollution-MOI (Schiermeier and Misra, 1983), the DOE-EPA/MAP3S evaluation (Shannon *et al.*, 1982), National Park Service (Stewart *et al.*, 1983), the EPRI evaluation (Ruff *et al.*, 1985), the International Sulfur Deposition Model Evaluation—ISDME (Clark *et al.*, 1989), and the NAPAP intercomparison (Dennis *et al.*, 1990). Each of these added insight into the performance of models, but the results were inconsistent and difficult to interpret with regard to the actual reliability of the models for strategic applications.

The MOI exercise was one of the first to try to compare models on a common, uniform basis.

The ideal relationship between an observation $C_{ok}$ and a prediction $C_{pk}$ at location $k$ of wet deposition is given by

$$\hat{C}_{ok} = \hat{C}_{pk}(x_{1k}, p) + \varepsilon_k \tag{6.5}$$

where the hats (ˆ) refer to transforming $C_{ok}$ to a normally distributed variable. In Eq. (6.5), $x_1$ refers to model inputs and $p$ denotes model parameters. The residuals associated with an ideal model should correspond to a "white noise" sequence $\varepsilon$, which means that $\varepsilon_k$ has a zero mean and constant variance. Furthermore, values of $\varepsilon_k$ are uncorrelated with each other or with any known input variable.

The objective of model testing is to determine whether the model calculations satisfy Eq. (6.5). A desirable model should meet the following criteria:

1. The bias, $\bar{\varepsilon}_k = 0$.
2. The standard deviation, $\sigma_\varepsilon$, is small compared with the variance of $C_p$.
3. $\varepsilon_k = f(x_1)$ or the cross-correlations $\overline{\varepsilon_k C_{pk}} = 0$ and $\overline{\varepsilon_1 x_{1k}} = 0$.

The first step in model evaluation is to plot $\varepsilon_k$ against $C_{pk}$. Desirable model behavior is indicated by a residual plot that looks like a band distributed about $\varepsilon = 0$. Plots that indicate the dependence of $\varepsilon$ on $C_p$ suggest that the model can be improved. Histograms of $\varepsilon_k$ also provide useful checks on a model's acceptability. Standard tests can determine the normality or log normality of the $\varepsilon$ distribution.

In summary, the following statistics provide basic information on the behavior of the model:

1. $\bar{\varepsilon} = \sum_k \varepsilon_k / n$
2. $\sigma_\varepsilon = [\sum_k (\varepsilon_k^2 - \bar{\varepsilon}^2/(n - 1)]^{1/2}$

It is important to note that model simulations can be interpreted only in terms of probabilities and that $\sigma_\varepsilon$ provides the key link between model simulations and observations. It is not meaningful to use a model that has not been tested against observations.

In the MOI study, eight models were compared for their performance in simulating monthly averaged ambient $SO_4^{2-}$ concentrations from the SURE data and long-term averaged $SO_4^{2-}$ deposition from AEI-CANSAP and EPRI networks. Table 6.5 illustrates the results for $C_p$ and $\varepsilon$. The columns labeled linear refer to computations in which $\varepsilon_k = (C_{ok} - C_{pk})$. The results of the logarithmic computations are stated as exponentials so that they become the familiar geometric means and standard deviations. The ideal values of the linear $\varepsilon$ and $\sigma\varepsilon$ are zero. The ideal values of the geometric mean and standard deviation of $C_o/C_p$ (logarithmic statistics) are unity. For an "adequate" model, $\overline{\varepsilon C_p}$ should be close to zero. The

TABLE 6.5 **Modeled Wet $SO_4^{2-}$ Deposition (kg/ha-yr) for Annual 1978**[a]

| | Linear | | | | | Logarithmic | | | | | |
|---|---|---|---|---|---|---|---|---|---|---|---|
| Model | $\bar{\varepsilon}$ | $\overline{C}_p$ | $\sigma_\varepsilon$ | $\sigma_{Cp}$ | $\overline{\varepsilon C}_p$ | $\overline{\varepsilon^g}$ | $\overline{C^g_p}$ | $\sigma_{\varepsilon g}$ | $\sigma_{C^g_p}$ | $\overline{\varepsilon C^g_p}$ | $n$ |
| AES[b] | 1.2 | 9.7 | 4.0 | 7.0 | −0.21 | 1.3 | 7.5 | 1.6 | 2.1 | 0.25 | 8 |
| MOE[b] | 0.8 | 10.1 | 2.8 | 6.0 | 0.03 | 1.1 | 8.8 | 8.8 | 1.4 | 0.20 | 8 |
| MEP | 4.4 | 6.5 | 2.5 | 4.2 | 0.79 | 1.8 | 5.4 | 1.6 | 1.9 | 0.27 | 8 |
| CAPITA | 4.5 | 6.4 | 3.5 | 3.6 | 0.70 | 1.8 | 5.4 | 1.5 | 1.8 | 0.67 | 8 |
| ASTRAP | 2.7 | 8.1 | 4.3 | 6.8 | 0.12 | 1.8 | 5.6 | 2.0 | 2.6 | 0.29 | 8 |

[a] From Schiermeier and Misra, 1983.
[b] A uniform background of 2 kg $SO_4^{2-}$/ha was added to predictions.

mean $\overline{C}_p$ and the standard deviation $\sigma_{Cp}$ provide information on the range of predictions for which the model has been tested. Note that the mean of observations can be obtained as follows:

$$\overline{C_o} = \overline{C_p} + \bar{\varepsilon} \text{ (linear)} \quad (6.6a)$$

$$\overline{C_o^g} = \overline{C_p^g}\,\overline{\varepsilon^g} \text{ (logarithmic)} \quad (6.6b)$$

A positive $\bar{\varepsilon}$(or $\bar{\varepsilon}_g > 1$) implies that the model calculated values are less (on average) than the observed values.

Table 6.5 gives annual (1978) model evaluation results for wet S deposition. For the calculations of annually averaged S deposition, $\sigma_\varepsilon$ is usually less than $\sigma_{Cp}$. Also, $\overline{\varepsilon C_p}$ is small for the linear statistics. For the monthly averages, $\bar{\varepsilon}$ and $\bar{\varepsilon}^g$ were close to their ideal values for most of the models. However, the variance of residuals $\sigma_\varepsilon$ in most cases was comparable to that of the calculations. Also, the correlations between $\varepsilon$ and $C_p$ are not negligible.

The models tested in the MOI exercise performed better for wet $SO_4^{2-}$ deposition than for airborne $SO_4^{2-}$ concentration calculation. This is surprising because wet S deposition is highly intermittent in nature, whereas airborne sulfate is ever present on a daily basis. Evidently the intermittent character of precipitation is lost in annual averaging. Because $SO_2$ concentrations are nearly always affected by local sources, these were not considered for the MOI evaluation exercise.

The $\sigma_\varepsilon$ and $\sigma_\varepsilon^g$ are measures of the combined uncertainties associated with the model predictions and the validation data. If the errors in the data form a random sequence or a white noise, then knowing the variance of these errors, one can compute the uncertainties associated with model predictions from $\sigma_\varepsilon$ and $\sigma_\varepsilon^g$.

The DOE and the EPA funded another comparison study under the MAP3S program, in which monthly predictions from three regional-scale models (BNL-AIRSOX, PNL-RAPT, and ASTRAP) were compared with observations from the SURE air quality and MAP3S precipitation chemistry networks (Shannon *et al.*, 1982). As in the previous study by the U.S./Canadian work group (MOI), the models were shown to perform similarly, agreeing modestly with observed $SO_2$ fields and reasonably with ambient $SO_4^{2-}$ concentrations. The AIRSOX model results agreed best with the wet deposition observations, although the limited number of observation sites made an objective evaluation difficult.

The U.S. National Park Service also sponsored a regional-scale model evaluation study. Three long-term regional-scale models (RTM-LT, ENAMAP-1, and the MIT model) have been evaluated using the SURE and MAP3S data bases (Stewart *et al.*, 1983b). ENAMAP-1 is an early version of ENAMAP-2S, which has already been described. The evaluation strategy was to use identical input data bases and preprocessing methods whenever possible. In addition, the specification on input parameters was governed by intermodel consistency to the extent possible. Both monthly and annually averaged concentrations and total wet deposition from the three models were compared with observations and each other. Predictions of the dry deposition of $SO_4^{2-}$ and $SO_2$ from the different models were also compared, although lack of measurements made an evaluation using observational data impossible.

In general, predictions from all three models were found to correlate well with observations when compared on an annual basis. All three models were found to reproduce the spatial distribution of annual average $SO_2$ concentrations in a qualitatively similar manner, with ENAMAP-1 calculating (on the average) higher $SO_2$ concentrations by the largest amount, followed by the MIT model. Similarly, all the models were found to reproduce the observed large-scale spatial variations of the annually averaged $SO_4^{2-}$ concentrations and, on the average, to calculate low $SO_4^{2-}$ concentrations. All the models calculated high annual $SO_4^{2-}$ wet deposition; ENAMAP-1 calculations were an order of magnitude greater than RTM-LT and MIT. All three models produced qualitatively similar spatial variations that matched the observed spatial patterns.

The monthly predictions from the three models did not correlate well with observations. The models calculated similar spatial variations of the $SO_2$ concentrations, with RTM-LT calculating slightly low and the ENAMAP-1 and MIT models calculating high by large amounts. The ranking of the three models varied from month to month when calculated monthly $SO_4^{2-}$ concentrations were compared with observations.

Three episodic regional transport models were evaluated in the National Park Service study described in Stewart *et al.* (1983b). The models evaluated were RTM-II, MESOPUFF (an early version of MESOPUFF-II), and ARL-ATAD (modified to incorporate linear chemistry). The models were used to simulate an 8-day $SO_4^{2-}$ episode that occurred in the eastern third of the United States during July 1978 based on the SURE observations. Because this episode was primarily a dry one, the wet deposition components of the models were not tested. Twenty-four-hour average $SO_2$ and $SO_4^{2-}$ concentrations calculated by the models were compared with observations. The evaluation showed that all models qualitatively predicted the location and approximate magnitude of the $SO_4^{2-}$ concentration during the last 3 days of the episode. In addition, all models qualitatively simulated the $SO_2$ concentration distribution during the episode, although with less skill than the $SO_4^{2-}$ predictions. Quantitatively, RTM-II showed the best overall performance for both $SO_4^{2-}$ and $SO_2$ of the three tested. MESOPUFF exhibited the lowest $SO_2$ and $SO_4^{2-}$ biases. ATAD tended to calculate higher concentrations to a greater extent than the other two models, particularly for $SO_2$. The performance of all models, particularly MESOPUFF, deteriorated for calculations of the highest concentrations. All models tested performed with less skill under lighter winds than under conditions of well-defined transport with modest wind speeds and uniform direction.

The results of a model evaluation study sponsored by EPRI have been reported (Ruff *et al.*, 1985). Both short-term (6-hour) and long-term (monthly) averages from three regional air quality models (RTM-II, ENAMAP-2S, and ACID) were compared with observations from the SURE and MAP3S data bases. Given the same input data and the absence of manipulation of the empirical model parameters, all three models performed poorly in their calculations of monthly averages, although the spatial correlation coefficients were considerably higher than those for the short-term averages.

Six-hour average $SO_2$ and $SO_4^{2-}$ ambient concentrations predicted by the models were compared with observations from the SURE data base. Poor correlations between predictions and observations for all three models were obtained. ACID and RTM-II calculated substantially lower $SO_2$ concentrations than those observed; ENAMAP-2S calculated high. An evaluation of the site-specific performance indicated that performance was poor at all sites. Differences between observation and calculation can exceed a factor of 10, and the measures of statistical representativeness show poor fidelity.

As a part of the NAPAP evaluation, the two comprehensive Eulerian models, ADOM and RADM, have been applied to the simulation of precipitation episodes that occurred in April 1981 (Venkatram *et al.* 1986; National Center for Atmospheric Research, 1986). These events were captured by a precipitation chemistry network set up for the OSCAR experiment, an intensive field investigation of the wet removal of pollutants by cyclonic storms (Easter *et al.*, 1984). As noted earlier, preliminary results from these simulations suggest that the models qualitatively reproduce observed $SO_4^{2-}$ and $NO_3^-$ wet deposition amounts on an event basis. The agreement was improved when the grid-averaged precipitation amounts used by the models match those at measuring stations (Venkatram *et al.*, 1986). However, both models performed poorly when predictions were compared with hourly observations.

A different scheme for comparing models was introduced by Clark *et al.* (1989) in the ISDME. They adopted measures for intercomparison based on ability to simulate 1980 spatial distributions of S wet deposition for annual conditions and the four seasons. Skill scores were determined from criteria involving (1) percentage of grid cells where predictions differed significantly from measurements (deemed most informative), (2) position of centroid of pattern, (3) orientation of pattern major axis, (4) location of deposition maximum, (5) maximum magnitude, and (6) seasonal distribution of annual deposition. Illustrative skill scores for some of the models compared are shown in Table 6.6. Performance according to the ISDME measures gives a somewhat inconsistent pattern. There are three distinct groupings of models. Those that exemplify relatively good performance included MOE and TVA models. Relatively poor performance was exemplified by RELMAP, RIVAD, RTM, and STATMOD. An in-between grouping included AES, ASTRAP, and (UM)ACID. The study indicated that the statistical models could perform as well as the more sophisticated Lagrangian models; the "spread" in skill was substantial with no overall "winner."

Dennis *et al.* (1990) report most recent model comparisons with models reflecting the current state of the art. Comparisons were made between RADM, ASTRAP, RELMAP, and MODEL T. This comparison was confined to annual averages, as reflected in the 1982–1985 ENA deposition networks observations. Equivalents of annual averages from RADM simulations were constructed as a composite made up of 30 cases of "representative" meteorological episodes in different seasons determined from cluster analysis of deposition data and 850-mb wind flow patterns (see, for example, Binkowski *et al.*, 1990). The results of the comparison are summarized in Table 6.7 for average annual wet deposition. These illustrative results are stated in terms of regression calculations between model predictions and observations. Here the $P$ values are the probability that the hypothesis ($H_0$) of $a$, the intercept, or $b$ (or $b - 1$), the slope of an observation vs. prediction plot, are the same value as the value stated. RADM and the linear models showed a similar degree of association with

**TABLE 6.6 Distribution of Skill Scores of Various Models for Seasonal, Overall, and Annual Spatial Patterns Reported from the ISDME[a,b]**

| | Summer | Spring | Autumn | Winter | Overall | Annual |
|---|---|---|---|---|---|---|
| | | AST | | | | |
| 9 | | MOE | | | | |
| | | TVA | | | | |
| | | | | | MOE | |
| 8 | | AES | | | TVA | MOE |
| | MOE | | MOE | | | |
| 7 | SERTAD | | UMACID | MOE | | |
| | TVA | | | | ASTRAP | |
| | | | RELMAP | | | |
| | | | RTM | TVA | | |
| 6 | AES | UMACID | TVA | UMACID | | |
| | ASTRAP | | | | UMACID | |
| 5 | | RELMAP | RIVAD | RELMAP | | TVA |
| | | STATMOD | STATMOD | RIVAD | | UMACID |
| | | | | | RIVAD | |
| | | | | | RTM | |
| | RIVAD | | | | RTM | ASTRAP |
| 4 | RTM | RIVAD | | ASTRAP | RELMAP | |
| | RIVAD | RTM | ASTRAP | | | |
| | | | | | | RELMAP |
| 3 | | | | RTM | | RIVAD |
| | | | | | | RTM |
| | | | | | STATMOD | |
| 2 | | | | | | |
| | | | | | | STATMOD |
| 1 | RELMAP | | | | | |
| | | | STATMOD | | | |
| 0 | STATMOD | | | | | |

[a] 0 Depicts the worst performance and 9 depicts the best.
[b] From Dennis *et al.*, 1990.

observations. ASTRAP, a linear model, seemed to be better than RADM in performance. Both RELMAP and TVA MODEL T gave biased estimates. RADM tended to underpredict large wet deposition values, but this was relatively small compared with RELMAP or MODEL T.

Comparison of seasonal patterns for precipitation concentration suggested no strong model bias. However, summer predictions were lower than observed—compensated by overestimation in winter.

Comparisons with empirically derived dry S deposition indicated that, overall, RADM more precisely estimated dry deposition than the linear models. Of the latter, only RELMAP approached RADM in performance. Both MODEL T and ASTRAP showed the majority of predictions differ-

**TABLE 6.7 Comparison of Model Predictions of Observed Average Annual Wet S Deposition from 1982 to 1985[a]**

| Coefficients | RADM | ASTRAP | RELMAP | TVA |
|---|---|---|---|---|
| Regression coefficients for linear model with intercept | | | | |
| Intercept ($a$) | 1.10 | −0.20 | −3.11 | −1.82 |
| Standard error | 0.71 | 0.95 | 1.04 | 1.32 |
| $P$ value ($H_0: a = 0$) | 0.131 | 0.832 | 0.004 | 0.173 |
| Slope ($b$) | 0.82 | 1.01 | 1.30 | 1.33 |
| Standard error | 0.08 | 0.11 | 0.12 | 0.15 |
| $P$ value ($H_0: b = 0$) | <0.001 | <0.001 | <0.001 | <0.001 |
| Multiple correlation coefficient ($R^2$) | 0.60 | 0.57 | 0.64 | 0.53 |
| Regression coefficients for zero-intercept model | | | | |
| Slope ($b$) | 0.94 | 0.99 | —[b] | 1.13 |
| Standard error | 0.02 | 0.03 | —[b] | 0.04 |
| $P$ value ($H_0: \|b-1\|<0.05$) | 0.589 | <0.999 | —[b] | 0.068 |

[a] From Dennis *et al.*, 1990.

[b] Slope estimate and regression statistics are not presented for this model because of the evidence that the intercept was significantly different from zero.

ent from observationally based estimates by at least a factor of 2. RADM tended to produce S dry deposition predictions that were about 20% low, compared with overestimates of the linear models.

Intercomparisons with $SO_2$ and $SO_4^{2-}$ concentration in the air indicated that RADM generally provided estimates that were biased at night but not in daytime. RADM produced less bias in $SO_2$ concentrations than either MODEL T or ASTRAP. None of the models appeared to be capable of reproducing airborne sulfate concentrations.

Figure 6.9 further illustrates the comparison between model calculations of grid cell averages and observations from the EMEFS network. In the figure are scatter plots for the relationship between 33-day average model and observed surface $SO_2$ concentrations, surface $SO_4^{2-}$ concentrations, wet sulfate concentration and wet sulfate deposition for the period 25 August, 1988 through 27 September, 1988. The calculations were made using RADM II (six layer version) (Barchet, 1991). The scatter diagrams show that the model overpredicts $SO_2$ concentrations, estimates surface $SO_4^{2-}$ concentrations, but underestimates wet sulfate concentration and wet deposition particularly at large values. The same comparisons using RADM II (one uniform layer version) showed similar results, except that the model tended to underpredict surface airborne $SO_4^{2-}$ concentration. The results were consistent for all geographic areas simulated; this suggested that an important sulfate-producing mechanism was absent in the

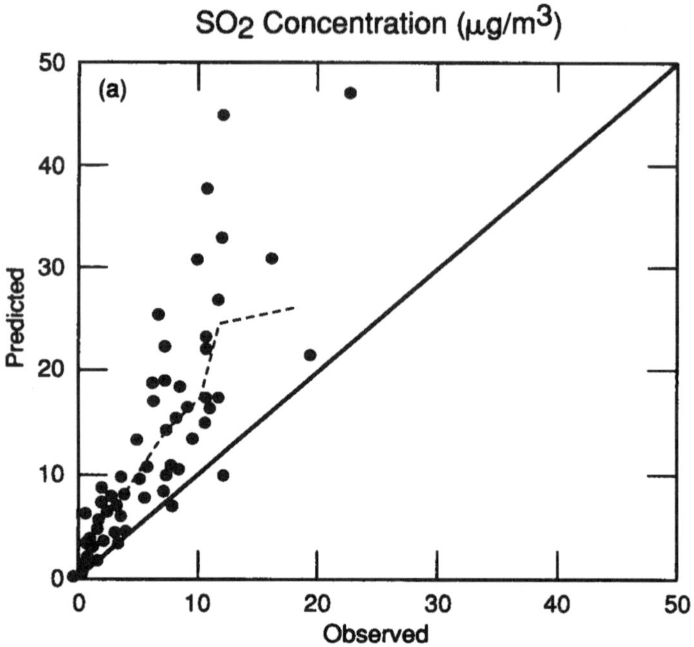


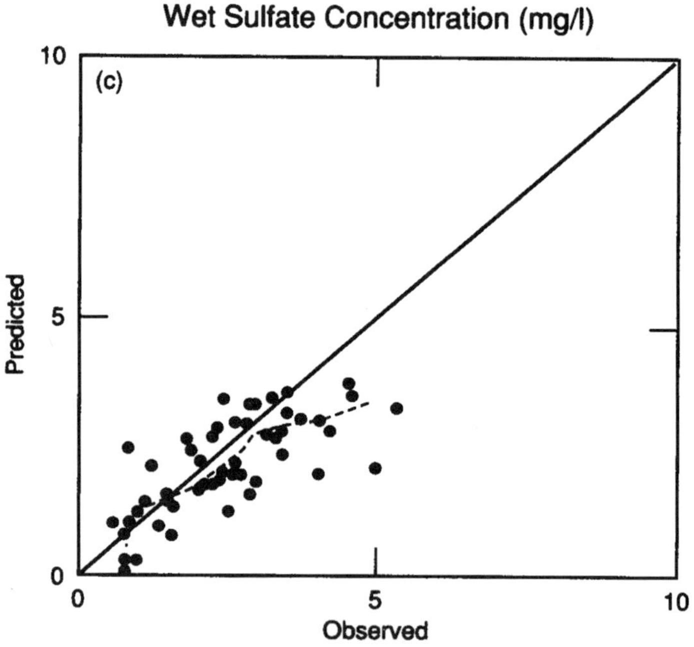


FIGURE 6.9 The relationship between the 33-day calculated and EMEFS observed averaged surface air concentrations of $SO_2$ (a), $SO_4^{2-}$ (b), precipitation weighted average wet $SO_4^{2-}$ concentration (c), and total wet $SO_4^{2-}$ deposition (d) for the eastern United States. Predictions are from the RADM II (six-layer) model. Broken lines are means of data points. (From Barchet, 1991).

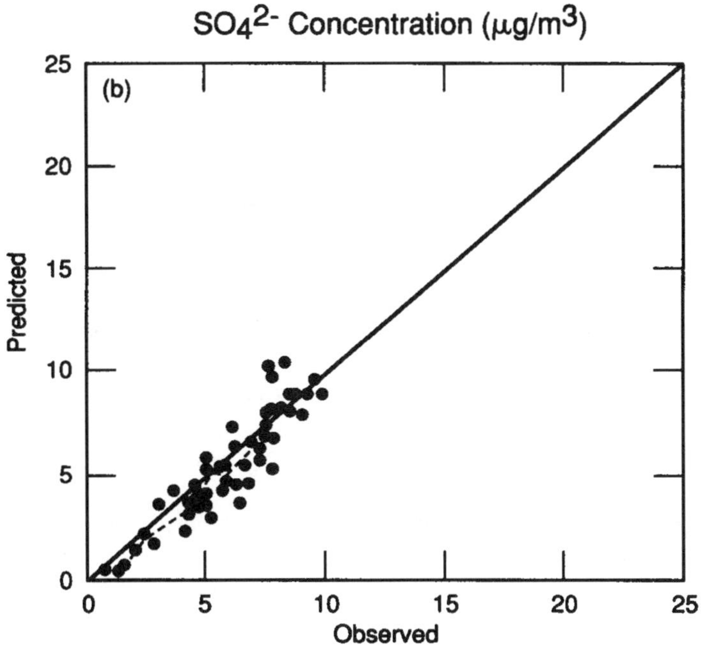


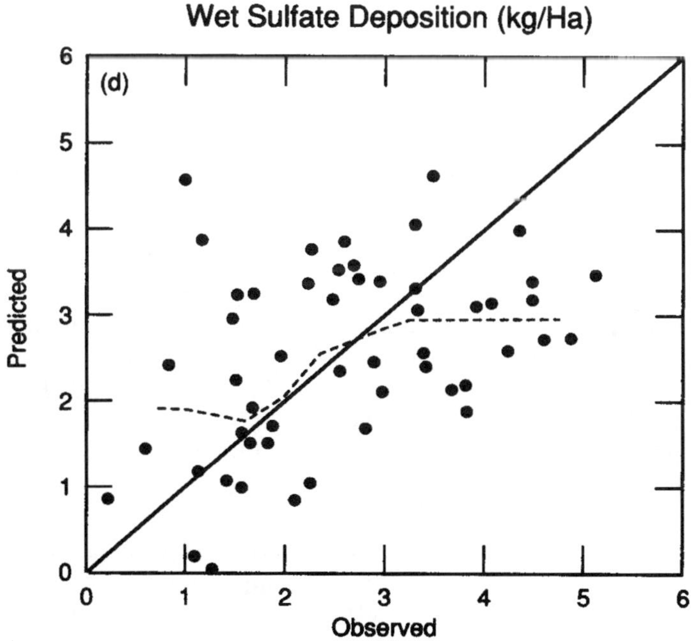


FIGURE 6.9 *(continued)*

single layer RADM version, presumably related to $SO_4^{2-}$ production in evaporating clouds.

The overprediction of $SO_2$ was accompanied by overestimation of $HNO_3$ concentrations. According to Barchet (1991), these biases appear to be related to the incommensurability between model estimates "averaged" for a model grid cell and measurements taken at a specific location. Subgrid-scale variations in air concentrations can be neither well simulated by the regional-scale models nor represented by measurements at a fixed station. The horizontal coarseness of the modeling grid causes unrealistic spatial smoothing of pollutants by the models. Vertical coarseness of the grid makes surface measurement a weak representation of the surface layer simulated by the models.

Unlike the linear models, RADM was found to be capable of reproducing wet $NO_3^-$ deposition data. These results were similar to wet $SO_4^{2-}$ deposition with essentially no bias for RADM calculation except a negative bias in the high deposition range.

Dennis *et al.* (1990) state that overall the operational evaluation of RADM shows improvement in regional modeling capability relative to the linear models. Improvement was found primarily in estimation of dry S deposition rather than wet deposition of $SO_4^{2-}$ and $NO_3^-$. Even with weaknesses in airborne $SO_4^{2-}$ predictions from RADM, the authors believe that RADM is the best model available at present for gaining insight into $SO_4^{2-}$ and $NO_3^-$ deposition for emission-change strategy development. However, there remain significant weaknesses in the RADM performance that will require continued effort for improvement. Historically, the complexity of RADM incorporating theoretically based process parameters does not as yet yield significant improvement over the linear models. However, at least the physical chemistry of $SO_x$ and $NO_y$ behavior in the presence of oxidants and the wet scavenging are explicit, to add insight into the regional behavior of these pollutants.

*Diagnostic Testing* Operational evaluation yields considerable information about the capability of models to simulate atmospheric conditions as characterized by limited observational data. However, this approach alone does not advance our understanding of the scientific foundation for the models. This understanding can be facilitated by specially designed studies that can isolate model response related to a particular process embodied in the model. For example, studies can be made to isolate dry, fair-weather conditions from precipitation; various forms of precipitation systems can also be examined. Programs employing special experiments to test model response are termed *diagnostic evaluation*. An adjunct to this approach is mechanistic testing, where a process component is isolated and evaluated—for example, (1) wind field reproduction through tracer release studies or (2) dry deposition with evaluation through boundary-layer flow characterization. Laboratory studies are also employed, as in the evaluation of gas flow chemical mechanisms and determination of aqueous-phase chemical oxidation rates. These supporting studies and experiments represent a major contribution to the development of air quality models over the last 30 years.

The meteorological components of the models have been evaluated extensively by comparison and weather observations, as well as special data characterizing the planetary boundary layer. Direct attempts to test the transport wind patterns for different weather situations have taken place. Two experiments have been conducted in the central and eastern United States: the Cross Appalachian Tracer Experiment, CAPTEX (e.g., Ferber and Heffter, 1984), and the Across North American Tracer Experiment, ANATEX (Draxler and Heffter, 1989).

In CAPTEX, seven releases of perfluorocarbon tracers were made from central Ohio and Sudbury, Ontario in fall 1983, with measurement of tracer by ground network and aircraft downwind to the Atlantic coast.

The capability of several trajectory models to simulate the CAPTEX data has been reported. For example, the ASTRAP model produced for 6-hour averages a minimally biased simulation, but the range of error in estimate is substantial for four of the seven cases examined (Argonne National Laboratory, 1986). As another example, Kahl and Samson (1986) found that beyond 24 hours of transport, the mean trajectory errors increased with time at a linear rate of 45 km per 12 hours. These researchers estimated that with a standard trajectory constant-pressure model, there was a 50% probability of exceeding a horizontal displacement error of 210 km in 24 hours and 350 km in 72 hours for the CAPTEX meteorology. Dennis *et al.* (1990) have summarized these and other CAPTEX results, noting that the uncertainty range of tracer concentration prediction could not be established on the basis of the results reported from CAPTEX analyses.

The ANATEX was a more elaborate tracer experiment involving the release of three different perfluorocarbon tracers on 33 occasions from two sites, Glasgow, MT and St. Cloud, MN. One tracer was released from St. Cloud over 3-hour periods at 2.5-day intervals, the second from St. Cloud at 5-day intervals, and a third from Glasgow, MT on the 3 h/2.5 day schedule. Samples were taken in a 77-station network extending into Alberta and Quebec, through ENA along the Gulf Coast to the Atlantic Ocean. Ground samples were taken for 24-hour periods over 84 days. In addition, 200-m tracer samples and aircraft samples were obtained.

Clark and Cohn (1990) compared the transport calculations with the ANATEX data from 11 LRTP models, three single-layer Lagrangian

(SLL), six multilayer Lagrangian (MLL), and two multilayer Eulerian (MLE) (including ADOM). This study indicated that the MLL models generally performed better than others. The performance of the SLL models depended on the method of defining single-layer transport vectors. Although the mean errors for ADOM and the Atmospheric Diffusion Particle-in-Cell (ADPIC) model were comparable with the MLL models for some transport times, their errors were 50 to 300% greater than those of the MLLs for other times.

The upper bounds of the model trajectory errors based on 66 ANATEX tracer releases exceeded those of the CAPTEX evaluations. Part of the difference lies in the wind flow regimes; the CAPTEX runs were chosen for simpler, nominally minimal curvilinear flow conditions. The ANATEX results are probably more representative than CAPTEX. Typically, the range of error was 225 to 650 km in 38-hour transport time to 250–900 km after 62-hour time; ADOM ranged between 200 km (13 hours) and 500 km (62 hours). These large errors in calculated trajectories reinforce the results gained from other studies qualifying the ability to link a specific source or a cluster of sources with deposition at specific receptors.

The gas-phase chemistry submodules have been tested against data from smog chambers. A variety of chamber data now exist that permit extensive testing for $SO_4^{2-}$ and oxidant production with different mixtures of $SO_2$, $NO_x$, and VOC. For example the SURADS chemistry (Mueller and Hidy, 1982) module simulates well the pattern of $O_3$ evaluation, $NO_2$ production and decay, and suspended $SO_4^{2-}$ production measured in the Battelle Institute chamber. The model simulates less well the HC changes. The performance also depends on the chamber experimental run and the ability to correct for chamber surface interaction. This kind of quality in comparisons for gas-phase chemistry is typical of submodules currently in use.

Dennis *et al.* (1990) have summarized the comparisons between chemical submodules of ADOM and RADM. Several versions of chemistry have been used for these models. Evaluation involves a variety of comparisons for different key species. Generally the simulations focused on (1) whether the chemistry simulates the specific maximum and the time of its occurrence and (2) whether the chemistry estimates the correct amount of material oxidized. Comparison of time plots indicates that the submodules can reproduce reasonably well the time variation of species concentration, but no quantitative measures of goodness of fit have been reported.

Intercomparisons between ADOM and RADM chemistry submodules indicate that for oxidation or oxidant production the ADOM scheme predicts responses to initial VOC and $NO_x$ conditions that are somewhat more intense than RADM. In both cases there is a complicated, nonlinear response to initial HC concentration and $NO_x$/VOC rates (see also Appendix A).

Laboratory experiments simulating cloud chemistry and scavenging have not been attempted in a manner comparable with the calculation. Therefore, tests of these components must rely on very limited cloud observations and on laboratory studies of hydrometeor and particle collision processes leading to scavenging. Only recently have detailed results of field studies of cloud processes become available. An example supported by Dickerson *et al.* (1987) provides a reasonably complete set of data for model component testing. Such tests have not yet been attempted. Testing of dry deposition or wet deposition has relied on field data taken at a single point. Essentially no areawide representative studies of the deposition detail for a wide range of cloud or precipitation types or surface conditions have been reported as yet.

Model evaluations to date have employed both operational and diagnostic evaluations to gain insight not only into the models themselves but also into the complex behavior of $SO_x$ and $NO_y$ and the troposphere. The evaluations to date indicate that the more sophisticated codes based on the best available physics and chemistry yield only a marginal improvement in operational performance compared with the simpler, linear process schemes. Without the benefit of insight from the laboratory studies and field studies described earlier, workers would be hard pressed to provide guidance for regional emission control strategies solely from the models. However, taken together, there is a broad body of information that can serve as a basis for interpreting productively mass balance and source–receptor analyses employing applications of the regional models which is vastly improved over knowledge in the 1970s.

Diagnostic testing has also included important investigations of the numerical methods used in models, as well as the influence of initial boundary conditions (e.g., Mueller and Hidy, 1982; Chang *et al*, 1990). Studies of SURADS, for example, found that numerical integrations using the SHASTA method were unacceptably susceptible to numerical errors, creating a *pseudodiffusion*. This resulted in adopting an improved cubic splice integration method that reduced this kind of error to acceptable levels. In studies of SURADS, and later in ADOM and RADM behavior, modeling of episodes was found to be dependent on initial conditions. This required development of methods to initialize calculations that would essentially eliminate the influence of initial conditions on the end results. Models have also been determined to be sensitive to boundary conditions and can experience significant error in the edge regions of the grid unless precautions are taken to minimize influx of chemicals. For the ENA this is accomplished by expanding the grid to encompass an area of negligible $SO_x$ and $NO_x$ sources.

## REGIONAL MODEL APPLICATIONS

Regional models can be adopted to a variety of applications that require estimation of SRRs and investigation of the role of atmospheric processes in distributing pollutants in space and time. Other applications include macroscopic S and N budget studies for ENA, investigations of emissions control strategies, and assessment of uncertainty in design for field experiments.

### Sulfur Budget for Eastern North America

Because of the magnitude of $SO_2$ emissions in the SURE region, it is of interest to construct at least some of the components of an S budget for ENA. Some insight into the sources, sinks, and inflow or outflow of S from ENA can be gained by comparing model results with S budget determined from coarse S-cycle estimates in the 1970s. The 1977 emissions inventory provides a direct estimate of anthropogenic S emissions (both $SO_2$ and $SO_4^{2-}$). The SURADS model simulations for 5 August 1977 estimate S removal by dry deposition and the net boundary flux (outflow minus inflow at the lateral boundaries). The SURE emissions, total dry deposition, and net boundary flux were converted to annual budget terms (simply by multiplying the daily total by 365) and compared with estimates of Galloway and Whelpdale (1980) for the United States and Canada east of 92°W longitude. Table 6.8 shows the comparison. Also included are MIT model balances of Fay *et al.* (1986). The emission estimates and dry deposition rates agree reasonably well. However, the net outflow from the SURADS estimate, 1.5 teragrams (Tg) S/yr, is less than half of Galloway and Whelpdale's EUS estimate of 3.5 Tg S/yr, even for a selected event day that exhibits evidence for long-range transport eastward.

Because comparison of an annual average with an estimate derived from a single day is difficult, estimates of the net-flow term were calculated from all the SURE model simulations. The resulting outflow loss of 1.5 Tg/yr is still considerably lower than the Galloway and Whelpdale (1980) estimate, but closer to the MIT estimate. The differences could be explained by noting that the SURADS simulations involved days with large anticyclones over the region. Those high-barometric-pressure areas produced easterly and southerly inflow winds on the eastern and southern sides of the region as well as the typical westerly inflow wind in the West and southwesterly outflow wind in the Northeast. In short, these days dominated by anticyclones may not represent typical days. In any event, even with a wet deposition term as high as or higher than Galloway and Whelpdale's, the S losses will not balance the sources. The balance would be better if a trans-Canadian boundary transport in excess of 1 Tg were

TABLE 6.8 **Estimated S Budget Terms (Tg S/yr)**

| | Eastern Canada | Eastern United States | Eastern North America[b] | SURE Region |
|---|---|---|---|---|
| *Source*[a] | | | | |
| Anthropogenic emissions | 2.1 (2) | 14 (10) | 16.1 (12) | 17.4 |
| Natural emissions | 0.36 | 0.47 | 0.83 | 0.14 |
| Inflow from oceans | 0.04 | 0.2 | 0.2 | 0.30 |
| Inflow from west | 0.1 | 0.4 | 0.5 | — |
| Transboundary flow | 2.0 (3.3) | 0.7 (0.4) | — | — |
| Totals | 4.6 | 15.8 | 17.6 | 17.5 |
| *Losses* | | | | |
| Transboundary flow | 0.7 (0.4) | 2.0 (3.3) | — | 1 |
| Wet deposition | 3.0 (1.8) | 1.5 (1.9) | 5.5 | ? |
| Dry deposition | 1.2 (1.6) | 3.3 (2.0) | 4.5 | 3.1 |
| Outflow to oceans | 0.4 (0.4) | 3.9 (1.8) | 4.3 | 1.5[c] |
| Totals | 5.3 (4.2) | 11.7 (9.0) | 14.3 | 5.6 |

[a] The first three columns are from Galloway and Whelpdale, 1980; numbers in parentheses are from Fay *et al.* (1986).
[b] United States and Canada east of 92°W.
[c] This term represents the net outflow, that is, the outflow minus the inflow. It should be compared with 4.3–0.2 or 3.6 Tg S/yr.

assumed (MIT gives a net flux of 2.9 Tg/yr). The practical budget calculation from the SURADS raises questions about the balancing terms and perhaps about the assumptions made by Galloway and Whelpdale (1980).

Dry deposition results are similar, but the net outflow to the oceans projected from the SURADS results are lower than the Galloway and Whelpdale estimates. The compensation may lie in wet deposition. Analogous calculations using EMEFS data and the newer models for long-term averages could be used to reexamine the ENA S budget; such analyses have not been reported. Even with its limitations, this example illustrates the potential utility of long-range transport models to determine the influence of source distributions and weather on transboundary flow and outflow over the ocean.

## Planning for Emission Control

The costs of controlling $SO_x$ and $NO_x$ emissions to reduce ENA deposition rates are potentially very great. Analyses of cost-effective control options using the early 1980 inventory of LRTAP models have been reported (e.g., OTA, 1984). These analyses have been used to rationalize more than 40 emission control bills introduced in Congressional committee between 1980 and 1987. After much debate, legislation in the form of the 1990 Clean Air Act Amendments (CAAA) was adopted in 1990 (U.S. Public Law 101-549). The new law calls for reduction of $SO_2$ emission in the United States by $10^7$ tons/yr from 1980 levels early in the next century. Ironically, the requirements of this legislation or its implementation strategy have little relationship to the geographically segregated analyses derived from SRR modeling. This is the case despite extensive analyses attempted over a decade and the insight gained from them.

The deposition models per se cannot project the benefits of returning to what is believed to be a natural state of ecosystem behavior. However, they can provide guidance on the change in exposure expected for target emissions reductions. Essentially all the evaluations prior to 1990 have used linear forms of LRTAP modeling, so that the potential associated with complex chemistry and deposition processes were excluded. The NAPAP assessment provided an extensive analysis of emission control options with the RADM that were compatible with the development of the 1990 CAAA provisions for $SO_x$ emission reduction. Although the RADM was used, the nonlinearities in chemistry driven by future projections of the VOC/$NO_x$ conditions were not considered; i.e., future VOC and $NO_x$ emission changes were not explicitly evaluated for various $SO_2$ emission reductions.

Two examples of $SO_x$ emission reduction scenarios are shown in Figure 6.10. The first case, in Figure 6.10a, represents for wet deposition patterns a test case referred to earlier that corresponds to annual $SO_4^{2-}$ deposition assuming a 50% uniform reduction of $SO_x$ from all sources. The result indicates a ‘‘marginal’’ nonlinear proportionality in $SO_4^{2-}$ deposition reduction, particularly near large source regions in the Ohio River Valley region. Figure 6.10b shows for comparison a case of hypothetical $SO_x$ emission reduction of electric power plants allocated by state to achieve the equivalent of a 50% total emissions reduction in ENA. Comparison of the two cases indicates that the utility reduction case differs from the uniform reduction case primarily in the border regions of southern Canada and the coastal parts of the United States. The utility reduction scenario would not be as useful for protecting remote receptor areas of the Adirondacks and southern Canada if it were employed.

Researchers at Systems Applications, Inc. (1984) calculated reduction of ambient $SO_4^{2-}$ concentrations and its effect on improving visibility using the RTM-LT model. Researchers tabulated 1980 visibility for several states, the projected 1995 visibility, and the percent improvement in summer median visibility assuming several $SO_2$ emission reduction scenarios, or achievement of the National Ambient Air Quality Standard (NAAQS) for $SO_2$. The emissions scenarios included a 24% mean reduction in 12 states, a 12-state scenario plus a 25% reduction of $SO_2$ emissions in Canada, an 8 million ton reduction in 31 states east of the Mississippi, and a 12 million ton reduction in 31 states. Except for the major reduction strate-

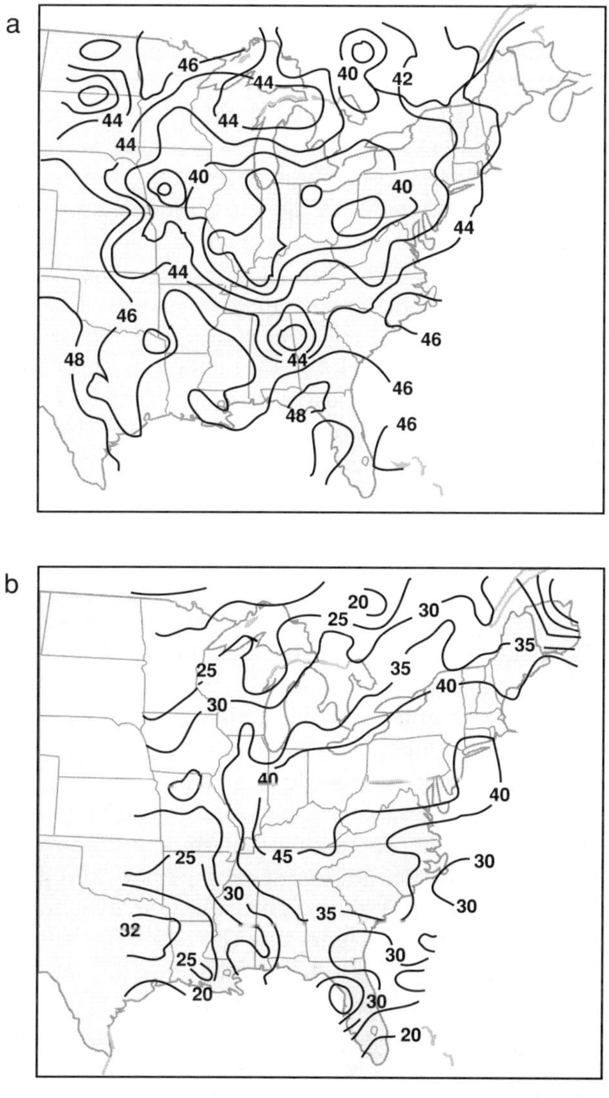


FIGURE 6.10 (a) Percentage change in annual wet sulfur deposition predicted by RADM for a 50% uniform reduction in $SO_2$ emissions from all sources (boundary and initial concentrations reduced by 50%). (b) Percentage change in annual wet sulfur deposition predicted by RADM for a hypothetical state-allocated utilities-only control program equivalent to 50% $SO_2$ reduction. Annual average is based on 14 cases, and initial and boundary conditions have not been changed. (From Dennis *et al.*, 1990).

gies involving the 31 eastern states, change in regional median summer visibility is expected to be marginal for these cases. Low visibility (10th percentile) would improve more substantially, however.

The calculations also placed options with the least tonnage and the least cost of control in perspective. The least-tonnage scenario considered sources by emission levels and by distance from a designated receptor. Least-cost scenarios included both the least tonnage and subregional fuel (or fuel switching) costs.

Using the MCARLO model, Husar (1983) has done such calculations for the Adirondacks. For the least-tonnage scenario, a 16% reduction in $SO_2$ emissions would result in a 50% reduction in S deposition for a radius of 500 km around the receptor. Husar's calculations showed that both the least-tonnage and least-cost scenarios are much more efficient in reducing deposition per ton of $SO_2$ removed than uniformly reducing emission everywhere. The least-cost scenario would involve control of sources. Because of the zone-of-influence relationship, the burden of cost and control is placed on nearby sources, as expected.

Husar's focus on receptor protection rather than source control was a creative conceptual approach to developing a cost-effective emission control strategy that distinguished between sensitive and less sensitive receptor areas. Even though this approach was widely discussed, it did not affect the rollback approach taken in the 1990 CAAA.

Other approaches coupled an air quality model with environmental effects predictions to compare alternative schemes for aggressive emission reduction with existing technology versus a more considered long-term approach using a mix of existing and newly demonstrated technologies (e.g., DOE, 1989). These calculations indicated that essentially the same environmental protection for sensitive lake and forest systems could be achieved by 2020 by both approaches. However, the new technology scenario was more cost effective (EPRI, 1991).

These examples show that modeling is useful for establishing guidelines on strategy. However, the reliability of available models for quantitative, detailed emission reduction studies remains undetermined. Continued refinement of these calculations with more reliable models should ensure better decision making, provided that model reliability can be established. Eventually, model reliability could be examined ideally by comparing over the years the $SO_x$ and $NO_x$ emission reductions versus exposure conditions at a number of receptor locations.

## Uncertainty in Experimental Design for SSRs

Because the reliability of theoretically based models is unproved for a range of $SO_2$, $NO_x$, and VOC emission changes, workers have sought alternatives for deducing SRRs with at least a defined uncertainty based on measurements. Model verification can be accomplished by detailed comparison with field data, as attempted in the next tracer experiments like CAPTEX and ANATEX for chemically inert species. Model computations can be (and have been) used to optimize field program design to test model performance (e.g., Hidy *et al.*, 1976, 1985; Hansen, 1989).

An alternative SRR estimation technique was the development of an empirical approach that would be derived solely from measurements. In effect, this technique would extend the tracer-of-opportunity concept further by use of deliberately released material. The MATEX was proposed, in which tracer release would be followed from many sources that would be linked with deposition patterns from all sources (e.g., Hidy *et al.*, 1985). The relationship would be most straightforward for tracers with chemical characteristics identical to those of S and N (e.g., naturally rare S and N isotopes). In principle, the link could also be established using chemically inert tracers that simulated transport and dispersion, provided that conversion and losses (including deposition) could be accounted for indirectly.

Because no experiments like the MATEX have been attempted, the designer has no guidance except directly from measurements of "summed" contributions from all sources, as in the SURE, or from model simulations. Because such an experiment would be very expensive, the uncertainties placed on the results are critical for its feasibility and design. LRTAP modeling can provide some information on this question.

An approach to an uncertainty analysis in S deposition was proposed by Henry (see Hidy *et al.*, 1985). His model describes the source–receptor deposition relationship by an algebraic expression involving basic uncertainties in elements of the source–receptor event chain. These uncertainties are parameterized compactly as a series of terms, each representing the relative percentage error in one link of the chain. The variables that enter are:

1. The total deposition of S at a receptor
2. The concentration of passive tracer observed over the receptor area
3. The ratio of actual $SO_2$ emissions to tracer emissions from the source area
4. The fractions of S emissions from the source area that were dry (wet) deposited or otherwise lost upwind as $SO_2$
5. The fraction of S emissions from the source area that were dry (wet) deposited or otherwise lost upwind as $SO_4^{2-}$
6. The fraction of $SO_2$ emissions converted to $SO_4^{2-}$ en route
7. The ratio of primary $SO_4^{2-}$ emissions to $SO_2$-S emissions
8. The fraction of total $SO_2$ removed over the receptor area by dry (wet) depositions

9. The fraction of total $SO_4^{2-}$ removed over the receptor area by dry (wet) depositions

These terms are combined into an algebraic "error" formula that relates the overall percentage error in total S deposition at a receptor to the percentage errors in passive tracer dispersion and in each of these other contributing terms. Deriving the basic error formula is tedious but straightforward and amounts to doing a conventional differential error analysis. The result for the case of a single source area and a single receptor area is

$$P^2 = \sum_i c_i P_i^2 \tag{6.7}$$

where $P$ is the expected percentage error in the source apportionment (that is, the error in the estimated amount of S deposited at the receptor by the source). The term $P_i$ is the percentage error in $i$ contributing error variables $x_i$; $c_i$ is a coefficient for $P_i^2$. The summation extends over all $i$ error variables. The coefficient $c_i(x_1, x_2, \ldots)$ is, in general, a simple algebraic function of one or more of the error variables.

As a first approximation, all the errors are taken as uncorrelated. (The derivation does not require any other assumptions about the size or statistical distribution of the errors. However, before a specific design and data interpretation scheme is prepared for such an experiment, there is no way to estimate how the error terms might be correlated.) Table 6.9 identifies the 14 terms that enter into the summation in Eq. (6.7), together with their coefficients. Again, if some of the errors are not independent, the basic formula will include many additional terms involving cross-products of the percentage errors and coefficients of correlation among the variables.

Without loss of generality (and again assuming uncorrelated error terms) the full error formula can be condensed and rewritten to emphasize the relative importance of different kinds of errors. With some rearrangement and substitutions

$$P^2 = P_{\mathrm{F}}^2 + P_{\mathrm{R}}^2 + P_{\mathrm{T}}^2 \tag{6.8}$$

where in the notation of Table 6.9

$$\begin{aligned} P_{\mathrm{F}}^2 &= \left(\frac{S_1}{F}\right)^2 [L_1^2 P_{\mathrm{L1}}^2 + C^2 P_{\mathrm{C}}^2 \\ &\quad + (1 - L_1 - C)^2 P_{\mathrm{S1}}^2] \\ &= \left(\frac{S_2}{F}\right)^2 [L_2^2 P_{\mathrm{L2}} + C^2 P_{\mathrm{C}}^2 + Y^2 P_{\mathrm{Y}}^2 \\ &\quad + (Y + C - L_1)^2 P_{\mathrm{S2}}^2] \end{aligned} \tag{6.9}$$

TABLE 6.9 **Coefficients of the Complete Error Analysis Formula**[a,b]

| Variable | Meaning | Coefficient ($c_i$) |
|---|---|---|
| $T$ | Mass tracer above the receptor | 1 |
| $R$ | Ratio of $SO_2$ emissions to tracer emissions | 1 |
| $D_1$ | Fraction of $SO_2$ lost to dry deposition upwind of the receptor area | $(S_1/F)^2\, D_1^2$ |
| $W_1$ | Fraction of $SO_2$ lost to wet deposition upwind of the receptor area | $(S_1/F)^2\, W_1^2$ |
| $V_1$ | Fraction of $SO_2$ lost to cloud venting or otherwise lost upwind of the receptor area | $(S_1/F)^2\, V_1^2$ |
| $D_2$ | Same as $D_1$, but for $SO_4^{2-}$ | $(S_2/F)^2\, D_2^2$ |
| $W_2$ | Same as $D_2$, but for $SO_4^{2-}$ | $(S_2/F)^2\, W_2^2$ |
| $V_2$ | Same as $V_1$, but for $SO_4^{2-}$ | $(S_2/F)^2\, V_2^2$ |
| $Y$ | Ratio of primary $SO_4^{2-}$ emissions | $(S_2/F)^2\, Y^2$ |
| $C$ | Fraction of $SO_2$ converted to $SO_4^{2-}$ the receptor area | $(S_2/F)^2 - (S_1/F)^2$ |
| $S_{W1}$ | Fraction of $SO_2$ over the receptor area that is wet deposited there | $(S_{W1}/F)^2\,(1 - D_1 - W_1 - V_1 - C)^2$ |
| $S_{D1}$ | Fraction of $SO_2$ over the receptor area that is dry deposited there | $(S_{D1}/F)^2\,(1 - D_1 - W_1 - V_1 - C)^2$ |
| $S_{W2}$ | Same as $S_{W1}$, except for $SO_4^{2-}$ | $(S_{W2}/F)^2\,(C + Y - D_2 - W_2 - V_2)^2$ |
| $S_{D2}$ | Same as $S_{D1}$, except for $SO_4^{2-}$ | $(S_{D2}/F)^2\,(C + Y - D_2 - W_2 - V_2)^2$ |

[a] This gives the coefficients in an error formula in Eq. (6.7).
[b] From Hidy *et al.*, 1985; reproduced with permission of the Electric Power Research Institute.

and

$F = [(1 - L_1 - C)S_1 + (C + Y - L_2)S_2]$ is the fraction of all emissions reaching the receptor that actually deposit there.

$S_{1(2)} - [S_{W1(W2)} + S_{D1(D2)}]$ is the fraction of all $SO_2$ ($SO_4^{2-}$) deposited on the receptor area.

$L_{1(2)} = [D_{1(2)} + W_{1(2)} + V_{1(2)}]$ is the total upwind loss of $SO_2$ ($SO_4^{2-}$).

$P_{L1}^2 = (\sigma^2(L_1)/L_1^2)$ is the percentage error in the term $L_1$, and $P_C^2$, $P_{S1}^2$, $P_{L1}^2$, $P_Y^2 P_{S2}^2 P_R^2$, $P_T^2$, are defined similarly.

Because the errors in $R$ and $T$ (i.e., $P_R^2$ and $P_T^2$) can be estimated or obtained directly, the error analysis reduces to a sensitivity analysis of $P_F^2$, the (normalized) error variance of $F$.

Equation (6.9) has a number of important implications for the design of any experiment attempting to trace $SO_x$ or $NO_x$ in air masses. First, the coefficients $S_1/F$ and $S_2/F$ dominate the growth of errors, partly because the other terms are smaller but mostly because the term $F^{-1}$ grows quickly with large downwind distances. The rapidity with which $F^{-1}$ grows to be very large is a key limitation to making source apportionment estimates from passive tracer data with empirical models.

To understand better how the various error processes contribute to the rapid growth of source apportionment errors (described by $F^{-1}$), the linear differential material balance model of acid deposition was used to generate arbitrary but realistic estimates of the relative magnitudes of the error coefficients. The model assumes 4-hour average rates of $SO_2$-to-$SO_4^{2-}$ transformation, dry deposition, wet deposition (if precipitation occurs during a 4-hour interval), and dilution by a mixture of clean air. An air parcel containing emissions from an initial source area (of order 100 $km^2$) is followed for 3 days as it travels downwind, during which it encounters additional sources and periods of precipitation.

The principal result is that after about 1 day's travel time the overall error (proportional to $F^{-1}$) becomes very large. As long as any realistic set of assumptions is used, the general conclusion is much the same and remarkably insensitive to the particular conditions imposed.

The reason becomes evident if one looks at how individual error terms grow with time. Figure 6.11 displays the growth of the error coefficients for dry and wet deposition of $SO_2$ and $SO_4^{2-}$ in one scenario for error propagation in source apportionment. The error terms vary as the squares of these coefficients. Note that the fractional uncertainties in $SO_2$ dry and wet deposition grow to unity or larger within 20–32 hours and exponentially thereafter. Even when the calculation is repeated with different but equally plausible assumptions about dry deposition and wet scavenging rates, these two curves behave in much the same way. One can conclude that at distances close to the source, errors due to $SO_2$ deposition upwind (plot a) will dominate those for $SO_4^{2-}$. As for errors in deposition *at* the receptor (plot b), $SO_2$-related errors are again dominant within about a day's travel, but for travel beyond about 36 hours the $SO_4^{2-}$-related errors become dominant (because there is little $SO_2$ left at these longer distances).

To examine the sensitivity of the source apportionment error with reference to the trajectory model, three nominal uniform estimates were made for the magnitudes of errors in each individual term of the generic error analysis for each step of the calculation. These estimates are low (20%), medium (50%), and high (100%). The following rationale was adopted for these choices.

The low estimate, 20%, is based on ideal conditions in the field; it is usually very difficult to make a single environmental field measurement to better than about 10%. But because each of the terms in the generic error analysis is itself a combination or interpolation from many single measurements, an error of only 20% in each generic error term is optimistic. The medium error estimate, 50%, may be more realistic for a well-defined achievable experiment. The high error estimate, 100%, is conservative for *some* of the terms; however, because of large uncertainties in complicat-

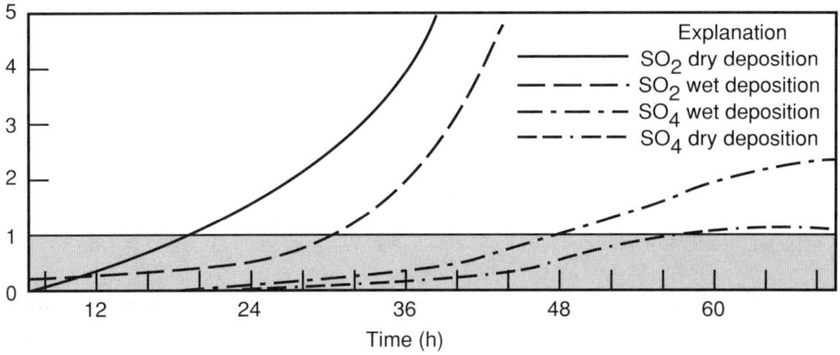


(a) Dimensionless error coefficients for deposition losses upwind of the receptor.

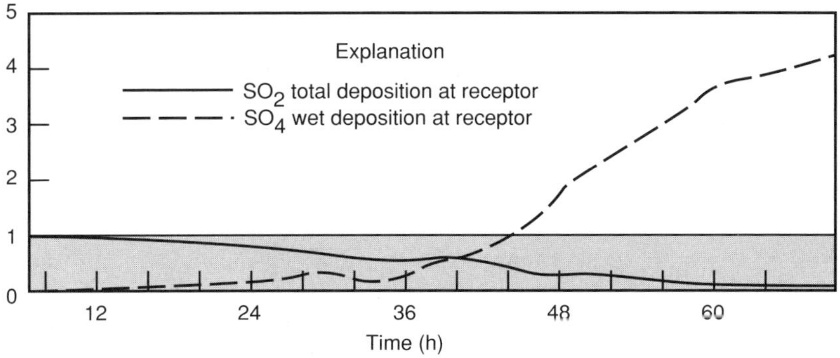


(b) Dimensionless error coefficents for total depositions of $SO_2$ and $SO^{2-}_4$ ($S_2$) over the receptor.

**FIGURE 6.11** Growth with time of error coefficients (% × $10^{-2}$) in overall source apportionment error formula. Values larger than unity lead to rapid overall growth of error in source apportionment. (From Hidy *et al.*, 1985; courtesy of the Electric Power Research Institute.)

ing factors, such as wind shear, convective dispersion, precipitation scavenging, and the like, errors of 100% (at least in the tracer-to-emission ratio term, $R$, or in the scavenging terms, S1 and S2, for example) are potentially the level expected from a state-of-the-art experiment.

In a generic material balance calculation for acid deposition, the terms are calculated incrementally along the entire trajectory of the air parcel (one example is the upwind loss of $SO_2$ due to dry deposition). The errors in these terms will accumulate with downwind distance, because the error at each step is influenced by the errors at all previous steps. Just how quickly the errors will grow in time is not at all obvious. One might guess, for example, that errors double roughly every 24 hours. Alternatively, one might suppose that the growth rate is proportional to a fractional power of travel time, say $t^{1/2}$, because some error terms are sums and cancellation of errors might cause these terms to grow less quickly with time. It is also likely that some errors will be correlated with each other or with the observations.

Estimating the net consequence of competing influences would require detailed Monte Carlo calculations, which the researchers have not attempted at this stage. Instead, to bound the error growth problem (without resorting to more elaborate and costly methods), they investigated three possibilities for how the errors might behave with time. The exercise was carried out with a simple deposition model. The three "generic" growth curves were

- Errors increasing exponentially with time (doubling in 24 hours)
- $t^{1/2}$ error growth rate
- Constant error (zero growth rate)

The last case was included as a reference against which to judge the other two cases.

Figure 6.12 displays the overall growth in the percentage error of source apportionment by the generic material balance approach for initial errors of 20%, 50%, and 100%. Each region of the figure corresponds to initial errors remaining constant. The upper boundary curve corresponds to initial errors growing like $t^{1/2}$. For times shorter than 24 hours the exponential growth case always leads to smaller overall error than does the $t^{1/2}$ growth case. Beyond about 36 hours the reverse is true. The regions shown are approximate and portions of small overlap have been ignored. The horizontal line drawn at 100% error level suggests a maximum "acceptable" apportionment error for an individual tracer experiment based on target specifications for a MATEX. (If repeated trials could be made under similar but truly independent conditions, the seasonal average error for source apportionment would be severalfold smaller.) Note that within 12–16 hours the overall error is already 100% (or greater), except in the implausible circumstances that initial errors are 20% or less *and* do not grow with time. Even in this most unlikely case, the overall error doubles within about 24 hours.

Figure 6.13 helps to explain why this is so by showing how primary sources of error change with travel time. The figure is a stacked chart for the fractions of the total error variance contributed by the five most important sources of error. It shows that 95% or more of the error variance can be explained by just five sources: (1) the ratio of tracer to $SO_2$ emissions at the source, (2) the measured mass of tracer over the receptor, (3) the fraction of $SO_2$ dry deposition upwind, (4) the fraction of $SO_2$

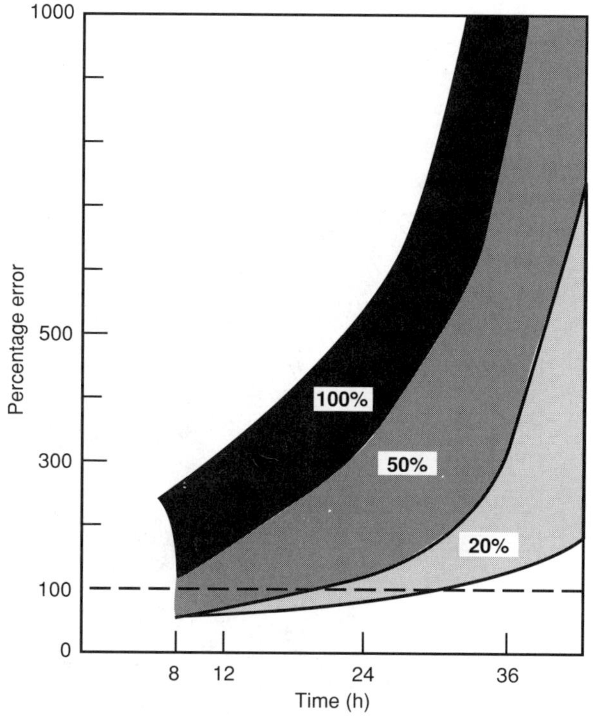


FIGURE 6.12 Fractional contributions of some individual error terms to overall error variance. Shading conventions are shown by error term. (From Hidy *et al.*, 1985; courtesy of the Electric Power Research Institute.)

wet deposition upwind, and (5) the fraction of $SO_2$ dry deposition over the receptor. For times less than about a day, the total error variance is attributable almost entirely to three terms:

1. The percentage error in the ratio of tracer emissions to S emissions at the source
2. The percentage error in the dry scavenging fraction for $SO_2$ over the receptor
3. The percentage error in the observed concentration of tracer over the receptor.

Beyond about a day the major sources of overall error are the percentage errors in the fractions of dry and wet deposition of $SO_2$ upwind of the source. Within about 1.5 days of travel time (which is already comparable to the characteristic travel times from the major $SO_2$ source regions to the major acid-sensitive receptor regions in ENA), the overall error in source apportionment becomes largely dominated by the uncertainty in the path-

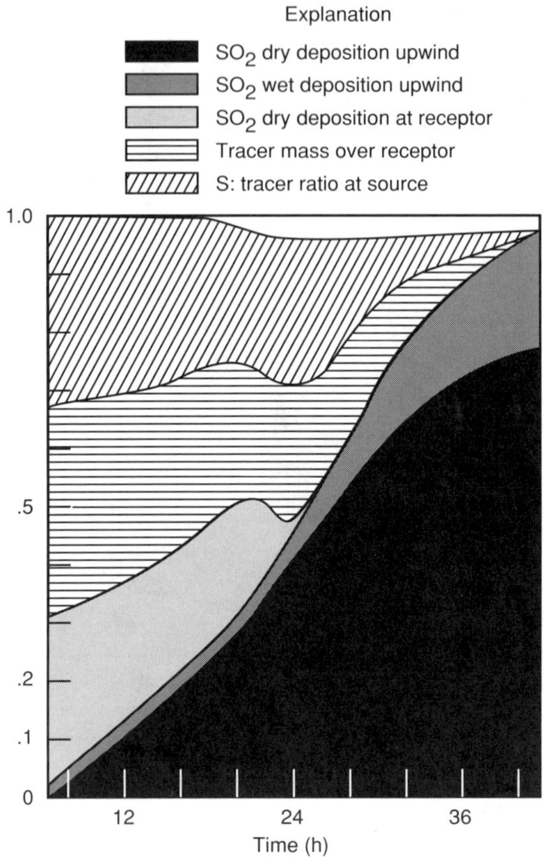


FIGURE 6.13 Fractional contributions of some individual error terms to overall error variance. Shading conventions are shown by error term. (From Hidy *et al.*, 1985; courtesy of the Electric Power Research Institute.)

integrated fraction of material lost upwind by dry deposition! This single error term by itself accounts for perhaps three-quarters of the total error variance in source apportionment. The large uncertainty in present knowledge of $SO_2$ dry deposition, especially in its space–time variations with conditions in the atmospheric surface boundary layer and ground properties, is apparently the single greatest impediment to the inert-tracer approach for source apportionment. The $SO_4^{2-}$ formation and deposition terms are unimportant to the error analysis because in the early history of the trajectory, little $SO_4^{2-}$ is formed, and in the later stages errors in $SO_4^{2-}$ deposition are overwhelmed by errors in the dry and wet $SO_2$ upwind deposition. The results are not particularly sensitive to reasonable variations in the $SO_2$ chemical conversion and wet and dry deposition rates, based on sensitivity analysis carried out with the trajectory LRTAP model.

## SUMMARY

Accessability to high-speed computers has created conditions for a proliferation of a large number of LRTAP models. The models serve as an important adjunct to the characterization of air chemistry by interpretation of field observation in the light meteorological experience and laboratory studies. The ability of the models to synthesize information on atmospheric dynamics facilitates the visualized interpretation of a wide variety of observations on spatial and temporal scales of continental proportions.

Progress in development of the models has been very substantial in the past decade. Their development has proceeded along two routes—the *Lagrangian* or air mass trajectory computation and the *Eulerian* or "fixed"-grid finite-difference approach. The Lagrangian calculation provides a straightforward link between specific course contributions to S and N concentrations or deposition and receptor conditions. However, the Eulerian framework provides the better mathematical structure to account for dynamic interactions that invoke air motion and exchange, cloud and precipitation and nonlinear air chemistry.

Progress on model development for ENA is marked by the construction and evaluation of two Eulerian models, ADOM and RADM. These contain the most sophisticated and elaborate descriptions of the various (nonlinear) processes involved in accounting for $SO_x$ and $NO_x$ emissions injected into the air. The comparison with simple models has shown the elaborate models to be only marginally superior to the simpler, linear codes when compared with long-term observations. The Eulerian codes are clearly a conceptual improvement over the simpler models for short-term, episode simulations. Although RADM and ADOM are not considered fully developed for many operational or regulatory purposes, they are believed to offer the most promise for future analyses because of their ability to accommodate explicitly improved data for initial and boundary conditions, as well as complex gas-phase and cloud processes and surface deposition. These models are the *only* means available to explore future conditions in which major differences in spatial $SO_x$, $NO_x$, and VOC emissions distribution may develop.

The evaluation of LRTAP models includes two basic methods—operational and diagnostic. The former relies on testing the reliability of the model estimates with field observations of pollutant concentrations and deposition patterns. The latter were special, elaborate field experiments or laboratory studies to test in relative isolation components of the models. The evaluation of LRTAP generally has focused on operational characteristics and has been hampered by data sets that are robust enough in variables and space or time dimensions. The intercomparisons with observations, to date, confirm qualitatively that our knowledge can explain the behavior of $SO_x$ and $NO_y$ over ENA. The reliability of quantitative estimations appears to have improved substantially in the past decade, such that there is reduced bias between calculations and observation, and the uncertainty in long-term averaged sulfate or nitrate concentrations or deposition is generally less than 100%.

Operational evaluations have strong limitations because all observations are basically either surface point measures or derived from snapshots taken from aircraft. These measures are not compatible with the grid or area averaging that must be done, given the spatial resolution of models and the resolution of their input and boundary conditions data.

Diagnostic evaluation of models has resulted in considerable refinement in the choice and use of semiempirical rate coefficients in linear models. More important, such testing has provided critical insight into explicit parameterizations used in describing chemical and deposition, as well as the reliability of numerical integrations used in the Eulerian schemes. Diagnostic evaluations have been hampered by the lack of detailed data from field experiments. ADOM and RADM, for instance, have been thoroughly compared with only approximately 2 months of data from EMEFS, although that experiment now has nearly 2 years of data available. Full evaluation of the models over the range of meteorological conditions from different seasons is continuing and should be reported in the next few years.

The LRTAP models can be used with the broad knowledge of regional air chemistry for emission control strategy development, provided the uncertainties in their performance qualify the results.

The models have been used for a variety of applications. Some of these have concerned scientific questions about the sulfur and nitrogen budgets, as well as the comparative importance of different physicochemical processes in regional-scale phenomena. Other applications have been aimed at development of air pollution control policy. This work over a period of a decade furnished a variety of studies exploring alternatives for $SO_x$ emission controls. However, the options to optimize strategies for minimum cost or other criteria do not appear to have influenced directly the final form of the U.S. Clean Air Amendments of 1990.

With the implementation of the $SO_x$ and $NO_x$ emissions reductions from the 1990 CAAA, the models will continue to have important application. They will give guidance about progress toward reduction in dry and wet deposition rates over ENA.

# 7 RETROSPECTIVE

Only since the early 1970s has substantial attention been devoted to atmospheric chemistry over continental dimensions. This interest was stimulated to a great extent by European concern about the deposition of acidifying species from the atmosphere and their environmental effects. Air pollution, particularly from $SO_x$ and $NO_y$ and their dispersion over western Europe, contributes to increased wet deposition of acid sulfate and nitrate over a natural level. Similar conditions have been found in areas of heavy industrialization and urbanization in eastern North America (ENA). Around such areas, elevated pollution levels, optically dense haze, and elevated wet deposition rates occur for distances extending over a thousand kilometers from pollutant sources. The regionwide phenomena involve the interaction between emissions sources and meteorological processes that disperse material near the ground on unexpectedly large spatial scales and temporal scales extending over several days. Concern about environmental influences has motivated studies to characterize subcontinental atmospheric chemistry. ENA has been a primary target for extensive study in the late 1970s and 1980s. Characterization of pollution behavior in this region corroborates conclusions of the European experience.

## STATUS OF KNOWLEDGE

As a result of the analysis of data from extensive monitoring programs and short-term measurement expeditions, knowledge of $SO_x$ and $NO_x$ behavior over ENA has improved dramatically. Air and wet deposition measurements have provided a reasonably complete picture of average pollutant distributions and precipitation chemistry, as well as their within-year and year-to-year variability. Measurements from the major experiments like the SURE and the EMEFS give a detailed picture of $SO_x$ behavior at the ground for periods between 1977 and 1990. Exploration of conditions aloft during the same period complemented the ground observations. A number of shorter-term experimental programs have added considerable insight into the links between $SO_x$ and $NO_y$ chemistry and oxidant chemistry, as well as the link to wet deposition and visibility impairment.

The observations are consistent with a regionwide condition of elevated $SO_x$ and $NO_y$ concentrations throughout much of ENA resulting from widespread dispersion of $SO_x$ and $NO_x$ emissions. Pollution levels are particularly high around areas of high emission density, notably the Ohio River Valley northeastward to southern Ontario and Quebec. High $SO_x$ and $NO_x$ concentrations and a chemically reactive atmosphere reduced by $NO_x$ and HC or VOC interactions create regional conditions characterized by high oxidant concentrations and reduced visibility. $SO_4^{2-}$ particles, which are products of $SO_2$ oxidation, play a significant role in regional visibility impairment. Atmospheric processes induce pollutant removal by continuous gas absorption and particle deposition at the ground, supplemented by intermittent cloud and precipitation scavenging. Dry and wet processes result in substantial acidic $SO_4^{2-}$ and $NO_3^-$ deposition throughout ENA, which has potentially significant adverse impacts on natural and managed ecosystems even in areas remote from sources.

Observations indicate that meteorological factors dominate the behavior of $SO_x$ and $NO_y$ over ENA, given the areal distributions of pollutant emissions. Emissions are predominantly anthropogenic. Natural S and N emissions occur, but they are estimated to be small compared with emissions from anthropogenic sources. The average distribution patterns can be explained, in part, by localized accumulation of airborne material, but much of the regionalization requires additional dispersal by airflow and mixing over distances exceeding 1000 km. Long-range transport of air pollutants (LRTAP) occurs surprisingly frequently over ENA, especially in summer. Injection of emissions above 1 km in height by tall stacks mixing at the ground, where removal can take place, facilitates long-range transport of pollutants. Researchers did not appreciate the regionwide significance of long-range transport and mixing processes until the work of the 1970s was complete. Also of great interest in the study of regional phenomena is the evidence of pollution accumulation and transport with deep mixing from the ground to heights exceeding 4 km. This deep mixing is facilitated by daytime convective processes, especially with summer cumulus cloud formation and evaporation.

Explanation of the main features of S and N distributions and their variations links emission distributions and variations with regional wind patterns and air mixing conditions. Quantification of concentration distributions and variability by reactant and product requires knowledge of atmospheric chemical conversion and surface removal processes. Seasonal changes in $SO_4^{2-}$ behavior are linked closely with chemical reactivity in both gas-phase photochemical reactions and aqueous-phase reactions. The presence of high oxidant concentrations, produced from photochemical reactions of VOC and $NO_x$, is critical to the chemistry of $SO_x$. The importance of gas-phase photochemistry to $SO_4^{2-}$ and $NO_3^-$ behavior in fair-weather conditions is well known from earlier urban experiments. Recent studies have produced evidence of accelerated *regional* photochemical activity even in remote, rural areas.

Particularly significant in chemical processing are the OH radical and $H_2O_2$, respectively, in gas-phase and aqueous $SO_2$ oxidation. The OH reaction with $SO_2$ is a dominant $SO_4^{2-}$ formation process in the gas phase. Unlike other reactions, the $H_2O_2$-$SO_2$ aqueous reaction is rapid and acid independent in a range commonly encountered in clouds and fog. Through the $H_2O_2$ reaction, supported at higher pH by $O_3$ and metal-catalyzed oxidation, absorbed $SO_2$ in hydrometeors can be oxidized readily to produce $SO_4^{2-}$. The seasonal change in airborne $SO_4^{2-}$ and precipitation $SO_4^{2-}$ appears to be a direct relation to the oxidant concentrations produced in $NO_x$/VOC-enriched air. Also important in scavenging are the differences in physicochemical mechanisms between species. For example, in contrast to $SO_4^{2-}$, $NO_3^-$ is not readily produced in aqueous media from NO or $NO_2$ but is found via efficient absorption of $HNO_3$ vapor. The behavior of aq $NO_3^-$ in precipitation apparently depends on gas-phase photochemical processes converting $NO_x$ to $HNO_3$, followed by acid absorption. $SO_4^{2-}$, on the other hand, is produced in gas-phase species in clouds as a function of $H_2O_2$ and other oxidant availability.

Although dry deposition occurs continuously, its dependence on surface character and wetness for both gases and particles has only recently been understood. The surface interaction, in particular, complicates considerably the capability to specify exposure conditions for ecosystems on a regional scale and adds large uncertainty to interpretation of variability in deposition patterns.

Wet deposition chemistry is perhaps best characterized by spatially and temporally representative observations. Interest in acid deposition has substantially improved knowledge about exposure. Wet deposition chemistry closely parallels the distribution of ambient $SO_x$ and $NO_y$ concentrations, with broad maxima in the subregions of highest emission density. Limited observations linking cloud scavenging with precipitation are consistent with long-established, simplified concepts of scavenging removal both in and below clouds. These concepts describe scavenging as a rate process linear in local pollutant concentration and in liquid water content. Researchers recognize, however, that the details of these processes still have to be measured for a range of cloud and storm conditions.

New information about visibility conditions over ENA is generally unavailable. Surveillance of the visual range, a coarse measure of visibility, has continued from airport weather records. However, no systematic observations of visibility in relation to air pollution have been conducted in ENA. Expeditionary experiments have been useful in relating light extinction conditions to aerosol particles. A complete, spatially representative data set for visual perception linked with pollution observations is still needed. Comparison with the visual range measurements available and particle chemistry tends to confirm that airborne $SO_4^{2-}$ particles account for an important fraction of the observed light extinction, a common index of visibility impairment. However, the efficiency of light scattering by $SO_4^{2-}$ particles appears to vary significantly by area. Information about the role of elemental (black) carbon in extinction over ENA is evidently limited regionally, as is information about (wavelength dependent) light absorption by $NO_2$.

Of considerable practical importance for cost-effective environmental management is the estimation of the effect of specific sources on regionally distributed receptors. Receptors can be affected both by exposure to ambient concentration and by surface deposition. Estimation of source–receptor relationships (SRRs) depends on detailed knowledge of meteorological and chemical factors determining exposure conditions.

Evidence of the contribution of sources to receptor exposure derives from two complementary sources: (1) field measurements of chemical parameters, coupled with meteorological observations, and (2) theoretically based modeling calculations. These two approaches have produced similar qualitative results, with some confidence in semiquantitative relationships for large subregions. Quantitative details and resolution of the impact of source groups on a $10^4$-$km^2$ scale or smaller have not been made, given the limitations of both methods.

Results delineating SRRs consistently link high exposures with source contributions such that their influence *decreases* in proportion to increasing distance from the source. The proportionality may be linear or nonlinear in character; the distinction is uncertain without direct link by tracer experiments. SRRs depend on prevailing wind direction, with sources contributing most to exposure directly downwind. Exposure also varies with local air mass stagnation conditions, but exposure extends to much greater distance when pollutants are transported by airflow.

A source's *zone of influence* has been defined in terms of mean ambient concentration patterns, wet deposition patterns, or intersite correlation analyzed using empirical orthogonal function analysis (EOFA) or Kriging. Modeling has also been used to define influence zone by calculating natural potential functions. The three approaches indicate that a ''significant–detectable'' influence area effectively extends 300–600 km radially from a source. This scale is common to both ambient concentrations and wet deposition of $SO_4^{2-}$ and $NO_3^-$, based on intersite correlation analysis.

SRR patterns in ENA depend on source and receptor location. For example, sources influencing northern Minnesota are very different from those influencing the Adirondacks. In the Adirondacks and the Northeast, in general, the prevailing westerly transport implicates local industry, as well as the industrial and urban regions to the west. This implication also applies to the Middle Atlantic states and, to some degree, to the Southeast. Ample evidence from both observations and modeling shows that a major fraction of pollution found in southeastern Canada comes from the United States. At the same time, Canadian sources have some effect on certain receptors in the northeastern United States.

The predictability of change in exposure given emission reduction at specific, targeted sources remains open to question. General regional patterns of pollutant distribution consistently lie over areas of high emission density. However, the significance of atmospheric chemistry and loss processes in modifying distribution patterns regionally influenced by transport and mixing appears to be identifiable in seasonal behavior involving oxidant limitations. For example, winter production of oxidant is less than summer production, which results in restrictions in winter $SO_4^{2-}$ production from reduced oxidant levels.

Historical changes in visibility and wet $SO_4^{2-}$ deposition appear to compare qualitatively with estimated regional $SO_x$ emission changes in ENA over a period of roughly 20 years. Beyond this conclusion, little is known specifically about SRRs and regional or local visibility impairment for ENA.

The quantification of SRRs has progressed beyond that known in the early 1980s primarily through improvements in LRTAP models. Representative field data on both S and N behavior from selected source groups necessary to complete a subregional material balance have not been available. However, recently acquired regionwide observations should facilitate renewed progress on SRR characterization ENA deposition conditions in the next few years.

## PERSPECTIVE AND UNCERTAINTIES

A comparison of 1976 knowledge of regional phenomena with what is known today shows that major advances have been made. Since the 1970s, our understanding of sulfur and nitrogen oxide chemistry has vastly improved. The regional-scale behavior of $SO_x$ and $NO_y$ in terms of source distributions and meteorological features is well described and explained by current scientific knowledge. Methods for the analysis of temporal and spatial variability have been applied, yielding substantial improvements in delineating empirical SRRs. Development of different theoretically based fluid dynamic models and analysis of these corroborate and amplify the conclusions reached from study of observations. Regional scale models can now be used with some confidence to estimate exposure patterns for $SO_x$ and related atmospheric chemical end products. SRRs can be estimated with sufficient confidence to develop broad regional control strategies for environmental protection.

Specific conclusions and uncertainties are:

- ENA is exposed to geographically widespread levels of ambient $SO_x$ and $NO_y$ concentrations that, on the average, exceed estimates of a continental baseline or background by as much as a factor of 10.
- ENA is exposed to wet $SO_4^{2-}$, $NO_3^-$, and acid deposition concentrations two to five times greater than that believed to represent a continental baseline.
- Visibility over ENA cannot be accounted for simply in terms of humidity or other meteorological factors. The areal distributions and temporal behavior of visibility are linked to air pollution levels.
- ENA visibility impairment can be ascribed with confidence to finely divided particles derived from air pollution; at least half of which are some form of $SO_4^{2-}$, probably ammonium salts or $H_2SO_4$.
- The suspected role of long-range transport over ENA has been verified for $SO_x$, $NO_y$, and $O_3$ behavior, especially in spatial distributions of these constituents.
- Ambient concentrations and wet deposition distributions are dominated by meteorological processes operating over a range of spatial scales from tens of kilometers to thousands of kilometers.
  —Air mass transport and mixing (or dilution) are major factors influencing these distributions.
  —Evidence that chemical conversion and deposition rates are also

important comes primarily from seasonal changes and from detailed study episodes of a few days duration. How these processes affect spatial distributions and introduce a nonlinear relation between source emissions and exposure changes remains to be detailed.

- Substantial evidence suggests that gas-phase photochemical processes, complemented by aqueous S and N chemistry involving $H_2O_2$ and $O_3$, are active in the ENA atmosphere. Oxidant concentrations appear to be important in accounting for variations in $SO_4^{2-}$ concentrations and deposition. Thus, the $NO_x$/VOC chemistry is important to S behavior in ENA.
- Dry deposition depends not only on aerodynamic factors but on surface resistance, which makes deposition highly variable. Deposition is difficult to estimate reliably from ambient concentration data and even more difficult to measure directly.
- Both wet and dry deposition occur in ENA. The dry process is believed to be dominant close to sources because of its proportionality to ambient concentration. At greater distances, evidence suggests that the total is evenly divided between the two processes. In very remote areas, wet deposition may predominate.
- Measurements and modeling have defined coarse subregional SRRs and ENA and have provided guidelines establishing zones of influence for significant source contributions.
- SRR characterization has progressed marginally since the early 1980s through significant improvements in theoretically based models. Neither old nor newer models have been verified thoroughly for reliability using field measurements.
- Information on SRRs for $NO_y$ has been severely hampered by incomplete and sparse data on nitrogen species distributions in the atmosphere. The material balance for $NO_y$ by species has not been completed for conditions in ENA. Data for spatial distributions in $NO_x$ ambient concentrations and their variability have only recently been acquired through EMEFS; they remain to be analyzed and interpreted.
- Ability to estimate quantitatively the consequences of emission changes on exposure remains inhibited by a lack of confidence in the reliability of projections from models or from inferences based on measurements of deposition. Reliability is most uncertain for conditions in which $NO_x$ and HC (VOC) emissions change substantially or spatial distributions would change. Input data for emissions and their variation have improved for $SO_x$ and $NO_x$, but remain uncertain for HCs.

Ongoing studies and new work must address significant limitations.

- Knowledge of the spatial and temporal character of dry deposition and ambient concentrations aloft to 6–7 km for both $SO_x$ and $NO_y$ needs to be enhanced. Without this knowledge, exposure will remain only partly specified and testing the reliability of models to estimate a material balance of emitted S and N species will be impossible.
- Spatial variations in chemical conversion need to be characterized more completely to estimate their significance in shaping distributions of pollutant concentration and deposition. Of particular concern are cloud processes during different seasons, where oxidant levels change and the ice-liquid phase shifts at cloud height change substantially.
- Knowledge of regional nitrogen oxide chemistry and species distributions needs to be expanded. Without completing an analysis of the EMEFS data, models cannot be satisfactorily verified for $NO_y$ behavior and exposure levels cannot be established reliably.
- The importance of oxidant chemistry is established, but routine HC or VOC measurements are inadequate to use reliably as input data for models. Such observations need to be made systematically and routinely along with organic oxygenates to better characterize the ENA photochemical environment.
- Field measurements with high spatial resolution characterizing $10^4$-$km^2$ areas have been sparse. Evidence indicates that spatial gradients are quite distinct for reactant species versus products. Systematic measurements are still needed with different grid resolution (spatial density) to ensure that processes are described in detail.
- The estimation of SRRs needs to be refined and made more quantitative for specific receptors or source groups with field data to directly test the model calculations.
  - —The transport and mixing component of atmospheric processing should be traced simultaneously and repetitively for several sources using release of inert tracers, as exemplified by CAPTEX and ANATEX.
  - —In principle, it is feasible to trace reactive species through atmospheric processing using stable rare isotope releases. Such experiments could be attempted to further test the hypothesis that transport and mixing dominate SRR characteristics for S and N.

## FUTURE PROSPECTS

Experiments planned for the 1990s along with analysis of the large EMEFS data set will provide a variety of new data for direct analysis and interpretation and for testing and verification of theoretical models. One can expect steady progress in knowledge about regional-scale processes over the next decade, but the prospect of dramatic breakthroughs is uncertain. Continuing research is needed to reinforce current understanding of atmospheric chemical behavior. The studies are particularly important for improving the SRR estimates, especially if the very high costs of reducing emissions to protect receptors from intense exposure are realized. Although measurements and theory show coarse quantitative relationships between sources and receptors, these results have not been considered reliable enough for rationalizing emission control strategies for individual sources. Completion of experimental programs now under way will expand the base of knowledge and increase confidence in results for state-level strategies, where interstate pollution behavior must be considered.

## CONTEXT OF REGULATORY POLICY

Over the past 20 years, substantial progress has been made toward formulating a scientific basis for effective emission management. Despite this progress, the experience of regional studies has had relatively little influence on regulatory policy except to clarify the nature and significance of major source areas to widespread, regional air quality determination. Frustration developed among North American scientists when the Clean Air Act Amendments (CAAA) were adopted by the U.S. Congress in 1990 (U.S. Public Law 101–549). This political action preempted the assimilation of new science from NAPAP into a new regulatory strategy. Despite 10 years of NAPAP and related effort beginning in 1980, the political strategy for $SO_x$ (and $NO_x$) emission reductions in the CAAA was largely formed by 1984. The remaining time to legislation passage involved political consensus building and a reversal of a "no-reduction" policy in the executive branch of the U.S. government. What was lost to the public were the choices for protection of vulnerable receptors given knowledge of source contributions as a function of distance and regional-scale transport direction. Crude analyses already indicated major differences in costs to the public of employing these options for the same level of receptor protections (e.g., Husar, 1983).

The acid rain provisions of the CAAA (Title IV) call for a national program to reduce $SO_x$ emissions of 10 million tons below 1980 levels by the year 2000, with a cap of 8.95 million tons. The Act calls for a reduction of 2 million tons of $NO_x$ from coal-fired power plants during the same time period. Reductions are not specified explicitly on a geographic basis, but the largest sources are targeted for early reduction strategies also call for $SO_2$ emissions credit targeting as a means of achieving reduction goals (Stensvaag, 1991). In its Title I, the CAAA would also require reductions of $NO_x$ (and probably VOC) to deal with the nonattainment of the present NAAQS for $O_3$ in many areas of the United States. At this time, such reductions have not been evaluated in terms of the Title IV requirements. The notes in Chapter 3, in combination with evaluation of geographic source distributions and the theory of $O_3$ production, indicate problems for NAAQS ($O_3$) attainment in the eastern United States by simple rollback (proportional reduction) of $NO_x$ and VOC emissions. Many analyses have shown that local ozone levels depend not only on ambient $NO_x$ and VOC concentrations but also on the VOC/$NO_x$ ratio. The ratio changes with distance from sources such that local $O_3$ reduction may exacerbate conditions downwind, as found in Southern California. The regional scale is even more complex and has not been analyzed carefully for ENA.

If the scientific community wants to play a significant role in strategic policy development for air pollution management, this community will have to be more effective in its communications to the decision maker. This will involve an appreciation and understanding of the timing and the criteria for consensus building, along with analyses that are relevant to informed decisions.

An improved ability to bring increasingly esoteric science to the lay public will be critical to the future impact of environmental science on uniformed public policy. Conversely, continued public support of environmental science on a scale of NAPAP and now climate change will depend on valuing science for improving decision making over the use of basic intuition or common sense. In this respect, I believe that continued investment in regional studies of atmospheric chemistry is inherently important. Without such studies, the public will not really know if their policies and judgments in the era of the late 20th century were truly effective in protecting the future of human and natural ecosystems.

The dimension of effective communications of *relevant information* to the lay public should be included in any topical list of shortcomings in scientific knowledge. Relevant information can no longer be confined to the debate about the nature and significance of environmental stress. It will increasingly be extended to provide options for improved management of environmental risk, taking into account benefits of protection versus cost of pollution reduction.

APPENDIX

# A CHEMICAL CONVERSION AND DEPOSITION PROCESSES

## INTRODUCTION

The interpretation of observations and the construction of theoretically based models depend on knowledge of atmospheric processes. Atmospheric transport and mixing from air motion have been studied extensively for many years in conjunction with weather phenomena. Perhaps the greatest recent expansion of knowledge relevant to regional-scale phenomena has come from research on the chemical processes affecting $SO_x$ and $NO_y$ and the production of oxidants, as well as dry and wet deposition.

The theory of these processes is summarized in this appendix. Comments are given about the application of such knowledge to the interpretation of regional-scale phenomena. The discussion on chemical conversion is largely derived from Calvert *et al.* (1985). Information on dry deposition of gas and particles comes from reviews of Hicks and Matt (1986) and Slinn (1986). Scavenging mechanisms in clouds and precipitation are described by Hidy (1984b), Hales (1986), and Slinn (1986).

## CHEMICAL CONVERSION

Much of the fundamental chemistry and physics related to the acid deposition issue is reasonably well established. Yet in the development of sophisticated transport and transformation models, it is difficult to incorporate accurately the many processes that lead to acid deposition in an integrated, systematic way. The key ingredients needed to address the problem are the amounts and locations of the emissions of both anthropogenic and natural S, N, and organic compounds; the transport and dispersion of the acids and their precursors over the hundreds of kilometers that may separate large source and sensitive receptor areas; the rates of chemical transformation of $SO_x$, $NO_x$, and organic compounds to acids; and the rates of dry and wet deposition of the acid-forming species and their emitted precursors. This discussion evaluates present knowledge of one important aspect of this problem: the chemical mechanisms leading to acid production in the troposphere. The detailed mechanisms of acid generation in the troposphere are important to transport and transformation models, the calculations of which are sensitive to the choice of transformation rates used. This sensitivity results largely from the differences in how readily the acid precursors and the acids themselves are removed from the atmosphere. Once $SO_2$ is oxidized to $H_2SO_4$ aerosol, for example, it is believed to be dry deposited less rapidly than is $SO_2$. If aerosols composed of $H_2SO_4$ and its salts (such as ammonium bisulfate and ammonium sulfate) are incorporated into precipitating clouds, then the deposition of these species can be faster than that of gaseous dry deposited $SO_2$. $HNO_3$ developed in the troposphere is much more rapidly deposited on the surface of the earth than are its precursors, NO and $NO_2$. Thus, the amounts and the chemical nature of the acids deposited at sites many kilometers from the source of the precursors are functions of the rates with which the acids are developed from the precursors. The link between chemical conversion and deposition is quantified through model calculations, the reliability of which depends on the fidelity of process description. There have been many reviews of the chemistry of acid production in the troposphere, and a consensus on the important reactions leading to acid generation is developing (Calvert, 1984).

## Gas-Phase Processes

Both gas- and liquid-phase chemical processes are important sources of acid generation within the troposphere. Gas-phase reactions not only generate $H_2SO_4$ and $HNO_3$ through oxidation of $SO_2$ and the oxides of nitrogen (NO, $NO_2$) but also lead to the generation of several different oxidants that are important reactants for acid generation in the condensed (cloud) phase.

To follow the chemistry of acid development, Calvert considered the results of computer simulations of the gas-phase chemistry expected to occur at the extremes of the seasons (1 January and 1 July) at 40°N latitude. The reaction scheme used is a version of Calvert and Stockwell (1984), in which the $NO_y$ chemistry has been updated in view of studies of Cantrell *et al.* (1985) and the zenith angle–dependent photochemical processes are treated more realistically. The chemistry was simulated in a model by means of four scenarios of initial concentrations believed to be representative of the nonurban atmospheres of ENA during these seasons. The rate constants of the chemical reactions were varied with chosen temperature profiles. The ranges of the temperatures used, 16° to 30°C for 1 July and −6° to 4°C for 1 January, are representative of those encountered during these days at stations in ENA near 40°N latitude.

The $J$ values (the apparent first-order rate constants for molecules undergoing photodecomposition) for the 16 photochemical processes included in the mechanism were calculated for elevations near sea level and in accord with the seasonal and diurnal variations in the solar irradiance and the known adsorption and quantum yield data for these species. Here $J(x) = (\sigma)_\lambda(\phi_x)_\lambda\ (I)_\lambda\ d\lambda$, where $(\sigma)_\lambda$, $(\phi_x)_\lambda$, and $(I)_\lambda$ represent the absorption cross section of the $x$ species, the primary quantum yield of the particular process involving $x$, and the solar irradiation for the small wavelength interval for $\lambda$ to $\lambda + d\lambda$, respectively. The integral is over all wavelengths for which the $\sigma\phi I$ product is nonzero. The July and January $J$ values for $NO_2$ and $O_3$ differ a great deal. In theory these differences play a large role in defining the seasonal variations observed in acid generation in the troposphere.

The mixture of HCs that make up the organic component of the reactive mixture is given a regional perspective based on measurements of Caldwell (1980) and Lonnemon (1979). The components identified in the air range from low-molecular-weight alkanes and alkenes to aromatics. The urban mixture is enriched with acetylene ($C_2H_2$), $n$-butane ($n$-$C_4H_{10}$), propylene, *iso*-pentane, toluene, and *m*-xylene. The rural sample is represented by a diluted urban mixture, which is relatively depleted in more reactive species. An index of depletion is sometimes identified in relation to acetylene, a low-reactivity HC. This approach is generally consistent with the mixtures cited, except for ethane ($C_2H_6$), which appears to be enriched in the rural sample relative to $C_2H_2$.

$SO_2$, $NO_2$, and HCs are oxidized to "secondary" species by several oxidizing agents that are formed in gas-phase reactions in the troposphere. The important reactive species include molecules such as $O_3$, $H_2O_2$, methyl hydroperoxide, and peroxyacetic acid and reactive free radicals such as HO, $NO_3$, and $HO_2$. The mechanisms of their formation, involving NO, $NO_2$, HCs, aldehydes, and sunlight, are now reasonably well characterized.

The concentration of $O_3$ is established in the troposphere through the photolysis of $NO_2$ [reaction (A.1)], followed by reaction (A.2) and, to an undefined extent, by the mixing of stratospheric $O_3$ through occasional stratospheric–tropospheric mixing near storm systems.

$$NO_2 + h\nu\ (\lambda < 430\ \text{nm}) \rightarrow O(^3P) + NO \quad (A.1)$$

$$O(^3P) + O_2(+M) \rightarrow O_3 + (+M) \quad (A.2)$$

where $\lambda$ and $\nu$ are the wavelength and frequency of light and M is any molecule as a collision body. $O_3$ is removed, in part, by reaction (A.3) with NO to re-form $NO_2$:

$$O_3 + NO \rightarrow NO_2 + O_2 \quad (A.3)$$

The cycle of reactions (A.1)–(A.3) that occurs during the day leads to a concentration of $O_3$ within the troposphere that is given approximately (usually within a factor of 2) by relation (A.4):

$$[O_3] = [NO_2]\,J(1)/[NO]k_3 \quad (A.4)$$

where $k_3$ is the rate constant for reaction (A.3). The apparent first-order photodissociation constant for reaction (A.1), $J(1)$, is proportional to the sunlight intensity present in the troposphere in a wavelength range that causes the dissociation of $NO_2$ (290–430 nm). Thus, the $O_3$ concentration usually grows as the sunlight intensity increases during the day, and the $[NO_2]/[NO]$ ratio climbs as NO is converted to $NO_2$. The observed NO-to-$NO_2$ conversion is promoted through chain reactions (A.5)–(A.9) involving the HO, $HO_2$, and organic peroxy radicals ($RO_2$) formed from the HCs and their oxidation products (such as aldehydes and CO).

$$HO + RH \rightarrow H_2O + R \quad (A.5)$$

$$R + O_2 \rightarrow RO_2 \quad (A.6)$$

$$RO_2 + NO \rightarrow RO + NO_2 \quad (A.7)$$

$$RO + O_2 \rightarrow R'COR'' + HO_2 \quad (A.8)$$

$$NO_2 + NO \rightarrow HO + NO_2 \quad (A.9)$$

R′COR″ represents a ketone or aldehyde; R′ and R″ are H atoms or organic radicals such as $CH_3$ and $C_2H_5$.

The important hydroxy radical HO that initiates this sequence of reactions is formed in the relatively clean atmosphere largely through the photodecomposition of $O_3$:

$$O_3 + h\nu(\lambda < 306\ \text{nm}) \rightarrow O(^1D) + O_2 \quad (A.10)$$

$$O(^1D) + H_2O \rightarrow 2HO \quad (A.11)$$

$$O(^1D) + O_2/N_2 \rightarrow O(^3P) + O_2N_2 \quad (A.12)$$

Thus, the HO radical concentration that develops in a tropospheric air mass is a complex function of the concentration and the nature of the hydrocarbons present, [NO], [$NO_2$], humidity, and the intensity of the solar radiation at wavelengths <306 nm, which controls the magnitude of $J(O_3)$, the apparent first-order rate constant for reaction (A.10). Seasonal differences in these major factors result in the lower rates of oxidant generation in the winter than in summer: $O(^1D)$ generation in reaction (A.10) occurs at a lower rate in winter, and the fraction of $O(^1D)$ atoms that generate HO radicals in reaction (A.11) decreases at lower [$H_2O$] characteristic of the winter months, as reaction (A.12) gains importance.

The reactions of the HO free radical with $SO_2$ and $NO_2$ are judged to be major gas-phase sources of $H_2SO_4$ and $HNO_3$, respectively:

$$HO + SO_2\ (+M) \rightarrow HOSO_2\ (+M) \tag{A.13}$$

$$HO + NO_2\ (+M) \rightarrow HONO_2\ (+M) \tag{A.14}$$

The $HOSO_2$ radical product of reaction (A.13) leads ultimately to $H_2SO_4$. The relative importance of the alternative pathways for the reactions of this free radical has been studied before (Cox, 1974; Davis *et al.*, 1979; Margatan, 1984). This work suggests strongly that reaction (A.15), a radical chain-carrying step, represents its dominant fate in the atmosphere (Stockwell and Calvert, 1983; Margatan, 1984); the $SO_3$ product of reaction (A.15) reacts rapidly to generate $H_2SO_4$ in reaction (A.16):

$$HOSO_2 + O_2 \rightarrow HO_2 + SO_3 \tag{A.15}$$

$$SO_3 + H_2O \rightarrow H_2SO_4 \tag{A.16}$$

Other gas-phase mechanisms of $SO_2$ oxidation are expected theoretically; reaction (A.17) can occur with $O(^3P)$ atoms and reaction (A.19) with "Criegee" intermediates (for example, $CH_2OO$), which are formed from $O_3$-alkene reactions such as reaction (A.18):

$$O(^3P) + SO_2\ (+M) \rightarrow SO_3(+M) \tag{A.17}$$

$$O_3 + C_2H_4 \rightarrow \underset{CH_2—CH_2}{\overset{O–O–O}{|\quad\ \ |}} \rightarrow CH_2OO + CH_2O \tag{A.18}$$

$$CH_2OO + SO_2 \rightarrow CH_2O + SO_3 \tag{A.19}$$

$$SO_3 + H_2O \rightarrow H_2SO_4 \tag{A.16}$$

Theory suggests that oxidation of $SO_2$ in reaction (A.17) can be important only in stack $NO_2$- and $SO_2$-rich gas plumes, whereas reaction (A.19) may be significant only in highly polluted urban atmospheres. The dominance of the HO-$SO_2$ reaction in the gas-phase generation of $H_2SO_4$ has been suggested in many studies involving computer simulations (Calvert *et al.*, 1983). For air masses with concentrations of impurities in the range common to the ENA, reaction (A.13) is believed to provide >98% of the pathway for $H_2SO_4$ generated in gas-phase reactions. It is instructive to compare the total rates of $SO_2$ oxidation from gas-phase reactions as estimated by computer calculations. The calculated rates exhibit both seasonal and diurnal variations that reflect largely the changing rates of generation of HO radicals in the photochemical processes involving $O_3$, reactions (A.10)–(A.12), and other species. In midsummer under clear skies, the maximum theoretical rate is about 9% $SO_2$/h for polluted conditions and somewhat lower (6% $SO_2$/h) for the cleaner, rural atmosphere. Previous simulations have shown that the rates of $SO_2$ oxidation expected for the highly polluted and very clean atmospheres are both somewhat lower than those for the air masses of intermediate levels of impurities. These midsummer rates are not insignificant when one considers that theoretically they will result in 34–49% conversion of an initial quantity of $SO_2$ to acid during a 24-hour clear period. Simulations for the midwinter cases show much lower rates, with only 3.1–4.7% conversion of $SO_2$ in a 24-hour period.

Current reaction rate data and computer simulations of tropospheric chemistry suggest that the gas-phase oxidation of $NO_2$ proceeds much more rapidly than that of $SO_2$, largely through reaction (A.14) during daylight. The rate constant for this HO radical reaction is ~10 times that for reaction (A.13) with $SO_2$. During the night an additional reaction sequence is expected to lead to $HNO_3$ as well; it is initiated by the $O_3$ molecule reactions with $NO_2$ (Calvert and Stockwell, 1984).

$$O_3 + NO_2 \rightarrow NO_3 + O_2 \qquad (A.20)$$

The $NO_3$ radical product of reaction (A.20) is a relatively unimportant source of $HONO_2$ during the daylight hours, as it is destroyed readily through its rapid photolysis in sunlight and by its rapid reaction with NO:

$$NO_3 + \nu(\lambda < 640\text{ nm}) \rightarrow O_2 + NO \qquad (A.21)$$

$$\rightarrow O(^3P) + NO_2 \qquad (A.22)$$

$$NO_3 + NO \rightarrow 2NO_2 \qquad (A.23)$$

The generation of NO through $NO_2$ photolysis in reaction (A.1) and $NO_3$ loss in reactions (A.21) and (A.22) cease at night, and much of the existing NO is oxidized to $NO_2$ by $O_3$ after sunset. Thus, the loss of $NO_3$ in reactions (A.21)–(A.23) should become unimportant, and the $NO_3$ concentration theoretically rises at sunset. Various acid-forming reactions of $NO_3$ can become important during the night. For example, $NO_3$ reacts with $NO_2$ to form $N_2O_5$ in reaction (A.24). $N_2O_5$ may react with water to form $HNO_3$ in reaction (A.25), or it may form the acid directly by H atom abstraction from aldehydes and other reaction compounds [reaction (A.26)]:

$$NO_3 + NO_2\ (+M) \rightleftharpoons N_2O_5\ (+M) \quad (A.24)$$

$$N_2O_5 + H_2O \rightarrow 2HONO_2 \quad (A.25)$$

$$NO_3 + RCHO \rightarrow HONO_2 + RCO \quad (A.26)$$

The rate constant for the gas-phase reaction (A.25) between $N_2O_5$ and $H_2O$ is uncertain, as undetected heterogeneous processes can lead to an overestimate of this rate. Tropospheric measurements of Platt *et al.* (1980) yield $k_{25} < (2.9 \pm 0.9) \times 10^{21}$ cm$^3$ per molecule per second. Other laboratory estimates of Tuazon *et al.* (1983) yield $k_{25} < (1.3 \pm 0.2) \times 10^{21}$ cm$^3$ per molecule per second. As the accuracy in $k_{25}$ estimates has improved, the value has tended toward zero; Calvert *et al.* (1985) assumed $k_{25} = 0$ in simulations. The gas-phase chemistry alone can, in theory, convert $NO_2$ to $HNO_3$ at very significant rates. For midsummer conditions, ~18 and 26%/h, respectively, of the $NO_2$ can give $HNO_3$ in gas-phase reactions; in midwinter, rates are much lower in theory, 0.8 and 1.9%/h for rural and urban conditions, respectively. Thus, if gas-phase chemistry alone were operative during a 24-hour period, 98.5 and 99.8% of an initial quantity of $NO_2$ could be oxidized to $HNO_3$ for midsummer cases and about 18 and 37% for the midwinter cases. If the values reported by Tuazon *et al.* (1983) are correct, the 24-hour average rates of $NO_2$ gas-phase oxidation are significantly higher than these estimates. Solution-phase reactions can also contribute, especially to the winter period rates, as discussed below.

Gas-phase reactions can also generate organic acids by various mechanisms (Calvert and Stockwell, 1983). Reactions of the Criegee intermediates with, for example, $H_2O$ vapor or $HO_2$ radicals with aldehydes can lead to formic acid, acetic acid, and the higher organic acids. The theoretical fraction of total acid that is composed of organic acids is relatively large. For tropospheric air masses with the typical range of compositions of the organic gases present, theory would estimate between 5 and 20% of the total acids to be organic acids.

## Aqueous-Phase Cloud Water Chemistry

***$H_2SO_4$ Generation*** Many of the efficient oxidizing agents formed from gas-phase chemistry in the troposphere, such as $O_3$, $H_2O_2$, peroxyacetylnitrate, methylhydroperoxide, and peroxyacetic acid, do not react with gaseous $SO_2$ at measurable rates, although the thermodynamic potential for reaction is great. When these species are dissolved in cloud water, they readily oxidize dissolved $SO_2$ (largely in the form of $HSO_3^-$ ions).

Calvert *et al.* (1985) estimated the relative significance of the various possible tropospheric oxidants for bisulfite in cloud water in a treatment is similar to that originally given by Penkett *et al.* (1979) and in subsequent evaluations by Martin (1984), Schwartz (1984a), and Calvert and Mohnen (1983). They assumed that the gaseous $SO_2$ concentration is 1 ppb and that this concentration is in equilibrium with $SO_2$ in cloud water. Pure water equilibrated with atmospheric $CO_2$ alone would have a pH of ~5.6. Charlson and Rodhe (1982) estimated that the pH of rain water uncontaminated by anthropogenic sources would range from 4.5 to 5.6 because of the variability of the S cycle alone. In the absence of basic materials such as $NH_3$ and $CaCO_3$, average pH values of 5 are expected to occur in pristine locations. For most of the oxidants considered, the rates are strongly pH dependent. For all but the reactions of $H_2O_2$, $CH_3O_2H$, and $CH_3COO_2H$, the rates decline sharply as cloud water becomes more acidic; this effect results largely from the decreased solubility of $SO_2$ as pH decreases.

For the typical conditions in nature, the highest rates of $HSO_3^-$ oxidation are observed for the reactant $O_3$ in solutions at pH>5. In these calculations it is assumed that the cloud water is in equilibrium with gaseous $O_3$ at 50 ppb, and the rate constants recommended by Hoffmann (1986) are used. The rates are somewhat slower for 15°C than for supercooled, liquid cloud water at −5°C. In this case, the enhancement of the rate at the lower temperature arises from increased solubility for the reactants that more than offsets the lower rate constant at the lower temperature. In the case of the $O_3$ oxidant, the rate decreases as the cloud water becomes more acidic; at −5°C, rates of $10^4$, $10^3$, $10^2$, 10, and 1% $SO_2$/h are expected for pH values of ~5.6, 5.1, 4.6, 4.1, and 3.5, respectively.

Gas absorption is estimated by thermodynamic projection in terms of Henry's law,

$$(\text{conc})_{\text{aq.}} = H(\text{conc})_{\text{atm}}.$$

Henry's law constants along with the negative logarithm of the ionization constant for acids (pK) are given in Table A.1.

For the lower pH values usually encountered in clouds and rainwater, the theoretical rates of oxidation of S(IV) by $H_2O_2$ are also very high and, more important, the rates are relatively insensitive to the pH over the entire range of values. With $H_2O_2$, as well as $CH_3O_2H$ and $CH_3COO_2H$, the decrease in rate expected to occur with increased acidity because of the lower $SO_2$ solubility is almost completely offset by the increase in rate caused by acid catalysis of the reaction. Hoffman (1986) has derived theoretical rates in this case, and the gas-phase concentration of $H_2O_2$

**TABLE A.1 Essential Thermodynamic Data, Henry's Law Constants (*H*), and Acidity Constants (pK)[a] at 25.0°C**

| Component | [C] (atm) | *H* (M atm$^{-1}$) | [C] (M) | p$K_{a1}$ | p$K_{a2}$ |
|---|---|---|---|---|---|
| HONO | $1.0 \times 10^{-9}$ | 49 | $4.9 \times 10^{-8}$ | 3.15 | — |
| $CH_3OOH$ | $2.0 \times 10^{-10}$ | $1.0 \times 10^{3}$ | $2.0 \times 10^{-7}$ | 11.5 | — |
| $CH_3COOOH$ | $1.0 \times 10^{11}$ | $1.0 \times 10^{4}$ | $1.0 \times 10^{-7}$ | 8.20 | — |
| PAN | $3.0 \times 10^{-8}$ | 3.6 | $1.1 \times 10^{-7}$ | — | — |
| $HO_2$ | $4.0 \times 10^{-11}$ | $5.0 \times 10^{4}$ | $2.0 \times 10^{-6}$ | 4.45 | — |
| HCHO | $3.0 \times 10^{-8}$ | $6.3 \times 10^{3}$ | $1.9 \times 10^{-4}$ | — | — |
| OH | $1.0 \times 10^{-13}$ | 200 | $2.0 \times 10^{-11}$ | 11.9 | — |
| $H_2O_2$ | $1.0 \times 10^{-9}$ | $1.6 \times 10^{5}$ | $1.6 \times 10^{-4}$ | 11.65 | — |
| $O_3$ | $5.0 \times 10^{-8}$ | $1.0 \times 10^{-2}$ | $5.0 \times 10^{-10}$ | — | — |
| $SO_2$ | $2.0 \times 10^{-8}$ | 1.24 | $3.2 \times 10^{-5}$ | 1.89 | 7.22 |

[a] For definitions of $K_{a1}$ and $K_{a2}$, $[S(IV)] = H_{SO_2}P_{SO_2}R_0^{-1}$; $R_0 = [SO_2 \cdot H_2O]/[S(IV)]$; $R_1 = [HS_3^-]/[S(IV)]$; $R_2 = [SO_3^{2-}]/S(IV)]$; pH = 5.0

$$R_0 = \frac{[H^+]^2}{([H^+]^2 + K_{a1}[H^+] + K_{a1}K_{a2})};$$

$$R_1 = \frac{K_{a1}[H^+]^2}{([H^+]^2 + K_{a1}[H^+] + K_{a1}K_{a2})};$$

$$R_2 = \frac{K_{a1}K_{a2}}{([H^+]^2 + K_{a1}[H^+] + K_{a1}K_{a2})};$$

where [S(IV)] = sulfuroxidation state ($SO_2$, $SO_2 \cdot H_2O$, $HSO_3^-$); $P_{SO_2}$ = partial pressure of $SO_2$ over water.

equilibrated with the cloud water is 1 ppb. This leads to a $[H_2O_2]$ in the cloud water of about $3.4 \times 10^{-5}$ M for the 15°C scenario, a value representative of the ambient measurements in cloud and rainwater in the nonurban atmosphere during the nonwinter seasons (Kadlecek *et al.*, 1983; Richards *et al.*, 1983).

In theory, other peroxy compounds such as peroxyacetic acid and methylhydroperoxide can lead to S(IV) oxidation in tropospheric cloud water. These and other organic peroxides, expected from the computer simulations, have not yet been identified in the troposphere. The Henry's law constraints and rate constants for $CH_3O_2H$ and $CH_3COO_2H$ reactions with S(IV) have been studied by Lind and Kok (1986). These data have been used to estimate the rates of S(IV) oxidation by these oxidants, assuming that 1 ppb (total gas and liquid phase) of each oxidant is present. The rates of oxidation of bisulfite ion by these species in cloud water at pH <5 are significantly lower than those for $H_2O_2$ at equivalent total gas- and liquid-phase concentrations. Although the rate constants for these oxidants with S(IV) are nearly equal to the rate constants for $H_2O_2$, the oxidants' lower solubilities in water lead to the lower rates.

In theory, a highly significant contribution to $HSO_3^-$ oxidation can arise from HO and $HO_2$ radicals formed in gas-phase reactions but transported to the cloud droplets as suggested by Chameides and Davis (1982). They assumed a cloud droplet radius of 12 $\mu$m and three different choices for the sticking coefficient of the free radicals on the droplet ($\alpha$), $10^{-1}$, $10^{-2}$ and $10^{-3}$; $\alpha$ is the fraction of the total collisions between the gaseous species and the surface that result in capture of the species by the droplet. Calvert *et al.* (1985) adopted the parameters used by Chameides and Davis for their standard model for this case: solar zenith angle, 30°; cloud transmissivity, 0.5; liquid water content of cloud, 1.5 g/m$^3$; [$O_3$], 30 ppb; and temperature, 7°C. Minimal reliability data are available on the actual size of $\alpha$, but it is in the range of $\geqslant 10^{-2}$. With such values, this reaction pathway could be an important source of acid in the troposphere during daylight hours.

Various other liquid-phase reactions may contribute to S(IV) oxidation, but they appear to be less important than the peroxides and $O_3$. The data of Penkett *et al.* (1979) show that the uncatalyzed oxidation of S(IV) by oxygen is unimportant for tropospheric conditions (Calvert *et al.*, 1985). However, Fe(III) and Mn(II) ions catalyze this oxidation, so this process can be an important source of acid for certain conditions. Calvert *et al.* (1985) chose a range of concentrations that are believed to be typical of the rainwater in the eastern United States: [Fe(III)] = $10^{-5}$, [Mn(II)] = $10^{-6}$ M. [Fe(III)] = $10^{-7}$, [Mn(II)] = $10^{-8}$ M. At the highest concentrations assumed here, the equivalent gas-phase rates of $SO_2$ conversion to acid could be > 100%/h in cloud water at pH>5. These rates decline substantially as the acidity of the cloud water increases: 100, 10, and 1%/h for pH values of 5, 4, and 3, respectively. For the conditions observed in Los Angeles fog water by Hoffman and Jacob (1984), $\leq 10^{-4}$ and $10^{-5}$ M for Fe(III) and Mn(II), respectively, this pathway of acid generation could dominate all others, especially during the night.

Oxidants such as peroxyacetylnitrate and its higher homologs, even when present at 5 ppb in the gas phase, are relatively unimportant oxidants for S(IV) (Lee, 1984).

The Calvert *et al.* (1985) study indicates that graphitic carbon (1 mg/L cloud water) is an unimportant catalyst for the S(IV) oxidation in clouds present in air masses with usual pollutant concentrations. For the carbon catalysis to compete with $H_2O_2$ in solutions with pH values <6, its concentration must exceed 0.1 g/L, a value that is probably encountered rarely in the troposphere. In some cases a very high carbon content has been observed in precipitation (Davis *et al.*, 1984), and graphite carbon–catalyzed oxidation of bisulfite ions could be significant for these seemingly rare conditions. $NO_2$ rates of S(IV) oxidation have also been shown to be unimportant for the commonly encountered tropospheric conditions (Schwartz, 1984a; Martin *et al.*, 1981).

Simulations of the gas-phase chemistry for case studies reported by Calvert *et al.* (1985) give insight into the expected concentrations of various oxidants that can be developed in theory for midsummer and midwinter conditions. The estimates of the concentrations of $CH_3OH$ and $CH_3COO_2H$ are qualitative indicators of these quantities, as the rate constants for the formation reactions of these species $CH_3O_2 + HO_2 \rightarrow CH_3O_2 + O_2$, $CH_3COO_2 + HO_2 \rightarrow CH_3COO_2H + O_2$ are uncertain. Calvert *et al.* have chosen the lower of current estimates for these reactions, $k = 1.5 \times 10^{-12}$ cm$^3$ per molecule per second (Kerr and Calvert, 1984). Neglecting loss processes at the ground in these simulations adds to the uncertainty of these estimates, but the estimates should be qualitatively useful. In the case of $H_2O_2$, $CH_3O_2H$, and $CH_3COO_2H$, the theoretical maximum concentrations can be well above the 1-ppb level assumed. There is a large difference between the concentrations of all the oxidants formed in summer and in winter. Theoretical midsummer maxima (ppb) for two summer cases (urban and rural) are $H_2O_2$, 12, 4.7; $CH_3O_2H$, 4.2, 2.0; $CH_3COO_2H$, 4.1, 0.008; and $O_3$, 150, 35. For midwinter cases, much lower values were calculated: $H_2O_2$, 0.002, 0.015; $CH_3O_2H$, $6.8 \times 10^{-5}$, $1.3 \times 10^{-3}$; $CH_3COO_2H$, $4.4 \times 10^{-5}$, $6.1 \times 10^{-4}$; $O_3$, 44, 27.

***$HNO_3$ Generation in Clouds*** Comparisons of clear air composition with that of cloud water in winter measurements suggested to Lazrus *et al.* (1983) that $HNO_3$ acid was produced in clouds because negligible amounts of $HNO_3$ were found in winter cloudy air. The most probable reactions involve the uptake of $N_2O_5$ and/or $NO_3$ into cloud water (Heikes and Thompson, 1981, 1984; Richards, 1983). The reaction sequence is the following:

$$O_3 + NO_2 \rightarrow NO_3 + O_2 \quad (A.20)$$

$$NO_3 + NO_2 \rightleftharpoons N_2O_5 \quad (A.27)$$

$$N_2O_5 + H_2O\ (\text{cloud water}) \rightarrow 2H^+ + 2\ NO_3^- \quad (A.28)$$

The $NO_3$ radicals may also be captured by cloud water and react, in part, with soluble organic matter, such as hydrated formaldehyde ($HOCH_2OH$), to generate $HNO_3$ as well as oxidation products of the organic species in reaction (A.29):

$$NO_3 + HOCH_2OH(aq) \rightarrow H^+(aq) + NO_3^-(aq) + HOCH_2O(aq) \quad (A.29)$$

$$HOCH_2O(aq) + O_2(aq) \rightarrow HCO_2H(aq) + HO_2(aq) \quad (A.30)$$

$$2HO_2(aq) \rightarrow H_2O_2(aq) + O_2(aq) \quad (A.31)$$

The theoretical studies of Heikes and Thompson (1981, 1984) suggest that the rates of $HNO_3$ formation in clouds by reactions (A.28) and (A.29) and their analogs can be significant (equivalent gas-phase conversion rates for $NO_2$ of hundreds of %/h), provided that the sticking coefficients for $N_2O_5$ and $NO_3$ on cloud droplets or moist aerosol particles are $>10^{-3}$. The nighttime generation of $HNO_3$ and nitrate aerosol ($NH_4NO_3$) in Los Angeles (Richards, 1983) and the depletion of $NO_3$ (and by inference its coupled $N_2O_5$ storage source) in nighttime atmospheres of high humidity or in the presence of clouds or rain (Platt, 1983) provide indirect field evidence for these reactions. In theory, the $N_2O_5$–cloud water reaction (A.28) can become important in winter when the lower temperatures favor the maintenance of the $N_2O_5$ product. Significant accumulation of this compound in the atmosphere can occur during the night, and $HNO_3$ can be formed as the $N_2O_5$ comes in contact with clouds, fogs, or moist aerosol surfaces common in the winter season. It seems probable that the diminished rate of $HNO_3$ formation from the HO-$NO_2$ reaction (A.14) during the winter months is supplemented by an $N_2O_5$–cloud water reaction during this period. It is also probable that the relative insensitivity of the aq $NO_3^-$ deposition data to season reflects the contribution of reaction (A.28). The $SO_2$ analog of the reaction sequences (A.20), (A.27), and (A.28) cannot lead to the generation of $H_2SO_4$ at night. Oxidation of $SO_2$ by $O_3$ in the gas phase is very slow (Calvert, 1984), and the midwinter rates of $SO_4^{2-}$ generation are reduced with the decline in generation of HO, $HO_2$, and other oxidants.

## Coupled Gas/Liquid Acid Generation

Theoretical considerations indicate that generally the two most important oxidants for $HSO_3^-$ in cloud water, $H_2O_2$ and $O_3$, arise in large part from gas-phase reactions. Calvert *et al.* (1985) observed that the generation of $O_3$ depends on a complex set of gas-phase reactions that are a function of the concentrations of $NO_x$, HCs, and their oxidation products, the relative humidity, and the sunlight intensity. In the gas phase, $H_2O_2$ arises from the interactions of hydroperoxy ($HO_2$) and hydrated hydroperoxy $H_2O \cdot HO_2$ radicals:

$$2HO_2(+M) \rightarrow H_2O_2 + O_2\ (+M) \qquad (A.32)$$

$$HO_2 + H_2O \rightleftharpoons H_2O + HO_2 \qquad (A.33)$$

$$H_2O + HO_2\ (+M) \rightarrow H_2O_2 + O_2 + H_2O(+M) \qquad (A.34)$$

The apparent second-order rate constant ($k$) for $H_2O_2$ formation $[R(H_2O_2) = k(HO_2)^2]$ is very sensitive to the concentration of water vapor; increasing the $H_2O$ pressure from 0 to 14 torr doubles the constant

(Kircher and Sander, 1984). In theory, other mechanisms can also contribute to $H_2O_2$ generation in the cloudy troposphere. For example, Zika *et al.* (1982) reported that the time dependence of $[H_2O_2]$ in rain during storms in southern Florida and the Bahamas did not follow that for $SO_4^{2-}$ and aq $NO_3^-$ in the rainwater. They concluded that aqueous-phase $H_2O_2$ generation may be important. Chameides and Davis (1982) have suggested from theoretical considerations that $HO_2$ radicals, generated in the gas phase, can diffuse to cloud water and that aqueous-phase reactions of $HO_2$ (and $O_2^-$ derived from $HO_2$) could account for a significant rate of $H_2O_2$ formation in clouds.

Schwartz (1984b) came to a similar conclusion on the basis of other theoretical considerations. Accepting that the sticking coefficient of $HO_2$ radicals of cloud droplets is $>10^{-3}$, the rates of $H_2O_2$ generation by this coupled gas-phase cloud process are in theory approximately equal to those of the gas-phase sequence of reactions (A.32)–(A.34). Calvert *et al.* (1985) and Hoffmann and Jacob (1984) have suggested an additional type of $H_2O_2$-forming reaction: photosensitized generation at the surface of certain sunlight-absorbing semiconductor solids suspended in cloud water. These processes can be important sources of $H_2O_2$ only if substantial amounts of efficient sensitizers are present in the cloud water. The limited concentration data available for solid photosensitizers in precipitation suggest that this mechanism of $H_2O_2$ formation is important in few circumstances. Kok (1982) has suggested that lightning is another potential source of $H_2O_2$ in clouds. However, the relatively large amounts of $H_2O_2$ predicted in simulations come entirely from the gas-phase processes (A.32)–(A.34), a preferred mechanism for $H_2O_2$ formation.

Field studies have suggested that one of the largest uncertainties in the current theory of acid generation in the troposphere has been the magnitude of the available supply of $H_2O_2$ present in the atmosphere. Hypothetically, $H_2O_2$ could be the limiting reagent for acid generation under many conditions. Measurements of the gas-phase $H_2O_2$ concentration in the nonurban troposphere of ENA have provided experimental estimates of this important constituent using airborne, $H_2O_2$ detection equipment. Lazrus and other investigators made flights over ENA during October and November 1984. Significant levels of $H_2O_2$ were observed in most of the air masses sampled. Theory suggests that the largest gas-phase $H_2O_2$ concentrations will be observed in summer and the smallest in winter, with an intermediate concentration in the spring and autumn, during which these first measurements were made. The $H_2O_2$ concentration in winter precipitation at Whiteface Mountain, NY, is typically much smaller than that in summer samples. This observation is expected if the reactions of the $HO_2$ radicals, generated in reaction sequences that are initiated photochemically, are the dominant source of $H_2O_2$. Findings confirm that gaseous $H_2O_2$ is present in the troposphere during the autumn in sufficient quantity to be a major oxidant for S(IV) in cloud water, and it is a reasonable extrapolation of theory that gas-phase $H_2O_2$ will be more abundant during summer in ENA.

In view of the considerations of Calvert *et al.* (1985), the observed seasonal dependence of $SO_4^{2-}$ deposition in ENA, largest in midsummer and lowest in midwinter, probably reflects the much greater rates of generation of oxidizing agents (such as HO radical, $O_3$, $H_2O_2$, and $CH_3O_2H$) in summer than in winter. All of the oxidant-forming reactions are triggered by the HO radical formation involving the $O_3$ molecule [reaction (A.10)]. The general shape of the distribution of seasonal average $SO_4^{2-}$ deposition shows a summer maximum corresponding to the maximum seasonal $J[O_3]$ values. Ambient concentrations of particulate $SO_4^{2-}$ also tend to show a summer seasonal maximum with the maximum in $J$ values. This suggests that both the gas- and liquid-phase chemistries of $H_2SO_4$ generation are strongly influenced by photochemistry. In contrast, there is no strong dependence of the aq $NO_3^-$ deposition on season. This can be explained in terms of the changing contribution with season from the two major $HNO_3$ sources. In summer, $HNO_3$ generation will probably be dominated by the fast reaction (A.14) with some acid generation in the reaction sequence (A.20), (A.27), and (A.28). In winter, when the $N_2O_5$ is maximal, the $HNO_3$ generation in reaction (A.28) can become more important as reaction (A.14) decreases in importance. The strong seasonal variations in $SO_4^{2-}$ production, coupled with the nearly constant aq $NO_3^-$ generation during the seasons, leads to the observed seasonal dependence on the $SO_4^{2-}/NO_3^-$ molar ratio in wet deposition (Calvert *et al.*, 1985). Current theories on acid development in the troposphere agree qualitatively with these limited observations. The results suggest that the seasonal variation in $SO_4^{2-}$ deposition results largely from an oxidant limitation during the winter when photochemical activity is at its minimum. The extent to which the summer acid deposition is regionally oxidant limited is uncertain. Furthermore, available data have not been analyzed for such oxidant limitations near sources, even though SRRs have been explained in this light.

## DRY DEPOSITION

The exchange of atmospheric trace chemicals with the surface is influenced by a wide variety of meteorological, chemical, and biological processes, any one of which constitutes an intriguing target for specialized research. Modeling atmosphere–surface exchange calls for a balanced consideration of all the relevant processes, without undue emphasis on any particular aspect of the overall phenomenon.

Turbulent transfer between the air and the surface is bidirectional. S species present an interesting and important example. $SO_2$ distributed throughout the lower atmosphere is carried to the surface by turbulence, causing some material to enter into the air space of plant canopies. Most of the $SO_2$ that enters plant biomass is transferred through stomata, but some is also absorbed by plant cuticle and other parts of the vegetation (Chamberlain, 1980). Once the deposited S is dissolved in mesophyllic fluid, biological processes can chemically reduce part of it, permitting subsequent emission as $H_2S$ (Taylor and Tingey, 1983). In addition, there can be a flux of $H_2S$ and other S species from the ground, especially if the area is rich in decaying matter. Thus, upward air motions can carry with them S species derived from surface biological activity, as well as the residual $SO_2$ left after incomplete surface capture of material transported downward by other eddies.

This description leads to a circumstance in which the surface plays an active role in the overall turbulent exchange process. If numerical modeling is concerned with the long-range transport and subsequent atmospheric concentrations of $SO_2$ alone, then it is the species-specific transfer of $SO_2$ to the surface that is of interest.

Both deposition and emission are subsets of a more general phenomenon—turbulent exchange between the atmosphere and the surface. Only in special cases can the net turbulent transfer be considered unidirectional. (Nevertheless, this is the conventional modeling assumption.) A frequently encountered problem is how to combine the wide range of contributing processes into a single model without detrimental simplification and in a manner that is suitable for field application in measurement programs as well as in conjunction with transport modeling.

## Deposition of Gases

***The Resistance Analogy*** The components of a multiple-resistance model used by many workers (Hicks *et al.*, 1985) combine separate formulations of meteorological, chemical, and biological processes and relate atmospheric fluxes F to concentrations $C_a$ in air near the surface. Separate resistances are identified with transfer through the air $r_a$, with uptake in the canopy (involving a resistance $r_s$), and with an intermediate canopy-level quasi-laminar flow resistance $r_b$. An additional resistance is shown to represent exchange with the surface itself.

In most applications of a resistance model to pollutant deposition, the various surfaces at which deposition occurs are considered to be ultimate sinks of the material in question. One could generalize exchange with vegetation at the same time as emissions are occurring from the underlying soil at a rate of $E$. Then manipulation of the appropriate relations indicates that the net flux observed by sensors mounted above the canopy would be

$$F = r_s E - C_a)/[(r_a + r_b) + r_s] \qquad (A.35)$$

where $(r_a + r_b)$ is the resistance associated with transfer through the air to the receptor surface and $r_s$ is the net resistance associated with uptake by the vegetation. If sensors mounted above the canopy detect a negative value of $F$, indicating net exchange toward the surface, then the apparent deposition velocity $V_d = -F/C_a$ can be written as

$$V_d = 1/[r_a + r_b + r_s)] - (E/C_a)$$

$$[r_s/(r_a + r_b) + r_s] \qquad (A.36)$$

If $E$ were constant with time, then a plot of $F$ against $C_a$ (or of $V_d$ against $C_a$) would be informative, since a negative intercept might be expected. However, errors of measurement and observational scatter will greatly influence any such analysis. The matter is further complicated by the fact that $E$ is a complicated function of many external variables.

*Aerodynamic Resistance* The formulas for aerodynamic resistance $r_a$ are derived from micrometeorological relationships. For the deposition of gases or very small particles on land and frozen surfaces,

$$r_a = [\ln(z/z_0) - \psi_c]/ku_* \qquad (A.37)$$

and the friction velocity,

$$u_* = \tau/\rho = k\overline{U}/[\ln(z/z_0) - \psi_m] \qquad (A.38)$$

Here $z$ is the height, $z_0$ is the roughness length, $k$ is the von Kármán constant (0.4), $\tau$ is the shearing stress at the surface, $\rho$ is the air density, $\overline{U}$ is the mean horizontal wind speed at reference height $z$, and $\psi_m$ or $\psi_c$ are air density stability correction factors for momentum and contaminant transport. $\psi_m$ and $\psi_c$ are functions of stability to mixing in air.

1. Stable atmospheric condition (cool air under warm air[1]

$$\psi_m \simeq \psi_c = -5z/L \qquad (0 < z/L < 1) \qquad (A.39)$$

[1] More precisely, stable conditions are defined in terms of a vertical temperature change that is less than the lapse rate for an adiabatic atmosphere. ($\Delta T/\Delta z < -10°C/km$). Unstable conditions refer to $\Delta T/\Delta z > -10°C/km$.

2. Unstable conditions (warm air over cool air)

$$\psi_m = \exp[0.032 + 0.448 \ln(-z/L) - 0.132(\ln(-z/L)^2)] \quad (A.40)$$

and

$$\psi_c = \exp[0.598 + 0.390 \ln(-z/L) - 0.090 [\ln(-z/L)^2] \quad (A.41)$$

Here the scale length $L$ is the Monin and Obukhov length. This parameter is often determined from a relation involving the Richardson number,

$$\mathrm{Ri} = (g/T_0)\frac{\partial\theta/\partial z}{(\partial\overline{U}/\partial z)^2}$$

where $g$ is the gravitational constant, $\theta$ is the mean potential temperature,[2] and $T_0$ is the surface temperature.

For deposition of gases on smooth water surfaces,

$$r_a\ (+\ r_b) = [\ln(ku^*z/D) - \psi_c]/ku_* \quad (A.42)$$

where $D$ is the molecular diffusivity of the gas. For wavy water surfaces, Eq. (A.37) is more appropriate.

*Quasi-Laminar Layer* The intermediate or quasi-laminar transfer resistance depends on the fluid (air) mixing near the surface. The nature of the surface roughness or the vegetation canopy plays an important role in this microscale flow and thus influences the mass transfer rate to the surface. For most trace gases, $r_b$ is not a critical property. It becomes important, however, for species with high molecular weight and consequently low molecular diffusivity, for particles, or in any circumstances in which the surface is smooth enough (or the atmosphere stable enough) that atmospheric turbulence is minimal and a significant quasi-laminar layer can develop.

For gases, the layer resistance $r_b$ is conventionally stated in terms of the Schmidt number, $\mu/\rho D$, which is the ratio of molecular transport of momentum and mass (where $\mu$ is the viscosity of the gas). A form of power law for $r_b$ is sometimes adapted:

$$r_b \sim u_*^{-1}\,\mathrm{Sc}^{-n} \quad (A.43)$$

where $n = \frac{1}{2}$ to $\frac{2}{3}$.

[2] The potential temperature is the temperature that a parcel of air would reach if it were brought adiabatically (without heat loss) to a pressure of 1000 millibars (mb).

*Surface Resistance* When characterizing regional-scale processes, it is necessary to use some procedure for simplifying surface exchange processes in a rational yet convenient way. Surface resistance for soluble gases is influenced to a great extent by surface wetness. Perhaps the single most important surface parameter for nonbiological material is moisture. Seasonally the surface can have distinctly different moisture characteristics. Even snow can change surface moisture with temperature. Ice is known to have a liquid water surface layer above temperatures of roughly −10 to −5°C. The deposition velocity changes dramatically with snow temperature. The break point is approximately −5°C.

In agrometeorology and forest meteorology, the surface is often viewed, from above, as being a uniform "big leaf," with stomatal, cuticular, mesophyll, and other exchange properties characteristic of the biological species of which the surface is constructed, but it is modified to represent the effective behavior of the plant canopy as a whole instead of individual biological surfaces. A single stomatal resistance is used to characterize the entire canopy, as is a single cuticular resistance, etc. Values of these resistances can be derived from two different kinds of approach: either directly from aerodynamic measurements made above the surface or by building up the gross canopy exchange properties from a more detailed subcanopy model. The first of these two methods is more common, because the second involves detailed consideration of transfer through air within the canopy and transfer with individual surface elements that will have different uptake characteristics depending on their location. (For example, see Baldocchi *et al.*, 1987, for a discussion of the stomatal resistance component.)

Following the work of Jarvis (1976), among others, stomatal resistance is expressed (for each biological species) as a function of several controlling variables, primarily photosynthetically active radiation, air temperature, humidity, and plant water stress. The first three of these variables can be measured using state-of-the-art instruments. Plant water stress is acknowledged to be important, but it cannot be measured easily; instead, it is necessary to infer water stress from other information. Every plant species has its own characteristic description in these terms. The techniques that are being used are based on information available in the agricultural, meteorological, and biological literature, with corrections being applied to account for the differences in molecular diffusion of different gaseous species (Bennett *et al.*, 1973). In essence, transfer through open stomata is controlled by molecular diffusion; known values of molecular diffusivity of gaseous species are used to adjust stomatal resistances originally derived for water vapor or $CO_2$ exchange.

Once gases enter stomatal cavities, they can be subjected to chemical reactions occurring in the cavity (involving the wide variety of natural HC species present there) or can follow the same uptake path as for $CO_2$ (involving absorption by mesophyll tissue). For many gases, the relative importance of these two pathways is uncertain. $O_3$, for example, is known to enter plant tissue through stomata, but it is sufficiently insoluble that dissolution into mesophyll tissue seems likely to be inefficient. Although chemical reactions within stomatal cavities may play an important role (Tingey and Taylor, 1982), these uncertainties and the fact that experiments show small mesophyll resistance to trace gas exchange of $SO_2$ and $O_3$, mesophyll resistances are not yet considered in detail in the multiple-resistance model.

Cuticular resistances are also of some importance, especially at night, when stomatal activity is reduced to zero but uptake across the leaf cuticle can continue. Values drawn directly from plant physiological literature are used with modification to account for the role of transfer through the canopy before encountering cuticle surfaces. Transfer directly to sub-canopy soil is being used to generate improved values of the biologically influenced resistance components used in the big leaf model (Baldocchi *et al.*, 1987).

Biologically influenced resistances depend on the chemical species. The biological nature of the surface also influences the aerodynamic resistance to transfer. However, chemical speciation does not influence aerodynamic resistance (provided the species in question does not react to release heat during its turbulent transport); the aerodynamic resistance to transfer is determined by atmospheric turbulence. The rougher the surface, the greater is turbulence intensity and the smaller is aerodynamic resistance. Turbulence intensity depends on several factors including friction at the surface, surface heating, and evaporative coding.

Voldner *et al.* (1986) estimated from the literature values of $r_s$ for the gases $SO_2$ and $NO_x$ for each surface in the land-use data bank and for each season. For $SO_2$, resistance $r_s$ to water is assumed to be zero, because the gas is quite soluble at the pH levels found in most surface waters. Values of $r_s$ for $SO_2$ range from 0.1 to 20 s/cm and for $NO_x$ from 1 to 70 s/cm. The assumption is consistent with the observations of Owers and Powell (1974). It does not conform to the anomalously high value found over a small artificial lake by Garland (1977), who found $r_s$ to water is assumed to be zero in 8 cases out of 14 and to average 0.45 s/cm in the remaining 6 cases. Garland speculated that the finite resistance may be due to an organic layer on the water surface. More experiments are needed over natural waters to resolve the uncertainty.

Most measurements of $SO_2$ deposition to forests have been made over pine stands. The measurements indicate that $r_s$ has a strong diurnal and seasonal variation, that it follows the transpiration rate of the plant, and that it is a complex function of the physical properties and physiological state of the plant (Hosker and Lindberg, 1982). The $r_s$ values higher than stomatal resistances estimated from transpiration occur most of the time, suggesting that there is a finite mesophyll resistance for $SO_2$ within the stomata. Low uptake during stomatal closure suggests that $r_s$ associated with cuticle and bark is high.

Evidence of a saturation effect after exposure to a high dose of $SO_2$ when plants are dry further complicates the situation; $r_s$ is generally lower initially for wet plants than for dry plants. However, as exposure to $SO_2$ continues, the wet plant resistance increases at a rate that depends on the oxidizing ability (metal content, etc.), the pH, and the organic content of the liquid film surrounding the leaf/needle. If the partial pressure of the gas decreases, uptake is reduced and may even result in release of $SO_2$. Considering the complexity of the gas–surface interaction, it is not surprising that measured or estimated values of $r_s$ have a large range.

In contrast to $SO_2$, relatively little information exists on the dry deposition of $NO_x$ to natural surfaces. The lack of knowledge is, in part, due to the difficulties in measuring concentrations of the various oxides of nitrogen. The values for forests are based primarily on the work of Grennfelt *et al.* (1983) and Hicks *et al.* (1983). Grennfelt *et al.* measured long-term uptake of $NO_x$ by pine needles in Sweden by the Kuvett method. A living branch of pine was enclosed in a chamber, and $NO_x$ was passed through the chamber, thus eliminating the aerodynamic resistance. The researchers found that the uptake followed the transpiration rate of the plant. It was a function of the size and number of stomata of the needles, the physiological activity of the plant, and the leaf area index. In the daylight hours of the growing season (May–October), $r_s$ was 1.3–2.5 s/cm. During nighttime, in fall and winter, it was much larger. A strong diurnal variation in $r_s$ was observed by Hicks *et al.* (1983) for deciduous forest. Average dry deposition velocities peaked at 1.2 s/cm in the morning and decreased to essentially zero at night.

Based on the foregoing observations, $r_s$ over pine forest for $NO_x$ is taken as 10 s/cm for winter, day or night. For the remainder of the year, it is set at 1 and 5 s/cm during day and night, respectively. For deciduous forest in winter $r_s$ was chosen to be 20 s/cm, similar to that for $SO_2$. It was chosen to be marginally higher than for coniferous forest during fall and early spring because of a lower leaf area index.

Over cultivated fields, grassland, swamps, and water bodies during winter, 50 s/cm was selected. This reflects low uptake by snow-covered surfaces observed by Gravenhorst and Böttger (1982) and Granat and Johansson (1983). They found that the deposition velocity to snow was less than 0.015 cm/s (67 s/cm) and 0.03 cm/s (33 s/cm), respectively.

For grassland, assuming active growth in spring, well-cropped pastureland in summer, and a somewhat revived growth in fall, daytime $r_s$ is chosen to be 2, 3, and 2 s/cm, respectively. Nighttime values are higher, 5, 4, and 10 s/cm, respectively. These choices are based on the micrometeorological observations of Duyzer *et al.* (1983) and Delany and Davis (1983), as well as on the chamber studies of Gravenhorst and Böttger (1982).

For urban areas a value of 10 s/cm is chosen for $r_s$. If one takes an urban area to be a mixture of buildings, parkland, and roads, this value is consistent with the field observations of Gravenhorst and Böttger (1982) over cement but is somewhat higher than initial resistance observed by Judeikis and Wren (1978). However, Judeikis and Wren also observed saturation after exposure.

A surface resistance for $HNO_3$ of 0.1 s/cm is chosen. Limited field observations over pastureland using the gradient method support this choice.

## Controlling Gas Processes

The many mechanisms that contribute to transfer between the atmosphere and the surface are sufficiently complicated that identification of dominant processes is often not easy. However, consideration of the major resistance components $r_a$, $r_b$, and $r_c$ is frequently enlightening. Field data taken recently ideally contain sufficient information to enable $r_a$, $r_b$, and $r_s$ to be determined independently. Analysis of controlling resistance involves (1) determination of the total resistance to transfer between the atmosphere and the surface as $r_t = -C_a/F$, (2) computation of the aerodynamic resistance $r_a$ from the supporting measurements of wind speed and the turbulent exchange rates of momentum, sensible heat, and water vapor (relevant expressions are summarized by Wesely and Hicks, 1977, and Hicks *et al.*, 1985); (3) computation of $r_b$ by assuming the applicability of the canopy-scale result of Garratt and Hicks (1973) and by applying a Schmidt number correction (see Wesely and Hicks, 1977); and (4) calculation of the effective canopy resistance $r_s$ as the residual $r_s = r_t - r_a - r_b$.

Hicks and Matt (1986) plotted scatter diagrams of eddy fluxes exemplified by $SO_2$ and $NO_2$ against the corresponding air concentrations (e.g., Figure A.1). The data were obtained during a field study conducted in Pennsylvania. The data correspond to a period of high atmospheric instability. In the case of the $SO_2$ flux measurements, the observations generally show a downward flux that is roughly proportional to $SO_2$ concentration measured near the surface. Similar characteristics were found for $O_3$. For both $SO_2$ and $O_3$, deposition dominates emission of gas at the surface, although a few positive flux values are noted. The case of $NO_2$ was more complicated. A flux out of the forest is frequent and larger on occasion than the downward flux. Additional studies have shown similar behavior

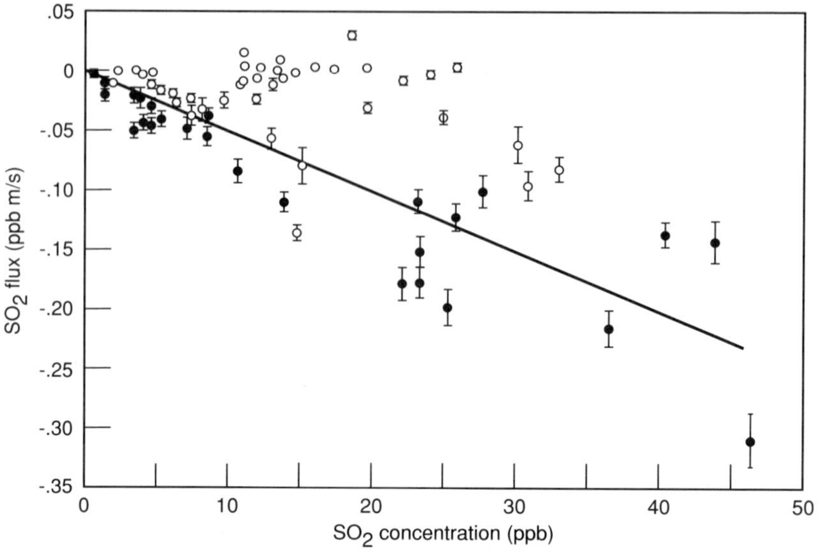


FIGURE A.1 A scatter diagram of half-hour average eddy fluxes of $SO_2$ against atmospheric $SO_2$ concentrations, as measured in a field experiment conducted in Pennsylvania (see Hicks, 1986). Bars correspond to standard errors on the average covariances between concentrations and airflow fluctuation. These are shown whenever large enough to be drawn. All data having errors of measurement smaller than the indicated average are shown, including both day and night values. Solid points indicate values obtained in thermally unstable air conditions (1100–1500 hours local standard time), emphasized to show the overall relevance of a daytime deposition velocity of about 0.5 cm/s as drawn. Note that negative fluxes indicate transfer toward the surface, i.e., deposition. (From Hicks and Matt, 1986).

for $NO_2$. Anomalous signals from instruments may be a factor. However, the soil condition in an agricultural area such as at this site, where nitrogen fertilizer is used, may also be important. Given the ambiguities in these and other data, Hicks and Matt concluded that the simple resistance model is not adequate to describe $NO_2$ surface interactions.

The resistance components can be estimated by applying the resistance model for $SO_2$ deposition. Hicks and Matt (1986) report that all the separate resistance components ($r_a$, $r_b$, $r_s$) have significant diurnal cycles and that the two major factors influencing $SO_2$ exchange are $r_a$ and $r_s$ (since these display the least conductance and hence the greatest resistance). The three components are similar in phase. Transfer through the air is most efficient when surface heating by solar radiation is the greatest, and it is also solar radiation that drives the photosynthesis, which results in trace gas exchange with biological tissue.

The canopy resistance appeared to be most important in determining rates of dry deposition of $SO_2$ in the Hicks and Matt study. The surface studied is on agricultural crop, primarily maize, alfalfa, and soybeans. The magnitudes of the conductances suggest that $r_b$ is likely to become important only for the case of particle exchange.

## Deposition of Particles

Particle deposition is believed to be less susceptible to surface conditions than are gases. Unlike gases, particles deposit without any known biological assimilation. However, surface "stickiness" (such as wetness) is undoubtedly a factor. Although some evidence of intermittent reentrainment of particles exists, particle deposition is generally assumed to be unidirectional to the surface, at least for particles <10 $\mu$m in diameter. A resistance model is often used, but the intermediate-layer $r_b$ is considered dominant in the process. For particles >10 $\mu$m in diameter, "bounce-off" has to be considered. There is little quantitative theory for estimating the reentrainment. The deposition velocity depends on surface characteristics, as well as on particle size. Large particle deposition approaches values determined by gravitational sedimentation (Milford and Davison, 1985). In this regime, the deposition velocity decreases with size in accordance with the sedimentation velocity, $V_s = d_p^2 \rho g F / 18\mu_g$, where $\rho$ is the particle density, and $F$ is a shape factor for nonspherical particles ($F = 1$ for spheres). In the small particle range (<0.1 $\mu$m), Brownian diffusion in the intermediate layer dominates transfer, such that $V_d$ decreases with increase in particle size. For particles in the range 0.1 to 1.0 $\mu$ diameter, neither sedimentation nor Brownian motion is important. In this intermediate regime, deposition is said to be controlled by impaction and interception but is small compared with the small- and large-particle regimes.

Slinn (1986) studied the effect of surface roughness and surface character or particle deposition. The more rough (and wet–rough) the surface, the less particle size is a factor. For very rough surfaces, such as conifer forests, little particle size dependence is expected, and the deposition velocity becomes large.

$SO_4^{2-}$ particles are predominantly in the range 0.1–1.0 $\mu$m diameter. Therefore, their deposition velocity is often taken to be between 0.1 and 1.0 cm/s, with a most likely value of 0.3–0.5 cm/s, independent of air speed, for modeling calculations.

A new correlation for the deposition velocity $[1/(r_a + r_b)]$ that does not explicitly estimate $r_b$ has been proposed by Schack *et al.* (1985). The two coefficients $\beta$ and $\gamma$ depend on the structure of rough surface gases and particulates over a wide range of particle sites and wind velocities.

TABLE A.2 **References for Values of the Correlation Factors $\beta$ and $\gamma$**[a]

| No. | Surface | Reference |
|---|---|---|
| (1) | Sticky artifical grass | (Chamberlain, 1966, 1967) |
| (2) | Rye grass | (Chamberlain, 1966, 1967; Clough, 1975) |
| (3) | 0.45–1.6 cm gravel | (Sehmel *et al.*, 1974) |
| (4) | Artificial grass | (Sehmel, 1972; Sehmel *et al.*, 1973) |
| (5) | Water ($u_* = 44$ cm s$^{-1}$) | (Sehmel, *et al.*, 1974) |
| (6) | Water ($u_* = 117$ cm s$^{-1}$) | (Sehmel, *et al.*, 1974) |
| (7) | Moss | (Clough, 1975) |

[a] From Schack *et al.*, 1985; reprinted with permission from Pergamon Press.

$$X = \frac{\beta d_p}{(\mu/\rho)^{1/2}} \left(\frac{u^*}{z_0}\right)^{1/2} \mathrm{Sc}^{1/3}, \qquad X = \frac{d_p u^{*1/2}}{(\mu/\rho)^{1/2}} \mathrm{Sc}^{1/3} \tag{A.44a}$$

$$X = \frac{\beta d_p}{(\mu/\rho)^{1/2}} \left(\frac{u^*}{z_0}\right)^{1/2} \mathrm{Sc}^{1/3}, \qquad Y = (V_d - V_s)\frac{\gamma d_p}{8D} \tag{A.44b}$$

$V_s$ is the particle sedimentation velocity and $D$ is the gas or particle diffusivity, which is $kTf/3\pi\mu d_p$, $k$ is Boltzmann's constant, $f = 1 + (2\lambda/d\rho)(1.257 + 0.400 \exp(0.555\, d\rho/\lambda)$, and $\lambda$ is the mean free path of the gas. Table A.2 lists values for $\beta$ and $\gamma$.

## WET SCAVENGING

The scavenging of gases and particles by hydrometeors takes place in clouds (rainout) or below clouds (washout) by diverse mechanisms. The gaseous processes involve gas absorption sometimes followed by chemical reaction, as discussed earlier. Particle collection processes involve "capture" resulting from nucleation of moisture on a small particle or collision processes. As in the case of dry deposition on surfaces, encounter and capture may take place as a result of Brownian diffusion, turbulent diffusion, or collisions resulting from differential gravitational settling, inertial impaction, or interception. Although there is no proof by field measurement, calculations suggest that in-cloud processes should be more effective in scavenging particles than below-cloud processes. The collection of material in rainwater is sometimes described in terms of a washout ratio $\xi$, defined as

$$\xi \approx \frac{\text{concentration of material in ground-level precipitation}}{\text{concentration of material in ground-level air samples}}$$

For gases, $\xi \approx H$, the Henry's law constant, if equilibrium is achieved between the hydrometeors and the air. The (wet) deposition rate at the ground, $k_w$, is $\xi P$, where $P$ is the precipitation rate. Under nonequilibrium conditions, scavenging will depend on chemical reactions in the hydrometeor. Examples of reactions in water were discussed previously for $SO_2$ and $NO_x$.

The scavenging rate of particles in or below clouds is often expressed in terms of a collection efficiency $\eta(a, R)$, which depends on the radius $a$ of the hydrometeors present and the aerosol particle radius $R$:

$$\eta(a, R) \approx \frac{\text{mass of particles collected during the hydrometeor's fall}}{\begin{pmatrix}\text{volume of air swept out}\\ \text{by the hydrometeor}\end{pmatrix} \times \begin{pmatrix}\text{mass concentration}\\ \text{of aerosol particles}\end{pmatrix}}$$

## Nucleation Scavenging

Condensation of moisture on initially dry or "moist" suspended particles, *nucleation*, is the mechanism for the production of cloud droplets. In the creation of the droplets, the nuclei are scavenged. As the droplets grow in a supersaturated environment, they accrete and eventually may become large enough to fall to the ground. They fall in a turbulent medium containing suspended material, which can be scavenged by collision or absorption. Nucleation is estimated to be an effective, perhaps dominant mechanism for scavenging hygroscopic aerosol particles, such as $H_2SO_4$, ammonium sulfate [$(NH_4)_2SO_4$] or ammonium nitrate ($NH_4NO_3$). These will deliquesce to form wet condensation nuclei at relative humidity above 65–70%, growing to larger equilibrium sizes as relative humidity approaches 100%. Above 100% relative humidity (supersaturation), particles will be "stable" or in equilibrium at increasingly high supersaturation as particle size decreases until a "critical" diameter is reached below which particles are "unstable" with respect to size. For soluble nuclei, the following proportionalities hold: the diameter of a particle at equilibrium with water vapor is inversely proportional to the ratio of the vapor pressure over the droplet and the equilibrium vapor pressure of the flat surface ($P_{sat}/P_{sato}$ or $S_v$). Achievement of a higher vapor pressure ratio in supersaturated air can lead to progressively active smaller nuclei. There is a critical vapor pressure ratio ($S_{cr}$) which is defined at the critical diameter of wet particles where the vapor pressure ratio is maximum. The presence of a nonsoluble fraction in particles of a given size can significantly raise the supersaturation, $S_{cr}$. The chemical composition of nuclei also plays a role, although the differences between aerosol particles composed of pure solute species are minor.

Figure A.2a shows an example of three hypothetical particle size distributions formed by superimposing the two modes often observed in three aerosol size spectra of $SO_4^{2-}$ (Hidy, 1984b). The smaller (mean diameter ≤ 0.5 μm) mode corresponds to particles formed by condensation of gas reaction products (e.g., $H_2SO_4$); the larger (mean diameter > 0.5 μm) is known as the accumulation mode, which is attributed primarily to coagulation of smaller particles and to emissions from combustion processes. A third, coarser fraction 1 μm is also routinely observed. These "giant"

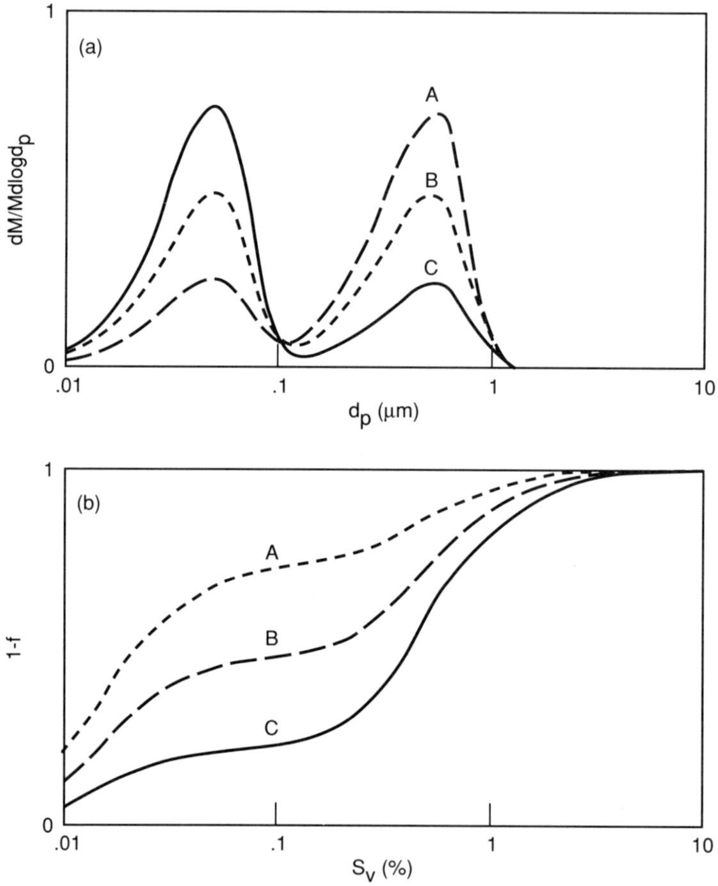


**FIGURE A.2** Hypothetical (a) fog nuclei mass size distributions and (b) scavenging efficiency $(1 - f)$ as a function of supersaturation $(S_v)$. Curves A, B, and C are for 75, 50, and 25% total mass concentration (M) in larger size mode, respectively. (Drawing courtesy of Hoffman, 1986; reprinted with permission from Pergamon Press.)

particles derive from coagulation, from repeated growth in droplets followed by evaporation, and from blowing dust or debris. Calculated activation of particles in the 1-$\mu$m distributions, assumed to be $(NH_4)_2SO_4$ particles, is shown in Figure A.2b as functions of $S_v$. The nonactivated fraction is expressed as $f = C_a/C$, the ratio of unactivated particles to total particle concentration. The ratio of nuclei concentration to total particle concentration $(1 - f)$, then represents the aerosol scavenging efficiency in a fog. In this situation, distribution (A) having the largest coarse particle fraction shows the highest scavenging efficiency, as expected from the $S_v$ versus $d_p$ relationship.

Few measurements of $S_v$ have been made in the atmosphere. Some observations indicate that oscillations of humidities occur in radiation fog with excursions about a mean relative humidity near 100%, but rarely exceed 0.1–0.2%. Hudson (1984) measured interstitial aerosol in fog and cloud environments and found systematic removal (i.e., activation) of large nuclei. These differences were used to calculate an effective supersaturation for the ambient atmosphere. In fogs and sampling of stratus clouds, Hudson's measurements indicated that $S_v$ is in the range 0.03–0.2%. Comparing these measured or calculated values to Figure A.2b shows that little of the smaller-size particle mode is likely to be scavenged in these environments. Conversely, solutes that are associated with a coarser (3 $\mu$m diameter) fraction can be readily activated.

At the same time, it is difficult to determine what fraction of the large nuclei will be activated or what droplet size they will achieve. The relative increase in droplet size is faster for smaller diameters, and modeling has shown it is possible for droplets of any size to grow from nuclei of any mass, depending on the $S_v$ history they have experienced, mixing, and the aerosol size spectrum (Lee and Pruppacher, 1977). Limited field data address this ambiguity. Measurements of solute mass for individual fog droplets have been attempted. Hudson and Rogers (1984) indirectly measured condensation nuclei within stratus cloud droplet spectra, selectively removing droplets above a certain size. They report an increasing fraction of low-$S_{cr}$ (i.e., large) nuclei in the larger droplets and a vanishing fraction of these nuclei in the interstitial aerosol spectrum. The size–composition relationship within droplet spectra may strongly affect the chemistry and deposition of fogborne material. Very little is known from field observations about this question.

Evidence that nucleation (and other scavenging processes) is effective in removing airborne $SO_4^{2-}$ in clouds comes from direct observations of interstitial particle behavior. For example, Schwartz (1987) found that virtually all of the airborne $SO_4^{2-}$ is removed from air by the cloud scavenging process.

## Collision Processes

For aerosol particles of a uniform radius $R<<a$, the rate of mass accumulation on hydrometeors ($W$) is given by

$$W = \pi M \int^{\infty} a^2 q(a)\eta(a, R)n_a(a)\, da \tag{A.45}$$

where $M$ is the total mass of particles per unit volume, $q(a)$ is the vertical speed of the raindrops of radius $a$, and $n_a(a)$ is the raindrop size spectrum. Given the parameters $\eta$, $q$, and $n_a$, in principle, one can calculate the removal rate of particles for conditions where Eq. (A.45) is applicable.

A washout rate coefficient can also be defined from Eq. (A.45) as

$$\Lambda = \pi \int^{\infty} a^2 q(a)\eta(a, R)n_a\, da \tag{A.46}$$

The collection efficiency $\eta$ can be written in terms of the different mechanisms of interest. For cases in which electrical, thermal, and diffusiophoretic forces are weak and are considered secondary (for such special cases, see Wang *et al.* 1978), Slinn (1977) has given expressions for mechanistic elements making up $\eta(a, R)$. These can be derived from appropriate expressions from Hidy (1984b). Slinn (1977) has specified three of the collision mechanisms and has suggested the following three formulas for computing the collection efficiency from the sum of the corresponding component efficiencies:

$$\eta_{\text{impaction}} = [(\text{Stk} - \text{Stk}_{\text{cr}})/(\text{Stk} + \text{Stk}_{\text{cr}} + 2/3)]^{3/2} \tag{A.47a}$$

$$\eta_{\text{interception}} = 3R/a \tag{A.47b}$$

$$\eta_{\text{diffusion}} = 4\text{Sh}/\text{Re Sc} \tag{A.47c}$$

Here the Stokes number (Stk) for the droplet is $2q\rho_p R^2/9\mu_g a$, where $q$ is the fall velocity of the hydrometeor of radius $a$. The critical Stokes number ($\text{Stk}_{\text{cr}}$) is the value of Stk below which collection does not take place because particles follow the curvilinear flow of gas around the hydrometeor. The Sherwood number (Sh) can be calculated from the Frössling equation, $\text{Sh} \simeq 2(1 + 0.4\text{Sc}^{1/3}\text{Re}^{1/2})$, when the Reynolds number is $qa/\nu$. More refined estimates of these component efficiencies are available in the literature (Slinn, 1981).

The corresponding numerical values of $\eta(a, R)$, obtained by summing the component efficiencies, become large for both very large and very small particle sizes and are small at intermediate sizes around 0.1 $\mu$m. Because contributions of secondary mechanisms are disregarded in Eqs. (A.47a–c), $\eta(a, R)$ values computed in this manner can be considered conservatively low estimates of actual behavior. An upper limit estimate of $\eta(a, R)$ of unity can be used. Because this practice leads in some cases to efficiency values 1000 times larger than those obtained from Eq. (A.47a–c), it is somewhat limited in value, at least in the present context, where particle scavenging by nucleation is assumed unimportant. Because of this discrepancy, Eqs. (A.47a–c) are recommended for practical calculation under the conditions where nucleation scavenging is small. Refinements for these expressions are available, for example, in Slinn (1981).

When summed, these expressions give a collection efficiency that will be large for both large and small particles, with the minimum at intermediate sizes around 0.5 $\mu$m particle diameter. The minimum, sometimes called the *Greenfield gap* for its discoverer, has been found in scavenging observations during precipitation. In three cases reported by Radke *et al.* (1980) the minimum in scavenging is observed for particles between 0.5 and 1.0 $\mu$m in diameter. There was also evidence that efficiencies in the Brownian diffusion regime tends to decrease for particles smaller than 0.1 $\mu$m in diameter. This result has not been explained yet, but it may be related to electrical charging effects.

The calculation of the washout rate coefficient requires estimation of the fall velocity of the hydrometeors. The fall velocity of raindrops, $q(a)$, depends on their shape as well as internal fluid circulation (Pruppacher and Klett, 1978). For practical estimates, empirical fits to measurements are available. Those of Dingle and Lee (1972) are often applied. For $q(a)$ in cm/s,

$$(0.05 \le a \le 0.7 \text{ mm}) \qquad q(a) = 27.2692 - 1206.2884a + 348.0768a^2$$

$$(0.7 \le a \le 2.9 \text{ mm}) \qquad q(a) = 155.6745 - 613.4914a + 123.3392a^2 \qquad \text{(A.47)}$$

The size spectrum of raindrops varies by cloud type. However, for practical calculations, the Marshall–Palmer distribution is often chosen (see Pruppacher and Klett, 1978),

$$n_a(a) = N_R/\Delta_a \exp(-\Delta_a a) \qquad \text{(A.49)}$$

where $\Delta_a = 8.2P^{-0.21}$ mm$^{-1}$ is a rainfall-dependent parameter ($P$ is the rainfall rate in mm/h). In this application, the raindrop concentration $N_R$ should be approximately 1950 $P^{-0.21}$ drops/m$^3$.

If an aerosol size distribution is polydispersed, the collection efficiency is more complicated, taking the form

$$\Lambda = \pi \int_0^\infty \int_0^\infty a^2 q(a) \eta(a, R) n_a(a) n(R)\, da\, dR \qquad \text{(A.50)}$$

Examples of washout coefficient curves for various rain and aerosol size distributions have been reported by Dana and Hales (1976).

Irregular and varied geometries of snow or ice particles complicate the calculation of the washout coefficient because of ambiguities in size distributions, fall velocities, and scavenging efficiencies. The relation for washout (scavenging) coefficient for snow particles is in the same form as for raindrops, but the drop radius is replaced by an equivalent radius $R_e$ and combined with a relation for the precipitation rate ($P$),

$$W \simeq \text{const} \times P\eta(d_e, a)/L \tag{A.51}$$

where the constant is of order unity and $L$ is a characteristic cross-sectional length for the hydrometeor. These have been reported by Slinn (1981) and are: graupel 0.014 cm, plates and stellar dendrites 0.0027 cm, powder snow and spatial dendrites 0.001 cm, plane dendrites $3.8 \times 10^{-4}$ cm, and needles $1.9 \times 10^{-3}$ cm. A semiempirical relation for $\eta$ or $\Lambda$ can be found in the work of Slinn (1981), or Knutson and Stockham (1977).

Assimilation of suspended material during the cloud formation and decay process involves the transport of trace gases and particles into the cloud by air motion, transport inside the cloudy air, then microscopic processes of interphase transfer. Analysis of the microphysics of interphase transport supports the notion that contamination of cloud particles intimately involves the water nucleation process. Junge (1963) has estimated that as much as 50–80% of the mass of aerosol may be active in the condensation process during a typical storm. The additional complexity of attachment by inertial forces or by diffusion along with collision and accretion of hydrometeors results in the net scavenging found in hydrometeors falling out of clouds as precipitation. The efficiency of both the nucleation process and the collisional processes in scavenging of material depends on the life of the cloud and the air flowing through the cloud system. A detailed mathematical description incorporating all mechanisms has not been attempted. However, the microphysical basis for attempting such a theoretical calculation with useful approximation is accessible (Mason, 1971; Pruppacher and Klett, 1978). The limitation now lies in the scarcity of actual in- and extra-cloud observations of air chemistry to confirm the validity of approximations required for a practical solution to such problems.

In the absence of detailed parameterization of rainout and washout processes, the use of collection rate theory accounting for the dynamic scavenging processes may be unwarranted. The scavenging of airborne material is often treated empirically in terms of material balances, which have been discussed in the literature using two analogous parameters, the washout ratio $\xi$ and the rainout efficiency $\varepsilon_R$, after Junge (1963).

For particles, the data available from nuclear bomb debris indicate that

$\xi \simeq 10^6$ (Gedeonov *et al.*, 1970). Slinn (1981) has reviewed a variety of additional data for washout ratio; the data show considerable scatter but tend to decrease with precipitation rate and increase with particle size. The washout ratios extend over the range $10^4$–$10^6$. Generally, the larger ratios are associated with particles greater than 1 $\mu$m in radius; smaller ratios appear with particles less than 1 $\mu$m.

The rainout efficiency is calculated assuming that the component mean concentration $\bar{\rho}_A$ in the air is uniform with height and scavenging by washout is small compared with rainout. The rainout efficiency $\varepsilon_R$ is defined by noting that the concentration of material A in rainwater, $C_A$, is proportional to the fraction in the air. In other words,

$$C_A = \rho_L \varepsilon_R \bar{\rho}_A / \text{LWC} \tag{A.52}$$

where LWC is the liquid water content in grams per cubic meter. For $C_A$ in milligrams per liter and $\rho_L$ in grams per cubic meter, $\bar{\rho}_A$ is in micrograms per cubic meter at standard temperatures and pressure, and $O \leq \varepsilon_R \leq 1$.

If the concentration ratio $C_A/\bar{\rho}_A$ is taken at the ground, the rainout efficiency calculated on a volume basis is equivalent to the product of the washout ratio and LWC/$\rho_L$.

In many cases, the liquid water content of precipitating clouds lies between 0.5 and 2.0 g/m$^3$. For rural conditions, cloud formation and precipitation may occur in relatively uniform pollutant concentrations. In such situations, the rainout efficiency may be represented by ground-level conditions. Table A.3 gives some values based on aerosol and rain data taken simultaneously, estimating $\varepsilon_R$ by precipitation event for $SO_4^{2-}$ and $NO_3^-$. If particulate $SO_4^{2-}$ and $NO_3^-$ only are used to calculate $\varepsilon_R$, values of $\varepsilon_1$ and $\varepsilon_3 > 1$ are found, which cannot occur by the definition of $\varepsilon_R$. The values for $SO_4^{2-}$ ($\varepsilon_3$) are much greater than unity, which is not possible. If the calculation includes $SO_2$, values of $\varepsilon$ for $SO_4^{2-}$ ($\varepsilon_2$) are reduced compared with values based on particulate $SO_4^{2-}$ only ($\varepsilon_1$). There is a suggestion of segregation of events in cases where $SO_2$ absorption is important compared with particulate $SO_4^{2-}$ scavenging. Particulate $SO_4^{2-}$ concentrations are low at ground level but were found to be substantially larger aloft in these experiments. Calculation of $\varepsilon_3$ based on values aloft provides lower values, which can be reduced further to less than unity by including the highly soluble gas $HNO_3$. Estimates of EUS $HNO_3$ concentrations aloft at 300 m are about 1 ppb. Using this value in addition to particulate $NO_3^-$ concentrations gives a rainout efficiency which suggests that more $NO_3^-$ found in rain may come from the gas phase rather than from particle scavenging per se.

**TABLE A.3 Calculated Values of the Rainout Efficiency Based on Data Taken during Summer 1977 and 1978 in the Northeastern United States[a,b]**

| Date | Aerosol composition ($\mu mol/m^3$) | | | Rainwater composition ($\mu mol/L$) | | Rain volume ($cm^3/h$) | $SO_4^{2-}$ rainout efficiency[b] | | $NO_3^-$ rainout efficiency[b] |
|---|---|---|---|---|---|---|---|---|---|
| | $SO_2$ | $SO_4^{2-}$ | $NO_3^-$ | $SO_4^{2-}$ | $NO_3^-$ | | $\varepsilon_1$ | $\varepsilon_2$ | $\varepsilon_3$ |
| 8/08/77 | 0.041 | 0.216 | 0.0032 | 35 | 25 | 222 | 0.16 | 0.14 | 7.8 |
| 8/09/77 | 0.041 | 0.080 | 0.0048 | 110 | 87 | 139 | 1.38 | 0.91[c] | 18.0 |
| 8/11/77 | 0.123 | 0.213 | 0.0032 | 86 | 70 | 148 | 0.40 | 0.26 | 21.0 |
| 8/15/77 | 0.041 | 0.024 | — | 46 | 29 | 238 | 1.90 | 0.71[c] | — |
| 8/18/77 | 0.041 | 0.021 | 0.0016 | 43 | 32 | 356 | 2.00 | 0.69[c] | 20.0 |
| 8/25/77 | 0.082 | 0.047 | 0.0016 | 65 | 92 | 55 | 1.40 | 0.50[c] | 57.0 |
| 8/27/77 | 1.68 | 0.134 | 0.0016 | 110 | 80 | 356 | 0.82 | 0.061 | 50.0 |
| 8/31/77 | 1.11 | 0.208 | 0.0016 | 64 | 65 | 255 | 0.31 | 0.048 | 41.0 |
| 7/04/78 | 0.082 | 0.040 | 0 | 25 | 13 | 32 | 0.62 | 0.20 | Large |
| 7/11/78 | 0.123 | 0.033 | 0.0064 | 69 | 77 | 196 | 2.09 | 0.43 | 12.0 |
| 7/15/78 | 0.082 | 0.224 | 0 | 95 | 68 | 46 | 0.42 | 0.31 | Large |
| 7/24/78 | 0.165 | 0.027 | 0.0016 | 42 | 20 | 1010 | 1.56 | 0.22 | 12.0 |
| 7/28/78 | 0.123 | 0.070 | 0.0016 | 81 | 47 | 270 | 1.20 | 0.42 | 29.0 |
| 7/30/78 | 0.123 | 0.053 | 0.0016 | 51 | 50 | 71 | 0.96 | 0.29 | 31.0 |

[a] From Hidy, 1982. Copyright © 1982 by D. Reidel Publishing Co. Dordrecht, Holland.

[b] The rainout efficiency $\varepsilon_1$ is based on the $SO_4^{2-}$ in rainwater and $SO_4^{2-}$ particle concentrations; $\varepsilon_2$ is based on rainwater $SO_4^{2-}$ and the sum of particulate $SO_4^{2-}$ an $SO_2$ (as $SO_4^{2-}$) in the aerosol. The efficiency $\varepsilon_3$ is based on rainwater $NO_3^-$ and particulate $NO_3^-$ sampled on a Teflon-coated glass-fiber filter.

[c] Hypothesized to be dominated by nucleation of water particle on a $SO_4^{2-}$ particle and particulate $SO_4^{2-}$ scavenging.

## Cloud Droplet Deposition on Trees

As discussed in Chapter 3, an important component of moist deposition is in-cloud or fog deposition recurring on high-elevation ecosystems. This case involves the collection of small, windblown droplets, snowflakes, or ice crystals on vegetation. The models developed to describe the process employ knowledge of aerodynamic collection processes described earlier. The particular model often cited is described by Lovett *et al.* (1982). Comparison of deposition mechanisms including in-cloud collection or occult deposition is included in Lovett and Kinsman (1990).

Basically, the model of Lovett *et al.* (1982) envisages the exchange of cloud (fog) droplets and water vapor between the forest canopy and the air as encompassing two aerodynamic regimes. Within the regimes, the fluxes are calculated as a ratio of potential concentration gradients and resistances as in the case of dry deposition. The appropriate gradients are the cloud liquid water content for cloud water deposition and the water vapor pressure for evaporation. Foliage surfaces are assumed to be saturated and perfect sinks for cloud droplets.

According to Lovett *et al.* (1982), the first regime is the air space of the forest canopy, where air is mixed by turbulence. Their analysis apportions this volume into layers 1 m high, calculating between layers the transfer resistances from the aerodynamic roughness of the canopy and the canopy-top wind speed.

The second regime covers the boundary layers surrounding the individual canopy surfaces. Here water vapor transport takes place mainly by molecular diffusion and droplets penetrated by inertial impaction. Resistances to vapor transfer in this regime have been determined for conifer samples. In wind tunnel experiments, Lovett *et al.* have measured resistances to droplet collection for six canopy components. At each height level, the boundary layer resistances for canopy components are summed in parallel. These are converted to a general area base by taking the ratio of the component surface area to ground area.

A similar form is used for evaporation. These consist of a set of linear, coupled equations relating fluxes to the ratio of concentration gradients and resistances for each canopy location. In the case of an example, a balsam fir stand is 10.3 m high. Model calculations of gross and net (gross minus evaporation) droplet deposition fluxes for a range of wind speeds and relative humidities were conducted. The predictions were found to be within 7% of the observed net deposition rates during four cloud events.

The Lovett *et al.* approach has been used to characterize the cloud droplet deposition rates from basic measurements taken at several ENA locations.

APPENDIX

# B ATMOSPHERIC VISIBILITY

High concentrations of suspended particles and $NO_2$ directly affect visibility or atmospheric discoloration. Reduced visibility depends on the amount of suspended material in the sight path of an observer. As an influencing factor, air pollution could be regulated on the basis of visibility determination. Indeed, a history of observations and theory have inferred the association, and visibility is often identified by the lay person as an index of pollution. With increased concern for preservation of visual aesthetics, research on visibility coupling atmospheric optical factors and human perception of optical changes has increased in recent years.

This appendix summarizes the current direction of visibility science. The discussion is based on an article by Richards (1988). Works by Middleton (1958) and Hidy (1984b) contain additional details on atmospheric optical properties and human perception. This appendix also discusses the measurement and prediction of air pollution effects on the light that enters the eye of an observer or the lens of an optical instrument and summarizes atmospheric optics related to visibility along sight paths close to horizontal.

## DEFINITIONS OF TERMS

When an observer views a scene, light from the observed objects travels through the atmosphere to the observer's eye. The straight line from an object being viewed to the observer's eye is a *sight path*. The scenic vista of an object being viewed is called a *target*. The viewer generally "sees" radiance from a target through a narrow range of wavelength on such a sight path. This range is considered centered in the green part of the light spectrum. The *radiance* of a ray of light is equal to the rate of energy transfer through an area perpendicular to the ray divided by the product of the area, the solid angle of view, and the wavelength range. The complete description of the radiance in the following includes information on the light wavelength, location of measurement, and direction of view. The notation uses a subscript to indicate the location of measurement but does not include symbols to indicate the wavelength and direction of view.

When a ray of light passes through the atmosphere, it is weakened by scattering and absorption. The rate of energy loss from the ray is proportional to the radiance of the ray, and the constant of proportionality is the *extinction coefficient* $b_{\mathrm{ext}}$, which has units of length$^{-1}$. The extinction coefficient is the sum of the *scattering coefficient* $b_{\mathrm{s}}$ and *absorption coefficient* $b_{\mathrm{a}}$. An *average extinction coefficient* between points a and b on a sight path is defined as

$$\bar{b}_{\mathrm{ext}} = \frac{1}{b - \mathrm{a}} \int_a^b b_{\mathrm{ext}}\, dx \tag{B.1}$$

Thus, the extinction coefficient and its components $b_{\mathrm{s}}$ and $b_{\mathrm{a}}$ are considered critical optical parameters of the atmosphere dependent on the suspended material in the air. The scattering coefficient is proportional to mass concentration of particles in the 0.1- to 1.0-$\mu$m-diameter range when scattering is most intense. Absorption depends on the absorptivity of material, such as black particles (carbon soot) or colored gas such as $NO_2$.

*Initial radiance* $I_0$ is the radiance measured at some point considered the initial point on the sight path, often measured close to or at the target. The *apparent radiance* $I$ is the radiance entering the eye or measuring instrument a distance $x$ from where the initial radiance is measured. The *transmittance* $T$ on the sight path is equal to the fraction of the initial radiance that traverses the sight path and reaches the location where the apparent radiance is measured. Therefore, the *transmitted radiance* is $I_0T$.

## OPTICAL PROPERTIES

The transmittance of the sight path is related to the average extinction coefficient $b_{\mathrm{ext}}$ by

$$T = \exp[-(\bar{b}_{ext} x)] \qquad (B.2)$$

In other words, the transmittance is a related (exponential) function of scattering on absorbing material suspended in the air. Equation (B.2) contains no approximations or assumptions, but it is subject to the restriction to monochromatic radiation because $b_{ext}$ is a function of wavelength. The equation can be generalized to incorporate a value of $b_{ext}$ for a range of wavelengths corresponding to visible light. Equation (B.2) shows that the transmittance of a sight path of given length is equivalent to knowing the average extinction coefficient for that sight path.

Atmospheric transmittance for a sight path is related to air quality through relationships that are well understood in principle but difficult to deal with in practice (Sloane and White, 1986).

Reporting visibility monitoring data to the public requires a simple index. Under foggy or very hazy conditions, the *visual range* has a practical connotation as an index. Reporting visual range for a clear atmosphere is less satisfactory because the complicated conceptual construct has an ambiguous relation to how far a human on the surface of the earth can see. One advantage it does have is that the visual range is proportional to the reciprocal of the extinction coefficient and, as a result, shows the nonlinear effect of increased extinction on visibility.

The literature in the field of air pollution has come to rely increasingly on the *green light extinction coefficient* as a measure of the visual air quality. For atmospheric conditions, this coefficient is often identified with its strongest contributor, light scattering from particles as given by the scattering coefficients $b_{sp}$; that is, $b_{ext}$ is assumed to be equivalent to $b_{sp}$ for ENA conditions, and

$$b_{sp} \approx \text{const} \times \text{mass concentration of fine particles} \qquad (B.3)$$

Units of reciprocal megameters ($Mm^{-1}$) have been recommended for technical communications. Richards (1988) has suggested, as an alternative to visual range, that publications for the lay public give information on visual air quality by reporting the green light extinction coefficient in units of percent per kilometer. As an example, a visual range of 50 km corresponds to an extinction coefficient of 78 $Mm^{-1}$ or 7.8% $km^{-1}$. Air molecules alone contribute 1% $km^{-1}$ to the green light extinction coefficient at sea level; values smaller than this occur only at high altitudes.

The fact that the extinction in a 1-km sight path is not equal to the extinction coefficient would cause confusion only when the visibility is quite limited. For example, when the visual range is 10 km, the extinction coefficient is 39% $km^{-1}$. In this case, the extinction in a 10-km sight path is 32%, which is not greatly different from the extinction coefficient.

## Apparent Radiance

For any sight path, some of the apparent radiance is due to light scattered into the direction of view, which is then transmitted to the location where the apparent radiance is measured. This portion of the apparent radiance is called the *path radiance* $I_p$. The magnitude of the path radiance depends on many factors, some of which are unrelated to air quality. The path radiance makes distant, dark objects appear lighter and usually somewhat blue.

The apparent radiance is equal to the sum of the transmitted radiance and the path radiance:

$$I - I_0 T + I_p = I_0 \exp(-b_{ext} x) + I_p \tag{B.4}$$

Light entering the eye from a sight path to a distant target comes from two sources: one source is the target itself, which contributes the transmitted radiance, and the other is the intervening atmosphere, which contributes the path radiance.

Except for light from extremely intense sources, such as lasers, the transmittance of the sight path and the path radiance do not depend on the initial radiance. Therefore, knowledge of the transmittance $T$ and path radiance $I_p$ for a sight path makes possible the calculation of the apparent radiance that would be measured for any initial radiance. The ability to determine values for the transmittance and path radiance of a sight path is therefore of great practical interest.

## Equilibrium Radiance

The concept of *equilibrium radiance* $I_e$ is sometimes introduced because it can be used to simplify the mathematics of light transmittance and it helps in the development of an intuitive understanding of atmospheric optics. The equation of radiative transfer governs the propagation of radiation through a medium that scatters and absorbs light. It can be written in the form (Chandrasekhar, 1960)

$$\frac{dI}{dx} = -b_{ext}(I - I_e) \tag{B.5}$$

where $I$ is the radiance in a narrow wavelength range for a given point on a sight path in a given direction of view and $I_e$ is defined by

$$I_e = \frac{w_0}{4\pi} \int_{4\pi} I(Z')P(Z, Z')\, dZ' \tag{B.6}$$

This integral is presented here for background information; the value of $I_e$ will be determined by simpler means than evaluating the integral. The quantity $I_e$ gives the radiance scattered into the sight path by an element of volume on the sight path divided by the extinction coefficient of the atmosphere in that element of volume. In Eq. (B.6), $I(Z')$ is the radiance traveling in direction $Z'$ incident on the element of volume in which the scattering takes place, $P(Z, Z')$ is the *phase function* that describes the efficiency with which light is scattered from the incident direction $Z'$ to the direction of the sight path $Z$, and $w_0$ is the *single scattering albedo* given by

$$w_0 = \frac{b_s}{b_{ext}} \tag{B.7}$$

Because particle scattering is stronger than gas molecule scattering, molecular scattering is disregarded. The single scattering albedo also contains a key measure of suspended particle influence ratios to total extinction. The integration in Eq. (B.6) is performed over all directions of incident radiation. The phase function is normalized so that

$$\int_{4\pi} P(Z, Z')\, dZ' = 4\pi \tag{B.8}$$

Figure B.1 shows an example of a normalized phase function for a moderately polluted atmosphere of the composition in Table B.1. The cosine of the scattering angle is used as the abscissa in the plot, so the areas under the curves are proportional to the contribution to the scattering coefficient. Scattering near zero degrees (i.e., scattering that changes the direction of travel of the light by only a small angle) is very much more efficient than scattering at greater angles. The value of the phase function at zero degrees is 73.

The value of $I_e$ for a given wavelength range at a given point on a sight path depends on the quantities in Eq. (B.6).

1. Illumination of the sight path
2. Light-scattering coefficient
3. Angular dependence of light scattering
4. Light extinction coefficient
5. Direction of view

A *uniform region* of the atmosphere is one in which the first four quantities do not depend on the distance $x$ along the sight path. Within a uniform region, $I_e$ is not a function of $x$, and Eq. (B.5) can be integrated to obtain

$$I - I_e = (I_0 - I_e)e^{-(b_{ext}x)} \tag{B.9}$$

This equation shows that the difference between $I$ and $I_e$ will decrease exponentially at a rate that depends on the extinction coefficient. If the

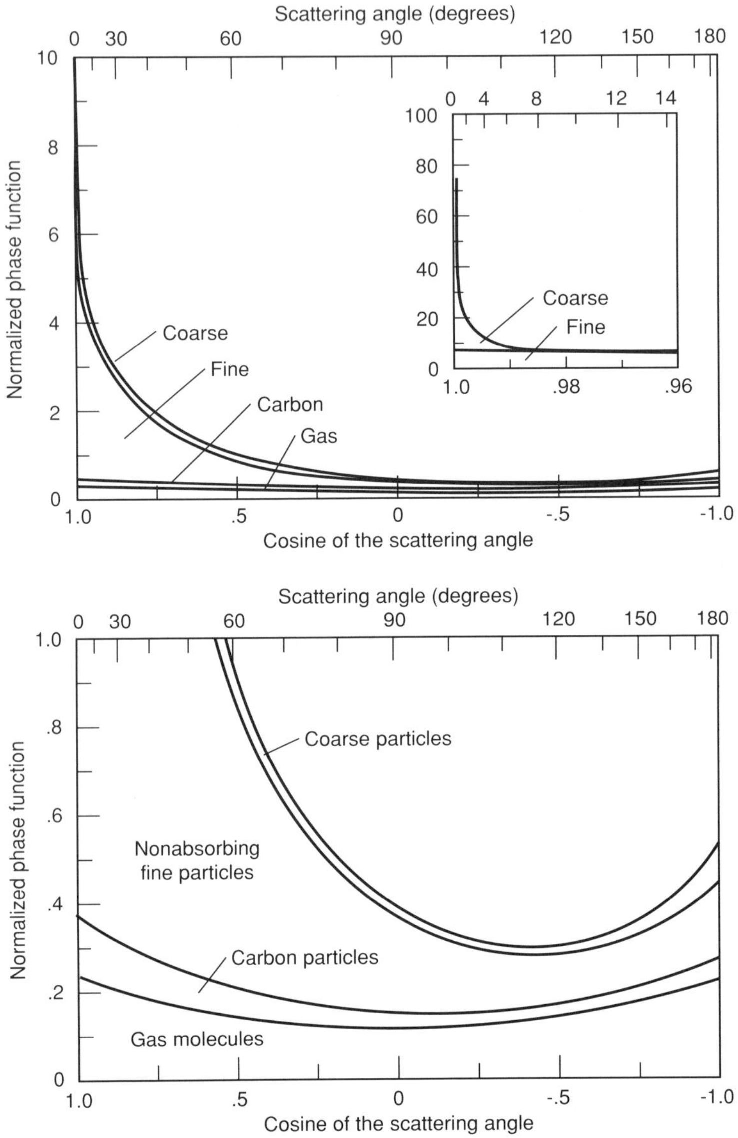


**FIGURE B.1** Normalized phase function for the atmospheric composition in Table B.1, showing the contribution of each constituent. (From L. W. Richards, unpublished, 1988.)

radiance is equal to $I_e$, both sides of Eq. (B.5) are equal to zero and the radiance $I$ will not change with increasing $x$. Thus, the radiance of any ray in a uniform region of the atmosphere tends toward the value of $I_e$ and,

TABLE B.1. **Composition of the Atmosphere Used to Calculate the Phase Function in Figure B.1**[a]

| **Component** | **Log-normal size distribution** | | **Index of refraction** | | **Volume conc.** | **Scattering coefficient** | **Extinction coefficient** |
|---|---|---|---|---|---|---|---|
| | $\bar{d}_p$ (μm) | | **real** | **imaginary** | ($\mu m^3/cm^3$) | ($Mm^{-1}$) | ($Mm^{-1}$) |
| Gas | | | | | | 10.5 | 10.5 |
| Carbon particles | 0.10 | 2.0 | 2.0 | 0.5 | 1. | 3.3 | 15.4 |
| Fine particles (<2.5 μm) | 0.30 | 2.0 | 1.5 | 0.0 | 9. | 46.0 | 46.0 |
| Coarse particles (>2.5 μm) | 6.0 | 2.1 | 1.5 | 0.0005 | 10. | 7.4 | 7.7 |
| Total | | | | | 20. | 67.3 | 79.6 |

[a] From Richards, 1988.

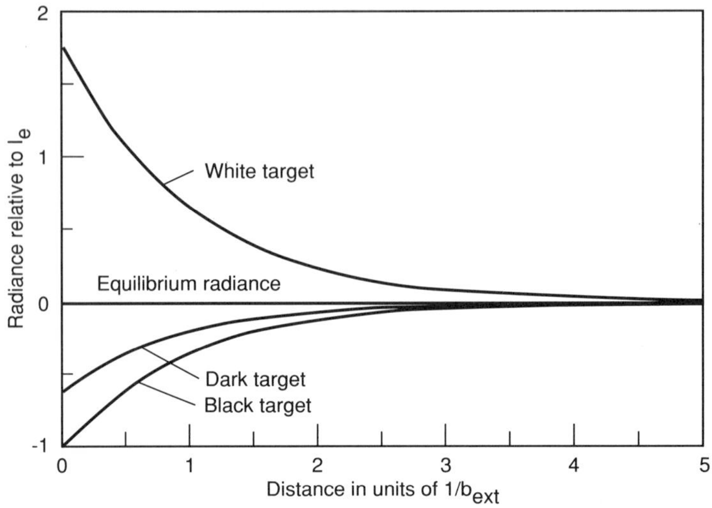


FIGURE B.2 Diagram showing how radiances approach the equilibrium radiance in a uniform region of the atmosphere. (From Middleton, 1958.)

when it reaches this value, does not change further. Figure B.2 illustrates the way radiances approach the equilibrium radiance asymptotically. Because of this behavior of radiances, $I_e$ is called the equilibrium radiance here. Many scientists, especially astronomers and others who work with the theory of radiative transfer, refer to $I_e$ as the *source function* (see, for example, Chandrasekhar, 1960). Another related quantity is the *path function*, which is the equilibrium radiance multiplied by the extinction coefficient.

Given this framework, one can derive simple and useful relationships between the equilibrium radiance and the path radiance for a sight path in a uniform region of the atmosphere. Imagine a sight path directed toward a target whose radiance can be varied. To provide an example, this target could be a sheet of translucent white plastic illuminated from the back under a hood that limits the amount of daylight falling on the plastic. If the radiance of the target is adjusted so that $I$ has the same value as $I_0$ for any distance $x$ within the uniform region, Eq. (B.5) requires that both $I$ and $I_0$ be equal to the equilibrium radiance $I_e$. When $I_e$ is substituted for $I$ and $I_0$ in Eq. (B.4), one obtains

$$I_e = I_e T + I_p \tag{B.10}$$

which can be solved to obtain the relationship

$$I_p = I_e(1 - T) \tag{B.11}$$

Because the path radiance does not depend on the initial radiance for initial radiances of reasonable intensity, this expression is valid for any target and any initial radiance that is not intense enough to cause physical changes in the scattering and absorbing medium. Some authors obtain this expression by more complicated derivations that involve integration along the sight path (Middleton, 1958).

Because the transmittance and path radiance for any sight path can be determined by measuring the transmittance of radiance differences, Eq. (B.11) makes it possible to calculate the equilibrium radiance for this sight path if it is in a uniform region of the atmosphere. The equation also makes it possible to calculate the path radiance for any sight path within a uniform region of the atmosphere for which the transmittance and the equilibrium radiance are known.

For sight paths within a uniform region of the atmosphere, Eq. (B.11) can be combined with Eq. (B.4) to obtain

$$I = I_0 T + I_e(1 - T) \tag{B.12}$$

Use of this equation to calculate the values of parameters related to human perception is described later.

When atmospheric conditions are so hazy that a horizontal sight path can be made sufficiently long to have a very small transmittance yet still be short enough to remain within a uniform region of the atmosphere, Eq. (B.12) shows that a horizon sky radiance will be equal to the equilibrium radiance. Because horizon sky radiance is frequently used in visibility monitoring, this result is an important one.

The curvature of the earth causes initially horizontal sight paths to increase in elevation with increasing distance. The increased elevation is 500 m (1640 ft) at 80 km (50 mi), 1000 m (3280 ft) at 113 km (70 mi), and 1500 m (5000 ft) at 140 km (87 mi). When the atmosphere is clear enough for sight paths to be longer than 100 to 150 km, the atmosphere will almost certainly not be uniform in a sight path toward the sky and the horizon sky radiance will differ from the equilibrium radiance. There is no simple way to calculate the horizon sky radiance for very long sight paths in the earth's atmosphere. Under conditions in ENA, the sight paths are generally small so that atmospheric uniformity is a good practical assumption.

*Approximation for Equilibrium Radiance* Knowledge of the equilibrium radiance facilitates calculation of how the radiance of a ray passing through a light-scattering and absorbing medium will be altered. Therefore, it is instructive to develop an approximate expression that gives information on the relative importance of various factors affecting the equilibrium radiance. Sight paths in which visibility reduction by air pollution is observed by the general public are almost always close to horizontal, so the approximate expression is restricted to this case.

Richards *et al.* (1985) reported a simple, approximate expression for the equilibrium radiance on a horizontal sight path:

$$I_e = \frac{w_0}{4\pi}[2F_+ + 2F_- + F_0P(Z, Z_0)] \tag{B.13}$$

This expression is derived by evaluating the integral in Eq. (B.6), assuming that within the upward and downward $2\pi$ steradian solid angles the upwelling and downwelling radiation are completely diffuse. $F_+$ is the flux of diffuse light reflected upward from the earth's surface, $F_-$ is the flux of diffuse skylight incident on the earth's surface; $F_0$ is the direct solar flux measured normal to the solar rays, $P(Z, Z_0)$ is the phase function for the scattering of light from the direction $Z_0$ of the solar rays to the direction $Z$ of the line of sight; and $w_0$ is the single scattering albedo in Eq. (B.7). The phase function $P(Z, Z_0)$ is normalized as described in Eq. (B.8). Near the surface of the earth, Eq. (B.13) can be simplified by the relation

$$F_+ = R(F_- + F_0 \cos\theta) \tag{B.14}$$

where $R$ is the *diffuse reflectance* (*albedo*) of the surface of the earth and $\theta$ is the angle between the sun's rays and the normal axis to the surface of the earth.

The most important assumption made in the derivation of Eq. (B.13) is that the upflowing and downward-directed diffuse fluxes are each perfectly diffuse. Gordon (1964) has shown that the radiance reflected from the earth's surface is not perfectly diffuse; the skylight is also not perfectly diffuse. Furthermore, poorest agreement occurs for conditions when the sun is near the horizon. Therefore, Eq. (B.13) is introduced as a useful approximation for exploring functional forms and orders of magnitude in relationships.

One important conclusion that can be drawn from Eq. (B.13) [or from Eq. (B.6)] is that increasing the absorption tends to decrease the equilibrium radiance. Thus, absorption has opposing effects on the extinction coefficient and the equilibrium radiance. If the ratio of scattering to absorption is quite variable, the equilibrium radiance, and hence the path radiance, need not correlate well with the extinction coefficient, even in cases where the illumination of the sight path is held constant. Therefore, it is necessary to measure two optical properties of the atmosphere to characterize the path radiance and visibility. Measuring only the sight path transmittance or extinction coefficient is not adequate. Middleton (1958) noted that under some conditions, the apparent contrast of a bright target against the horizon sky can be increased by adding absorption to the atmosphere.

*Factors Affecting the Equilibrium Radiance* Mariners have long known that land can be detected over the horizon by looking for changes in sky color. The reason for this phenomenon is that the equilibrium radiance is strongly affected by the reflectance of the surface of the earth. The data in Table B.2, which were calculated using Eq. (B.13) and (B.14), show this relationship. The first two columns give the input data assumed in the calculations. These results apply to cases in which the aerosol absorption is negligible. The third column gives the ratio of the equilibrium radiance to the solar flux, which is also equal to the ratio of the horizon sky radiance to the solar flux when the conditions described in the previous discussion of the horizon sky radiance are satisfied. The equilibrium radiance and hence the path radiance and horizon sky radiance are highly sensitive to the reflectance of the ground. A snowfall or ground fog can increase these radiances by a factor of 2 or 3, and seasonal foliage changes can have a significant effect.

TABLE B.2. **Relative Importance of Light from Ground, Sky, and Sun in Contributing to the Equilibrium Radiance When the Absorption Is Negligible and the Phase Function Has a Value of 0.4**[a]

| Conditions | | | Percentage of equilibrium radiance due to light from the | | |
|---|---|---|---|---|---|
| Ground reflectance $R$ | Ratio of skylight (F) to sunlight ($F_0$) ($F_-/F_0$) | Ratio of equilibrium to sunlight flux (%) $100I_e/F_0$ | Ground | Sky | Sun |
| 0.10 | 0.10 | 6.53 | 26.8 | 24.4 | 48.8 |
| 0.15 | 0.10 | 7.40 | 35.5 | 21.5 | 43.0 |
| 0.20 | 0.10 | 8.28 | 42.3 | 19.2 | 38.5 |
| 0.40 | 0.10 | 11.8 | 59.5 | 13.5 | 27.0 |
| 0.80 | 0.10 | 18.8 | 74.6 | 8.47 | 16.9 |
| 0.10 | 0.20 | 8.28 | 23.1 | 38.5 | 38.5 |
| 0.15 | 0.20 | 9.23 | 31.0 | 34.5 | 34.5 |
| 0.20 | 0.20 | 10.19 | 37.5 | 31.3 | 31.3 |
| 0.40 | 0.20 | 14.0 | 54.5 | 22.7 | 22.7 |
| 0.80 | 0.20 | 21.6 | 70.6 | 14.7 | 14.7 |

[a] From Richards, 1988.

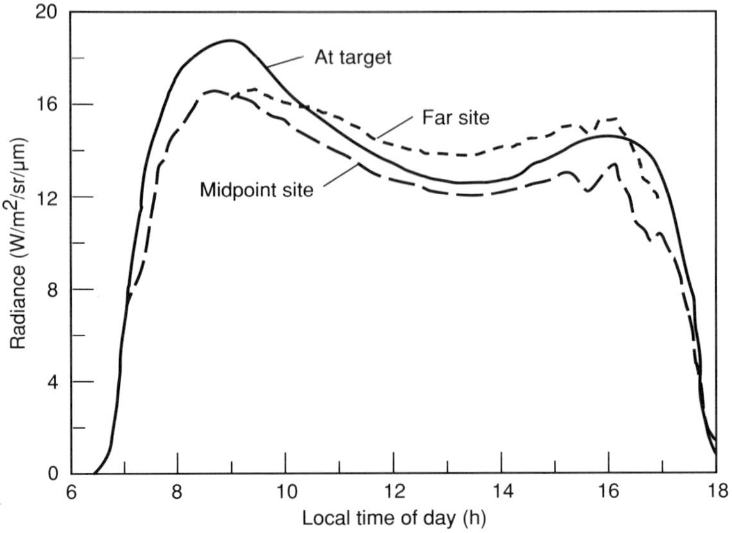


FIGURE B.3 Sky radiances measured at several distances along one sight path with an azimuth angle of 304°. (From Richards, 1988.)

The last three columns in Table B.2 give the percentage of the equilibrium radiance (or path radiance or horizon sky radiance) resulting from

1. Light reflected from the ground and then scattered into the sight path
2. Diffuse sky light scattered into the sight path
3. Direct sun light scattered into the sight path

The contribution of light reflected from the ground is important even over dark landscapes and is dominant when the ground reflectance (albedo) is greater than 40%. It is apparent from the data in Table B.2 that each of the terms in Eq. (B.13) makes a significant contribution to the equilibrium radiance and that variations in any of them will have an important effect on the equilibrium radiance.

Figure B.3 shows an example cited by Richards (1988). This figure shows sky radiance data measured in a southwestern U.S. location at various points along a sight directed toward the dark mountains from a *far site* surrounded by light-colored desert. The sky radiance was measured at the mountains (at target) looking at the sky directly away from the far site. The sky radiances labeled *midpoint site* and *far site* were measured 30 and 63 km from the mountains, respectively. Between the mountains at the initial point of the sight path and the midpoint site, the sight path was mostly over light-colored desert sparsely covered with grass and sagebrush. The instruments were aligned to come as close as possible to making all radiance measurements along the same sight path that passed over the top of the mountains. However, the data in Figure B.3 are from a cloud-free day during which the sky radiances were uniform enough that any instrument misalignment should not have significantly weakened the following argument.

One feature of these curves is expected. During the middle of the day the smallest sky radiance was measured at the midpoint site and the largest at the far site. This observation shows that the sky radiance decreased in the part of the sight path over dark forest and then increased over the light-colored desert, which clearly shows the important effect of the ground reflectance on the equilibrium radiance and hence on the apparent radiance of the sky.

Another feature is not expected. The sky radiances are greatest in the morning when the instruments are looking away from the sun. Equation (B.13) and the functional form of the phase function shown in Figure B.1 predict that the sky radiance should be greatest in the afternoon when the angle for scattering of the sun's rays into the sight path is small.

Richards' (1988) explanation of the large sky radiances in the early morning is that the earth's surface tends to reflect light back in the direction of the illumination (see also Gordon, 1964). In the morning, every rock, tree, and hillside has a sunlit side and a shaded side, which causes much of the light reflected from the surface of the earth to be directed back toward the sun. Atmospheric scattering through a small angle, which is very efficient, can deflect the light toward the measuring instruments. In these data, this direction of the surface reflectance has a larger effect on the horizon sky radiance than does the direction of the scattering of the sun's rays into the sight path. The sky radiance measured at the far site shows a weaker maximum in the morning. This result is reasonable, because the desert surrounding this site was rather flat and the grass and brush were not very opaque. Therefore, the light reflected from the surface of the earth at low sun angles was more nearly diffuse in the desert than in the hilly and tree-covered areas.

These examples illustrate that details of the reflectance of the ground have a strong effect on the equilibrium radiance, which, in turn, controls path and sky radiances. As a consequence, determination of the phase function as an index for atmospheric light scattering is of limited use in providing information on path and horizon sky radiances.

These results also show that the equilibrium radiance and hence the path radiance are not simple quantities. Both depend on the illumination of the atmosphere, which is typically uneven and variable. Therefore, the equilibrium and path radiances will undergo capricious variations caused by changing clouds in the sky and the nonuniform reflectance of the ground. Variations in the path radiance contain precisely the information that must be known to deal with the variation in the parameters that are likely to be useful in characterizing the human visual response. Therefore, in visibility monitoring it is desirable to make measurements sensitive to the variations in the equilibrium and path radiance when human perception is of interest. In general, this has not been done; instead, historic observations rely either on nephelometer measurements of $b_s$ or on human estimates of visual range.

The discussion here leads to another important conclusion. *Visibility is site specific*. The magnitude and direction of the reflectance of the ground under a sight path affect the equilibrium radiance, hence the measured radiances.

These points are supported and amplified by a discussion of the properties parameters related to human vision.

## Contrast Measurements

In air pollution research, the interest in atmospheric optics arises from a desire to infer the intuitive relationship between air pollution and the human perception of the resulting haze. Therefore, it is important that optical data from air pollution research and monitoring studies provide information on optical parameters (ideally) related to human perception.

*Contrast* is one of the more important parameters (Blackwell, 1946), and contrast measurements are widely used for visibility monitoring. When an object or target of apparent radiance $I_t$ is viewed against a background (such as the sky) of apparent radiance $I_b$, the *contrast* $C$ is defined as

$$C = \frac{I_t - I_b}{I_b} = \frac{I_t}{I_b} - 1 \tag{B.15}$$

The *initial contrast* $C_0$ of a target against its background is the contrast measured at the start of the sight path, usually at or close to the target. The *apparent contrast* $C$ is the contrast measured at the distance $x$ from the location where $C_0$ was measured.

There is some variability in the name used for the contrast of a target measured at the start of the sight path. Middleton (1958) and Dzubay and Clubb (1981) use the term *intrinsic contrast*; Duntley (1948) and Malm (1979) use the term *inherent contrast*. The U.S. Environmental Protection Agency's Report to Congress (1979) uses the term *initial contrast*, which is adopted here.

Equation (B.15) can be rewritten considering the initial radiance and the background apparent radiances as

$$C = C_0 \frac{I_{bo}}{I_b} \exp(-\bar{b}_{ext} x) \tag{B.16}$$

This is the basis for using *contrast teleradiometry*. This technique has become a widely used approach for inferring changes in visibility in relation to a visual impression of impairment of a scenic vista. Changes in contrast basically describe the transmittance of radiance differences through the atmosphere (Richards, 1988).

Equation (B.16) is rarely applied in practice. Often, the horizon sky is used as the background, and it is assumed that the radiance of the horizon sky is the same at the target and at the observer. If this assumption is made, $I_b$ and $I_{bo}$ are equal, and Eq. (B.16) can be simplified to the commonly used equation

$$C = C_0 e^{-(\bar{b}_{ext} x)} \tag{B.16a}$$

Nearly all discussions of the use of Eq. (B.16a) contain several assumptions that must be satisfied. In fact, according to Richards (1988), the necessary and sufficient assumption is that the background radiances are equal at each end of the sight path. Although they do not say so, workers have the objective of making this assumption valid by restricting measurements to situations where the sight path is in a uniform region of the atmosphere and the background radiance is equal to the equilibrium radiance. When these assumptions are satisfied, it is said that *standard viewing conditions* exist (Allard and Tombach, 1981). Note that very few detailed lists of assumptions about the uniformity of pollutant concentrations along the sight path, cloud conditions, and curvature of the earth contain any mention of the strong influence of the reflectance of the ground on the equilibrium radiance. This omission is an important one. The data in Figure B.3 clearly show that variations in the reflectance of the ground can invalidate the assumptions underlying Eq. (B.16a). Such reflectance variations are expected to be relatively moderate in the East but a more frequent phenomenon in the desert Southwest, where the sight paths can extend from the light-colored terrain to dark, tree-covered mountains.

In common practice, the initial contrast $C_0$ is not measured, and its value is assumed or estimated (Malm *et al.*, 1982). The initial contrast is sometimes assumed to be constant, but it is more common to assume that the initial contrast depends only on the sun angle (i.e., the time of day and the time of year). In either case, the values of the initial contrast are obtained through estimation procedures whose accuracy is difficult to determine. At a minimum, the accuracy of these procedures is limited because factors other than the sun angle affect the illumination of the target and horizon sky; hence, they affect the initial contrast. Errors in the initial contrast cause errors in the average extinction coefficient, which can be calculated from Eq. (B.16a).

Numerical examples can examine the relative importance of various factors that determine the contract of distant objects. If a target is far enough away, the path radiance can be shown to be almost always much greater than the transmitted radiance. For simplicity, in this derivation the sight path is assumed to be in a uniform region of the atmosphere and the horizon sky radiance is assumed equal to the equilibrium radiance. Equation (B.15) can be rearranged to give the apparent radiance $I_t$ of a target with contrast $C$ against a background of the horizon sky,

$$I_t = I_e(C + 1) \tag{B.17}$$

When the equation is evaluated close to the target, it gives a value for the initial radiance in terms of the initial contrast. If this initial radiance is substituted in Eq. (B.12), one obtains

$$I = I_e(C_0 + 1)T + I_e(1 - T) \tag{B.18}$$

where the first term on the right is the transmitted radiance and the second term is the path radiance. The two terms on the right of Eq. (B.18) are equal when the transmittance has the value $T_{eq}$ given by

$$T_{eq} = \frac{1}{2 + C_0} \tag{B.19}$$

When the sight path is long enough that the transmittance is less than $T_{eq}$, the path radiance is greater than the transmitted radiance. For a relatively dark target with an initial contrast of $-0.8$ (such as a sunlit, wooded hillside), $T_{eq} = 0.83$. Scattering by gas molecules alone will reduce the transmittance of green light to this value for a 17-km sight path. (Scattering by gas molecules alone attenuates green light by about 1% per kilometer.) If the day is relatively clear and the aerosol extinction is equal to the molecular scattering, the path radiance is larger than the transmitted radiance for any sight path longer than 9 km. Similar calculations are easily done for other initial contrasts.

The teleradiometer targets used for visibility studies are often selected to be dark and at a greater distance than 17 km. In such cases, the great majority of the radiance reaching the teleradiometers used in these studies is typically due to the path radiance. Richards (1988) notes that the path radiance is a highly variable quantity that is quite sensitive to variations in the illumination of the atmosphere. This variability explains why it is so difficult to extract useful extinction data from contrast teleradiometer measurements. Most of the light being measured is from capricious origins in the atmosphere, especially when the sight paths are so long that the probability of uniform illumination for the full length is quite small. The unwanted variability in the extinction coefficient calculated from contrast measurements enters the data through the denominator of Eq.

(B.15). This source of variability can be eliminated by using radiance differences in the data analysis.

These results can be extended to an observer viewing a scenic vista. For most objects in a scene more distant than 10 or 15 km, most of the light entering the observer's eye is due to the path radiance even on quite clear days. The path radiance cannot be ignored in perception studies.

In a nonuniform region of the atmosphere, the data from a given sight path may be valid only for that sight path. Even so, data for the transmittance and path radiance from the monitoring data base can be used in Eq. (B.4) to calculate the apparent radiance $I$ corresponding to any initial radiance $I_0$ (which is not large enough to modify physically the atmosphere in the sight path by heating the air). Such calculations are valid for the sight path, times, and wavelengths of the monitoring data. Within these constraints, it is possible to calculate the apparent contrast of any target if the initial radiance of the target and its background are known. Thus, the contrast degradation caused by the atmosphere in the sight path can be calculated from radiance difference monitoring data.

If the monitoring data are obtained for a sight path that is representative of a uniform region of the atmosphere, then similar calculations can be done for any sight path within that region. If, as is often the case when the sun is shining, the equilibrium radiance depends on the direction of view, the calculations can be done only for sight paths with the same direction of view as the monitoring sight path.

The first step of the calculations for a uniform region is to evaluate the extinction coefficient from the monitoring data using Eq. (B.2) or (B.7). Next is to calculate the equilibrium radiance from the path radiance and transmittance using Eq. (B.11). Then for any other sight path meeting the criteria in the previous paragraph, the transmittance of the sight path can be calculated from the path length and extinction coefficient using Eq. (B.2). Finally, for any initial radiance $I_0$ (which is not large enough to modify the atmosphere physically), the apparent radiance $I$ can be calculated from Eq. (B.12).

This procedure is particularly powerful for characterizing visibility. It is the one usually required to perform calculations for computer-generated photographs that illustrate the visual effects of air pollution. It can also be used to derive from radiance difference monitoring data the optical information necessary for perception studies. As noted previously, these calculations are restricted to the wavelengths for which monitoring data are available and to the times and sight paths for which the monitoring data are representative.

The foregoing discussion assumes that the initial radiances of the targets are known or can be determined from other data. For some types of studies, it may be necessary to estimate the effect of the atmospheric optical properties on the initial radiance of a target. A complete solution of this problem requires a radiative transfer calculation for the atmosphere above the target. In the typical case (clouds in the sky) lack of information about the vertical profile of the haze and the variability and angular dependence of the reflectance of the ground makes anything more exact than rough estimates impractical. One case that is relatively simple is a sunlit surface on a cloud-free day with long sight paths. Under these conditions, the attenuation of the direct sunlight by haze is approximately balanced by the increased diffuse sky radiance caused by the haze with the result that variations in the atmospheric haze have a relatively small effect on the radiance of the target.

## PARAMETERS CHARACTERIZING HUMAN PERCEPTION

The techniques described in the previous two sections are straightforward. The results of such calculations should provide to researchers interested in human perception the data they require to link visual experience with atmosphere optics that markedly improves historical observations.

A visual parameter that can be calculated is the *modulation M*:

$$M = \frac{I_{\max} - I_{\min}}{I_{\max} + I_{\min}} \tag{B.20}$$

which differs from contrast in that the denominator is the sum of the two radiances in the numerator instead of being just one of them. The modulation parameter is, in effect, a measure of human ability to discriminate texture in a broad target. As with contrast, it is possible to define an *initial modulation* at the start of the sight path and an *apparent modulation* at a distance $x$. These quantities are defined for a limited field of view, or target area. If the spatial variation of radiances on the target is coarse enough that atmospheric shimmer does not blur them when observed at a distance $x$, then the calculation methods presented in the previous section can be used to calculate the apparent modulation of any target from monitoring data if the initial values of the maximum and minimum radiances within the target area are known.

When the spatial variation of the radiances of the target is fine enough that they are blurred by atmospheric shimmer, these calculations are no longer applicable. The apparent modulation will be a function of the fineness of the spatial variation of the target radiance. This information can be described by a modulation transfer function, which is the apparent modulation divided by the initial modulation as a function of the degree of fineness of the spatial variation of the target radiance. To the extent that atmospheric shimmer is not of interest in air quality studies, the modulation transfer function is also not of interest. Air pollution tends not to blur images, so only the modulation transfer for coarse features on a target should be considered in air pollution studies (Duntley, 1948).

Visual parameters, such as contrast and modulation, are also limited in that they do not provide a basis for calculating the value of other optical parameters. If only the initial and apparent contrast and path length are known for a sight path, it is not possible to calculate from this information other parameters such as the transmittance of the sight path or the modulation. For example, use of Eq. (B.16a) requires the additional information that the measurements were made in a uniform region of the atmosphere and that the background radiance was equal to the equilibrium radiance. Because of this limitation, the radiance required for radiance difference calculations needs to be monitored. Then the values of all parameters of interest can be calculated from the monitoring data.

The measurement methods discussed here also provide useful information related to the visibility of plumes from point sources. White *et al.* (1986) have shown that plume contrast is largely determined by the difference between the equilibrium radiance of the plume and the surrounding air. This difference provides an additional incentive for measuring equilibrium radiances in visibility research studies.

The equilibrium radiance, the path radiance, depend strongly on how the atmosphere is illuminated and hence is related rather poorly to air quality. Any visual parameter, such as contrast, contrast transmittance, modulation, modulation transfer function, or blue–red ratio, that depends on the path radiance has a similar dependence on the illumination of the atmosphere and is poorly related to air quality. Because of their dependence on the illumination of the atmosphere, none of the visual parameters just mentioned provides a useful index of air quality or a complete basis for regulations.

In the absence of precise monitoring methods for relating air quality and visibility, workers continue to rely on nephelometric measurements of scattering coefficient as an index for particle concentration and visual range. This simple monitoring device is increasingly supplemented with teleradiometry and photographic records even though both of these are difficult to relate to air pollution measures.

# REFERENCES


Adams, D. F., Farwell, S. O., Robinson, E., and Pack, M. R. (1980). Biogenic Sulfur Emissions in the SURE Region. Report EA-1561. Electric Power Research Institute, Palo Alto, California.

Allard, D., and Tombach, I. (1981). *Atmos. Environ.* **15,** 1847.

Allwine, K. J., and Whiteman, C. D. (1985a). *MELSAR: A Mesoscale Air Quality Model for Complex Terrain*, Vol. 1. PNL-5460-Vol. 1. Pacific Northwest Laboratory, Richland, Washington.

Allwine, K. J., and Whiteman, C. D. (1985b). *MELSAR: A Mesoscale Air Quality Model for Complex Terrain*, Vol. 2, Appendices. PNL-5460-Vol. 2. Pacific Northwest Laboratory, Richland, Washington.

Anthes, R. A., Hsie, E.-Y., and Kuo, Y.-H. (1987). Description of the Penn State/NCAR Mesoscale Model Version 4(MM4). Tech. Note NCAR/TN-282 +STR. National Center for Atmospheric Research, Boulder, Colorado, 66 p.

ApSimon, H. M., Goddard, A. J. H., and Wrigley, J. (1985a). *Atmos. Environ.* **19,** 99–111.

ApSimon, H. M., Goddard, A. J. H., Wrigley, J., and Crompton, S. (1985b). *Atmos. Environ.* **19,** 113–125.

Argonne National Laboratory. (1986). Environmental Research Division Technical Progress Report 1984–1985. Report ANL-86-15. Argonne National Laboratory, Argonne, Illinois, p. 2–42.

Artz, R., Pielke, R. A., and Galloway, J. (1985). *Atmos. Environ.* **19,** 47–63.

Baldocchi, D. D., Hicks, B. B., and Camara, P. (1987). *Atmos. Environ.* **21,** 91–101.

Ball, R. H. (1981). Matrix Methods to Analyze Long-Range Transport of Air Pollutants. DOE/EV-0127. Office of Environmental Assessments, Regional Impacts Division, Washington, D.C.

Barchet, W. R. (1991). The EMEFS Model Evaluation: An Interim Report. Report PNL-7914. Pacific Northwest Laboratory, Richland, Washington.

Barnes, C. R., Schoeder, R. A., and Peters, N. E. (1982). *Northeast Environ. Sci.* **1,** 187–197.

Barrie, L. A., and Hales, J. (1984). *Tellus* **36B,** 333–335.

Benkley, C. W., and Bass, A. (1978). MESOPAC: A Mesoscale Meteorology Package. Report EPA-600/7-78. U.S. Environmental Protection Agency, Research Triangle Park, North Carolina.

Benkley, C. W., and Schulman, L. L. (1979). *J. Appl. Meteorol.* **19,** 772–780.

Bennett, J. H., Hill, A. C., and Gates, D. M. (1973). *JAPCA* **23,** 957–962.

Bhumralkar, C. M., Johnson, W. B., Mancuso, R. L., Thiullier, R. A., Wolf, D. E., *et al.* (1980). Interregional Exchanges of Airborne Sulfur Pollution and Deposition in Eastern North America. In: *Second Joint Conference on Applications of Air Pollution Meteorology*, pp. 225–231. American Meteorology Society, New Orleans, March 24–27, 1980.

Bhumralkar, C. M., Mancuso, R. L., Wolf, D. E., and Johnson, W. B. (1981). *Tellus* **33,** 142–161.

Bhumralkar, C. M., Endlich, R. M., Nitz, K., Brodzinsky, R., Mastinez, J. R., *et al.* (1983). Refined Air Pollution Model for Calculating Daily Regional Patterns and Transfrontier Exchanges of Airborne Sulfur in Central and Western Europe. In: *Air Pollution Modeling and Its Application*, II, C. De Wispelaere (ed.), pp. 265–276. Plenum Publishing, New York.

Bhumralkar, C. M., Endlich, R. M., Nitz, K. C., Brodzinsky, R., and Mayerhofer, P. (1984). Lagrangian Long-Range Air Pollution Model for Eastern North America. In: *Air Pollution Modeling and Its Application*, III, C. De Wispelaere (ed.), pp. 35–56. Plenum Publishing, New York.

Bilonik, R. (1985). *Atmos. Environ.* **19,** 1829–1845.

Bilonik, R., and Nichols, D. (1983). *Atmos. Environ.* **17,** 1063–1072.

Binkowsky, F. S., Chang, J. S., Dennis, R. L., Reynolds, S., Samson, P. J., and Shannon, J. D. (1990). Regional Acid Deposition Modeling. SOS/T Report 3. In: *Acid Deposition: State of Science and Technology, Vol. 1.* National Acid Precipitation Program, Washington, D.C.

Blackwell, H. R. (1946). *J. Opt. Soc. Am.* **36,** 624.

Blifford, I. B., Jr., and Meeker, G. D. (1967). *Atmos. Environ.* **1,** 709–712.

Blumenthal, D. L., Keifer, W. S., and McDonald, J. A. (1981). Aircraft Measurements of Pollutants and Meteorological Parameters during the Sulfate Regional Experiment (SURE) Program. Report EA-1909. Electric Power Research Institute, Palo Alto, California.

Blumenthal, D. L., McDonald, J. A., Keifer, W. S., Tommerdahl, J. B., Saeger, M. L., and White, J. H. (1984). *Atmos. Environ.* **18,** 733–749.

Bolin, B., and Persson, C. (1975). *Tellus* **27,** 281–310.

Bottenheim, J., and Gallant, A. (1986). *Trans AGU* **67,** 902.

Bottenheim, J. W., Brice, K. A. and Anlauf, K. G. (1984). *Atmos. Environ.* **18,** 2609–2619.

Brady, P. (1977). *J. Appl. Meterol.* **17,** 12–21.

Brodzinsky, R., Cantrell, B. K., Endlich, R. M., and Bhumralker, C. M. (1984). *Atmos. Environ.* **18,** 2361–2366.

Bruck, R. I., Miller, P., and Laut, J. (1985). Investigation into the Health of Forests in the Vicinity of Gothic, Colorado. Document No. EPA-908/9-85-001. Environmental Protection Agency, Region VIII, Denver, Colorado.

Bryson, R., and Hare, F. K. (1974). *Climate of North America*, Vol. 11. *World Survey of Climatology*, R. A. Bryson and F. K. Hare (eds.). Elsevier, New York.

Caldwell, G. (1980). Regional Air Pollution Study, Gas Chromatography Laboratory Operation. Report EPA-600/4-80-006, Research Triangle Park, North Carolina, U.S. Environmental Protection Agency, January 1980.

California Air Resources Board. (1983). Summary of 1983 Air Quality Data for Gaseous Particulate Pollutants, Vol. XV.

California Air Resources Board. (1985). Third Annual Report to the Governor and the Legislature on the Air Resources Board's Acid Deposition Research and Monitoring Program, December.

California Air Resources Board (1991). The Atmospheric Acidity Protection Program: Annual Report to the Governor and the Legislature. CARB, Sacramento, California, 93 pp.

Calvert, J., (ed). (1984). *$SO_2$, NO and $NO_2$, Oxidation Mechanisms: Atmospheric Considerations*. Butterworth, Boston.

Calvert, J. G., and Mohnen, V. (1983). Chemistry of Acid Formation. In: *Acid Deposition: Atmospheric Processes in Eastern North America*, pp. 155–201, National Academy of Sciences, Washington, D.C.

Calvert, J. G., and Stockwell, W. R. (1983). *Environ. Sci. Technol.* **17,** 428A-443A.

Calvert, J. G., and Stockwell, W. R. (1984). Mechanism and Rates of the Gas-Phase Oxidations of Sulfur Dioxide. In: *$SO_2$, NO and $NO_2$ Oxidation Mechanisms: Atmospheric Considerations* J. G. Calvert (ed.), pp. 1–62. Butterworth, Boston.

Calvert, J. G., Heikes, B. G., Stockwell, W. R., Mohnen, V. A., and Kerr, J. A. (1983). Some Considerations of the Important Chemical Processes in Acid Deposition. In: *Chemistry of Multiphase Atmospheric Systems*, W. Jaeschke (ed.), pp. 615–647. Reidel, Dordrecht.

Calvert, J. G., Lazrus, A., Kok, G. L., Heikes, B., Whan, J., Lind, J., and Cantrell, C. (1985). Chemical Mechanisms of Acid Generation in the Troposphere. *Nature* **317,** pp. 27–35.

Caneill, J. Y., Granier, J. P., Racher, P., and Rosset, R. (1985). Finite-Element Formulation of a Variational Procedure of Wind-Field Adjustment Over Complex Terrain. EUR-9678-EN. Commission of the European Communities, Luxembourg.

Cantrell, C. A., Stockwell, W. R., Anderson, L. G., Busarow, K., Perner, L. D., Schmeltekopf, A., Calvert, J. G., and Johnston, H. S. (1985). *J. Phys. Chem.* **89,** 139–146.

Carmichael, G. R., and Peters, L. K. (1980). The Transport, Chemical Transformation, and Removal of $SO_2$ and Sulfate in the Eastern United States. In: *Studies in Environmental Science*, Vol. 8, M. M. Benarie (ed.), pp. 31–36. Elsevier, Amsterdam.

Carmichael, G. R., and Peters, L. K. (1984a). *Atmos. Environ.* **18,** 937–951.

Carmichael, G. R., and Peters, L. K. (1984b). *Atmos. Environ.* **18,** 953–967.

Carmichael, G. R., Kitada, T., and Peters, L. K. (1984). A Second Generation Combined Transport/Chemistry Model for the Regional Transport of $SO_x$ and $NO_x$ Compounds. In: *Air Pollution Modeling and Its Application*, III, C. De Wispelaere (ed.), pp. 525–538. Plenum Publishing, New York.

Carmichael, G. R., Peters, L. K., and Kitada, T. (1986). *Atmos. Environ.* **20,** 173–188.

Chamberlain, A. C. (1966). *Proc. R. Soc. Lond.* **A290,** pp. 236–265.

Chamberlain, A. C. (1967). *Proc. R. Soc. Lond.* **A296,** pp. 45–70.

Chamberlain, A. C. (1980). In: *Dry Deposition of Sulfur Dioxide, Atmospheric Sulfur Deposition,* D. S. Shriner, C. R. Richmond, and S. E. Lindberg (eds.), pp. 185–198. Ann Arbor Science, Ann Arbor, Michigan.

Chameides, W. L., and Davis, D. D. (1982). *J. Geophys. Res.* **87,** 4863–4877.

Chandrasekhar, S. (1960). *Radiative Transfer*. Dover Publications, New York.

Chang, J. S., Binkowski, F. S., Seaman, N., *et al.* (1990). The Regional Acid Deposition Model and Engineering Model, NAPAP. SOS/T Report 4. In: *Acidic Deposition: State of Science & Technology*, Vol. 1. National Acid Precipitation Assessment Program, Washington, D.C.

Charlson, R. J., and Rodhe, H. (1982). *Nature* **295,** 683–685.

Charlson, R. J., Lanagner, J., Radhe, H., Leouy, C. B., and Warren, S. G. (1991). *Tellus* **43AB,** 152–163.

Clapp, P. F. (1964). *Mon. Weather Rev.* **92,** 495–507.

Clark, T. L., and Cohn, R. D. (1990). The Across North America Tracer Experiment, Model Evaluation Study. Report EPA 600/3-90/051. U.S. Environmental Protection Agency, Research Triangle Park, North Carolina, 141 pp.

Clark, T. L., Dennis, R. L., Voldner, E. C., Olson, M. P., Seilkop, S. K., and Alvo, M. (1987). International Sulfur Depositiion Model Evaluation. Report EPA 600/3-87/008. U.S. Environmental Protection Agency, Research Triangle Park, North Carolina.

Clark, T. L., Voldner, E. C., Dennis, R. L., Seilkop, S. K., Alvo, M., and Olson, M. P., (1989). *Atmos. Environ.* **23,** 2449–2462.

Cleveland, W. S., and Guarino, R. (1976). *Technometrics* **18,** 401–409.

Clough, W. S. (1975). *Atmos. Environ.* **9,** 1113–1119.

Clutton-Bock, M. (1967). *Technometrics* **9,** 261–269.

Coffey, P. E. (1984). *JAPCA* **34,** 905–906.

Cogbill, C. V., and Likens, G. E. (1974). *Water Resour. Res.* **18,** 2261–2270.

Cogbill, C. V., Likens, G. E., and Butler, T. J. (1984). *Atmos. Environ.* **18,** 2261–2270.

Cooper, J. A., and Watson, J. G., Jr. (1980). *JAPCA* **30,** 1116–1125.

Court, A., and Gerston, R. (1966). *Geogr. Rev.* **56,** 543–550.

Covert, D., Waggoner, A., Weiss, K., Ahlquista, N., and Charlson, R. (1980). Atmospheric Aerosols, Humidity, and Visibility. In: *The Character and Origin of Smog Aerosols*, G. M. Hidy *et al.* (eds.), p. 559. Wiley (Interscience), New York.

Cowling, E. B. (1980). A Historical Resume of Progress in Scientific and Public Understanding of Acid Precipitation and Its Biological Consequences. Report 18/80. SNSF Project, Oslo, Norway.

Cox, R. A. (1974). *J. Photochem.* **3,** 291–304.

Dana, T., and Hales, J. M. (1976). *Atmos. Environ.* **10,** 45–50.

Danard, M. (1977). *Mon. Weather Rev.* **105,** 572–581.

Daum, P. H., Kelly, T., Schwartz, S. E., and Newman, L. (1984). *Atmos. Environ.* **18,** 2671–2684.

David, M. (1977). *Geostatistical Ore Reserve Estimation.* Elsevier, New York.

Davis, C. G., Bunker, S. S., and Mutschlecner, J. P. (1984). *J. Climatol. Appl. Meteorol.* **23,** 235–238.

Davis, D. D., Ravishankara, A. R., and Fischer, S. (1979). *Geophys. Res. Lett.* **6,** 113–116.

Davis, W. E. (1984). The Effect of Vertical Wind Shear on Regional-Scale Trajectories in Near-Frontal Conditions. In: *Fourth Joint Conference on Applications of Air Pollution Meteorology*, pp. 204–207. American Meteorological Society, Portland, Oregon, October 16–19, 1984.

Dawson, G. A., and Farmer, J. C. (1984). *J. Geophys. Res.* **89,** 4779–4787.

Delany, A. C., and Davies, T. D. (1983). *Atmos. Environ.* **17,** 1391–1394.

Dennis, R., Barchet, W. R., Clark, T. L., and Seilkop, S. K. (1990). Evaluation of Regional Acidic Deposition Models and Selected Applications of RADM. SOS/T Report 5. In: *Acid Deposition: State of Science and Technology, Vol. 1.* National Acid Precipitation Assessment Program, Washington, D.C.

Department of Energy (DOE) (1989). Clean Coal Technology Demonstration Program. DOE/FE-0125 Office of Fossil Energy, Washington, D.C.

Desautels, G. (1981). Trajectories of Air Parcels Reaching Quebec in 1978. Scientific Services Division, Atmospheric Environment Service, Quebec Region, Montreal.

Diaz, H. (1983). *J. Climatol. Appl. Meteorol.* **22,** 3–16.

Diaz, H., and Quayle, R. (1980). *Mon. Weather Rev.* **108,** 249–266.

Dickerson, M. H. (1978). *J. Appl. Meteorol.* **17,** 241–253.

Dingle, A. N., and Lee, Y. (1972). *J. Appl. Meteorol.* **11,** 877–879.

Dittenhoefer, A. C., and Ferullo, A. F. (1983). A Dual-Model Regional Air Back-Trajectory Model. In: *The Meteorology of Acid Deposition*, Perry J. Samson (ed.), pp. 312–323. Air Pollution Control Association, Pittsburgh, Pennyslvania.

Draxler, R. R. (1977). A Mesoscale Transport and Diffusion Model. ERL ARL-64. Air Resources Laboratory, Silver Spring, Maryland.

Draxler, R. R. (1979). *Atmos. Environ.* **13,** 1523–1533.

Draxler, R. R. (1980). Error Estimates of Calculated Air Concentrations from a Mesoscale Trajectory Model. In: *Second Joint Conference on Applications of Air Pollution Meteorology*, pp. 275–278. American Meteorological Society, New Orleans, Louisana, March 24–27.

Draxler, R. R. (1982a). User's Guide for a Long-Range Multi-Layer Atmospheric Transport and Dispersion Model. ERL ARL-112. Air Resources Laboratories, Silver Spring, Maryland.

Draxler, R. R. (1982b). *Atmos. Environ.* **16,** 2763–2776.

Draxler, R. R., and Heffter, J. L. (1989). Across North America Tracer Experiment (ANATEX). NOAA Tech. Memo., ERL/ARL-167 (Vol. 1). National Oceanic and Atmospheric Adminstration, Silver Spring, Maryland, 83 pp.

Draxler, R. R., and Taylor, A. D. (1982). *J. Appl. Meteorol.* **21,** 367–372.

Draxler, R. R., Ferber, A. J., Heffter, J. L., Cordella, R. H., Jr., and Miller, J. M. (1982). Atmospheric Transport and Dispersion of Pollutants and Related Meteorological Studies. DOE/EV/10081-T1. National Oceanic and Atmospheric Administration, Rockville, Maryland.

Dredgen, A., Thornton, D., and Bandy, A. (1986). *Trans. AGU* **67,** 902.

Dumont, G., Vervliet, F., De Saeger E., and Verduyn, G. (1985). Application of a Mesoscale Lagrangian Puff-Model to the Measurements of $SO_2^-$ Pollution Transport over Belgium. In: *Air Pollution Modeling and Its Application*, IV, C. De Wispelaere (ed.), pp. 453–470. Plenum Publishing, New York.

Duntley, S. Q. (1948). *J. Opt. Soc. Am.* **38,** 179.

Duyzer, J. H., Meger, G. M., and van Aalst, R. M. (1983). *Atmos. Environ.* **17,** 2117–2120.

Dzubay, T. G., and Clubb, K. R. (1981). *Atmos. Environ.* **15,** 2617.

Easter, R. C., Dana, M. T., Thorp, J. M., Busness, K. M., Hales, J. M., Raynor, G. S., Benkovitz, C. M., Tanner, R. L., and Shannon, J. D. (1984). Overview of the Oxidation and Scavenging Characteristics of April Rains (OSCAR) Experiment. PNL-4869. Pacific Northwest Laboratory, Richland, Washington.

Eddy, A., and Samson, P. J. (1986). The Use of Acid Deposition Models on Designing Long-Range Tracer Experiments. In: *Air Pollution Modeling and Its Application,* V, C. De Wispelaere, F. A. Schiermeier, and N. V. Gillani (eds.) pp. 501–518. Plenum Publishing, New York.

Eddy, A., Avara, E. P., and Panturat, S. (1984). Statistical Design and Analysis for MATEX. Unpublished report. Environmental Research and Technology, Newbury Park, California.

Electric Power Research Institute. (1991). Analysis of Alternative $SO_2$ Reduction Strategies. Report EN/GS-7132. Electric Power Research Institute, Palo Alto, California.

Eliassen, A. (1978). *Atmos. Environ.* **12,** 479–487.

Eliassen, A. (1980). *J. Appl. Meteorol.* **19,** 231–240.

Eliassen, A. (1984). Aspects of Lagrangian Air Pollution Modelling. In: Air Pollution Modeling and Its Application, III, C. De Wispelaere (ed.), pp. 3–21. Plenum Publishing, New York.

Eliassen, A., and Saltbones, J. (1975). *Atmos. Environ.* **9,** 425–429.

Eliassen, A., and Saltbones, J. (1983). *Atmos. Environ.* **17,** 1457–1473.

Eliassen, A., Hov, O., Isaksen, I. S. A., Saltbones, J., and Stordal, F. (1982a). *J. Appl. Meteorol.* **21,** 1645–1661.

Eliassen, A., Hov, O., Isaksen, I. S. A., Saltbones, J., and Stordal, F. (1982b). A Lagrangian Long-Range Transport Model with Atmospheric Boundary Layer Chemistry. In: *Air Pollution by Nitrogen Oxides,* T. Schneider and L. Grant (eds.), pp. 347–356. Elsevier, Amsterdam.

Ellenton, G., Ley, B., and Misra, P. K. (1983). Treating Exponential Mass Decay When Wet Scavenging Varies Discretely Within an Expanding Gaussian Dispersed Puff. In: *The Meteorology of Acid Deposition,* P. J. Samson (ed.), pp. 528–536. Air Pollution Control Association, Pittsburgh, Pennsylvania.

Ellenton, G., Ley, B., and Misra, P. K. (1985). *Atmos. Environ.* **19,** 727–737.

Endlich, R. M., Ludwig, F. L., Bhumralkar, C. M., and Estoque, M. A. (1982). *J. Appl. Meteorol.* **21,** 1441–1454.

Endlich, R. M., Ludwig, F. L., Bhumralkar, C. M., and Estoque, M. A. (1983). ENAMAP-2 Air Pollution Model for Long-Range Transport of Sulfur and Nitrogen Compounds. EPA 600/7-83-059. SRI International, Menlo Park, California.

Endlich, R. M., Ludwig, F. L., and Bhumralkar, C. M. (1984). *Atmos. Environ.* **18,** 2345–2360.

Endlich, R. M., Eynon, B. P., Ferek, R. J., Valdes, A. D., and Maxwell, C. (1986). Statistical Analysis of Precipitation-Chemistry Measurements over the Eastern United States. In: *Utility Acid Precipitation Program (UAPSP).* Report 112. EPRI, Palo Alto, California.

Environmental Research and Technology. (1984a). *ADOM/TADAP Model Development Program,* Vols. 6 and 7. ERT, Newbury Park, California.

Environmental Research and Technology. (1984b). *ADOM/TADAP Model Development Program,* Vols. 1–5. ERT, Concord, Massachusetts.

Environmental Research and Technology. (1985). *ADOM/TADAP Model Development Program,* Vol. 8. ERT, Newbury Park, California.

Fay, J. A., and Rosenzweig, J. J. (1980). *Atmos. Environ.* **14,** 355–365.

Fay, J. A., Golomb, D., and Kumar, S. (1985). *Atmos. Environ.* **19,** 1773–1782.

Fay, J., Kumar, S., and Golomb, D. (1986). *Atmos. Environ.* **20,** 1497–1500.

Fay, J., Golomb, D., and Kumar, S. (1987). *Atmos. Environ.* **21,** 61–68.

Fehsenfeld, F. C., Bollinger, M. J., and Lin, S. C. (1983). *J. Atmos. Chem.* **1,** 87–105.

Ferber, G. J., and Heffter, J. L. (1984). Cross Appalachian Tracer Experimental (CAPTEX '83) with Evaluation Workshop Information. National Oceanic and Atmospheric Administration, Air Resources Lab., Rockville, Maryland.

Ferman, M., Wolff, G. and Kelly, N. A. (1981). *JAPCA* **31,** 1074–1082.

Fernau, M. E., and Samson, P. J. (1990). *J. Appl. Meteorol.* **29,** 735–750 (Part I), 751–761 (Part II).

Fisher, B. E. A. (1978). *Atmos. Environ.* **12,** 489–501.

Fisher, B. E. A. (1982). *Atmos. Environ.* **16,** 2725–2734.

Fisher, B. E. A. (1983). *Atmos. Environ.* **17,** 1865–1880.

Fisher, B. E. A., and Clark, P. A. (1985). Testing a Statistical Long-Range Transport Model on European and North American Observations. In: *Air Pollution Modeling and Its Application,* IV, C. De Wispelaere (ed.), pp. 471–486. Plenum Publishing, New York.

Foley, G., and Hilst, G. (1983). An Experimental Determination of Sources of Acidic Deposition in the Eastern U.S. and Southeastern Canada. Unpublished informal memorandum. U.S. Environmental Protection Agency/Electric Power Research Institute, Washington, D.C./Palo Alto, California.

Fox, D. G. (1981). *Bull. Am. Meteorol. Soc.* **62,** 599–609.

Frank, A., and Posseill, N. (1976). Seasonality and Regional Trends in Atmospheric Sulfates. Paper presented at the meeting of the American Chemical Society, San Francisco, California.

Galloway, J. N., and Whelpdale, D. (1980). *Atmos. Environ.* **14,** 409–418.

Galloway, J. N., Likens, G. E., Keene, W. C., and Miller, J. M. (1982). *J. Geophys. Res.* **87,** 8771–8786.

Garland, J. A. (1977). *Proc. R. Soc. Lond.* **A354,** 245–268.

Garratt, J. R., and Hicks, B. B. (1973). *Q. J. R. Meteorol. Soc.* **99,** 680–687.

Gatz, D. F. (1980). Associations and Mesoscale Spatial Relationships among Rainwater Components. *J. Geophys. Res.* **85,** 5588.

Gatz, D. F. (1982). *J. Appl. Meteorol.* **17,** 600–608.

Gedeonov, L. I., Gritchenko, Z. G., Flyontov, F. M., and Zhilkana, M. I. (1970). Coefficient of Radioactive Aerosol Concentration in Atmospheric Precipitation. In: *Atmospheric Scavenging of Radioisotopes,* B. Styra, C. Garbaliauskas, and V. Lujanas, (eds.), pp. 150–156. U.S. Department of Commerce, Springfield, Virginia.

Gillani, N. V. (1985). Project PEPE-NEROS (1980). Measurements and Data Base—An Overview. Unpublished report to U.S. Environmental Protection Agency (EPA Cooperative Agreement CR-809713). Washington University, St. Louis, Missouri.

Gillani, N. V., and Pooler, F. (1980). Project PEPE-NEROS (1900) Measurements and Data Base—An Overview. Unpublished report (EPA Cooperative Agreement CR-809713). U.S. Environmental Protection Agency, Research Triangle Park, North Carolina.

Godden, D., and Lurmann, F. (1983). Development of the PLMSTAR Model and Its Application to Ozone Episode Conditions in the South Coast Air Basin. P-A702-200. Environmental Research and Technology, Newbury Park, California.

Golomb, D., Fay, J. A., and Kumar, S. (1985). Seasonal Control of Sulfate Deposition. Paper No. 85–1B.8R. Presented at the 78th Air Pollution Control Association Meeting, Detroit, Michigan, June.

Goodin, W. R., McRae, G. J., and Seinfeld, J. H. (1979). *J. Appl. Meteorol.* **18,** 761–771.

Goodin, W. R., McRae, G. J., and Seinfeld, J. H. (1980). *J. Appl. Meteorol.* **19,** 98–108.

Gordon, J. I. (1964). *Appl. Opt.* **3,** 556.

Granat, L., and Johansson, C. (1983). *Atmos. Environ.* **17,** 191–192.

Gravenhorst G., and Böttger, A. (1982). Field Measurements of NO and $NO_2$ Fluxes to and from the Ground. In: *Acid Deposition*, S. Beilke and A. J. Elshout (eds.), pp. 172–184. Reidel, Dordrecht.

Grennfelt, P., Bengtsson, C., and Skarby, L. (1983). Dry Deposition of $NO_2$ to Scots Pine Needles. In: *Precipitation Scavenging, Dry Deposition and Resuspension*, H. Pruppacher, R. Semonin, and G. Slinn (eds.), pp. 753–761. Elsevier, New York.

Hales, J. M. (1986). The Mathematical Characterization of Precipitation Scavenging and Precipitation Chemistry. In: *Handbook of Environmental Chemistry*, Vol. 4/A, O. Hutzinger (ed.), pp. 150–207. Springer-Verlag, Berlin.

Hansen, D. A. (1989). Project Plan for the Acid Deposition Eulerian Model Evaluation and Field Study. Unpublished report, Electric Power Research Institute, Palo Alto, California.

Hansen, D. A., and Hidy, G. M. (1982). *Atmos. Environ.* **16,** 2107–2172.

Harmon, H. H. (1976). *Modern Factor Analysis*. rev. ed. University of Chicago Press, Chicago.

Heffter, J. L. (1980). Transport Layer Depth Calculations. In: *Second Joint Conference on Applications of Air Pollution Meteorology*, pp. 787–791. American Meteorological Society, New Orleans, Lousiana, March 24–27.

Heffter, J. L. (1981). Air Resources Laboratory Atmospheric Transport and Dispersion Model (ARL-ATAD). Technical Memorandum ERL ARL-81. Air Resources Laboratory, National Oceanic and Atmospheric Administration, Silver Spring, Maryland.

Heffter, J. L. (1983). Branching Atmospheric Trajectory (BAT) Model. ERL ARL-121. Air Resources Laboratory, National Oceanic and Atmospheric Adminstration, Rockville, Maryland.

Hegg, D. A., and Hobbs, P. V. (1982). *Atmos. Environ.* **16,** 2613–2668.

Heikes, B. G., and Thompson, A. M. (1981). Nitric Acid Formation Mechanisms in-Cloud: Gas Phase, Photochemical and/or Heterogeneous Paths. *EOS Trans.* **62,** 884.

Heikes, B. G., and Thompson, A. M. (1984). Effects of Heterogeneous Processes on $NO_3$, HONO, $HNO_3$ Chemistry in the Troposphere [Correction to the article in *J. Geophys. Res.* **88,** 10883–10895 (1983)] *J. Geophys. Res.* **89,** 11829.

Heikes, B. G., Kok, G. L., Walega, J. G., and Lazrus, A. L. (1987). *J. Geophys. Res.* **92**(D1), 915–931.

Heisler, S. L., Collins, H., Collins, J., Hayden, P., Lurmann, F., Yaransian, T., and Young, J. (1988). Interim Emissions Inventory for Regional Air Quality Studies, EA6070. Electric Power Research Institute, Palo Alto, California.

Henderson, R. G., and Weingartner, K. (1982). *Atmos. Environ.* **16,** 1657–1665.

Henry, R. C. (1982). Stability Analysis of Receptor Models that Use Least Squares Fitting. In: *Receptor Models Applied to Contemporary Pollution Problems*, p. 141. SP-48, Air Pollution Control Association, Pittsburgh, Pennsylvania.

Henry, R. C., and Hidy, G. M. (1979). *Atmos. Environ.* **13,** 1581–1596.

Henry, R. C., and Hidy, G. M. (1982). *Atmos. Environ.* **16,** 927–9294.

Henry, R. C., Wang, Y. N., and Gebhart, K. A. (1991). *Atmos. Environ.* **25A,** 503–509.

Hershfield, D. (1962). *J. Appl. Meteorol.* **1,** 575–578.

Hicks, B. B. (1986). Combining Biology, Chemistry, and Meteorology in Modeling and Measuring Dry Deposition. Presented at the 2nd Internationl Symposium on Biosphere Atmosphere Exchange, Mainz, Federal Republic of Germany.

Hicks, B. B., and Matt, D. (1986). Combining Biology, Chemistry and Meteorology in Modeling and Measuring Dry Deposition. Presented at the 2nd International Symposium on Biosphere Atmosphere Exchange, Mainz, Germany.

Hicks, B. B., and McMillen, R. T. (1986). Application of Similarity Methods to Deduce Dry Deposition Covariances Using Imperfectly Responding Sensors. In: Proc. 5th Annual National Symposium on Recent Advances in Pollutant Monitoring of Ambient Air and Stationary Sources. Report EPA/600/9-85/029. U.S. Environmental Protection Agency, Research Triangle Park, North Carolina.

Hicks, B. B., Wesley, M. I., and Durham, J. L. (1980). Critique of Methods to Measure Dry Deposition. Report EPA-600/9-80-050 (NTIS PB81-126443). U.S. Environmental Protection Agency, Research Triangle Park, North Carolina.

Hicks, B. B., Matt, D. R., McMillen, R. T., Womack, J. D., and Shetter, R. E. (1983). Eddy Fluxes of Nitrogen Oxides to a Deciduous Forest in Complex Terrain. In: *The Meteorology of Acid Deposition.* P. J. Samson (ed.), pp. 189–201. Air Pollution Control Association, Pittsburgh, Pennsylvania.

Hicks, B. B., Baldocchi, D. D., Hosker, R. P. Jr., Hutchison, G. A., Matt, D. R., McMillen, R. T., and Satterfield, L. C. (1985). On the Use of Monitored Air Concentrations to Infer Dry Deposition, NOAA Technical Memorandum ERL ARL-141, p. 65.

Hicks, B. B., Matt, D., and McMillan, R. (1985). Operation and Results of the CORE Network. Unpublished report NOAA/ER21. Air Resources Laboratory, Oak Ridge, Tennessee.

Hidy, G. M. (1982). *Water Air, Soil Pollut.* **18,** 191–198.

Hidy, G. M. (1984a). *JAPCA* **34,** 518–531.

Hidy, G. M. (1984b). *Aerosols, an Industrial and Environmental Science*, Chapter 8. Academic Press, New York.

Hidy, G. M., Tong, E. Y., and Mueller, P. K. (1976). Design of the Sulfate Regional Experiment. Report EC-125 (1). Electric Power Research Institute, Palo Alto, California.

Hidy, G. M., Mueller, P. K., and Tong, E. Y. (1978). *Atmos. Environ.* **12,** 735–754.

Hidy, G. M., Hansen, D. A., Henry, R. C., Ganesan, K., and Collins, J. (1984). *JAPCA* **34,** 333–354.

Hidy, G. M., Hansen, D. A., and Bass, A. (1985). Feasibility and Design of the Massive Aerometric Tracer Experiment (MATEX). Report EA-4305(2)(3). Electric Power Research Institute, Palo Alto, California.

Hilst, G. R. (1992). *Atmos. Environ.* **26A,** 1413–1421.

Hilst, G. R., and Chapman, E. G. (1990). *Atmos. Environ.* **24A,** 1889–1901.

Hoffmann, M. R. (1986) *Atmos. Environ.* **20,** 1145–1154.

Hoffmann, M. R., and Jacob, D. J. (1984). Kinetics and Mechanisms of the Catalytic Oxidation of Dissolved Sulfur Dioxide in Aqueous Solution: An Application to Nighttime Fog Water Chemistry. In: *$SO_2$, NO and $NO_2$, Oxidation Mechanisms: Atmospheric Considerations.* J. G. Calvert (ed.), pp. 101–172, Butterworth, Boston.

Holzworth, G. (1972). Mixing Heights, Wind Speeds, and Potential for Urban Air Pollution through the Contiguous U.S. Report AP-101. U.S. Environmental Protection Agency, Research Triangle Park, North Carolina.

Hopke, P. K., Gladney, E. S., Gordon, G. E., Zoller, W. H., and Jones, A. G. (1975). *Atmos. Environ.* **10,** 1015–1025.

Hosker, R. P., and Lindberg, S. E. (1982). *Atmos. Environ.* **16,** 889–910.

Hudson, J. G. (1984). *J. Climatol. Appl. Meteorol.* **23,** 42–51.

Hudson, J. G., and Rogers, C. F. (1984). Interstitial CCN Measurements Related to Mixing in Clouds. Abstracts of the Proceedings, 9th International Cloud Physics Conference, Tallinn, Estonian SSR, USSR, 21–28 August.

Huebert, B. J., and Lazrus, A. L. (1980). *J. Geophys. Res.* **85,** 7322–7328.

Husain, L., Webber, J., and Canelli, E. (1983) *JAPCA* **33,** 1185.

Husar, R. B. (1983). Possible Remedies to Acid Rain. Unpublished report. Center for Air Pollution Impact and Trend Analysis (CAPITA) Washington University, St. Louis, Missouri.

Husar, R. B., and Patterson, D. (1980). *Ann. N.Y. Acad. Sci.* **338,** 399–417.

Interagency Task Force on Acid Precipitation (ITFAP). (1983). National Acid Precipitation Assessment Program, Annual Report to the President and Congress, pp. 35–40. Interagency Task Force-NAPAP, Washington, D.C.

Irving, P. (ed.) (1991). Acid Deposition: State of Science and Technology. Summary Report of the U.S. National Acid Precipitation Assessment Program. U.S. National Acid Precipitation Asssessment Program, Washington, D.C.

Jarvis, P. G. (1976). *Philos. Trans. R. Soc. Lond.* **B273,** 593–610.

Johannes, A. H., Altwicker, E. R., and Clesceri, N. L. (1981) Characterization of Acidic Precipitation in the Adirondack Region. Report EA 1826. Electric Power Research Institute, Palo Alto, California.

Johnson, W. B., Wolf, D. E., and Mancuso, R. L. (1978). *Atmos. Environ.* **12,** 511–527.

Journel, A. G., and Huijbugts, C. (1978). *Mining Geostatistics.* Academic Press, New York.

Judeikis, H. S., and Wren, A. G. (1978). *Atmos. Environ.* **12,** 2315–2319.

Junge, C. E. (1963). *Air Chemistry and Radioactivity.* Academic Press, New York.

Junge, C. E., and Werby, R. T. (1958). *J. Meteorol.* **15,** 417–425.

Kadlecek, J., McLaren, S., Camarota, N., Mohnen, V., and Wilson, J. (1983). Cloud Water Chemistry at Whiteface Mountain. In: *Precipitation Scavenging, Dry Deposition and Resuspension,* Vol. 1, H. Pruppacher, R. Semonin, and W. G. N. Slinn (eds.), pp. 103–114. Elsevier, New York.

Kadlecek, J. A., McLaren, S., and Mohnen, V. A. (1986). Cloudwater Chemistry, Winter–Summer Comparison at Whiteface Mountains. Unpublished report. Atmospheric Sciences Research Center, State University of New York, Albany.

Kahl, J. D., and Samson, P. J. (1986). *J. Climatol. Appl. Meteorol.* **25,** 1816–1831.

Kahl, J., Keeler, G., Fernau, M., and Samson, P. J. (1985). Evaluation of a Regional-Scale Acid Deposition Model. Paper No. 85–5.2. Presented at the 78th Annual Meeting of the Air Pollution Control Association, Detroit, Michigan, June 16–21.

Kelly, T. J., Stedman, D. H., and Ritter, J. A. (1980). *J. Geophys. Res.* **85,** 7417–7425.

Kerr, J. A., and Calvert, J. G. (1984). Chemical Transformation Modules for Eulerian Acid Deposition Models, Vol. 1. Available from NTIS (PB173714) Gas Phase Chemistry.

Kimball, K. D., Likens, G. E., Eaton, J. S., Weathers, K. C., Siccama, T. G., Borman, F. H., and Cogbill, C. V. (1986). A Five-Decade Comparison of Cloud Water Acidity on Mt. Washington, New Hampshire. Unpublished manuscript.

King, W., and Vukovich, F. (1982). *Atmos. Environ.* **16,** 1171–1182.

Kircher, C. C., and Sander, S. P. (1984). *J. Phys. Chem.* **88,** 2082–2091.

Kleinman, L. I. (1981). Concentration and Deposition of Sulfur Oxides in the Northeastern United States: A Comparison of AIRSOX Model Results with Monitoring Data for July 1978. BNL-29855. Brookhaven National Laboratory, Upton, New York.

Kleinman, L. I., Cederwall, R. T., and Meyers, R. E. (1979). Modeling Sulfur Oxide Concentrations: An Assessment of Energy Utilization Scenarios. BNL-26369. Department of Energy and Environment, Brookhaven National Laboratory, Upton, New York.

Klemm, H., and Brennan, R. J. (1981). Emissions Inventory for the SURE Region. Report EA-1913. Electric Power Research Institute, Palo Alto, California.

Kneip, T., Lioy, P., and Wolff, G. (1983). *Environ. Forum* **2,** 8.

Knutson, E. O., and Stockham, J. D. (1977). "Aerosols Scavenging by Snow: Comparison of Single-Flake and Entire Snowfall Results," In: Precipitation Scavenging (1974): Proceedings ERDA Symp. Series No. 41, National Technical Information Service, U.S. Department of Commerce, pp. 195–207.

Kok, G. L. (1982). *EOS Trans.* **63,** 891.

Kok, G. L., Lind, J. A., Gitlin, S. N., Heikes, B. G., and Shetter, R. E. (1986). *Anal. Chem.* **58,** 594–597.

Korshover, J. (1976). Climatology of Stagnating Anticyclones East of the Rocky Mountains, 1936–1975. Technical Memorandum, ERL-ARL-55. National Acid and Atmospheric Administration, Silver Spring, Maryland.

Lazaro, M. A. (1985). Meso- and Synoptic-Scale Pollutant Transport. Presented at the 78th Annual Meeting of the Air Pollution Control Association, Detroit, Michigan, 16–21 June.

Lazrus, A. L., Haagenson, P. E., Kok, G. L., Huebert, B. J., Kreitzberg, C. W., Likens, G. E., Mohnen, V. A., Wilson, W. E., and Winchester, J. W. (1981). Acidity in Air and Water in a Case of Warm Frontal Precipitation. Proceedings, 74th Annual Meeting Air Pollution Control Association Abstract 81-13.5.

Lazrus, A. L., Haagenson, P. E., Kok, G. L., Huebert, B. J., Kreitzberg, C. W., *et al.* (1983). *Atmos. Environ.* **17,** 581–591.

Lee, I. Y., and Pruppacher, H. R. (1977). *Pure and Appl. Geophys.* **115,** 523–545.

Lee, Y., and Kinsley, M. (1986). *Trans. AGU* **67,** 898.

Lee, Y. N. (1984). Gas–Liquid Chemistry of Natural Waters, *Abstr.*, Vol. 1, pp. 20–21. Brookhaven National Laboratory, New York.

Likens, G. E. (1976). *Chem. Eng. News* **54,** 19.

Likens, G. E., and Butler, T. J. (1981). *Atmos. Environ.* **15,** 1103–1109.

Likens, G. E., Bormann, F., and Eaton, J. (1980). Variations in Precipitation and Stream Water Chemistry at the Hubbard Brook Experimental Forest during 1964–1977. In: *Effects of Acid Precipitation on Terrestrial Ecosystems*, T. C. Hutchinson and Mittaras (eds.), pp. 443–464. Plenum Publishing, New York.

Liljestrand, H. M. (1982). Acidic Precipitation Source Identification by Chemical Mass Balance Methods Employing Fractionation Factors. In: *Receptor Models Applied to*

*Contemporary Pollution Problems*, p. 212. SP-48. Air Pollution Control Association, Pittsburgh, Pennsylvania.

Lind, J. A., and Kok, G. L. (1986). *J. Geophys. Res.* **91,** 7889–7895.

Lindberg, S. E., Lovett, G. M., Richter, D., and Johnson, D. (1986). *Science* **231,** 141–145.

Lioy, P. J., and Samson, P. J. (1979). *Environ. Int.* **2,** 77.

Lioy, P. J., Mallon, R. P., Lippman, M., Kneip, T. J., and Samson, P. J. (1982). *JAPCA* **32,** 1043–1047.

Liu, M. K., Stewart, D. A., and Henderson, D. (1982). *J. Appl. Meteorol.* **21,** 859–873.

Lodge, J. P., Jr. (1968). Chemistry of U.S. Precipitation. Final report on the National Precipitation Sampling Network. National Center for Atmospheric Research, Boulder, Colorado.

Logan, J. (1985). *J. Geophys. Res.* **90,** 10463–10482.

Lonneman, W. (1979). Hydrocarbon Measurements in Urban and Nonurban Sites Made during Past Field Studies by the Gas Kinetics and Photochemistry Branch Mobile Laboratory by EPA Contract Personnel, Research Triangle Park, North Carolina, U.S. Environmental Protection Agency, 9 May.

Lovett, G. R., and Kinsman, J. D. (1990). *Atmos. Environ.* **24A,** 2767–2786.

Lovett, G. R., Reiners, W. A., and Olson, R. K. (1982). *Science* **218,** 1303–1304.

Ludwig, F. L., and Byrd, G. (1980). *Atmos. Environ.* **14,** 585–587.

Ludwig, F. L., Endlich, R. M., and Nitz, K. C. (1986). A Model for Online Description of Air Flow Over Complex Terrain. Unpublished.

Lurmann, F. W., Godden, D. A., and Collins, H. M. (1985). User's Guide to the PLMSTAR Air Quality Simulation Model. M-2206-100. Environmental Research and Technology, Newbury Park, California.

Lyons, W., Dooley, J. C., Jr., and Whitby, K. T. (1978). *Atmos. Environ.* **12,** 633–640.

Malinowski, E. R., and Howery, D. G. (1980). *Factor Analysis in Chemistry*. Wiley, New York.

Malm, W. C. (1979). *JAPCA* **29,** 1042.

Malm, W. C., Pitchford, M., and Pitchford, A. (1982). *Atmos. Environ.* **16,** 2323.

MAP3S/RAINE Research Community. (1982). *Atmos. Environ.* **16,** 1603–1631.

Margatan, J. J. (1984). *J. Phys. Chem.* **88,** 3314–3318.

Martin, L. R., Damschen, D. E., and Judeikes, H. S. (1981). Sulfur Dioxide Oxidation Reactions to Aqueous Solutions. Available from NTIS (PB81-196834).

Martin, R. (1984). Kinetic Studies of Sulfite Oxidation in Aqueous Solution. In: *$SO_2$, NO and $NO_2$ Oxidation Mechanisms: Atmospheric Considerations*, J. G. Calvert (ed.), pp. 63–100. Butterworth, Boston.

Mason, B. J. (1971). *Physics of Clouds*, 2nd Ed. Oxford University Press, London.

Mass, C. F., and Dempsey, D. P. (1985). *Mon. Weather Rev.* **113,** 1211–1227.

Maul, P. R. (1980). Atmospheric Transport of Sulphur Compound Pollutants. MID/SSD/80/0026/R. Central Electricity Generating Board, Scientific Services Department, Midlands Region, Imperial College, London.

Maul, P. R. (1982). *Environ. Pollut.* **B4,** 1–25.

McDonald, G. C., and Schwing, R. C. (1973). *Technometrics* **15,** 463–481.

McNaughton, D. J. (1980). *Atmos. Environ.* **14,** 55–63.

McNaughton, D. J. (1981). *Atmos. Environ.* **15,** 1075–1079.

McNaughton, D. J., Benkowtiz, C. M., and Williams, R. C. (1981). *J. Appl. Meteorol.* **20,** 795–801.

Meyers, T. P., Hicks, B. B., Hosker, R. P., Jr., Womack, J. D., and Satterfield, L. C. (1991). *Atmos. Environ.* **25A,** 2361–2360.

Michael, P., and Hales, J. M. (1984). Air Pollutant Trajectories in Precipitating Storms. In:

*Meteorological Aspects of Acid Rain*, John I. Teasley (ed.), pp. 111–118. Butterworth, Stoneham, Massachusetts.

Michael, P., and Raynor, G. S. (1983). Modifications of Trajectory Models Needed for Pollutant Source-Receptor Analysis. In: *Air Pollution Modeling and Its Application*, II, C. De Wispelaere (ed.), pp. 403–414. Plenum Publishing, New York.

Middleton, W. E. K. (1958). *Vision through the Atmosphere*. University of Toronto Press, Toronto.

Milford, J. G., and Davidson, C. I. (1985). *JAPCA* **35,** 1249–1260.

Miller, M., and Everett, A. G. (1981). History and Trend of Atmospheric Nitrate Deposition in the Eastern U.S. In: *Formation and Fate of Atmospheric Nitrates*. EPA-600-19-81-025. U.S. Environmental Protection Agency, Washington, D.C.

Miller, M. S., Friedlander, S. K., and Hidy, G. M. (1972). A Chemical Element Balance for the Pasadena Aerosol. In: *Aerosols and Atmospheric Chemistry*, G. M. Hidy (ed.), pp. 301–311. Academic Press, New York.

Minnis, P. and E. F. Harrison (1984). *J. Clim. Appl. Meteorol.* **23,** 993–1051.

Mohnen, V. (1984). "Acid Rain: Problem Definition." In: *Acid Rain Source Book*, T. C. Elliott and R. G. Shwieger (eds.), pp. 14–26. McGraw-Hill, New York.

Mohnen, V. (1986). Estimated Deposition Levels of Pollutants in the Eastern U.S. Unpublished manuscript.

Mohnen, V. (1988). Mountain Cloud Chemistry Project: Wet, Dry and Cloud Water Deposition. EPA Contractors Report CR-813934-01-2. USEPA AREAL, Research Triangle Park, North Carolina.

Mohnen, V. and Kadlecek, J. (1989). Cloud Chemistry Research at Whiteface Mountain. *Tellus* **41B,** 79–91.

Moody, J., and Samson, P. J. (1986). The Influence of Meteorology on the Chemical Composition of Precipitation. Presented at ACS Symposium on Acid Rain, New York. (See also Samson and Moody, 1986.)

Moody, J. L., Kahl, J., and Samson, P. J. (1983). Initial Investigation of the Sensitivity of Source–Receptor Relationships to Non-Linear Atmospheric Transformations. In: *The Meteorology of Acid Deposition*, Perry J. Samson (ed.), pp. 346–357. Air Pollution Control Association, Pittsburgh, Pennsylvania.

Moyers, J. L., Ranweiler, L. E., Hopf, S. B., and Korte, N. E. (1977). *Environ. Sci. Technol.* **11,** 789–794.

Mueller, P. K., and Hidy, G. M. (1982). Sulfate Regional Experiment: SURADS Model Documentation. Report EA-1907. Electric Power Research Institute, Palo Alto, California.

Mueller, P. K., and Hidy, G. M. (1983). The Sulfate Regional Experiment: Report of Findings. Report EA-1901(2,3). Electric Power Research Institute, Palo Alto, California.

Mueller, P. K., and Watson, J. (1983). Eastern Regional Air Quality Measurements. Report EA-1914(1). Electric Power Research Institute, Palo Alto, California.

Mueller, P. K., Jansen, J. J., and Allan, M. A. (1984). Utility Acid Precipitation Study Program—Second Annual Summary Report. UAPSP 109. Edison Electric Institute, Washington, D.C.

Murphy, B. D. (1981). PHENIX: A Pseudo-Spectral Technique in the Computation of Seasonal Average Concentrations and Its Comparison with the Trajectory Technique. In: *Air Pollution Modeling and Its Application*, II, C. De Wispelaere (ed.), pp. 285–302. Plenum Publishing, New York.

National Academy of Sciences. (1986). *Acid Deposition—Long-Term Trends*. NAS Press, Washington, D.C.

National Acid Precipitation Assessment Program (1990a). *Acid Deposition: State of Science and Technology*, Vol. 1, SOS/T Report 1, U.S. Government Printing Office, Washington, D.C.

National Acid Precipitation Assessment Program (1990b). *1990 Integrated Assessment Report*. U.S. Government Printing Office, Washington, D.C.

National Center for Atmospheric Research. (1985). The NCAR Regional Acid Deposition Model. NCAR/TN-256+STR. NCAR, Boulder, Colorado.

National Center for Atmospheric Research. (1986). Preliminary Evaluation Studies with the Regional Acid Deposition Model (RADM). NCAR/TN-265+STR, NCAR, Boulder, Colorado.

National Oceanic and Atmospheric Administration. (1974). *Climatic Atlas of the United States*. U.S. Department of Commerce, Washington, D.C.

Nieman, B. L. (1980). Air Quality and Acid Rain Results for the Ohio River Basin Energy Study (ORBES). Report to the Environmental Protection Agency, Office of Environmental Engineering and Technology, Washington, D.C.

Oden, S. (1968). The Acidification of Air and Precipitation and Its Consequences in the Natural Environment. Ecological Committee Bull. No. 1, Swedish National Research Council, Stockholm.

Oehlert, G. W. (1984). A Statistical Analysis of the Trends in the Hubbard Brook Bulk Precipitation Chemistry Data Base. Technical Report 260, Series 2. Department of Statistics, Princeton University, Princeton, New Jersey.

Office of Technology Assessment, 1984. Acid Rain and Transported Air Pollutants: Implications for Public Policy. Report OTA-204. U.S. Congress, Office of Technology Assessment, Washington, D.C.

Olson, M. P., and Oikawa, K. K. (1983). A Preliminary Evaluation of the Sulfate Concentration Forecasts for the PEPE Project. In: *The Meteorology of Acid Deposition*, P. J. Samson (ed.), pp. 367–386. Air Pollution Control Association, Pittsburgh, Pennsylvania.

Organization of Economic Cooperation and Development (1977). The OECD Programme on Long Range Transport of Air Pollutants, Measurements and Findings. OECD, Paris.

Owers, M. J., and Powell, A. W. (1974). *Atmos. Environ.* **8,** 63–67.

Padmanabhamurty, B. (1975). *Atmos. Environ.* **9,** 365–366.

Parkhurst, W. J., Crawford, T., Lee, N. T., and Lokey, J. (1984). The Effects of Large-Scale Utility Sulfur Dioxide Emissions Control Program on Local and Regional Sulfur Measurements. Paper presented at the 77th APCA Annual Meeting, San Francisco, California.

Patterson, D. E., Husar, R. B., Wilson, W. E., and Smith, L. F. (1981). *J. Appl. Meteorol.* **20,** 404–420.

Penkett, S. A., Jones, B. M. R., Brice, K. A., and Eggleton, A. E. J. (1979). *Atmos. Environ.* **13,** 123–137.

Peters, N. E., and Bonelli, J. E. (1982). Chemical Composition of Bulk Precipitation in the North-Central and Northeastern U.S. Circular 874. U.S. Geological Survey, Washington, D.C.

Peters, N. E., Schroeder, R. A., and Troutman, D. E. (1982). Temporal Trends in Acid of Precipitation and Surface Waters. Water Supply Paper 2188, p. 35. U.S. Geological Survey, Washington, D.C.

Peterson, J. T. (1970). *Atmos. Environ.* 4, 501–518.

Peterson, J. T. (1972). *Atmos. Environ.* **6,** 433–442.

Pielke, R. A. (1984). *Mesoscale Meteorological Modeling*. Academic Press, Orlando, Florida.

Pierson, W. R., Brachaczek, W., Truex, T., Butler, J., and Korniski, T. (1980). *Ann. N.Y. Acad. Sci.* **338,** 145–173.

Pierson, W. R., Brachaczek, W., Gorse, R., Jr., Japar, S., and Norbeck, J. (1987). *Environ. Sci. Technol.* **21,** 679–691.

Platt, U. (1983). Origin of Nitrous and Nitric Acid in the Atmosphere. In: *Chemistry of Multiphase Atmospheric Systems* W. Jaeschke (ed.), pp. 299–319. Reidel, Dordrecht.

Platt, U., Perner, D., Winder, A. M., Harris, G. W., and Pitts, J. N., Jr. (1980). *Geophys. Res. Lett.* **7,** 89–92.

Pleim, J., Yamartino, R., Lung, W., and Coffee, P. E. (1983). Development of an Acid Rain Impact Assessment Model. In: *The Meteorology of Acid Deposition*, P. J. Samson (ed.), pp. 400–419. Air Pollution Control Association, Pittsburgh, Pennsylvania.

Porch, W. M. (1980). Numerical Wind Field Model Validation in Complex Terrain with Application to Pollutant Transport. In: *Second Joint Conference on Applications of Air Pollution Meteorology*, pp. 608–613. American Meteorological Society, New Orleans, Lousiana, 24–27 March.

Porch, W. M., and Lange, R. (1983). Spatially Averaging Cross-Wind Sensors and Numerical Model Results for Nocturnal Drainage Winds in Complex Terrain. In: *Sixth Symposium on Turbulence and Diffusion*, American Meteorological Society, pp. 181–183. Boston, Massachusetts, 22–25 March.

Prahm, L. P., and Christensen, O. (1977). *J. Appl. Meteorol.* **16,** 896–910.

Prandtl, L., and Tietjens, O. G. (1934). *Fundamentals of Hydro- and Aerodynamics*, p. 73 et seq. Dover, New York.

Prinz, B. and Stratmann, H. (1968). The Possible Use of Factor Analysis in Investigating Air Quality. *Staub* **28,** 33–42.

Pruppacher, H. (1986). The Role of Cloud Physics in Atmospheric Multiphase Systems: Ten Basic Statements. In: *Chemistry of Multiphase Atmospheric Systems*, W. Jaeschke (ed.), pp. 133–190. NATO ASI Series G6. Springer-Verlag, Berlin.

Pruppacher, J. R., and Klett, J. D. (1978). *Microphysics of Clouds and Precipitation.* Reidel, London.

Radke, L., Hobbs, P. V., and Eltgroth, M. (1980). *J. Appl. Meteorol.* **19,** 715–722.

Rahn, K. (1983). *Environ. Forum* **2,** 14.

Rahn K., and Lowenthal, D. (1984). *Science* **223,** 132.

Rahn, K., and Lowenthal, D. (1985). *Science* **228,** 275–284.

Rahn, K., Wunschel, K. R., and Lowenthal, D. H. (1985). Elemental Tracers Applied to Transport of Aerosol from Midwest to Northeast. Final Report, Cooperative Agreement CR-810903. Atmospheric Sciences Research Laboratory, Office of Research and Development, U.S. Environmental Protection Agency, Research Triangle Park, North Carolina.

Ramsdell, J. V. (1981). Atmospheric Transport Modeling. PNL-SA-10050. Pacific Northwest Laboratory, Richland, Washington.

Rao, S. T., Pleim, J., and Czapski, U. (1983). *JAPCA* **33,** 32–41.

Raynor, G. S. (1984). Preliminary Evaluation of Mesoscale Precipitation Chemistry Studies. Unpublished report, Brookhaven National Laboratory, Upton, New York.

Raynor, G. S., and Hayes, J. V. (1982). *Atmos. Environ.* **16,** 1647.

Richards, L. W. (1983). *Atmos. Environ.* **17,** 397–402.

Richards, L. W. (1988). *JAPCA* **38,** 784–791.

Richards, L. W., Anderson, J. A., Blumenthal, D. L., McDonald, J. A., Kok, G. L., and Lazrus, A. L. (1983). *Atmos. Environ.* **17,** 911–914.

Richards, L. W., Bergstrom, R. W., and Ackerman, T. P. (1985). *Atmos. Environ.* **20,** 387.

Roberts, J. M., Fehsenfeld, F. C., and Liu, S. C. (1984). *Atmos. Environ.* **18,** 2421–2432.

Ruff, R. E., Nitz, K. C., Ludwig, F. L., Bhumralkar, C. M., Shannon, J. D., Sheih, C. M., Lee, I. Y., Kumar, R., and McNaughton, D. J. (1985). *Atmos. Environ.* **19,** 1103–1115.

Samson, P. J. (1980). *J. Appl. Meteorol.* **19,** 1382.

Samson, P. J., and Moody, J. (1986). The Influence of Meteorological Factors on Precipitation Chemistry. Report UAPSP III, Electric Power Research Institute, Palo Alto, California.

Samson, P. J., and Small, M. J. (1984). Atmospheric Trajectory Models for Diagnosing the Sources of Acid Precipitation. In: *Modeling of Total Acid Precipitation Impacts*, J. L. Schnoor (ed.), pp. 1–24. Butterworth, Stoneham, Massachusetts.

Samson, P. J., Fox, R. J., and Foltman, R. A. (1983). Sensitivity Analysis of a Climatological Air Trajectory Model. In: *Air Pollution Modeling and Its Application*, II, C. De Wispelaere (ed.), pp. 415–428. Plenum Publishing, New York.

Samson, P. J., Moody, J. L., Kahl, J., and Keeler, G. (1986). Uncertainties in Quantifying Source-Receptor Relationships for Atmospheric Acids. In: *NATO ASI Series*, Vol. G6, *Chemistry of Multiphase Atmospheric Systems*, W. Jaeschke (ed.), pp. 727–740. Springer-Verlag, Berlin.

Schack, C. J., Jr., Pratsinis, S. E., and Friedlander, S. I. (1985). *Atmos. Environ.* **19,** 953–960.

Schiermeier, F. A., and Misra, P. K. (1983). Evaluation of Eight Linear Regional-Scale Sulfur Models by the Regional Modeling Subgroup of the United States/Canadian Memorandum of Intent Work Group 2. In: *The Meteorology of Acid Deposition*, P. J. Samson (ed.), pp. 330–345. Air Pollution Control Association, Pittsburgh, Pennsylvania.

Scholtz, M. T., and Weisman, B. (1985). A Multi-Layered, Long-Range Transport, Lagrangian Trajectory Model: Comparison with Fully Mixed Single Layer Models. In: *Air Pollution Modeling and Its Application*, IV, C. De Wispelaere (ed.), pp. 51–70. Plenum Publishing, New York.

Schwartz, S. E. (1984a). Gas-Aqueous Reactions of Sulfur and Nitrogen Oxides in Liquid-Water Clouds. In: *$O_2$, NO and $NO_2$ Oxidation Mechanisms: Atmospheric Considerations*, Calvert, J. G. (ed.), pp. 173–208. Butterworth, Boston.

Schwartz, S. E. (1984b). *J. Geophys. Res.* **89,** 11589–11598.

Schwartz, S. E. (1987). *Ann. N.Y. Acad. Sci.* **502,** 83–144.

Science Applications International Corporation. (1985a). Southern California Edison Company Rock Creek Lake Meteorological and Air Quality Monitoring Program Quarterly Data Report, July through September 1984. SAI Corp., La Jolla, California.

Science Applications International Corporation. (1985b). Southern California Edison Company Rock Creek Lake Meteorological and Air Quality Monitoring Program Quarterly Data Report, October through December 1984. SAI Corp., La Jolla, California.

Scire, J. S., and Lurmann, F. W. (1983). Development of the MESOPUFF II Dispersion Model. In: Sixth Symposium on Turbulence and Diffusion, pp. 32–35. Boston, Massachusetts, 22–25 March.

Scire, J. S., Lurmann, F. W., Bass, A., and Hanna, S. R. (1983a). Development of the MESOPUFF II Dispersion Model. Contract No. 68–02–3733. Environmental Sciences Research Laboratory, U.S. Environmental Protection Agency, Research Triangle Park, North Carolina.

Scire, J. S., Lurmann, F. W., Bass, A., and Hanna, S. R. (1983b). User's Guide to the MESOPUFF II Model and Related Processor Programs. Contract No. 68-02-3733. Environmental Sciences Research Laboratory, U.S. Environmental Protection Agency, Research Triangle Park, North Carolina.

Scott, B. C. (1978). *J. Appl. Meteorol.* **17,** 1375–1389.

Sehmel, G. A. (1972). Turbulent Deposition of Monodisperse Particles on Simulated Grass. In: *Assessment of Airborne Particles*, T. T. Mercer, P. E. Morrow, and W. Stöber (eds.), pp. 18–42. Charles C Thomas, Springfield, Illinois.

Sehmel, G. A., Sutter, S. L., and Hanna, M. T. (1973). Dry Deposition Processes. Pacific Northwest Laboratory Annual Report for 1972, Atmospheric Sciences, NBWL-1751, Pt. 1, Battelle, Pacific Northwest Laboratory, Richland, Washington, pp. 43–49.

Sehmel, G. A., Hodgson, W. H., and Sutter, S. L. (1974). Dry Deposition of Particles. Pacific Northwest Laboratory Annual Report for 1973, Atmospheric Sciences, BNWL-1850, Pt. 3, Battelle, Pacific Northwest Laboratory, Richland, Washington, pp. 157–162.

Seigneur, C., Saxena, P., and Mirabella, V. (1984). Development and Evaluation of an Acid Rain Trajectory Model. Presented at the 77th Annual Meeting of the Air Pollution Control Association, San Francisco, California, 24–29 June.

Seigneur, C., Hudischewskyj, A. B., and Saxena, P. (1985). Mathematical Modeling of the Concentration, Chemical Composition, and Size Distribution of Aerosols in Plumes. SYSAPP-85/026. Systems Applications, Inc., San Rafael, California.

Seigneur, C., Hudischewskyj, A. B., and Saxena, P. (1986). Further Development and Application of a Mathematical Model of Sulfate and Nitrate Formation in Plumes. SYSAPP-86/028. Systems Applications, Inc., San Rafael, California.

Semb, A. (1978). *Atmos. Environ.* **12,** 455–460.

Shannon, J. D. (1981). *Atmos. Environ.* **15,** 689–701.

Shannon, J. D. (1985a). Estimation of North American Anthropogenic Sulfur Deposition as a Function of Source/Receptor Separation. In: *Air Pollution Modeling and Its Application*, IV, C. De Wispelaere (ed.), pp. 119–128. Plenum Publishing, New York.

Shannon, J. D. (1985b). User's Guide for the Advanced Statistical Trajectory Regional Air Pollution (ASTRAP) Model. EPA 600/8-85-016. Argonne National Laboratory, Illinois.

Shannon, J. D., Kleinman, L., Benkovitz, C., and Berkowitz, C. (1982). Intercomparison of MAP3S Models of Long-Range Transport and Deposition. In *Proceedings, Third Joint Conference on Applications of Air Pollution Meteorology*, pp. 29–32. San Antonio, Texas, 12–15 January.

Sheih, C. M. (1983). An Evaluation of Wind Field Interpolation Schemes Used in Studies of Regional-Scale Pollutant Transport. In: *The Meteorology of Acid Deposition*, Perry J. Samson (ed.), pp. 324–329. Air Pollution Control Association, Pittsburgh, Pennsylvania.

Sherman, C. A. (1978). *J. Appl. Meteorol.* **17,** 312–319.

Sirois, A. V., and Fricke, W. (1992). Regionally Representative Daily Air Concentration of Acid-Related Substance in Canada: 1983–1987. *Atmos. Environ.*, **26A,** 1983–1987.

Sisterson, D. L., Borwersox, V. C., Olson, A. R., Meyers, T. P., and Vong, R. L. (1990). Deposition Monitoring Methods and Results. State-of-Science/Technology Report 6, National Acid Precipitation Program, Washington, D.C.

Slinn, W. G. N. (1977). *Water Air Soil Pollut.* **7,** 513–543.

Slinn, W. G. N. (1981). Precipitation Scavenging, In: *Atmospheric Science and Power Production*, pp. 466–532, D. Randerson (ed.), Chapter 11. U.S. Department of Energy, Technology Information Center, Washington, D.C.

Slinn, W. G. N. (1986). Deposition of Atmospheric Pollution. Unpublished lecture notes, American Meteorological Society Adv. Study Inst. on the Chemical and Meteorological Aspects of Atmospheric Source-Receptor Relationships, Oak Ridge, Tennessee.

Sloane, C. S. (1984). *Atmos. Environ.* **18,** 1217–1229.

Sloane, C. S., and White, W. H. (1986). *Environ. Sci. Technol.* **20,** 760.

Stensland, G. J., and Semonin, R. (1982). *Bull. Am. Meteorol. Soc.* **63,** 1277–1284.

Stensvaag, J.-M. (1991). *Clean Air Act 1990 Amendments: Law and Practice.* Wiley Law, Somerset, New Jersey.

Sternberg, S. (1985). Montana Air Quality Data and Information Summary for 1983. Air Quality Bureau, Environmental Sciences Division, Department of Health and Environmental Sciences, Helena, Montana.

Stevens, R. K., Dzubay, T., Lewis, C., and Shaw, R., Jr. (1984). *Atmos. Environ.* **18,** 261–272.

Stewart, D. A., Morris, R. E., and Liu, M. K. (1983). Evaluation of Long-Term Regional Transport Models of Interest to the National Park Service. SYSAPP 83/215. Systems Applications, Inc., San Rafael, California.

Stewart, D. A., Morris, R. E., Liu, M. K., and Henderson, O. (1983a). *Atmos. Environ.* **17,** 1225–1252.

Stewart, D. A., Morris, R. E., Hudischewskj, A. B., and Liu, M. K. (1983b). Evaluation of Episodic Regional Transport Models of Interest to the National Park Service. SYSAPP 83/020. Systems Applications, Inc., San Rafael, California.

Stockwell, W. R., and Calvert, J. A. (1983). *Atmos. Environ.* **17,** 2231–2235.

Stockwell, W. R., and Calvert, J. G. (1983). *J. Geophys. Res.* **88,** 6673–6682.

Streets, D. G., Lesht, B. M., Shannon, J. D., and Veselka, T. D. (1985). *Environ. Sci. Technol.* **19,** 887–893.

Sweden's Case Study for the United Nations Conference in the Human Environment (1972). Air Pollution Across National Boundaries. The Impact on the Environment of Sulfur in Air and Precipitation. Stockholm.

Swindel, B. F. (1974). *Am. Statist.* **82,** 63–65.

Systems Applications, Inc. (1984). Visibility and Other Air Quality Benefits of Sulfur Dioxide Emission Controls in the Eastern U.S., Vol. I. Report SYSA-84/165, September. Prepared for the U.S. Environmental Protection Agency.

Systems Applications, Inc. (1985). Modeling Regional Haze in the Southwest: A Preliminary Assessment of Source Contributions, Vol. 1. SYSAPP/85-038. SAI, San Rafael, California.

Takacs, J. F., and Starheim, F. J. (1983). Intercomparison of Meteorological Trajectory Techniques Available for Use in Long-Range Transport Modeling. In: *The Meteorology of Acid Deposition*, P. J. Samson (ed.), pp. 302–311. Air Pollution Control Association, Pittsburgh, Pennsylvania.

Tanner, R. (1984). Specification of Chemical Measurement Capabilities Required for MATEX. Unpublished manuscript, Brookhaven National Laboratory, Upton, New York.

Taylor, G. E., and Tingey, D. T. (1983). *Plant Physiol.* **72,** 237–244.

Tingey, D. T., and Taylor, G. E., Jr. (1982). Variation in Plant Response to Ozone: A Conceptual Model of Physiological Events. In: *Effects of Gaseous Air Pollution in Agriculture and Horticulture*, M. H. Unsworth, and D. D. Ormrod (eds.), pp. 113–138. Butterworth Scientific Press, London.

Tombach, I. H., and Allard, D. (1983). Comparison of Visibility Measurement Techniques: Eastern U.S. Report EA-3292. Electric Power Research Institute, Palo Alto, California.

Tombach, I. H., Fitz, D. R., Drake, R. L., and Muller, B. (1985). Aerosols in the Rural West 1981–1982. Paper No. 85-16.8. Presented at the 78th Annual Meeting of the Air Pollution Control Association, Detroit, Michigan, 16–21, June.

Topol, L. (1984). Plan for Controlling the Quality of Measurements and Data Base during the Utility Acid Precipitation Study Program. UAPSP unpublished report.

Topol, L. (1989). UAPSP Precipitation Data Displays for 1979 through 1987. Utility Acid Precipitation Program, UAPSP Report 116. Electric Power Research Institute, Palo Alto, California.

Trijonis, J. (1982). *Atmos. Environ.* **16,** 2431–2445.

Trijonis, J., and Yuan, K. (1978). Visibility in the Northeastern U.S. Report EPA-600/3-78-075. U.S. Environmental Protection Agency, Research Triangle Park, North Carolina.

Tuazon, E., Atkinson, A., Plum, C. N., Winer, A. M., and Pitts, J. N., Jr. (1983). *Geophys. Res. Lett.* **10,** 953–956.

Tuncel, S. G., Olmez, I., Parrington, J. R., Gordon, G. E., and Stevens, R. K. (1985). *Environ. Sci. Technol.* **19,** 529–537.

U.S. Environmental Protection Agency (1975). Position Paper on Regulation of Atmospheric Sulfate. Report EPA-450/2-75-002. Office of Planning and Standards, Research Triangle Park, North Carolina.

U.S. Environmental Protection Agency. (1979). Protecting Visibility. An EPA Report to Congress. Report No. EPA-450/5-79-008, NSEPA, Washington, D.C.

U.S. Environmental Protection Agency. (1984). The Acidic Deposition Phenomenon and Its Effects. Critical Assessment Review Papers. Report EPA-600/8-83-016BF(1). Office of Research and Development, Washington, D.C.

Van Dop, H., and De Haan, B. J. (1983a). *Atmos. Environ.* **17,** 1449–1456.

Van Dop, H., and De Haan, B. J. (1983b). A Mesoscale Air Pollution Transport Model: Outline and Preliminary Results. In: *Air Pollution Modeling and Its Application*, II. C. De Wispelaere (ed.), pp. 303–316. Plenum Publishing, New York.

Van Egmond, N. D., and Kesseboom, H. (1983a). *Atmos. Environ.* **17,** 257–266.

Van Egmond, N. D., and Kesseboom, H. (1983b). *Atmos. Environ.* **17,** 267–274.

Venkatram, A. (1984). Stochastic Long-Range Transport Models. AMS/EPA Workshop, Sources and Evaluation of Uncertainties in Long Range Transport Models, Woods Hole, Massachusetts, September 1984.

Venkatram, A., and Pleim, J. (1985a). *Atmos. Environ.* **19,** 659–667.

Venkatram, A., and Pleim, J. (1985b). An Examination of Linear Parameterizations in a Statistical Long-Range Transport Model. In: *Air Pollution Modeling and Its Application*, IV, C. De Wispelaere (ed.), pp. 37–50. Plenum Publishing, New York.

Venkatram, A., Ley, B. E., and Wong, S. Y. (1982). *Atmos. Environ.* **16,** 249–257.

Venkatram, A., Karamchandani, P., and Scire, J. S. (1986). Data Base Priorities Identified by Regional Modeling Studies. Paper No. 86-29.4. Presented at the 79th Annual Meeting of the Air Pollution Control Association, Minneapolis, Minnesota, 22–27 June.

Voldner, E. C., and Sirois, B. (1985). *Water Air Soil Pollut.* **30,** 179–186.

Voldner, E. C., Olson, E. C., Oikawa, K., and Loiselle, M. (1981). *J. Geophys. Res.* **86**(C6), 5339–5346.

Voldner, E. C., Barrie, L. A., and Sirois, A. (1986). *Atmos. Environ.* **20,** 2101–2123.

Vong, R. J. (1989). Network Measurement of Droplet Deposition of Atmospheric Pollutants. SOS/T Report 6 Section 4, National Acid Precipitation Assessment Program, Washington, D.C.

Vukovich, F. (1979). *Atmos. Environ.* **13,** 255–265.

Vukovich, F., and Fishman, J. (1986). *Atmos. Environ.* **20,** 2423–2433.

Walcek, C. J., and Taylor, G. R. (1986). *J. Atmos. Sci.* **43,** 339–355.

Walcek, C. J., Brost, R. A., Chang, J. S., and Wesely, M. (1986). *Atmos. Environ.* **20,** 949–964.

Walmsley, J. L., and Mailhot, J. (1983). On a Method of Evaluation of Performance of a Trajectory Model for Long-Range Transport of Atmospheric Pollutants. In: *Air Pollution Modeling and Its Application*, II, C. De Wispelaere (ed.), pp. 277–284. Plenum Publishing, New York.

Wang, P. K., Grover, S., and Pruppacher, H. R. (1978). *J. Atmos. Sci.* **35,** 1735–1743.

Ward, J. H. (1963). *J. Am. Statist. Assoc.* **58,** 236–244.

Weisman, B. (1980). Long Range Transport Model for Sulphur. Paper No. 80-54.6. Presented at the 73rd Annual Meeting of the Air Pollution Control Association, Montreal, Quebec, 22–27 June.

Weisman, B. (1983). Application of Trajectory Model to Regional Transport Characterization. In: *Air Pollution Modeling and Its Application*, II, C. De Wispelaere (ed.), pp. 245–264. Plenum Publishing, New York.

Wesely, M. L., and Hicks, B. B. (1977). *JAPCA* **27,** 1110–1116.

Whelpdale, D., and Barrie, L. A. (1982). *Water Air Soil Pollut.* **18,** 7–23.

White, W. H., Seigneur, C., Heinold, D. W., Richards, L. W., Wilson, W. E., and Roberts, P. T. (1986). *Atmos. Environ.* **20,** 2243–2257.

Williamson, H. J., and DuBose, A. (1983). Receptor Model Technical Series, Volume III, User's Manual for Chemical Mass Valance Model. Report EPA-450/4-83-014, U.S. Environmental Protection Agency, Research Triangle Park, North Carolina.

Wilson, J. (1984). Feasibility and Design of the Massive Aerometric Tracer Experiment (MATEX). Informal report. Environmental Research and Technology, Newbury Park, California.

Wilson, J., Mohnen, V., and Kadlecek, J. (1982). *Atmos. Environ.* **16,** 1667–1676.

Winkler, P. (1967). Diploma thesis. Meteorol. Inst., University of Mainz, Mainz, Germany.

Wolff, G. T., and Lioy, P. (1980). *Environ. Sci. Technol.* **14,** 1257–1260.

Wolff, G. T., Lioy, P., Wright, G., Meyers, R., and Cedewall, R. (1977). *Atmos. Environ.* **11,** 797.

Wolff, G. T., Ferman, M., Kelly, N., Strongs, D., and Ruthkosky, M. (1982). *JAPCA* **32,** 1216–1220.

Wolff, G. T., Morrissey, M. L., and Kelly, N. A. (1984). *J. Climatol. Appl. Meteorol.* **23,** 1333–1341.

Wright, S. C., Grabenstetter, J. E., and Holdsworth, J. D. (1984). Ontario Hydro's Application of Fisher's Statistical Long-Range Transport Model. Paper No. 84-107.4. Presented at the 77th Annual Meeting of the Air Pollution Control Association, San Francisco, California, 24–29 June.

Wyoming Department of Environmental Quality (WDEQ). (1984). Wyoming's Air Quality Ambient Air Monitoring Data—1984. Air Quality Division, Cheyenne, Wyoming.

Yamartino, R., Pleim, J., and Lung, W. (1985). Development of an Acid Rain Impact Assessment Model. In: *Air Pollution Modeling and Its Application*, IV, C. De Wispelaere (ed.), pp. 91–118. Plenum Publishing, New York.

Yocke, M. A. (1981). A Three-Dimensional Wind Model for Complex Terrain. Ph.D. thesis, University of California, Berkeley.

Young, J. R. (1982). Data Inventory Report for the Chevron Baseline Visibility Study, Clear Creek Properties, CO. ERT Doc. No. P-B220-260, Environmental Research and Technology, Newbury Park, California.

Young, J. R., Ellis, E. C., and Hidy, G. M. (1988). *J. Environ. Qual.* **17,** 1–26.

Zannetti, P., Tombach, I. H., and Crencek, S. J. (1989). *An Analysis of Visual Range in the Eastern United States under Different Meteorological Regimes.*" *JAPCA* **39,** 200–203.

Zeldin, M. D., and Ellis, E. C. (1984). Trends in Precipitation Chemistry in Southern California. Paper No. 84-19.6. Presented at the 77th Annual Meeting of the Air Pollution Control Association, San Francisco, California, 24–29 June.

Zika, R., Saltzman, E., Chameides, W. L., and Davis, D. D. (1982). *J. Geophys. Res.* **87,** 5015–5017.

# INDEX

ISBN 0-12-347255-5

90065

9 780123 472557